Smart Innovation, Systems and Technologies

Volume 43

About this Series

The Smart Innovation, Systems and Technologies book series encompasses the topics of knowledge, intelligence, innovation and sustainability. The aim of the series is to make available a platform for the publication of books on all aspects of single and multi-disciplinary research on these themes in order to make the latest results available in a readily-accessible form. Volumes on interdisciplinary research combining two or more of these areas is particularly sought.

The series covers systems and paradigms that employ knowledge and intelligence in a broad sense. Its scope is systems having embedded knowledge and intelligence, which may be applied to the solution of world problems in industry, the environment and the community. It also focusses on the knowledge-transfer methodologies and innovation strategies employed to make this happen effectively. The combination of intelligent systems tools and a broad range of applications introduces a need for a synergy of disciplines from science, technology, business and the humanities. The series will include conference proceedings, edited collections, monographs, handbooks, reference books, and other relevant types of book in areas of science and technology where smart systems and technologies can offer innovative solutions.

High quality content is an essential feature for all book proposals accepted for the series. It is expected that editors of all accepted volumes will ensure that contributions are subjected to an appropriate level of reviewing process and adhere to KES quality principles.

More information about this series at http://www.springer.com/series/8767

Atulya Nagar · Durga Prasad Mohapatra
Nabendu Chaki
Editors

Proceedings of 3rd International Conference on Advanced Computing, Networking and Informatics

ICACNI 2015, Volume 1

Editors
Atulya Nagar
Department of Computer Science
Liverpool Hope University
Liverpool
UK

Nabendu Chaki
Department of Computer Science
and Engineering
University of Calcutta
Kolkata, West Bengal
India

Durga Prasad Mohapatra
Department of Computer Science
and Engineering
National Institute of Technology Rourkela
Rourkela, Odisha
India

ISSN 2190-3018 ISSN 2190-3026 (electronic)
Smart Innovation, Systems and Technologies
ISBN 978-81-322-3416-6 ISBN 978-81-322-2538-6 (eBook)
DOI 10.1007/978-81-322-2538-6

Springer New Delhi Heidelberg New York Dordrecht London

Softcover re-print of the Hardcover 1st edition 2016

Printed on acid-free paper

Springer (India) Pvt. Ltd. is part of Springer Science+Business Media (www.springer.com)

Preface

It is indeed a pleasure to receive overwhelming response from researchers of premier institutes of the country and abroad for participating in the 3rd International Conference on Advanced Computing, Networking, and Informatics (ICACNI 2015), which makes our endeavor successful. The conference organized by School of Computer Engineering, KIIT University, India during 23–25 June 2015 certainly marks a success toward bringing researchers, academicians, and practitioners in the same platform. We have received more than 550 articles and very stringently have selected through peer review 132 best articles for presentation and publication. We could not accommodate many promising works as we tried to ensure the quality. We are thankful to have the advice of dedicated academicians and experts from industry to organize the conference in good shape. We thank all people participating and submitting their works and having continued interest in our conference for the third year. The articles presented in the two volumes of the proceedings discuss the cutting edge technologies and recent advances in the domain of the conference.

We conclude with our heartiest thanks to everyone associated with the conference and seeking their support to organize the 4th ICACNI 2016.

Conference Committee

Chief Patron
Achyuta Samanta, Founder, Kalinga Institute of Industrial Technology and Kalinga Institute of Social Sciences, India

Patrons
N.L. Mitra, Chancellor, Kalinga Institute of Industrial Technology, India
Premendu P. Mathur, Vice Chancellor, Kalinga Institute of Industrial Technology, India

General Chairs
Amulya Ratna Swain, Kalinga Institute of Industrial Technology, India
Rajib Sarkar, Central Institute of Technology Raipur, India

Program Chairs
Manmath N. Sahoo, National Institute of Technology Rourkela, India
Samaresh Mishra, Kalinga Institute of Industrial Technology, India

Program Co-chairs
Ashok Kumar Turuk, National Institute of Technology Rourkela, India
Umesh Ashok Deshpande, Visvesvaraya National Institute of Technology, India
Biswajit Sahoo, Kalinga Institute of Industrial Technology, India

Organizing Chairs
Bibhudutta Sahoo, National Institute of Technology Rourkela, India
Madhabananda Das, Kalinga Institute of Industrial Technology, India
Sambit Bakshi, National Institute of Technology Jamshedpur, India

Technical Track Chairs

Computing

Umesh Chandra Pati, SMIEEE, National Institute of Technology Rourkela, India
Debasish Jena, International Institute of Information Technology Bhubaneswar, India

Networking

Shrishailayya Mallikarjunayya Hiremath, National Institute of Technology Rourkela, India
Daya K. Lobiyal, Jawaharlal Nehru University, India

Informatics

Korra Sathya Babu, National Institute of Technology Rourkela, India
Hrudaya Kumar Tripathy, Kalinga Institute of Industrial Technology, India

Industrial Track Chairs

Debabrata Mahapatra, Honeywell Technology Solutions, India
Binod Mishra, Tata Consultancy Services, India

Session and Workshop Chairs

Bhabani Shankar Prasad Mishra, Kalinga Institute of Industrial Technology, India
Ramesh K. Mohapatra, National Institute of Technology Rourkela, India

Special Session Chair

Ajit Kumar Sahoo, National Institute of Technology Rourkela, India

Web Chairs

Soubhagya Sankar Barpanda, National Institute of Technology Rourkela, India
Bhaskar Mondal, National Institute of Technology Jamshedpur, India

Publication Chair

Savita Gupta, Vedanta Aluminum Limited

Publicity Chairs

Ram Shringar Rao, Ambedkar Institute of Advanced Communication Technologies and Research, India
Himansu Das, Kalinga Institute of Industrial Technology, India

Registration Chairs

Sachi Nandan Mohanty, Kalinga Institute of Industrial Technology, India
Sital Dash, Kalinga Institute of Industrial Technology, India

Technical Program Committee

Adam Schmidt, Poznan University of Technology, Poland
Akbar Sheikh Akbari, University of Gloucestershire, UK
Al-Sakib Khan Pathan, SMIEEE, International Islamic University Malaysia (IIUM), Malaysia
Andrey V. Savchenko, National Research University Higher School of Economics, Russia

Annappa, SMIEEE, National Institute of Technology Karnataka, Surathkal, India
B. Narendra Kumar Rao, Sree Vidyanikethan Engineering College, India
Biju Issac, SMIEEE, FHEA, Teesside University, UK
C.M. Ananda, National Aerospace Laboratories, India
Ch. Aswani Kumar, Vellore Institute of Technology, India
Dhiya Al-Jumeily, Liverpool John Moores University, UK
Dilip Singh Sisodia, National Institute of Technology Raipur, India
Dinabandhu Bhandari, Heritage Institute of Technology, Kolkata, India
Ediz Saykol, Beykent University, Turkey
Jamuna Kanta Sing, SMIEEE, Jadavpur University, India
Jerzy Pejas, Technical University of Szczecin, Poland
Joel J.P.C. Rodrigues, Instituto de Telecomunicacoes, University of Beira Interior, Portugal
Koushik Majumder, SMIEEE, West Bengal University of Technology, India
Krishnan Nallaperumal, SMIEEE, Sundaranar University, India
Kun Ma, University of Jinan, China
Laszlo T. Koczy, Széchenyi István University, Hungary
M. Murugappan, University of Malayesia, Malayesia
M.V.N.K. Prasad, Institute for Development and Research in Banking Technology (IDRBT), India
Maria Virvou, University of Piraeus, Greece
Mihir Chakraborty, Jadavpur University, India
Natarajan Meghanathan, Jackson State University, USA
Patrick Siarry, SMIEEE, Université de Paris, Paris
Pradeep Singh, National Institute of Technology Raipur, India
Rajarshi Pal, Institute for Development and Research in Banking Technology (IDRBT), India
Soumen Bag, Indian School of Mines Dhanbad, India
Takuya Asaka, Tokyo Metropolitan University, Japan
Tuhina Samanta, Bengal Engineering and Science University, India
Umesh Hodeghatta Rao, SMIEEE, Xavier Institute of Management, India
Vivek Kumar Singh, SMIEEE, MACM, South Asian University, India
Yogesh H. Dandawate, SMIEEE, Vishwakarma Institute of Information Technology, India
Zahoor Khan, SMIEEE, Dalhouise University, Canada

Organizing Committee

Arup Abhinna Acharya, Kalinga Institute of Industrial Technology, India
C.R. Pradhan, Kalinga Institute of Industrial Technology, India
Debabala Swain, Kalinga Institute of Industrial Technology, India
Harish Kumar Patnaik, Kalinga Institute of Industrial Technology, India
Mahendra Kumar Gourisaria, Kalinga Institute of Industrial Technology, India
Manas Ranjan Lenka, Kalinga Institute of Industrial Technology, India

Manjusha Pandey, Kalinga Institute of Industrial Technology, India
Manoj Kumar Mishra, Kalinga Institute of Industrial Technology, India
Pinaki Sankar Chatterjee, Kalinga Institute of Industrial Technology, India
R.N. Ramakant Parida, Kalinga Institute of Industrial Technology, India
Santwana Sagnika, Kalinga Institute of Industrial Technology, India
Sharmistha Roy, Kalinga Institute of Industrial Technology, India
Siddharth Swarup Rautaray, Kalinga Institute of Industrial Technology, India
Subhasis Dash, Kalinga Institute of Industrial Technology, India
Sujoy Datta, Kalinga Institute of Industrial Technology, India
Tanmaya Swain, Kalinga Institute of Industrial Technology, India

External Reviewer Board
Ahmed Farouk Metwaly, Mansoura University, Egypt
Ankit Thakkar, Nirma University, India
Arun Jana, Centre for Development of Advanced Computing, India
Bharat Singh, Indian Institute of Information Technology Allahabad, India
Biswapratap Singh Sahoo, National Taiwan University, Taiwan
Chiheb-Eddine Ben N'Cir, University of Tunis, Tunisia
Gaurav Sharma, Galgotias University, India
Gudikandhula Narasimha Rao, Andhra University, India
Gunter Fahrnberger, University of Hagen, Germany
Hamidreza Khataee, Griffith University, Australia
Jitendra Agrawal, Rajiv Gandhi Proudyogiki Vishwavidyalaya, India
Manuj Darbari, Babu Banarasi Das University, India
Mithileysh Sathiyanarayanan, University of Brighton, UK
Monish Chatterjee, Asansol Engineering College, India
Musheer Ahmad, Jamia Millia Islamia University, India
N. Bhalaji, SSN College of Engineering, India
K. Nandhini, Sastra University, India
Nisarg Gandhewar, SB Jain Institute of Technology, Management and Research Nagpur, India
P. Victer Paul, Sri Manakula Vinayagar Engineering College, Puducherry, India
R. Rajavel, SSN College of Engineering, India
Rajasekar Karthik, Oak Ridge National Laboratory, USA
Rajesh Duvvuru, Andhra University, India
Roberto Herrera Lara, National Polytechnic School, Ecuador
Shahriar Shahabuddin, University of Oulu, Finland
Sobhana N.V., Rajiv Gandhi Institute of Technology, India
Sonali Agarwal, Indian Institute of Information Technology Allahabad, India
Soren Atmakuri Davidsen, Sri Venkateswara University, India
Sreedevi K. Menon, Amrita Vishwavidyapeetham, India
Srikumar Panda, Indian Institute of Technology Ropar, India
Subrata Biswas, Netaji Subhash Engineering College, India

V. Rajinikanth, St. Joseph's College of Engineering, India
Vijay Kumar, Manav Rachna International University, India
Waad Bouaguel, University of Tunis, Tunisia
Wu-Chen Su, Kaohsiung Chang Gung Memorial Hospital, Taiwan
Yilun Shang, Hebrew University of Jerusalem, Israel
Yogendra Narain Singh, Uttar Pradesh Technical University, India

Contents

About the Editors

Prof. Atulya Nagar holds the Foundation Chair as Professor of Mathematical Sciences at Liverpool Hope University where he is the Dean of Faculty of Science. He has been the Head of Department of Mathematics and Computer Science since December 2007. A mathematician by training he is an internationally recognized scholar working at the cutting edge of applied nonlinear mathematical analysis, theoretical computer science, operations research, and systems engineering and his work is underpinned by strong complexity-theoretic foundations. He has extensive background and experience of working in universities in the UK and India. He has edited volumes on Intelligent Systems and Applied Mathematics; he is the Editor-in-Chief of the International Journal of Artificial Intelligence and Soft Computing (IJAISC) and serves on editorial boards for a number of prestigious journals such as the Journal of Universal Computer Science (JUCS). Professor Nagar received the prestigious Commonwealth Fellowship for pursuing his Doctorate (D.Phil.) in Applied Non-Linear Mathematics, which he earned from the University of York in 1996. He holds B.Sc. (Hons.), M.Sc., and M.Phil. (with Distinction) from the MDS University of Ajmer, India.

Dr. Durga Prasad Mohapatra received his Ph.D. from Indian Institute of Technology, Kharagpur. He joined the Department of Computer Science and Engineering at the National Institute of Technology, Rourkela, India in 1996, where he is presently serving as Associate Professor. His research interests include software engineering, real-time systems, discrete mathematics, and distributed computing. He has published more than thirty research papers in these fields in various international journals and conferences. He has received several project grants from DST and UGC, Government of India. He has received the Young Scientist Award for the year 2006 from Orissa Bigyan Academy. He has also received the Prof. K. Arumugam National Award and the Maharashtra State National Award for outstanding research work in Software Engineering for 2009 and 2010, respectively, from the Indian Society for Technical Education (ISTE), New Delhi. He is to receive the Bharat Sikshya Ratan Award for significant contribution to academics awarded by the Global Society for Health and Educational Growth, Delhi.

Prof. Nabendu Chaki is a Professor in the Department Computer Science and Engineering, University of Calcutta, Kolkata, India. Dr. Chaki did his first graduation in Physics at the legendary Presidency College in Kolkata and then in Computer Science and Engineering, University of Calcutta. He completed his Ph.D. in 2000 from Jadavpur University, India. He shares two US patents and one patent in Japan with his students. Professor Chaki is active in developing international standards for Software Engineering. He represents the country in the Global Directory (GD) for ISO-IEC. Besides editing more than 20 book volumes in different Springer series including LNCS, Nabendu has authored five text and research books and about 130 peer-reviewed research papers in journals and international conferences. His areas of research include distributed computing, image processing, and software engineering. Dr. Chaki has served as Research Assistant Professor in the Ph.D. program in Software Engineering in U.S. Naval Postgraduate School, Monterey, CA. He has strong and active collaborations in US, Europe, Australia, and other institutes and industries in India. He is a visiting faculty member for many universities in India and abroad. Dr. Chaki has been the Knowledge Area Editor in Mathematical Foundation for the SWEBOK project of the IEEE Computer Society. Besides being on the editorial board of several international journals, he has also served on the committees of more than 50 international conferences. Professor Chaki is the founder Chair of the ACM Professional Chapter in Kolkata.

Part I
Soft Computing Techniques for Advanced Computing

Fuzzy Logic Based UPFC Controller for Voltage Stability and Reactive Control of a Stand-Alone Hybrid System

Asit Mohanty, Meera Viswavandya, Sthitapragyan Mohanty and Dillip Mishra

Abstract This paper is mainly focused on the implementation of fuzzy logic based UPFC controller in the isolated wind-diesel-micro hydro system for management of reactive power and enhancement of voltage stability. For better analysis a linearised small signal transfer function system is considered for different load inputs. The fuzzy based UPFC controller has been tunned to improve the reactive power of the off grid system. Simulation in MATLAB environment has been carried out and the effectiveness of fuzzy tuned controller is established..

Keywords Reactive power compensation · Standalone wind-diesel-micro hydro system · UPFC

1 Introduction

Renewable energy sources by nature are intermittent and non predictable though they are plentily available in the nature. Hybrid energy sources combining multiple energy sources mitigate this problem to a great extent as because shortfall due to one source is replenished by other. Generally standalone hybrid models exist near the place of consumption and can be connected to the main grid. One or more renewable sources are combined to form a Hybrid system where the inadequacy of generation of power because of one source is met by the other source [1, 2]. Wind Diesel Micro hydro system is quite popular choice of combined energy source where a combined network of wind, micro hydro and diesel system work to provide

A. Mohanty (✉) · M. Viswavandya · S. Mohanty
College of Engineering & Technology, Bhubaneswar, India
e-mail: asithimansu@gmail.com

D. Mishra
EAST, Bhubaneswar, India

A. Nagar et al. (eds.), *Proceedings of 3rd International Conference on Advanced Computing, Networking and Informatics*, Smart Innovation, Systems and Technologies 43, DOI 10.1007/978-81-322-2538-6_1

continuous power supply to the load, Turbines used in this particular model are Induction generators and they use the Synchronous generator based Diesel engine as back up [3–5]. Turbines having Induction generators have some operational difficulties as they require reactive energy for its smooth working. The much needed reactive energy is provided by the Synchronous machine to a small extent but the real compensation is impossible without the FACTS devices who not only manages the reactive power but enhances the overall stability of the system.

Capacitor banks [6–8] which is mentioned in many articles fail to assist in the compensation of reactive energy and improve voltage stability. Its because the wind is intermittent and load change is unavoidable. The voltage problem and reactive power mismatch are mitigated by the FACTS devices [9–11].

UPFC is one of the important member of FACTS devices which has been utilized like SVC and STACOM for compensation of reactive power. UPFC acts as a better compensator and increases the voltage stability and angle stability. Management of reactive energy is extremely essential as shortfall of reactive power. makes the system voltage varying. The importance of reactive power can be accessed from the fact that the shortfall of reactive power makes the whole system unstable. UPFC like other FACTS members has proven its ability in compensating reactive energy and improving the stability margin. Furthermore a Fuzzy logic controller can be added to the UPFC Controller to tune the parameters and improve the stability status of the system to a great extent.

A simulink based Fuzzy logic tuned wind diesel micro hydro system is discussed with reactive power compensator like UPFC with step change in load, for better transient stability. For better analysis the mathematical model has been derived. The proportional and integral constants are finely tuned by the Fuzzy controller.

2 System Configuration and Its Mathematical Modeling

The wind-diesel-micro hydro hybrid system essentially takes the generating devices like wind turbine and micro hydro turbine to supply power to the loads. The backup is provided by the Diesel generator. The synchronous based Diesel genset also helps the system in improvement of reactive power. The single line system block with UPFC controller is shown in Fig. 1. The system parameter table is shown in this paper and is mentioned as Table 3 (Fig. 2).

The balanced equation of Reactive power of (SG, UPFC, IG, and LOAD) is expressed as [13].

The reactive power balance equation of the system for uncertain load ΔQ_L is

$$\Delta Q_{SG} + \Delta Q_{UPFC} = \Delta Q_L + \Delta Q_{IG} + \Delta Q_{IGH} \quad (1)$$

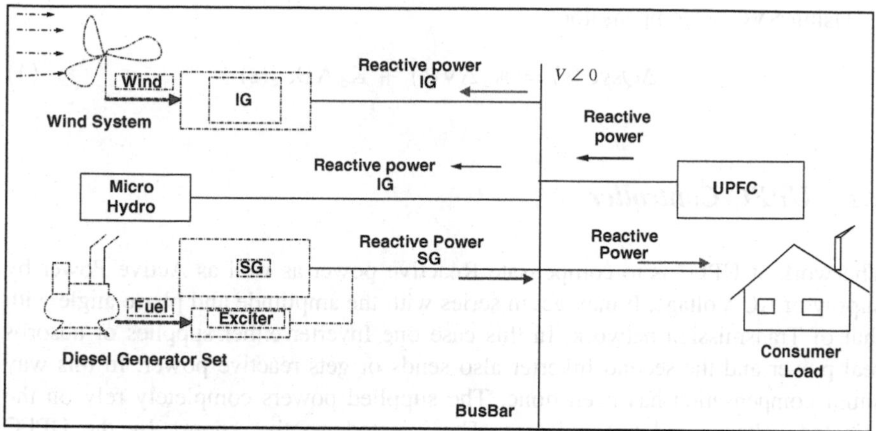

Fig. 1 Wind diesel micro hydro hybrid system block

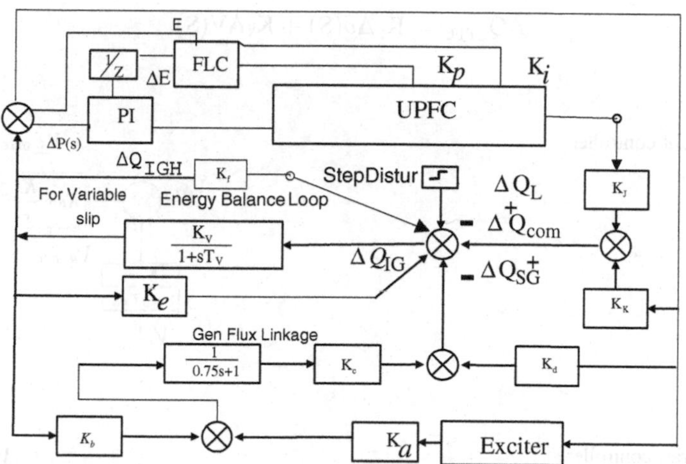

Fig. 2 Transfer function of fuzzy logic based UPFC controller

$\Delta Q_{SG} + \Delta Q_{UPFC} - \Delta Q_L - \Delta Q_{IG} - \Delta Q_{IGH}$ deviates the system output voltage.

$$\Delta V(S) = \frac{K_V}{1 + ST_V}\left[\Delta Q_{SG}(S) + \Delta Q_{COM}(S) - \Delta Q_L(S) - \Delta Q_{IG}(S) - \Delta Q_{IGH}\right] \quad (2)$$

$$Q_{SG} = \frac{\left(E'_q V \cos\delta - V^2\right)}{X'd} \text{ and } \Delta Q_{SG} = \frac{V\cos\delta}{X'd\Delta E'q} + \frac{E'q\cos\delta 2V}{X'd\Delta V} \quad (3)$$

Using SVC as compensator

$$\Delta Qsvc\,(s) = K_c\,\Delta V(s) + K_d\,\Delta\,B_{svc}(s) \tag{4}$$

2.1 UPFC Controller

The work of UPFC is to compensate Reactive power as well as Active Power by supply of AC Voltage. It may act in series with the amplitude and phase angle with that of Transmission network. In this case one Inverter either supplies or absorbs real power and the second Inverter also sends or gets reactive power. In this way shunt compensation has been done. The supplied powers completely rely on the injected voltages and bus voltages. The injected reactive energy by the UPFC Controller depends on V_m and angle δ (Figs. 3 and 4). So the small change in reactive power is equal to

$$\Delta Q_{UPFC} = K_j\,\Delta\delta(S) + K_k\Delta V(S) \tag{5}$$

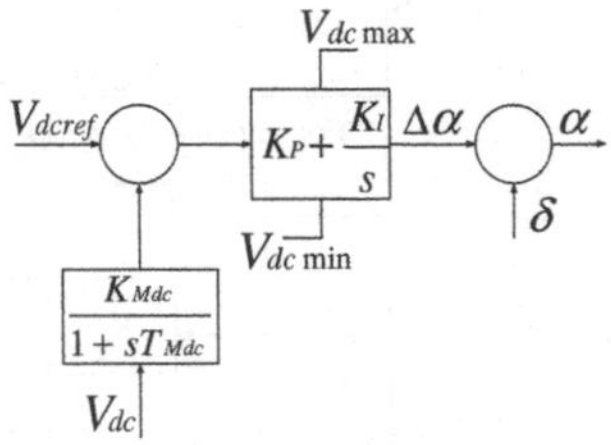

Fig. 3 Shunt controller

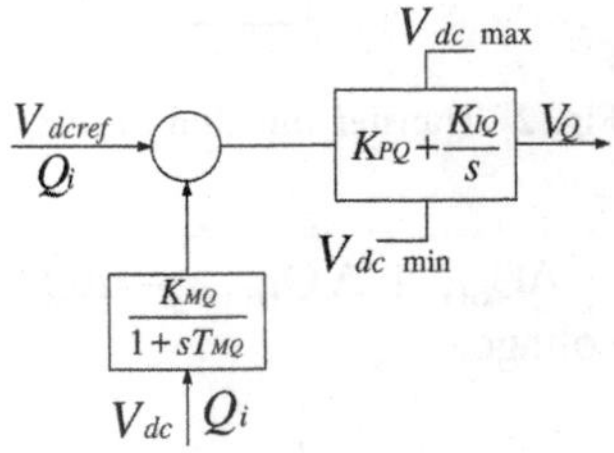

Fig. 4 Series controller

3 Fuzzy Logic Controller (FLC)

A fuzzy set has much more generalised approach than the ordinary sets. The proposed self tuned Fuzzy controller operates in 3 stages i.e. Fuzzification inference engine and defuzzification. These three stages with the help of membership functions tune the parameters of UPFC controller (Fig. 5).

3.1 *Self-Tuning Fuzzy PI Controller*

As power system is non linear a self tuned Fuzzy controller is designed to tune the proportional and integral controllers.of the PI controller. A PI controller can be explained as U(s) = KpE(s) + Ki ∫ E (s) and E*Kp + Ki * ∫ E = U. The design of Fuzzy controller is such that the inputs and outputs are taken into account. Kp and Ki of the PI controller are tuned by the inference engine of FLC. For this Fuzzy controller error and change in error work as inputs and the output as proportional and integral constants.

$$\mu Pi = \min(\mu(E), \mu(\Delta E)), \mu Ii = \min(\mu(E), \mu(\Delta E))$$

The fuzzy controller uses seven linguistic variables so that there are 49 rules and it uses triangular membership function for tuning the Kp and Ki. An auto tuned fuzzy controller has input membership functions as error (E) and change in error (Del E) and in the output it gives tuned values of Kp and Ki (Tables 1 and 2).

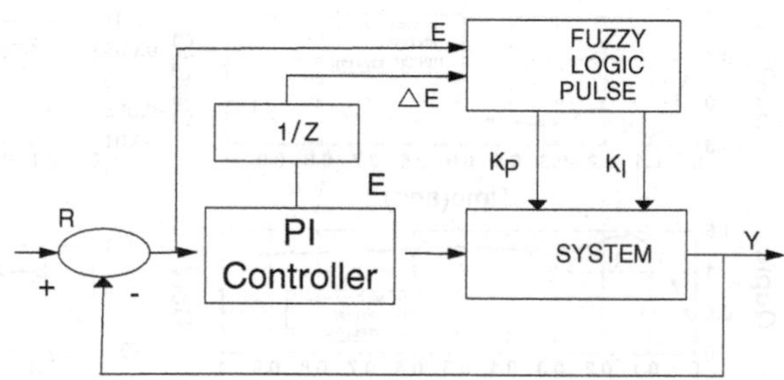

Fig. 5 Fuzzy PI controller block

Table 1 Fuzzy rule for Kp

E/ΔE	NL	NM	NS	Z	PS	PM	PL
NL	VL	VL	VB	VB	MB	M	M
NM	VL	VL	VB	MB	MB	M	MS
NS	VB	VB	VB	MB	M	M	MS
Z	VB	VB	MB	M	MS	VS	VS
PS	MB	MB	M	MS	MS	VS	VS
PM	VB	MB	M	MS	VS	VS	Z
PL	M	MS	VS	VS	VS	Z	Z

Table 2 Fuzzy rule for Ki

E/ΔE	NL	NM	NS	Z	PS	PM	PL
NL	Z	Z	VS	VS	MS	M	M
NM	Z	Z	VS	MS	MS	M	M
NS	Z	VS	MS	MS	M	MB	MB
Z	VS	VS	MS	M	MB	VB	VB
PS	VS	MS	M	MB	MB	VB	VL
PM	M	M	MB	MB	VB	VL	VL
PL	M	M	MB	VB	VB	VL	VL

4 Simulation Results

From the simulation based on UPFC controller with fuzzy logic performances of different parameters are noticed. The wind diesel micro hydro hybrid power system is simulated in MATLAB/Simulink environment. With a step load change of (1–5) % and variable input parameters like wind and hydro energy, Variation of all existing system parameters are noticed and plotted as shown (Fig. 6). During observation it is found that UPFC provides good performance with increase size of synchronous generator than induction generator. Vital parameters such as settling time and peak overshoot are found reduced in case of fuzzy controller with respect to traditional PI Controller. The Control signal is noticed after fuzzification, rules creation and defuzzification. It provides good damping but it has some Negatives like creation of membership function, making of rules and suitability of scaling factors which is done by trial and error method (Fig. 7).

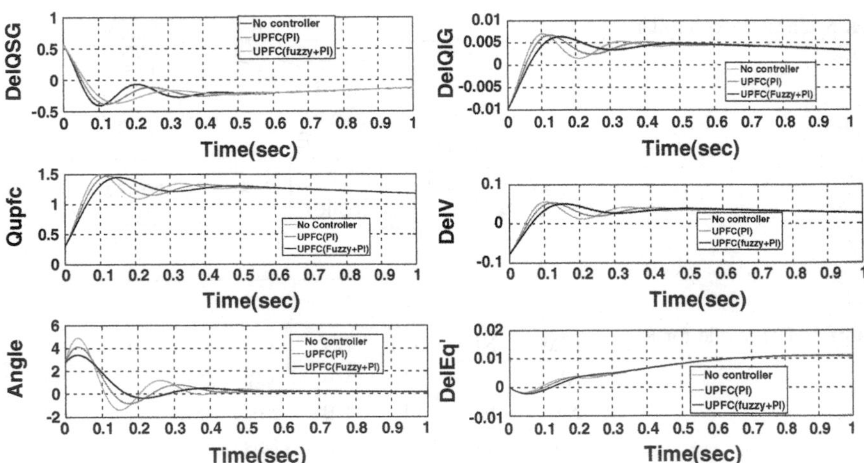

Fig. 6 Comparative results of wind diesel micro hydro hybrid system

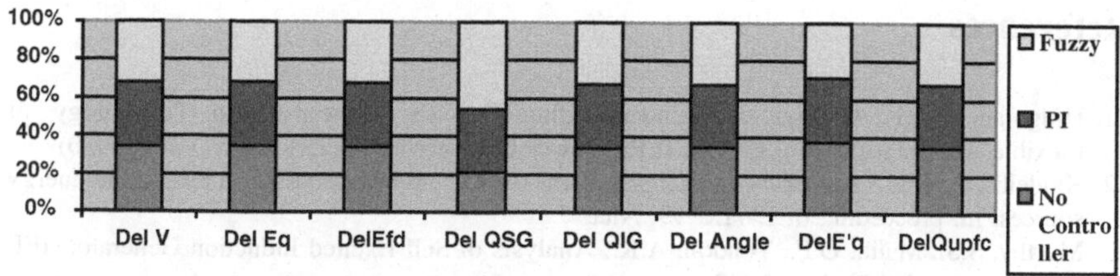

Fig. 7 Settling time of different parameters

5 Conclusion

This work has described a fuzzy logic tuned UPSC controller in the DG based hybrid system and discusses its impact in the management of reactive power and voltage stability. The parameters of the wind-diesel-micro hydro system perform in a better manner after proper tuning of the PI controller is done by Fuzzy controller. The simulated results show less settling duration and better overshoots. This particular hybrid system after the inclusion UPFC controller with soft computing approach performs robustly and takes the entire model to a more stable level.

Appendix

Table 3 Parameters of wind-diesel system

System Parameter	WindDieselSystem
Wind (IG)Capacity (Kilowatt) Diesel(SG) Capacity Load Capacity	100 KW 100 KW 200 KW
Base Power in KVA	200KVA

$P_{SG,=}$ 0.4 in KW $P_{IG,=}$ 0.6 in Kilowatts $P_{L}(pu)$= 1.0 in KW
Q_{SG} = 0.2 in KW Q_{IG},pu=0.189 in KVAR $Q_{L}(pu)$= 0.75 in KVAR
T'_{do} = 5.044
$E_{q}(pu)$ = 1.113 P_{IN},pu in Kilowatts =0.75 α in Radian= 2.44
X'_{d} =0.3
$E_{q}'(pu)$ = 0.96 r1=r2(pu)=0.19 $X_{d}(pu)$=1.0
V(pu) = 1.0
X1=X2(pu) = 0.56 T'_{do},s= 5 $T_{\alpha}(S)$= 0.05

References

1. Hingorani, N.G., Gyugyi, L.: Understanding FACTS, Concepts and Technology of Flexible AC Transmission System. IEEE Power Engineering Society, New York (2000)
2. Kaldellis, J., et al.: Autonomous energy system for remote island based on renewable energy sources. In: proceeding of EWEC 99, Nice
3. Murthy, S.S., Malik, O.P., Tondon, A.K.: Analysis of Self Excited Induction Generator, IEE Proceeding, vol. 129, Nov 1982
4. Padiyar, K.R.: FACTS controlling in power transmission system and distribution. New Age International Publishers (2008)
5. Tondon, A.K., Murthy, S.S., Berg, G.J.: Steady state analysis of capacitors excited induction generators. IEEE Transaction on Power Apparatus & Systems, March 1984
6. Padiyar, K.R., Verma, R.K.: Damping torque analysis of static VAR system controller. IEEE Trans. Power Syst. **6**(2), 458–465 (1991)
7. Bansal, R.C.: Three phase self excited induction generators, an over view. IEEE Transaction on Energy Conversion (2005)
8. Saha, T.K., Kasta, D.: Design optimisation & dynamic performance analysis of electrical power generation system. IEEE Transaction on Energy Conversion Device (2010)
9. Gould, I.E.: Baring: Wind Diesel Micro Hydro Power Systems Basics and Examples. National Renewable Energy Laboratory, United States Department of Energy
10. Fukami, T., Nakagawa, K., Kanamaru, Y., Miyamoto, T.: Performance analysis of permanent magnet induction generator under unbalanced grid voltages. Translated from Denki Gakki Ronbunsi, vol. 126, no. 2, pp. 1126–1133 (2006)
11. Mohanty, Asit, Viswavandya, Meera, Ray, Prakash, Patra, Sandeepan: Stability analysis and reactive power compensation issue in a microgrid with a DFIG based WECS. Electr. Power Energy Syst. **62**, 753–762 (2014)

APSO Based Weighting Matrices Selection of LQR Applied to Tracking Control of SIMO System

S. Karthick, Jovitha Jerome, E. Vinodh Kumar and G. Raaja

Abstract This paper employs an adaptive particle swarm optimization (APSO) algorithm to solve the weighting matrices selection problem of linear quadratic regulator (LQR). One of the important challenges in the design of LQR for real time applications is the optimal choice state and input weighting matrices (Q and R), which play a vital role in determining the performance and optimality of the controller. Commonly, trial and error approach is employed for selecting the weighting matrices, which not only burdens the design but also results in non-optimal response. Hence, to choose the elements of Q and R matrices optimally, an APSO algorithm is formulated and applied for tracking control of inverted pendulum. One of the notable changes introduced in the APSO over conventional PSO is that an adaptive inertia weight parameter (AIWP) is incorporated in the velocity update equation of PSO to increase the convergence rate of PSO. The efficacy of the APSO tuned LQR is compared with that of the PSO tuned LQR. Statistical measures computed for the optimization algorithms to assess the consistency and accuracy prove that the precision and repeatability of APSO is better than those of the conventional PSO.

Keywords APSO · LQR · Inverted pendulum · Riccati equation · Tracking control

S. Karthick (✉) · J. Jerome · G. Raaja
Department of Instrumentation and Control Systems Engineering,
PSG College of Technology, Coimbatore 641004, India
e-mail: skpattitude@gmail.com

J. Jerome
e-mail: jovithajerome@gmail.com

G. Raaja
e-mail: raaja1393@gmail.com

E. Vinodh Kumar
School of Electrical Engineering, VIT University, Vellore 632014, India
e-mail: vinothmepsg@gmail.com

A. Nagar et al. (eds.), *Proceedings of 3rd International Conference on Advanced Computing, Networking and Informatics*, Smart Innovation, Systems and Technologies 43, DOI 10.1007/978-81-322-2538-6_2

1 Introduction

Linear Quadratic Regulator, a corner stone of modern optimal control, has attracted considerable attention in the recent years due to its inherent robustness and stability properties [1]. A minimum phase margin of (-60°, 60°) and a gain margin of (-6, ∞) db provided by LQR enable the system to yield satisfactory response even during the small perturbations. Moreover, by minimizing a quadratic cost function which consists of two penalty matrices, namely Q and R matrices, LQR yields an optimal response between the control input and speed of response. Hence, the LQR techniques have been successfully applied to a large number of complex systems such as vibration control system [2], fuel cell systems [3] and aircraft [4]. Nevertheless, one of the major issues of LQR design for real time applications is the choice of Q and R weighting matrices. Even though, the performance of LQR is highly dependent on the elements of Q and R matrices, conventionally the matrices have been tuned either based on the designer's experience or via trial and error approach. Such approach is not only tedious but also time consuming. Hence, in this paper the conventional LQR design problem is reformulated as an optimization problem and solved using particle swarm optimization algorithm.

In literature, several results have been reported on PSO based state feedback controller design. For instance, in [5] selection of weighting matrices of LQR controller for tracking control of inverted pendulum has been solved using PSO. In [6] the performances of GA and PSO for FACTS based controller design have been assessed and reported that both the convergence and time consumption of PSO are less than those of the GA based feedback controller design. PSO based variable feedback gain control design for automatic fighter tracking problems have been investigated in [7] and it has been reported that PSO based LQR design yields better tracking response than the LMI based methods. However, the standard PSO has two important undesirable dynamical properties that degrade its exploration abilities. One of the most important problems is the premature convergence. Due to the rapid convergence and diversity loss of the swarm, the particles tend to be trapped in the local optima solution when solving multimodal tasks. The second problem is the ability of the PSO to balance between global exploration and local search exploitation. Overemphasize of the global exploration prevents the convergence speed of swarm, while too much search exploitation causes the premature convergence of swarm. These limitations have imposed constraints on the wider applications of the PSO to real world problems [8]. Hence, to better the convergence rate and speed of conventional PSO, we propose an adaptive PSO, whose inertia weight is varied adaptively according to the particle's success rate. The key aspect of the proposed APSO is that an adaptive inertia weight parameter (AIWP), whose weights are varied adaptively according to the nearness of the particles towards the optimal solution, is introduced in the velocity update equation of conventional PSO to accelerate the convergence of the algorithm. To assess the performance of the

APSO based LQR control strategy, simulation studies have been carried out on a benchmark inverted pendulum, which is a typical single input multi output system (input: motor voltage, output: cart position and pendulum angle).

2 Problem Formulation

Consider a linear time invariant (LTI) multivariable system,

$$\dot{X}(t) = AX(t) + Bu(t) \tag{1}$$

$$Y(t) = CX(t) + Du(t) \tag{2}$$

The conventional LQR design problem is to compute the optimal control input u^* by minimizing the following quadratic cost function.

$$J(u^*) = \frac{1}{2}\int_0^\infty \left[X^T(t)QX(t) + u^T(t)Ru(t)\right]dt \tag{3}$$

where $Q = Q^T$ is a positive semi definite matrix and $R = R^T$ is a positive definite matrix. By solving the following Lagrange multiplier optimization technique, the optimal state feedback gain matrix (K) can be computed.

$$K = R^{-1}B^T P \tag{4}$$

where P is the solution of following ARE.

$$A^T P + PA - PBR^{-1}B^T P + Q = 0 \tag{5}$$

The elements of Q and R matrices play a vital role in determining the penalty on system states and control input when the system deviates from the equilibrium position. Normally, the Q and R matrices are chosen as diagonal matrices such that the quadratic performance index is a weighted integral of squared error. The sizes of Q and R matrices depend on the number of state variables and input variables respectively. As an alternate to conventional trial and error based manual tuning of these weighting matrices, in the following section, a bio-inspired evolutionary algorithm, an adaptive PSO, has been incorporated in the LQR control strategy for the optimal selection of Q and R.

3 Adaptive PSO

In the last decade, several variants of PSO have been put forward to enhance the performance of conventional PSO. All the proposed variations are mainly to improve the convergence and exploration-exploitation capabilities of PSO. One of the variations incorporated in the PSO is the use of inertia weight parameter to accelerate the convergence of particles towards optimum value. As the inertia weight not only determines the contribution rate of a particle's previous velocity to its current velocity but also yields the required momentum for the particles to move across the solution space, it is important to control the inertia weight to strike a balance between the global search and local exploitation. The larger value of inertia weight concentrates more on global search, while the smaller inertia weight focuses highly on fine tuning the current search space. A comprehensive survey on the use of inertia weight schemes in PSO algorithms is given in [9]. In this paper, we extend the idea of adaptive inertia weight strategy to solve the LQR optimization problem.

To implement an adaptive inertia weight strategy, it is important to evaluate the position of the swarm during every iteration step. Hence, the success percentage (SP) of particles is used to update the velocity adaptively. Large value of SP indicates that the particles have reached the best value and the particles are slowly progressing towards the optimum, whereas a small value of SP implies that the particles are fluctuating around the optimum value with very less improvement. Hence, the success rate can be used to modify the inertia weight adaptively. If the fitness of the current iteration is less than that of the previous iteration the success count (SC) is set to 1, else it is set to zero. Computing the ratio of the SC to the number of iterations, the SP value is computed and used to arrive at the adaptive inertia weight parameter (AIWP) as given below. Table 1 gives the pseudo code of an adaptive PSO algorithm.

$$w(t) = (w_{\max} - w_{\min})SP + w_{\min} \tag{6}$$

Table 1 APSO pseudo code

Randomly initialize Particle swarm, minimum and maximum values of inertia weight $(w_{\min}, w_{\max})$
for i <=100
Set Success Count (SC) = 0
Evaluate the fitness of particle swarm using $f = ISE = \int e^2(t)dt$
for i = 1 to 30
if $f < f_{pbest_i}$
SC = SC + 1
$f_{pbest_i} \leftarrow f$
$x_{pbest_i} \leftarrow x_i$
end if
if $f < f_{gbest_i}$
$f_{gbest_i} \leftarrow f$
$x_{gbest_i} \leftarrow x_i$
end if
for d = 1 to dimensions
$v_i^d(t+1) = w * v_i^d(t) + c_1 * rand_1 \left(pbest_i^d(t) - x_i^d(t)\right) + c_2 * rand_2 * \left(pbest_g^d(t) - x_i^d(t)\right)$

(continued)

Table 1 (continued)

$x_i^d(t+1) = x_i^d(t) + v_i^d(t+1)$
`end for`
`PS = SC/100;`
$w = (w_{\max} - w_{\min}) * SP + w_{\min}$
`end for`
`end for`

4 Single Inverted Pendulum

Single inverted pendulum is used as a typical benchmark system to evaluate effectiveness of various control schemes due to its highly nonlinear and inherently unstable properties. It consists of a DC motor and a pendulum, which is attached to the shaft of the motor. Two encoders are used to measure the position of the cart and the angle of the pendulum. Figure 1 shows the schematic diagram of a single inverted pendulum.

Two control schemes, namely swing up control and stabilization control, are used to meet the control objective of maintaining the pendulum angle at zero degree while the cart tracks the reference trajectory. The stabilization control is implemented using LQR due to the practical limitation on control input (motor voltage) given to the cart system. Using Euler-Lagrangian energy based approach the nonlinear equation of motion of pendulum can be written as

$$(M_c + M_p)\ddot{x}_c(t) + B_{eq}\dot{x}_c(t) - (M_p l_p \cos(\alpha(t)))\ddot{\alpha}(t) + M_p l_p \sin(\alpha(t))\dot{\alpha}^2(t) = F_c(t) \tag{7}$$

and

$$-M_p l_p \cos(\alpha(t))\ddot{x}_c(t) + \left(I_p + M_p l_p^2\right)\ddot{\alpha}(t) + B_p\dot{\alpha}(t) - M_p g l_p \sin(\alpha(t)) = 0 \tag{8}$$

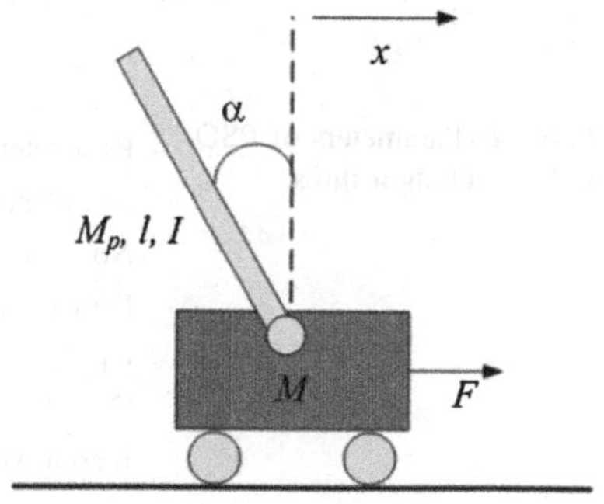

Fig. 1 Schematic diagram of single inverted pendulum

To obtain the state model, four variables namely, cart position, pendulum angle, cart velocity and pendulum velocity are taken as state variables and the state space model is obtained by linearizing the model around the equilibrium point ($\sin(\alpha) \cong \alpha$, $\cos(\alpha) \cong 1$). The following numerical state space model of inverted pendulum system is borrowed from [10] for controller design.

$$\begin{bmatrix} \dot{x}_c \\ \dot{\alpha} \\ \ddot{x}_c \\ \ddot{\alpha} \end{bmatrix} = \begin{bmatrix} 0 & 0 & 1 & 0 \\ 0 & 0 & 0 & 1 \\ 0 & 2.2643 & -15.8866 & -0.0073 \\ 0 & 27.8203 & -36.6044 & -0.0896 \end{bmatrix} \begin{bmatrix} x_c \\ \alpha \\ \dot{x}_c \\ \dot{\alpha} \end{bmatrix} + \begin{bmatrix} 0 \\ 0 \\ 2.2772 \\ 5.2470 \end{bmatrix} u \tag{9}$$

$$y = \begin{bmatrix} 1 & 0 & 0 & 0 \\ 0 & 1 & 0 & 0 \\ 0 & 0 & 1 & 0 \\ 0 & 0 & 0 & 1 \end{bmatrix} \begin{bmatrix} x_c \\ \alpha \\ \dot{x}_c \\ \dot{\alpha} \end{bmatrix} \tag{10}$$

5 Results and Discussion

The APSO based LQR tracking control algorithm is implemented in MATLAB 2013b. Table 2 gives the parameters used for PSO and APSO algorithms. The dimension of the optimization algorithms are chosen to be 3 as the number of variables to be optimized in the LQR design is 3 (q_{11}, q_{22} and r). Moreover, the number of iterations, particle size and cognitive acceleration constants in both PSO and APSO are same except the inertia weight. In case of conventional PSO inertia weight is linearly varied, whereas in APSO the inertia weight is adaptively varied according to the particle's success rate as given in (6). According to the fitness function ISE, the optimization algorithms are executed for the specified number of iterations and the global best of the particles, the weights of LQR, are obtained. Figure 2 illustrates the fitness function of both PSO and APSO algorithms.

Table 2 Parameters of PSO and APSO algorithms

Parameters	PSO	APSO
No. of population (N)	30	30
No. of iterations (i)	100	100
Dimensions (d)	3	3
C_1	0.9	0.9
C_2	1.2	1.2
Inertia weight (w)	0.9	AIWP

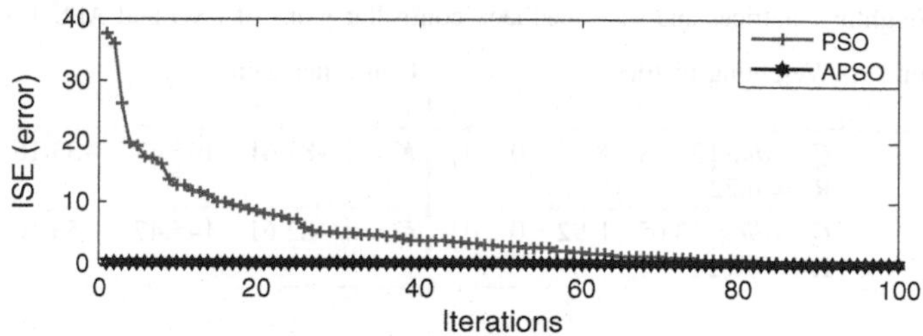

Fig. 2 Fitness function of PSO and APSO

From Table 3, it can be inferred that the minimum fitness function of APSO is less than that of the PSO, which accentuates that the accuracy of the APSO is better than that of PSO. Moreover, the convergence speed of APSO is faster than that of PSO. Figure 3 shows the surface plot of the optimization algorithms. It can be noted

Table 3 Statistical analysis of PSO and APSO

Statistical parameter	PSO	APSO
Mean (μ)	0.1011	0.0316
Standard deviation (σ)	0.2123	0.0367
Minimum (m_x)	0.00032	0.0020
Maximum (M_x)	0.6962	0.1122
Range (R)	0.6942	0.1119

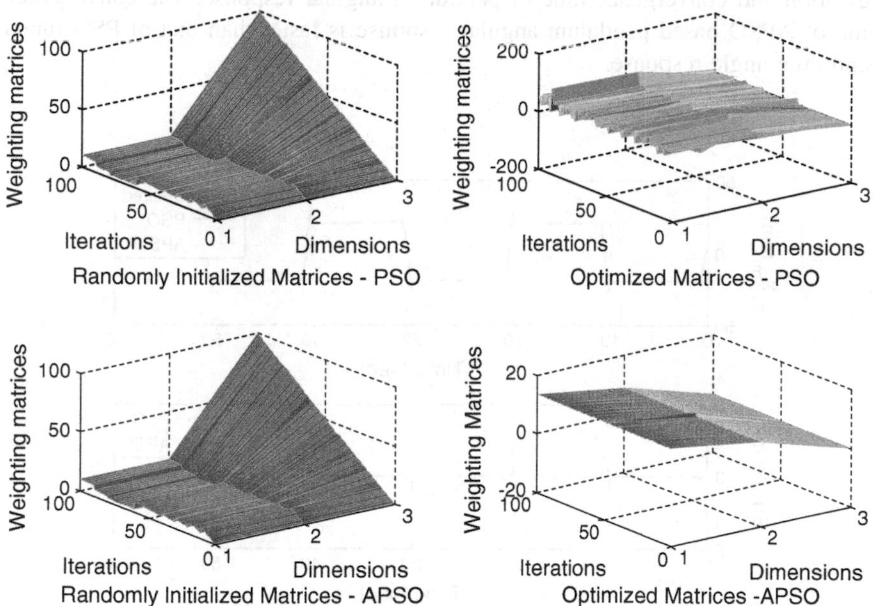

Fig. 3 Surface plots of PSO and APSO

Table 4 Weighting matrices and state feedback controller gains of PSO and APSO

Optimization algorithm	Weighting matrices	Controller gain
PSO	$Q = diag[31.88 \quad 8.97 \quad 0 \quad 0]$ $\mathrm{R} = 0.22$	$K = [-82.61 \quad 145.47 \quad -53.16 \quad 18.85]$
APSO	$Q = diag[13.65 \quad 8.92 \quad 0 \quad 0]$ $\mathrm{R} = 0.002$	$K = [-82.61 \quad 145.47 \quad -53.16 \quad 18.85]$

that the smoothness of the convergence is significantly better in APSO compared to PSO. Table 4 gives the corresponding Q and R matrices and controller gain of LQR obtained using the PSO and APSO algorithms.

5.1 Trajectory Tracking Response

To assess the tracking response of the APSO tuned LQR controller framework, a square test signal of 0.05 Hz with 40 cm (peak to peak) displacement amplitude is given and the response is illustrated in Fig. 4.

From Table 5, which gives the time domain specifications of the cart position response, it is worth to note that both the settling time and the dead time of the APSO tuned LQR is better than those of PSO tuned LQR. The pendulum angle response and its corresponding motor voltage are shown in Fig. 5. Table 6 gives the deviation and convergence time of pendulum angular response. The convergence time of APSO based pendulum angular response is faster than that of PSO tuned pendulum angle response.

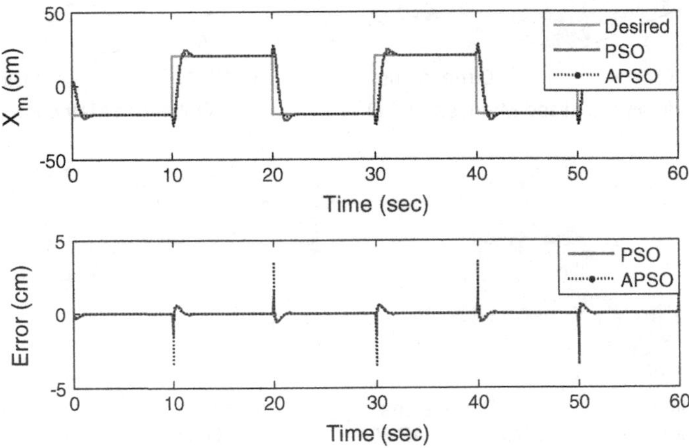

Fig. 4 Cart position and tracking error for square trajectory

Table 5 Comparison of cart position response

Optimization method	Time domain parameters			Performance index ISE
	t_d	t_s	$\%M_p$	
PSO	0.4	3.5	20	0.412
APSO	0.25	3	10	0.376

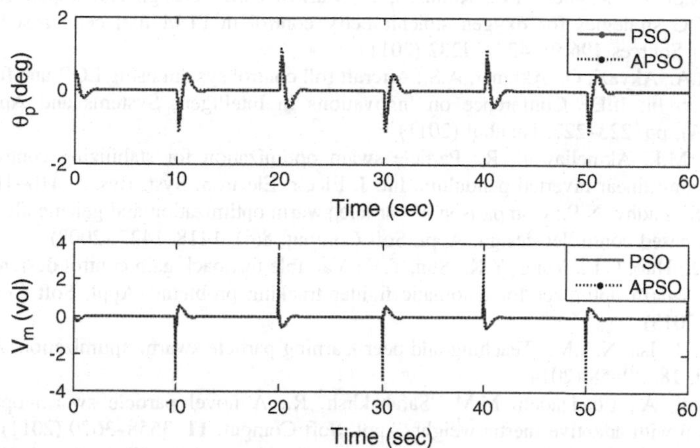

Fig. 5 Pendulum angle and motor voltage for square trajectory

Table 6 Pendulum angle response

Optimization method	Angle deviation (°)	Convergence time (s)
PSO	1.3	3.2
APSO	1.2	3.0

6 Conclusions

In this paper, the weight selection problem of LQR has been solved using the APSO algorithm and the efficacy of the controller has been tested on a benchmark inverted pendulum. To increase the convergence speed and precision of the conventional PSO, an AIWP has been introduced in the velocity update equation of PSO. Statistical measures calculated for the optimization algorithms prove that the introduction of AIWP significantly increases both the accuracy and consistency of the conventional PSO. Moreover, the trajectory tracking response of inverted pendulum accentuate that compare to PSO tuned LQR, the APSO tuned LQR controller framework can result in not only improved tracking response but also reduced tracking error.

References

1. Wang, L., Ni, H., Zhou, W., Pardalos, P.M., Fang, J., Fei, M.: MBPOA-based LQR controller and its application to the double-parallel inverted pendulum system. Eng. Appl. Artif. Intell. **36**, 262–268 (2014)
2. Ang, K.K., Wang, S.Y., Quek, S.T.: Weighted energy linear quadratic regulator vibration control of piezoelectric composite plates. J. Smart Mater. Struct. **11**(1), 98–106 (2002)
3. Niknezhadi, A., Miguel, A.F., Kunusch, C., Carlos, O.M.: Design and implementation of LQR/LQG strategies for oxygen stoichiometry control in PEM fuel cells based systems. J. Power Sources **196**(9), 4277–4282 (2011)
4. Usta, M.A., Akyazi, O., Akpinar, A.S.: Aircraft roll control system using LQR and fuzzy logic controller. In: IEEE Conference on Innovations in Intelligent Systems and Applications (INISTA), pp. 223–227. Istanbul (2011)
5. Solihin, M.I., Akmeliawati, R.: Particle swam optimization for stabilizing controller of a self-erecting linear inverted pendulum. Int. J. Electr. Electron. Syst. Res. **3**, 410–415 (2010)
6. Panda, S., Padhy, N.P.: Comparison of particle swarm optimization and genetic algorithm for FACTS-based controller design. Appl. Soft Comput. **8**(4), 1418–1427 (2008)
7. Tsai, S.J., Huo, C.L., Yang, Y.K., Sun, T.Y.: Variable feedback gain control design based on particle swarm optimizer for automatic fighter tracking problems. Appl. Soft Comput. **13**, 58–75 (2013)
8. Lim, W.H., Isa, N.A.M.: Teaching and peer-learning particle swarm optimization. Appl. Soft Comput. **18**, 39–58 (2014)
9. Nickabadi, A., Ebadzadeh, M.M., Safabakhsh, R.: A novel particle swarm optimization algorithm with adaptive inertia weight. Appl. Soft Comput. **11**, 3658–3670 (2011)
10. Vinodh Kumar, E., Jovitha, J.: Stabilizing controller design for self erecting single inverted pendulum using robust LQR. Aust. J. Basic Appl. Sci. **7**(7), 494–504 (2013)

Application of Fuzzy Soft Multi Sets in Decision-Making Problems

Anjan Mukherjee and Ajoy Kanti Das

Abstract Alkhazaleh and Salleh presented a fuzzy soft multi set theoretic approach to solve decision-making problems using the Roy-Maji Algorithm, which has some limitations. In this research work, we have proposed an algorithm to solve fuzzy soft multi set based decision making problems using the Feng's algorithm, which is more stable and more feasible than the Alkhazaleh–Salleh Algorithm. The feasibility of our proposed algorithm in practical applications is illustrated by a numerical example.

Keywords Decision making · Soft set · Level soft set · Fuzzy soft set · Fuzzy soft multi set · Fuzzy soft multi set part

1 Introduction

In 1999, Molodstov [12] initiated the notion of soft set theory as a general mathematical tool for dealing with vagueness, uncertainties and not clearly defined objects. Research works on the soft set theory are progressing rapidly. Some new algebraic operations and results on soft set theory defined in [2, 10]. Adding soft sets [12] with fuzzy sets [15], Maji et al. [9] defined fuzzy soft sets and studied their basic properties. As a generalization of soft set, Alkhazaleh and others [1, 4–6, 14] proposed the notion of a soft multi set and its basic algebraic and topological structures were studied. Alkhazaleh and Salleh [3] initiated the notion of fuzzy soft multi set theory and presented its application in decision making using Roy-Maji

A. Mukherjee (✉)
Department of Mathematics, Tripura University, Agartala, Tripura 799022, India
e-mail: anjan2002_m@yahoo.co.in

A.K. Das
Department of Mathematics, Iswar Chandra Vidyasagar College,
Belonia, Tripura 799155, India
e-mail: ajoykantidas@gmail.com

A. Nagar et al. (eds.), *Proceedings of 3rd International Conference on Advanced Computing, Networking and Informatics*, Smart Innovation, Systems and Technologies 43, DOI 10.1007/978-81-322-2538-6_3

Algorithm [13]. Maji et al. [11] first proposed the application of soft sets for solving the decision making problems and thereafter in 2007, they also presented an application on fuzzy soft sets based decision making problems in [13]. Kong et al. [8] mentioned that the Roy-Maji algorithm [13] was wrong and they introduced a revised algorithm. Feng et al. [7] studied the validity of the Roy-Maji algorithm [13] and mentioned that the Roy-Maji Algorithm [13] has some limitations. Also, they proposed an adjustable approach to fuzzy soft sets based decision making problems by using thresholds and choice values.

In fact, all these concepts have a good application in some real life problems. But, it is seen that all these theories have their own difficulties that is why in this paper, we are going to propose an algorithm to fuzzy soft multi set based decision making problems using Feng's algorithm, which is another one new mathematical tool for solving some real life applications of decision making problems. The feasibility of our proposed algorithm in practical applications is illustrated by a numerical example.

2 Preliminary Notes

In this current section, we briefly recall some basic notions of soft sets, fuzzy soft multi sets and level soft sets.

Definition 2.1 [12] Suppose that, U be an initial universe and $\hat{E}$ be a set of parameters. Also, let $\tilde{P}(U)$ denotes the power set of the universe U and $\hat{A} \subseteq \hat{E}$. A pair $(\tilde{F}, \hat{A})$ is said to be a soft set over the universe U, where $\tilde{F}$ is a mapping given by $\tilde{F} : \hat{A} \to \tilde{P}(U)$.

Definition 2.2 [3] Suppose $\{U_i : i \in \Lambda\}$ be a set of universes, such that $\cap_{i\in\Lambda} U_i = \phi$ and let for each $i \in \Lambda$, E_i be a sets of decision parameters. Also, let $\tilde{U} = \prod_{i\in\Lambda} FS(U_i)$ where $FS(U_i)$ is the set of all fuzzy subsets of U_i, $\hat{E} = \prod_{i\in\Lambda} E_{U_i}$ and $\hat{A} \subseteq \hat{E}$. A pair $(\tilde{F}, \hat{A})$ is said to be a fuzzy soft multi set over the universe $\tilde{U}$, where $\tilde{F}$ is a function given by $\tilde{F} : \hat{A} \to \tilde{U}$.

Definition 2.3 [3] For any fuzzy soft multi set $(\tilde{F}, \hat{A})$, where $\hat{A} \subseteq \hat{E}$ and $\hat{E}$ is a set of parameters. A pair $\left(e_{U_i,j}, \tilde{F}_{e_{U_i,j}}\right)$ is said to be a U_i—fuzzy soft multi set part of $(\tilde{F}, \hat{A})$ over U, $\forall e_{U_i,j} \in a_k$ and $\tilde{F}_{e_{U_i,j}} \subseteq \tilde{F}(\hat{A})$ is an approximate value set, for $a_k \in \hat{A}$, k $\in \{1, 2, 3, .., \text{m}\}$, $i \in \{1, 2, 3, .., \text{n}\}$ and $j \in \{1, 2, 3, .., \text{r}\}$.

Definition 2.4 [7] Let $\varpi = (\tilde{F}, \hat{A})$ is a fuzzy soft set over the universe U, where $\hat{A} \subseteq \hat{E}$ and $\hat{E}$ is a set of parameters. For $t \in [0, 1]$, the t-level soft set of ϖ is a crisp soft set $L(\varpi; t) = (\tilde{F}_t, \hat{A})$ defined by $\tilde{F}_t(e) = \left\{u \in U : \mu_{\tilde{F}(e)}(u) \geq t\right\}$, $\forall e \in \hat{A}$.

Definition 2.5 [7] Suppose $\varpi = (\tilde{F}, \hat{A})$ be a fuzzy soft set over U, where $\hat{A} \subseteq \hat{E}$ and $\hat{E}$ is the parameter set. Let $\lambda : \hat{A} \to [0, 1]$ be a fuzzy set in $\hat{A}$, which is called a threshold fuzzy set. The level soft set of the fuzzy soft set ϖ with respect to the fuzzy set λ is a crisp soft set $L(\varpi; \lambda) = \left(\tilde{F}_\lambda, \hat{A}\right)$ defined by $\tilde{F}_\lambda(e) = \left\{u \in U : \mu_{\tilde{F}(e)}(u) \geq \lambda(e)\right\}, \forall e \in A$.

Definition 2.6 [7] Let $\varpi = (\tilde{F}, \hat{A})$ is a fuzzy soft set over a finite universe U, where $\hat{A} \subseteq \hat{E}$ and $\hat{E}$ is the set of parameters. The mid-threshold of the fuzzy soft set ϖ define a fuzzy set $mid_\varpi : \hat{A} \to [0, 1]$ by $\forall e \in \hat{A}$, $mid_\varpi(e) = \frac{1}{|U|} \sum_{u \in U} \mu_{\tilde{F}(e)}(u)$ and the level soft set of ϖ with respect to the mid-threshold fuzzy set mid_ϖ, namely $L(\varpi; mid_\varpi)$ is said to be mid-level soft set of ϖ.

Definition 2.7 [7] Let $\varpi = (\tilde{F}, \hat{A})$ be a fuzzy soft set over a finite universe U, where $\hat{A} \subseteq \hat{E}$ and $\hat{E}$ is the parameter set. The max-threshold of the fuzzy soft set ϖ define a fuzzy set $\max_\varpi : \hat{A} \to [0, 1]$ by $\forall e \in \hat{A}$, $\max_\varpi(e) = \max_{u \in U} \mu_{\tilde{F}(e)}(u)$ and the level soft set of ϖ with respect to the max-threshold fuzzy set $\max_\varpi$, namely $L(\varpi; \max_\varpi)$ is said to be top-level soft set of ϖ.

3 An Adjustable Approach Based on Feng's Algorithm

3.1 Feng's Algorithm Using Choice Values

The details of Feng's Algorithm [7] for solving a decision-making problem based on a fuzzy soft set are as follows:

Algorithm 1 (*Feng's Algorithm*)

1. Input the fuzzy soft set $\varpi = (\tilde{F}, \hat{A})$
2. Input a threshold fuzzy set $\lambda : \hat{A} \to [0, 1]$ (or select a threshold value $t \in [0, 1]$ or select mid-level decision criterion or select top-level decision criterion) for solving decision making problem.
3. Obtain the level soft set $L(\varpi; \lambda)$ of ϖ with respect to the threshold fuzzy set λ (or $L(\varpi; t)$ or $L(\varpi; mid)$ or $L(\varpi; \max)$).
4. Present the level soft set $L(\varpi; \lambda)$ (or $L(\varpi; t)$; or $L(\varpi; mid)$; or $L(\varpi; \max)$) as in tabular form and also, obtain the choice value s_i of $u_i \in U, \forall i$.
5. The final optimal decision to be select u_k if $s_k = \max_i s_i$.
6. If k has more than one value, then any one of u_k may be chosen.

3.2 Application of Fuzzy Soft Multi Sets in Decision-Making Problems

In this section, we propose an algorithm (Algorithm 2) for fuzzy soft multi sets based decision making problems, using Feng's Algorithm [7], as described above. In the following, we have to show our algorithm (Algorithm 2):

Algorithm 2

(1) Input the (resultant) fuzzy soft multi set $(\tilde{F}, \hat{A})$
(2) Apply the Algorithm 1 (Feng's Algorithm) to the first fuzzy soft multi set part in $(\tilde{F}, \hat{A})$, to obtain the decision S_{k_1}.
(3) Modify the fuzzy soft multi set $(\tilde{F}, \hat{A})$, by taking all values in each row, where the choice value of S_{k_1} is maximum and changing the values in the other rows by 0 (zero), to get $(\tilde{F}, \hat{A})_1$.
(4) Apply the Algorithm 1 (Feng's Algorithm) to the second fuzzy soft multi set part in $(\tilde{F}, \hat{A})_1$, to obtain the decision S_{k_2}
(5) Modify the fuzzy soft multi set $(\tilde{F}, \hat{A})_1$, by taking the first two parts fixed and apply the method as in step (3) to the next part, to get $(\tilde{F}, \hat{A})_2$
(6) Apply the Algorithm 1 (Feng's Algorithm) to the third fuzzy soft multi set part in $(\tilde{F}, \hat{A})_2$, to obtain the decision S_{k_3}.
(7) Finally, we have the optimal decision for decision maker is $(S_{k_1}, S_{k_2}, S_{k_3})$.

3.3 Application in Decision-Making Problems

Let us consider three universes $U_1 = \{h_1, h_2, h_3, h_4\}$, $U_2 = \{c_1, c_2, c_3\}$ and $U_2 = \{v_1, v_2, v_3\}$ are sets of houses, cars and hotels respectively and let $E_{U_1} = \{e_{U_1,1}, e_{U_1,2}, e_{U_1,3}\}$, $E_{U_2} = \{e_{U_2,1}, e_{U_2,2}, e_{U_2,3}\}$, $E_{U_3} = \{e_{U_3,1}, e_{U_3,2}, e_{U_3,3}\}$ be the sets of respective decision parameters related to the above three universes.

Let $\tilde{U} = \prod_{i=1}^{3} FS(U_i)$, $\tilde{E} = \prod_{i=1}^{3} E_{U_i}$ and $\hat{A} \subseteq \hat{E}$, such that

$$\begin{aligned}\hat{A} =\{ & a_1 = (e_{U_1,1}, e_{U_2,1}, e_{U_3,1}), a_2 = (e_{U_1,1}, e_{U_2,2}, e_{U_3,1}), a_3 = (e_{U_1,2}, e_{U_2,3}, e_{U_3,1}), a_4 = (e_{U_1,3}, e_{U_2,3}, e_{U_3,1}), \\ & a_5 = (e_{U_1,1}, e_{U_2,1}, e_{U_3,2}), a_6 = (e_{U_1,1}, e_{U_2,2}, e_{U_3,2}), a_7 = (e_{U_1,2}, e_{U_2,3}, e_{U_3,3}), a_8 = (e_{U_1,3}, e_{U_2,3}, e_{U_3,3})\}.\end{aligned}$$

Assume that, Mr. X wants to buy a house, a car and rent a hotel with respect to the three sets of decision parameters as in above. Suppose the resultant fuzzy soft multi set be $(\tilde{F}, \hat{A})$ given in Table 1.

First, we apply the Algorithm 1 (Feng's Algorithm) to the U_1—fuzzy soft multi set part in $(\tilde{F}, \hat{A})$ to obtain the decision from the first fuzzy soft multi set part U_1. Now we represent the U_1—fuzzy soft multi set part in $(\tilde{F}, \hat{A})$ as in Table 2.

Table 1 The tabular representation of the fuzzy soft multi-set $(\tilde{F}, \hat{A})$

U_i		a_1	a_2	a_3	a_4	a_5	a_6	a_7	a_8
U_1	h_1	0.3	0.8	1	0.8	0.4	0.9	1	0.8
	h_2	0.4	0.9	0.8	0.6	0.6	0.6	0.9	0.7
	h_3	0.9	0.3	0.7	0.1	0.8	0.7	0.8	1
	h_4	0.7	0.8	0	0.5	0.7	0.5	0.4	0.9
U_2	c_1	0.8	0.8	0.8	0.5	1	0.8	0.8	0.8
	c_2	0.6	0.8	0.6	0.3	0.9	0.8	0.8	0.8
	c_3	0.6	0.5	0.3	0.1	0.9	0.5	0.5	0.5
U_3	v_1	0.9	0.7	0.5	0.5	0.8	0.8	0.5	0.8
	v_2	0.7	0.6	0.5	0.3	0.5	0.8	0.6	0.9
	v_3	0.9	0.5	0.7	0.4	0.4	1	0.8	0.9

Table 2 Tabular representation of U_1—fuzzy soft multi set part of $(\tilde{F}, \hat{A})$

		a_1	a_2	a_3	a_4	a_5	a_6	a_7	a_8
U_1	h_1	0.3	0.8	1	0.8	0.4	0.9	1	0.8
	h_2	0.4	0.9	0.8	0.6	0.6	0.6	0.9	0.7
	h_3	0.9	0.3	0.7	0.1	0.8	0.7	0.8	1
	h_4	0.7	0.8	0	0.5	0.7	0.5	0.4	0.9

In the Table 3, we see that the maximum choice value (s_k) is $s_3 = 6$ and scored by h_3. So we modify the fuzzy soft multi set $(\tilde{F}, \hat{A})$, by taking all values in each row are fixed, where the choice value of h_3 is maximized and changing the values in other rows by 0 (zero), to get $(\tilde{F}, \hat{A})_1$ as in Table 4.

We apply the Algorithm 1 (Feng's Algorithm) to the U_2—fuzzy soft multi set part in $(\tilde{F}, \hat{A})_1$, to obtain the decision from U_2—fuzzy soft multi set part in $(\tilde{F}, \hat{A})$. Now we represent the U_2—fuzzy soft multi set part in $(\tilde{F}, \hat{A})_1$ as in Table 5.

In Table 6, we see that the maximum choice value (s_k) is $s_1 = 3$ and scored by c_1. Therefore, we modify the fuzzy soft multi set $(\tilde{F}, \hat{A})_1$ by taking all values in each row are fixed, where the choice value of c_1 is maximized and changing the values in other rows by 0 (zero), to get $(\tilde{F}, \hat{A})_2$ (Table 7).

Similarly, we apply the Algorithm 1 (Feng's Algorithm) to the U_3—fuzzy soft multi set part in $(\tilde{F}, \hat{A})_2$, to obtain the decision from U_3—fuzzy soft multi set part in $(\tilde{F}, \hat{A})$. Now we represent the U_3—fuzzy soft multi set part in $(\tilde{F}, \hat{A})_2$ as in Table 8.

In Table 9, we see that the maximum choice value (s_k) is 2, scored by v_1 and v_2.

Thus, the final optimal decision for decision maker Mr. X is (h_3, c_1, v_1) or (h_3, c_1, v_2), i.e. Mr. X may chose (h_3, c_1, v_1) or (h_3, c_1, v_1).

Remark 1 In the step (7) of our algorithm (Algorithm 2), if there are too many optimal choices obtained, then decision maker may go back to the step (2) as in our algorithm (Algorithm 2) and replace the level soft set (decision criterion) that he/she once used to adjust the final optimal decision in the fuzzy soft multi set based decision making problems.

Table 3 Mid-level soft set of the U_1—fuzzy soft multiset part in $(\tilde{F},\hat{A})$, with choice values

		a_1	a_2	a_3	a_4	a_5	a_6	a_7	a_8	Choice value (s_k)
U_1	h_1	0	1	1	1	0	1	1	0	$s_1 = 5$
	h_2	0	1	1	1	0	0	1	0	$s_2 = 4$
	h_3	1	0	1	0	1	1	1	1	$s_3 = 6$
	h_4	1	1	0	1	1	0	0	1	$s_4 = 5$

Table 4 The tabular representation of the fuzzy soft multi-set $(\tilde{F},\hat{A})_1$

U_i		a_1	a_2	a_3	a_4	a_5	a_6	a_7	a_8
U_1	h_1	0.3	0.8	1	0.8	0.4	0.9	1	0.8
	h_2	0.4	0.9	0.8	0.6	0.6	0.6	0.9	0.7
	h_3	0.9	0.3	0.7	0.1	0.8	0.7	0.8	1
	h_4	0.7	0.8	0	0.5	0.7	0.5	0.4	0.9
U_2	c_1	0.8	0	0	0	1	0	0	0.8
	c_2	0.6	0	0	0	0.9	0	0	0.8
	c_3	0.6	0	0	0	0.9	0	0	0.5
U_3	v_1	0.9	0	0	0	0.8	0	0	0.8
	v_2	0.7	0	0	0	0.5	0	0	0.9
	v_3	0.9	0	0	0	0.4	0	0	0.9

Table 5 Tabular representation of the U_2—fuzzy soft multi set part in $(\tilde{F},\hat{A})_1$

		a_1	a_2	a_3	a_4	a_5	a_6	a_7	a_8
U_2	c_1	0.8	0	0	0	1	0	0	0.8
	c_2	0.6	0	0	0	0.9	0	0	0.8
	c_3	0.6	0	0	0	0.9	0	0	0.5

Table 6 Mid-level soft set of the U_2—fuzzy soft multi set part in $(\tilde{F},\hat{A})_1$, with choice values

		a_1	a_2	a_3	a_4	a_5	a_6	a_7	a_8	Choice value (s_k)
U_2	c_1	1	0	0	0	1	0	0	1	$s_1 = 3$
	c_2	0	0	0	0	0	0	0	1	$s_2 = 1$
	c_3	0	0	0	0	0	0	0	0	$s_3 = 0$

Table 7 Tabular representation of the fuzzy soft multi set $(\tilde{F},\hat{A})_2$

U_i		a_1	a_2	a_3	a_4	a_5	a_6	a_7	a_8
U_1	h_1	0.3	0.8	1	0.8	0.4	0.9	1	0.8
	h_2	0.4	0.9	0.8	0.6	0.6	0.6	0.9	0.7
	h_3	0.9	0.3	0.7	0.1	0.8	0.7	0.8	1
	h_4	0.7	0.8	0	0.5	0.7	0.5	0.4	0.9
U_2	c_1	0.8	0	0	0	1	0	0	0.8
	c_2	0.6	0	0	0	0.9	0	0	0.8
	c_3	0.6	0	0	0	0.9	0	0	0.5
U_3	v_1	0.9	0	0	0	0.8	0	0	0.8
	v_2	0.7	0	0	0	0.5	0	0	0.9
	v_3	0.9	0	0	0	0.4	0	0	0.9

Table 8 Tabular representation of U_3—fuzzy soft multi set part in $(\tilde{F},\hat{A})_2$

		a_1	a_2	a_3	a_4	a_5	a_6	a_7	a_8
U_3	v_1	0.9	0	0	0	0.8	0	0	0.8
	v_2	0.7	0	0	0	0.5	0	0	0.9
	v_3	0.9	0	0	0	0.4	0	0	0.9

Table 9 Mid-level soft set of the U_3—fuzzy soft multi set part in $(\tilde{F},\hat{A})_2$, with choice values

		a_1	a_2	a_3	a_4	a_5	a_6	a_7	a_8	Choice value (s_k)
U_3	v_1	1	0	0	0	1	0	0	0	$s_1 = 2$
	v_2	0	0	0	0	0	0	0	1	$s_2 = 1$
	v_3	1	0	0	0	0	0	0	1	$s_3 = 2$

Advantages of our algorithm (Algorithm 2) are as follows:

(1) From the above illustration, we have seen that our algorithm (Algorithm 2) is too simple and less computation than Alkhazaleh–Salleh Algorithm [3]. Because instead of computing comparison tables and calculating scores as in Alkhazaleh–Salleh Algorithm [3], we have to consider only choice values of objects form the level soft sets of fuzzy soft multi set parts in the fuzzy soft multi set.

(2) Also, our algorithm (Algorithm 2) is an adjustable algorithm, because the level soft set (decision rule) used by decision makers, which are changeable. For example, if we take top-level decision criterion in step (2) of our algorithm (Algorithm 2), then we have the choice value of each object in the top-level soft set of fuzzy soft multi set parts in the fuzzy soft multi set, if we take another decision rule such as the mid-level decision criterion, then we have choice values from the mid-level soft set of fuzzy soft multi set parts in the fuzzy soft multi set.

4 Conclusion

In [8], Kong et al. mentioned that the Roy-Maji Algorithm [13] was wrong and Feng et al. [7] mentioned that the Roy-Maji Algorithm [13] has some limitations and Alkhazaleh and Salleh [3] presented an application of fuzzy soft multi set based decision-making problems using Roy-Maji Algorithm [13], so Alkhazaleh–Salleh Algorithm [3] is not sufficient to solve fuzzy soft multi set based decision making problems. In this study, we have proposed an algorithm for fuzzy soft multi set based decision making problems using Feng's algorithm [7], which is more stable and more feasible than the Alkhazaleh–Salleh Algorithm [3] for solving decision-making problems based on fuzzy soft multi sets.

References

1. Alhazaymeh, K., Hassan, N.: Vague soft multiset theory. Int. J. Pure Appl. Math. **93**, 511–523 (2014)
2. Ali, M.I., Feng, F., Liu, X., Min, W.K., Shabir, M.: On some new operations in soft set theory. Comput. Math Appl. **57**, 1547–1553 (2009)
3. Alkhazaleh, S., Salleh, A.R.: Fuzzy soft multi sets theory. abstract and applied analysis, vol. **2012**, Article ID 350603, 20 p, Hindawi Publishing Corporation (2012). doi:10.1155/2012/350603
4. Alkhazaleh, S., Salleh, A.R., Hassan, N.: Soft multi sets theory. Appl. Math. Sci. **5**, 3561–3573 (2011)
5. Babitha, K.V., John, S.J.: On soft multi sets. Ann. Fuzzy Math. Inf. **5**, 35–44 (2013)
6. Balami, H.M., Ibrahim, A.M.: Soft multiset and its application in information system. Int. J. Scientific Res. Manag. **1**, 471–482 (2013)
7. Feng, F., Jun, Y.B., Liu, X., Li, L.: An adjustable approach to fuzzy soft set based decision making. J. Comput. Appl. Math. **234**, 10–20 (2010)
8. Kong, Z., Gao, L.Q., Wang, L.F.: Comment on A fuzzy soft set theoretic approach to decision making problems. J. Comput. Appl. Math. **223**, 540–542 (2009)
9. Maji, P.K., Biswas, R., Roy, A.R.: Fuzzy soft sets. J. Fuzzy Math. **9**, 589–602 (2001)
10. Maji, P.K., Biswas, R., Roy, A.R.: Soft set theory. Comput. Math Appl. **45**, 555–562 (2003)
11. Maji, P.K., Roy, A.R., Biswas, R.: An application of soft sets in a decision making problem. Comput. Math Appl. **44**, 1077-1083 (2002)
12. Molodtsov, D.: Soft set theory-first results. Comput. Math Appl. **37**, 19–31 (1999)
13. Roy, A.R., Maji, P.K.: A fuzzy soft set theoretic approach to decision making problems. J. Comput. Appl. Math. **203**, 412–418 (2007)
14. Tokat, D., Osmanoglu, I.: Soft multi set and soft multi topology. Nevsehir Universitesi Fen Bilimleri Enstitusu Dergisi Cilt. **2** (2011) 109–118
15. Zadeh, L.A.: Fuzzy sets. Inf. Control **8**, 338–353 (1965)

Fuzzy kNN Adaptation to Learning by Example in Activity Recognition Modeling

Vijay Borges and Wilson Jeberson

Abstract Activity Recognition is a complex task of the Human Computer Interaction (HCI) domain. k-Nearest Neighbors (kNN) a non-parametric classifier, mimics human decision making, using experiences for segregating a new object. Fuzzy Logic mimics human intelligence to make decisions; but suffers from requiring domain expertise to propose novel rules. In this paper a novel technique is proposed that comes with efficient fuzzy rules from the training data. The kNN classifier is modified by incorporating fuzzification of the feature space by learning from the data and not relying solely on domain experts to draw fuzzy rules. Additional novelty is the efficient use of the Fuzzy Similarity Relations and Fuzzy Implicators for hybridization of the kNN Classifier. The proposed hybridized fuzzy kNN classifier is shown to perform 5.6 % better than the classical kNN counterpart.

Keywords k-Nearest neighbors · Human activity recognition · Smart environments · Ubiquitous computing · Fuzzy sets

1 Introduction

Ubiquitous Computing unobtrusively gathers data from various devices/sensors, processes it; as to control physical processes and interact with human beings. Environments having Ubiquitous Computing facilities are called as Smart Environments [1]. This work focuses on Activity Recognition (AR), that would use

Authors would like to thank Computer Society of India for their research grant vide grant number 1-14/2013-09 dated 21/03/2013. Author$ would like to thank Government of Goa, India for deputational grant for research.

V. Borges (✉) · W. Jeberson
Sam Higginbottom Institute of Agriculture, Technology & Sciences (SHIATS), Allahabad, Uttar Pradesh 211 007, India
e-mail: vijayborges@gmail.com

A. Nagar et al. (eds.), *Proceedings of 3rd International Conference on Advanced Computing, Networking and Informatics*, Smart Innovation, Systems and Technologies 43, DOI 10.1007/978-81-322-2538-6_4

the underlying smart environment to infer/reason the activities of living entities (mostly humans) in an integrated manner. AR has varied applications like; security surveillance, activity recognition in smart infrastructures, assisted living etc.

Human Activities are very complex, as there may be multiple ways to do the same activity either concurrently/interleaved or in a shared fashion. The corresponding data in Smart Environments is also redundant, overlapping and seldom missing/ambiguous. kNN classifier maps a test object to the most common decision class among its k closest neighbors. But kNN classifier suffers loss in accuracy in the domain of human AR due to the stated complexities. Fuzzy set theory [2], is a natural, linguistic computing paradigm attempting to characterize imperfect data/information in a human-like form, by modeling vague (not clearly understood or expressed) information, with a degree of belonging to a certain concept. Extending fuzzy sets to kNN classifier, allows partial membership of an object and shows measures of its importance (strength) to different decision classes.

It is shown here that using the hybridized kNN with fuzzy extensions, the complex task of human Activity Recognition gives better classification accuracy over traditional kNN classifier. The novelty is learning fuzzy memberships from the k-nearest data samples; eliminating the need for *Domain Experts* to propose fuzzy memberships to get discriminating *Fuzzy Rules* [2]. Another novelty is the efficient use of fuzzy similarity relations and fuzzy implicators in the hybridization of the kNN Classifier. This paper is organized as follows. Section 2, introduces necessary theoretical background for fuzzy sets. In Sect. 3, the proposed hybridized Fuzzy kNN is presented: various fuzzy similarity relations and fuzzy implicators are shown. Experimentation is performed on the proposed hybridized Fuzzy kNN using real-life dataset and is reported in Sect. 4. The paper concludes in Sect. 5.

2 Fuzzy Set Theory

Due to human psychoanalytical classification mechanism, traditional set theory falls short in capturing humans vague decision boundaries. Fuzzy set theory, allows to associate a degree of membership to the belongingness of an object to a subset. This degree of membership allows vague decision boundaries to crop in, capturing humans psychoanalytical classification mechanism. For example, human activities like, *sleeping, toileting, eating*, cannot be aptly portrayed with crisp boundaries but rather vague boundaries using Fuzzy Set Theory which plays a pivotal role in decision making.

Given a universe of objects $\mathbb{U}$, and a class C. An object $x \in U$ (also called *input pattern*, *feature vector*), is said to belong to C, depending on the characteristic function $\mu_C : U \to [0, 1]$. This characteristic function defines Fuzzy Sets; which associates to every object x an associated degree of membership to the class C, as $\mu_C(x)$, and formally shown as $\{x, \mu_C(x)\}$. The advantages of fuzzy set theory is critical to the domain of Activity Recognition, where *human interactions* (vague, overlapping) in smart environments have to be clearly classified to recognizable *activities* by associating *human interactions* to various degrees of memberships to the *activity classes.*

2.1 Fuzzification of Feature Space by Learning from Data

The criticality to Fuzzy Set theory is in getting truly discriminating Fuzzy Membership Function (FMF), so that the input feature space can be correctly classified to appropriate output decisions/classes. The domain expert is assigned to come with these discriminating FMF's. Getting FMF's have their own challenges, like: (a) FMF's are problem dependent, there are hardly any *"one-fit-all"* FMF's for a class of problems; (b) FMF's deteriorates in the presence of absurd/noisy input data; (c) for dynamic problem spaces, static FMF's from the domain expert may be difficult to obtain. Hence a novel model is proposed by building these FMF's by learning from the input data itself, and thus eliminating the need of a domain expert.

Let $\mathbb{X} = \{\underline{x}_1, \underline{x}_2, .., \underline{x}_N\}$ where $\underline{x}_i \in \mathbb{R}^p$, be a set of input patterns (feature vectors) [human activity interactions in smart environment] and $\mathbb{A} = \{a_1, a_2, .., a_M\}$, be a set of pre-defined output classes [human activities/activity classes]. The need is to generate fuzzy rules using a mapping, $f: \{\underline{x}_i, .., \underline{x}_j\} \rightarrow a_k$. Let $\underline{x}_i[k]$, with interval range $\left[\underline{x}_i[k]^-, \underline{x}_i[k]^+\right]$, be divided into three fuzzy regions, *low, medium, high*. The $\underline{x}_i[k]$, is then assigned a membership grade corresponding to the fuzzy region/fuzzy set, $F = \{low, medium, high\}$, using the membership function, $\mu_F(\underline{x}_i[k])$, where $\mu_F(\underline{x}_i[k]) \in [0, 1]$. In this work, $\mu_F(.)$, is defined as a $\pi(\underline{x}_i[k], c, \lambda)$, and given as:

$$\pi(\underline{x}_i[k], c, \lambda) = \begin{cases} 2(1- \parallel \underline{x}_i[k] - c \parallel / \lambda); \text{if } \lambda/2 \leq \parallel \underline{x}_i[k] - c \parallel \leq \lambda \\ 1 - 2(\parallel \underline{x}_i[k] - c \parallel / \lambda); \text{if } 0 \leq \parallel \underline{x}_i[k] - c \parallel < \lambda/2 \\ 0; \text{otherwise} \end{cases} \quad (1)$$

where λ, is the scaling factor of the function $\pi(.)$, c is the central point and $\|.\|$, is the L^p*-norm*. The Euclidean norm is used for experiments (refer Sect. 4).

To generate c, λ, for the fuzzy membership function, $\pi(.)$, over fuzzy regions F, for input range $\underline{x}_i[k]$, where $i = 1, 2, \ldots, N$; $k = 1, 2, \ldots, p$; N: number of input patterns (vectors) and p: features of each pattern; the following methodology is employed. Let $r_{medium} = avg(\underline{x}_i[k]^-, \underline{x}_i[k]^+)$, be used to sub-divide the interval range into $[\underline{x}_i[k]^-, r_{medium})$, $(r_{medium}, \underline{x}_i[k]^+]$. Corresponding to the three fuzzy ranges, we define $H_{min_{medium}} = H_{min_{low}} = \underline{x}_i[k]^-$, $H_{max_{low}} = H_{min_{high}} = r_{medium}$, $H_{max_{medium}} = H_{max_{high}} = \underline{x}_i[k]^+$, $c_{medium} = r_{medium}$. To calculate λ_{medium}, we define two points $r1 = c_{medium} - (H_{max_{medium}} - H_{min_{medium}})/2$ and $s1 = c_{medium} + (H_{max_{medium}} - H_{min_{medium}})/2$; which would give $\lambda_{medium} = s1 - r1$. We set $r_{low} = r1$, $r_{high} = s1$. Using $c_{low} = r_{low}$, we next define two new points $r2 = c_{low} - (H_{max_{low}} - H_{min_{low}})/2$ and $s2 = c_{low} + (H_{max_{low}} - H_{min_{low}})/2$; which would give $\lambda_{low} = s2 - r2$. Similarly, setting $c_{high} = r_{high}$, we next define two new points $r3 = c_{high} - (H_{max_{high}} - H_{min_{high}})/2$ and $s3 = c_{high} + (H_{max_{high}} - H_{min_{high}})/2$; which would give $\lambda_{high} = s3 - r3$. These quantities are then used in Eq. (1), to fuzzify the feature space.

3 Modified Fuzzy k-Nearest Neighbor Classifier

In the domain of Activity Recognition in smart environments, the user (actor) interacts with the environment to do a pre-defined set of activities. An *Annotator* (observer) would collect these interactions and allocate a set of these interactions with the correct activity label (class label). These annotated collections of interactions are called as the training set. In the near future whenever the user interacts with the smart environment, the interactions are recorded and the resultant collection set is called as the test set. It is the task of the classification algorithm to provide the correct activity label to collections in the test set by using the training set.

A *k-Nearest Neighbors* (kNN) algorithm is one of the best non-parametric classification algorithm, that assigns an activity label to an input pattern based on the majority of the k-closest (based on distance metrics) neighbors it finds in the training set. kNN has accuracy sometimes matching the ideal Bayes Classifier. kNN's limitations are; the choice of k, that either slows-down/speeds-up the classifier; it always provide a classification even to absurd input patterns; when input pattern is vague (like in the case of human activities) kNN classifier fails to capture the vagueness in the pattern and its accuracy dips.

Here a modified Fuzzy kNN classifier is proposed, that hybridizes the traditional kNN, by adapting to problems where input patterns are vague, carry insufficient knowledge and contain noisy/absurd data. To address these problems, the fuzzy memberships functions are learnt from the training data itself. From that, fuzzy similarity measures are defined to quantize the degree to which two input patterns are similar for every feature in the patterns. Finally, the resultant fuzzy rules are connected using various fuzzy residuum t-norms and implicators [3], to get reduced and most discriminating fuzzy rules (rules with the best classification criteria).

3.1 Theory on Fuzzy Similarity and Rules Reductions via t-Norms

From Sect. 2.1, for an input pattern $\underline{x}_i \in \mathbb{R}^p$, the fuzzified feature space would be $\underline{x}_i = [\pi(\underline{x}_i[1], c_{low}, \lambda_{low}), \pi(\underline{x}_i[1], c_{medium}, \lambda_{medium}), \pi(\underline{x}_i[1], c_{high}, \lambda_{high}), \pi(\underline{x}_i[2], c_{low}, \lambda_{low}), \ldots, \pi(\underline{x}_i[p], c_{high}, \lambda_{high})]$. Traditional *Fuzzy kNN* classifier [4], would generally take the test object $\underline{t} \in \mathbb{R}^p$ and find the FMF to the k nearest neighbors training data (say $\mathbb{N}$, be set of k nearest neighbors), using the formula, $Z_{\underline{t}}(j) = \frac{\sum_{n=1}^{p} \underline{x}_i[n](1/L^p(\underline{t}(n), \underline{x}_i[n])}{\sum_{n=1}^{p}(1/L^p(\underline{t}(n), \underline{x}_i[n])}$; *where $L^p(.)$: Is the conventional L^p norm.* The fuzzy rule to get the correct class (activity) label would simply be: *Class* = $\text{argmax}_{\forall j \in \mathbb{N}} Z_{\underline{t}}(j)$.

The conventional *Fuzzy kNN* classifier is modified, by partitioning the input feature space to determine the fuzzy equivalence classes. These fuzzy equivalence classes are determined using fuzzy similarity relations for every feature between the *test object $\underline{t}$* and the *Nearest Neighbors $\underline{x}$*, that are returned by the traditional kNN classifier. Various fuzzy similarity relations R, over all the features a; given by Eqs. (2) and (3), are experimented with:

$$\mu_{R_a}(\underline{x}, \underline{t}) = 1 - \frac{|a(\underline{x}) - a(\underline{t})|}{|a_{max} - a_{min}|} \tag{2}$$

$$\mu_{R_a}(\underline{x}, \underline{t}) = exp\left(\frac{-(a(\underline{x}) - a(\underline{t}))^2}{2\sigma_a^2}\right) \tag{3}$$

To consolidate the resulting FMF's using either Eqs. (2) and (3), over the subset of features $w \subseteq p$, the following induction is used: $\mu_{R_w}(\underline{x}, \underline{t}) = \bigcap_{a \in w} \mu_{R_a(\underline{x}, \underline{t})}$ Resulting in induced similarity relation over all the subset of features and is analogous to the *antecedent* part of the fuzzy rule. To get the *consequence* part of the fuzzy rule, a novelty is introduced. Let the consequence part [i.e. the decision class(s)/activity label(s)], of the fuzzy rule be itself a fuzzy concept, $C \subseteq \mathbb{A}$. A T-transitive fuzzy similarity relation for a fuzzy concept C, for a given test object $\underline{x}$, is got as: $\mu_{R_w C}(\underline{x}) = \inf_{\underline{y} \in \mathbb{U}} \mathbb{I}\left(\mu_{R_w}(\underline{x}, \underline{y}), \mu_C(\underline{y})\right)$; where $\mathbb{I}$, is the fuzzy implicator and are given in many ways:

$$\text{Largest } I_l(x, y) = \begin{cases} 0; \text{ if } x = 1 \text{ and } y = 0 \\ 1; \text{ otherwise} \end{cases} \tag{4}$$

$$\text{Łukasiewicz } I_Ł(x, y) = min(1 - x + y, 1) \tag{5}$$

Finally, correct decision class (fuzzy concept) C, to the test object $\underline{y}$, is allocated using, $Class = \sup_{C \subseteq \mathbb{A}} \mu_{R_w C}(\underline{y})$

3.2 Proposed Algorithm of the Modified Fuzzy kNN by Learning from Data (FNN-LD)

Algorithm 1 : Fuzzy kNN by Learning from Data (FNN-LD)

Input: The annotated training set $\mathbb{L} = \{l_1, l_2, ..., l_N\}$, where $l_i \in \mathbb{R}^p$.
K : Number of Nearest Neighbors
C: Set of decision Classes. Such that $l_i \rightarrow C_j$.
$\underline{t}$: Test pattern without decision Class
Output: Allocate decision Class C_j for $\underline{t}$.
BEGIN:
1: $\mathbb{X} \leftarrow$ getNearestNeighbors$(\mathbb{L}, K, \underline{t})$ ▷ Traditional kNN Algorithm
2: **for each** $\underline{x_i} \subseteq \mathbb{X}$, $\underline{x_i} \in \mathbb{R}^p$ **do**
3: $\underline{x}_i = [\pi(\underline{x}_i[k], c_{low}, \lambda_{low})\underline{x}_i[1], ..., ...]$ ▷ Gives Fuzzified Regions
4: **end for**
5: **for each** $a \in p$ in fuzzified feature space $\underline{x}_i$ over F **do**
6: $\mu_{R_a}(\underline{x}, \underline{t}) = 1 - \frac{|a(\underline{x}) - a(\underline{t})|}{|a_{max} - a_{min}|}$ ▷ Fuzzy Similarity Relation R, $a \in p$
7: **end for**
8: $\mu_{R_w C}(\underline{t}) = \inf_{\underline{y} \in \mathbb{U}} \mathbb{I}\left(\mu_{R_w}(\underline{t}, \underline{y}), \mu_C(\underline{y})\right)$ ▷ T-transitive relation for Class C
9: **Output** *Class* $\leftarrow \sup_{C} \mu_{R_w C}(\underline{t})$ ▷ Allocate decision class to test pattern $\underline{t}$
END

The proposed Algorithm 1, takes training set $\mathbb{L}$, of human interactions in smart environment which are correctly labeled from set C by the *Annotator*; and, a test pattern $\underline{t}$, for which the decision class is to be provided; as its inputs. At line (1), the function $\mathbb{X} \leftarrow$ getNearestNeighbors$(\mathbb{L}, K, \underline{t})$, makes a call to the traditional Nearest Neighbor algorithm, which returns K closest neighbors to the test pattern $\underline{t}$, using some distance metric. At line (3) the input feature space is fuzzified by using the input data itself. The Fuzzy Similarity Relation R, is determined between the nearest neighbors $\underline{x}$ got from line (1), and the test patten $\underline{t}$, at line (6). This is innovative because many of the human activities being vague, overlapping, concurrent etc. (discussed in Sect. 1); are aptly captured by the fuzzy similarity relation. There are other alternative fuzzy similarity relations in Eqs. (2) and (3), either of which could be used. At line (8), the *consequence* part of the Fuzzy Rule is constructed for every decision Class C; via the T-transitive relation between the test pattern $\underline{t}$ belonging to the considered decision class $\left(\mu_C(\underline{y})\right)$, and previously found induced fuzzy similarity relation at line (6). Line (9), finds the decision class to be allocated to the test pattern $\underline{t}$. Algorithm 1, has quadratic time complexity, $O(Np)$, where N: number of input pattern and p: features of each pattern.

4 Experimental Results and Discussion

For testing and validating the proposed model, a publicly available *Cairo* dataset, available from the Washington State Universities *CASAS* testbed [5] is used. This dataset has two dementia effected senior residents and a pet, it has 27 motion sensors (other sensors are ignored), 600 activities that were recorded for a period of 3 months. It has 10 self-explaining macro-activity labels $\mathbb{A}$, that were used for annotating the datasets by the *Annotator*; *Breakfast, Lunch, Laundry, Dinner, Leave_home (L_Home), Taking_medicine (R_Med), Night_wandering (N_Wand), Go_to_Work (C_work), Bed, Bed_to_toilet (B_Toilet).* This dataset was selected since activities are vague like *Breakfast, Lunch, Dinner*; as all these are *eating* related activities; occurring around the *dining room.* These vague activities happen at different times in the day, but at exactly what time; would be of concern. This concern is addressed by using fuzzy set theoretic concepts. The dataset has multiple residents, having interactions which are concurrent, overlapping, sequential; these would have to be correctly classified by the proposed Algorithm 1.

A *k-fold cross validation*, with $k = 7$, was used to validate the accuracy of the proposed Algorithm 3.2. Fuzzy similarity relation Eq. (2), and the Łukasiewicz Fuzzy Implicator Eq. (5), were used for the setup. The feature set used were, **SENSOR[27]**: corresponding to 27 motion sensors, **DOW**[7]: corresponding to 7 days in a week, **TOD**[7]: where the time in a day was discretized to seven bins {[0–5),[5–10),[10–12),[12–15),[15–18),[18–21),[21–0)}. The K of the kNN Algorithm when set to $K = 11$, was seen to be best performing, for the *Cairo* dataset. The accuracy of the proposed Algorithm 3.2 (FNN-LD) was compared to other classifiers like Naive Bayesian(NB), Improved Naive Bayesian (INB) [6], Traditional kNN (kNN) and the result is tabulated in Table 1. The Improved Naive Bayesian (INB) [6], is a heuristic way of improving the Naive Bayes', skewed data biases, systematic and weighted magnitude errors. The FNN-LD overall accuracy was the highest at 79.71 %; outperforming the traditional kNN by over 5.60 %, INB by over 2.52 % and NB by 16.47 %.

Table 1 Accuracy of the proposed FNN-LD algorithm versus the naive bayesian, improved naive bayesian, traditional kNN approaches for different validation rounds

Val.#	# Test Obj.	Acc. NB (%)	Acc. INB (%)	Acc. kNN (%)	Acc. FNN-LD (%)
1	86	83.95	91.55	85.98	93.03
2	92	70.42	70.93	79.78	88.68
3	88	64.91	70.58	70.40	74.32
4	93	69.23	75.03	76.54	70.18
5	87	58.96	74.88	61.72	68.63
6	91	61.91	76.20	74.44	75.69
7	63	72.73	85.08	79.48	87.42
Avg. (%)	**100**	**68.44**	**77.75**	**75.48**	**79.71**

5 Conclusion and Future Direction

The proposed FNN-LD Algorithm, fuzzifies the input feature space by learning from the data. This eliminates the need of a domain expert in coming with discriminating fuzzy membership functions. Using these fuzzified regions, a hybridization of the traditional kNN algorithm is made possible. The proposed algorithm was tried on a smart environment dataset that had vague, overlapping, concurrent human activities, which had to be properly classified. The FNN-LD Algorithm gave an improved accuracy of 79.71 % augmenting its robustness. At every round of validation, the Algorithm 1, gets called afresh. As such the fuzzy similarity relation R at step (6), get re-calculated from scratch. This is reflected in the validation rounds where the accuracy of the FNN-LD varies from 93.03 % to 68.63 %. In the future updated fuzzy similarity relation R would be used across validation rounds, instead of re-calculating from scratch. Another limitation is the lack of experimental results of missing and absurd data on the efficiency of the FNN-LD. The FNN-LD can be further improved by using Fuzzy Granular Computing concepts [2], in the future.

References

1. Borges, V., Jeberson, W.: Survey of context information fusion for sensor networks based ubiquitous systems. Comput. Sci. Inf. Technol. **2**(3), 165–178 (2014). doi:10.13189/csit.2014.020306
2. Pedrycz, W., Skowron, A., Kreinovich, V.: Handbook of Granular Computing. Wiley, Chichester (2008)
3. Mundici, D.: Advanced Lukasiewicz Calculus and MV-Algebras, 35th edn. Springer, New York (2011)
4. Keller, J.M., Gray, M.R., Givens, J.A.: A fuzzy k-nearest neighbor algorithm. IEEE Trans. Syst. Man Cybern. **4**, 580–585 (1985)
5. WSU, CASAS Project. http://ailab.wsu.edu/casas/. Accessed 24 March 2014
6. Borges, V., Jeberson, W.: Fortune at the Bottom of the Classifier Pyramid. A Novel approach to Human Activity Recognition, Elsevier Procedia Computer Science. **46**, 37–44 (2015)

Plant Leaf Recognition Using Ridge Filter and Curvelet Transform with Neuro-Fuzzy Classifier

Jyotismita Chaki, Ranjan Parekh and Samar Bhattacharya

Abstract The current work proposes an innovative methodology for the recognition of plant species by using a combination of shape and texture features from leaf images. The leaf shape is modeled using Curvelet Coefficients and Invariant Moments while texture is modeled using a Ridge Filter and some statistical measures derived from the filtered image. As the features are sensitive to geometric orientations of the leaf image, a pre processing step is performed to make features invariant to geometric trans-formations. To classify images to pre-defined classes, a Neuro fuzzy classifier is used. Experimental results show that the method achieves acceptable recognition rates for images varying in texture, shape and orientation.

Keywords Curvelet transform · Invariant moment · Ridge filter · Neuro fuzzy

1 Introduction

Plants play a crucial role in Earth's ecology by providing sustenance, shelter and maintaining a healthy breathable atmosphere. Plants also have important medicinal properties and are used for alternative energy sources like bio-fuel. Building a plant database for quick and efficient classification and recognition is an important step towards their conservation and preservation. This is especially significant as many plant species are at the brink of extinction due to incessant de-forestation to pave the way for modernization. In recent years computer vision and pattern recognition techniques have been utilized to prepare digital plant cataloging systems. Most of these techniques rely on extraction of visual properties from plant leaf images and representing them as computer recognizable features using data modeling

J. Chaki (✉) · R. Parekh
School of Education Technology, Jadavpur University, Kolkata, India
e-mail: jyotismita.c@gmail.com

S. Bhattacharya
Department of Electrical Engineering, Jadavpur University, Kolkata, India

A. Nagar et al. (eds.), *Proceedings of 3rd International Conference on Advanced Computing, Networking and Informatics*, Smart Innovation, Systems and Technologies 43, DOI 10.1007/978-81-322-2538-6_5

techniques. Properties like shape and texture [1] and shape, texture and color [2] have been used for discrimination. Different data modeling techniques like fractal dimensions [3], Fourier analysis [4], and wavelets [5] have been applied. The current work proposes an innovative scheme of a plant recognition system based on shape and texture of the leaf. Shape is modeled using Curvelet Transform and Invariant Moments, while texture by using a Ridge Filter and some statistical measures derived from it. Experiments using a neuro-fuzzy classifier demonstrate acceptable recognition accuracies. The organization of the paper is as follows: Sect. 2 outlines the proposed approach, Sect. 3 provides details of the dataset and experimental results obtained, Sect. 4 compares the proposed approach vis-à-vis some contemporary approaches, Sect. 5 brings up the overall conclusions and scopes for future research.

2 Proposed Approach

As mentioned above, the proposed approach uses a combination of shape and texture features. The feature values are however sensitive to the size and orientation of the leaf image. To make them invariant to translation, rotation and scaling, a pre-processing step is used to standardize these parameters before feature calculation.

2.1 *Pre-Processing (PP)*

The objective of the pre-processing step is to standardize the scale and orientation of the image before feature computation. The raw image (*I*) is typically a color image oriented at a random angle and having a random size. See Fig. 1. The image is first converted to binary (*bw*) and grayscale (*gs*) forms. To make features rotation-invariant, the angle of the major axis of the leaf is extracted from the image and used to rotate it so that the major axis is aligned with the horizontal line (*rg* and *rb*). If visible, the white bounding rectangle is removed to superimpose the leaf over a homogeneous background (*cg*). At this point even though the leaf is horizontal, it can have varying translation factors with respect to the origin. To make the features translation-invariant, the background is shrunk until the leaf just fits within the bounding rectangle (*pg* and *pb*).

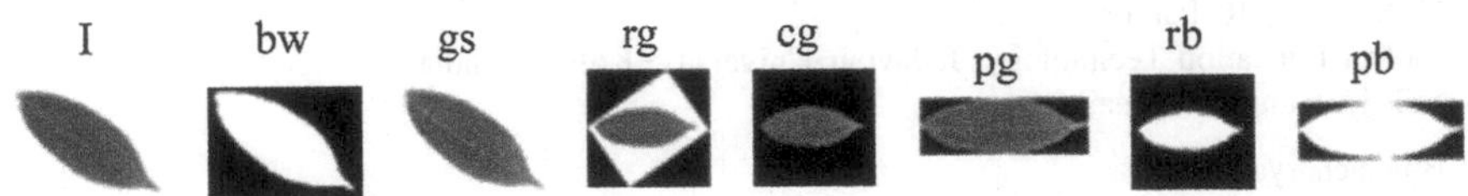

Fig. 1 PP steps of the original image I

Table 1 Pre defined scaling dimensions mapped to segments

Segment	R	Size (M × N)	Segment	R	Size
1	R ≤ 1.4	300 × 300	4	2.4 < R ≤ 3	100 × 300
2	1.4 < R ≤ 2	200 × 300	5	3 < R ≤ 5	65 × 300
3	2 < R ≤ 2.4	140 × 300	6	5 < R ≤ 13	45 × 300

To make the features scale-invariant, the image is rescaled to standard dimensions, called 'segments'. Since the ratio of major axis to minor axis, henceforth called 'aspect ratio' (*R*) is different for different leaf types, rescaling to a single size will produce distortions due to non-uniform scaling. Hence a scheme is devised so that the leaf can be scaled to one of 6 pre-determined segments based on different values of this ratio, with no or minor distortions. The output of the pre-processing block for each leaf image, is its segment number (*s*), and the grayscale and binary versions (*pg* and *pb*). The segment number assigned to each class along with the aspect ratio is tabulated below in Table 1. If $R > 13$, subsequent feature values were found to give inconsistent results.

2.2 Curvelet Transform (CT)

Curvelets were first introduced in [6]. Subsequently a faster form was developed called Fast Discrete Curvelet Transform (FDCT) which had two variants: unequally spaced Fast Fourier Transforms (USFFT) and wrapping function. The current work utilizes the wrapping function, which involves several sub-bands at different scales consisting of different orientations and positions in the frequency domain. An image with dimensions $M \times N$ is subjected to FDCT which generates a set of curvelet coefficients C indexed by scale a, orientation b and spatial location parameters p and q. Here $0 \leq m \leq M$, $0 \leq n \leq N$ and $\varphi_{a,b,p,q}$ is the curvelet waveform.

$$C(a,b,p,q) = \sum_{m=0,n=0}^{M,N} f(m,n)\varphi_{a,b,p,q}(m,n). \tag{1}$$

2.3 Invariant Moment (IM)

For a digital image, the moment m of pixel $P(x,y)$ is defined as: $m = x.y.P(x,y)$. The moment of the entire image is the summation of moments of all its pixels. More generally the moment of order (p,q) of an image is given by

$$m_{pq} = \sum_x \sum_y x^p y^q P(x, y). \tag{2}$$

Hu [7] proposed 7 moment features that can be used to describe images and are invariant to rotation. The first of these φ_1 is given by the following:

$$\varphi_1 = m_{20} + m_{02}. \tag{3}$$

To make the moments invariant to translation the image is shifted such that its centroid coincides with the origin of the coordinate system. The centroid of the image in terms of the moments, is given by:

$$x_c = \frac{m_{10}}{m_{00}}, y_c = \frac{m_{01}}{m_{00}}. \tag{4}$$

Then the central moments are defined as

$$\mu_{pq} = \sum_x \sum_y (x - x_c)^p (y - y_c)^q P(x, y). \tag{5}$$

It can be verified that

$$\mu_{00} = m_{00}, \mu_{10} = 0 = \mu_{01}. \tag{6}$$

To make the moments invariant to scaling, the moments are normalized by dividing by a power of μ_{00}

$$\gamma_{pq} = \frac{\mu_{pq}}{(\mu_{00})^{\omega}}, \omega = 1 + \frac{p+q}{2}. \tag{7}$$

The normalized central Hu moments are defined by substituting m terms in Eq. (3) by γ terms. The first normalized central invariant moment of an image I is:

$$M_1(I) = \gamma_{20} + \gamma_{02}. \tag{8}$$

2.4 *Ridge Filter (RF)*

In computer vision algorithms, particularly those dealing with image analysis, edge detection forms an important step for recognizing the shape, location and orientation of image objects. In some cases however we might be more interested in gaining information about the nature of surfaces on such objects. Ridge detection is a step towards understanding the corrugatory nature of these surfaces. Ridge filters have been mostly used to enhance the ridge and valley structures of fingerprint images as part of minutiae extraction modules [8]. In this paper we use a ridge filter to enhance vein structures of a leaf to model its texture content. To compute ridge information, a grayscale image is first partitioned into a grid of non-overlapping blocks. The standard deviation within each block is evaluated and a threshold is used to determine whether the region is a part of the object or the background. This generates a mask showing the pattern of ridge lines on the surface. The mask is normalized to have zero mean and unit standard deviation. From the mask image gradients are calculated and the local ridge orientation at each point is estimated. The ridge frequency is obtained within each block by rotating the block using the orientation values and finding peaks in projected grey values along the ridges. The spatial frequency of the ridges is determined by dividing the distance between the first and last peaks by number of peaks. The frequency and orientation values are used to generate a ridge filter to enhance ridge lines of the original image.

2.5 *Statistical Measures*

A normalized histogram p is calculated from the image data using a specified number of bins N. Three statistical measures calculated from the normalized histogram are uniformity (U), entropy (E) and third moment (T). If $p = [x(1), x(2), \ldots, x(N)]$ depicts the histogram and μ the mean of the distribution then:

$$U = \sum_{i=1}^{N} [x(i)]^2 \tag{9}$$

$$E = -\sum_{i=1}^{N} [x(i).\log_2 x(i)] \tag{10}$$

$$T = \frac{1}{(N-1)^2} \sum_{i=1}^{N} [x(i) - \mu]^3 \tag{11}$$

2.6 *Features*

The binary pre-processed image (pb) is subjected to a Curvelet Transform. From all possible coefficients C, the one with the highest energy (CC) is retained as it contains the most significant information pertaining to the image shape. See Fig. 2. The shape feature (FS) is formulated by computing M_1 from CC as per Eq. (8).

Fig. 2 Curvelet coefficient CC and ridge structures of a leaf

$$FS = M_1(CC) \tag{12}$$

The grayscale pre-processed image (*pg*) is subjected to a Ridgelet Filter to enhance the ridge structures of the leaf. See Fig. 2. Three statistical measures as defined in Eqs. (9)–(11) computed from the filtered image form the texture feature (*FT*).

$$FT = \{U, E, T\} \tag{13}$$

To study the joint effects of shape and texture features, a combined feature vector *FC* is used for class discrimination and recognition

$$FC = \{FS, FT\} \tag{14}$$

2.7 Classification

A leaf class consists of a set of member images. Each class is characterized by a collection of the *FC* vectors obtained during a training phase. A test image with its computed vector is said to belong to a specific class if the probability of its feature values being a member of that class is maximum. Since there is no prior mathematical model based on which data samples may be classified, classification is done solely on the basis of a number of observations which is used to train a Neuro Fuzzy Classifier (NFC) so as to combine the advantages of a fuzzy classification scheme and the automatic adaptation procedure of a neural network.

3 Experimentations and Results

To study validity of the proposed scheme, experimentations done using 600 images from Flavia [9] involving 30 classes having 10 images per class for training and 10 for testing. Each image is of size 300 × 225 and in JPG format. Figure 3 shows samples of the dataset. Overall accuracy obtained is 97 %. Details are shown in Table 2.

Fig. 3 Samples of the dataset

Table 2 Percentage recognition accuracies

Class	Acc (%)	Class	Acc (%)	Class	Acc (%)	Class	Acc (%)	Class	Acc (%)
1	100	7	100	13	100	19	90	25	100
2	100	8	100	14	100	20	100	26	100
3	90	9	100	15	100	21	100	27	90
4	100	10	100	16	90	22	90	28	100
5	100	11	90	17	100	23	90	29	100
6	100	12	100	18	90	24	90	30	100

4 Analysis

To put the current work in perspective with other contemporary works, their approaches were applied to the current dataset. Color, texture and shape features in [10] produce an accuracy of 74.3 %. Due to the large number of features, resource overheads were also high. The LBP based method of [11] was sensitive to noise, and different patterns of LBP were seen to produce incorrect classifications giving an accuracy of 37 %. Color information of [12] produces incorrect classifications due to small variations of green shades between leaves giving a 31 % accuracy. Fourier basis functions [4] produce an accuracy of 16 % as these are sinusoidal in nature with infinite lengths, and cannot suitably model transient signals with sharp changes, as is often encountered along leaf contours.

5 Conclusions

This article discusses a method of characterizing plant leaves by using a ridge filter and statistical measures to model texture information, together with Curvelet coefficients and invariant moments to model shape. Prior to feature extraction, the leaf image is made invariant to transformations through a pre-processing stage. To avoid distortion for leaves having different aspect ratios, a set of 6 predefined segments have been proposed. Classes are categorized using neuro fuzzy classifiers. Experimental results demonstrate that the proposed approach is effective in discriminating between 30 classes of leaf images having a variety of textures and shapes. Future work would involve research along two directions: (1) combination of other shape, texture and color based features with the current method. (2) Using other classifiers like k-Nearest Neighbor, Support Vector Machine etc.

References

1. Beghin, T., Cope, J.S., Remagnino, P., Barman, S.: Shape and texture based plant leaf classification. In: International Conference on Advanced Concepts for Intelligent Vision Systems (ACVIS), pp. 345–353 (2010)
2. Kebapci, H., et al.: Plant image retrieval using color, shape and texture features. Comput. J. **53** (1), 1–16 (2010)
3. Du, J.-X., Zhai, C.-M., Wang, Q.-P.: Recognition of plant leaf image based on fractal dimension feature. Neurocomputing **116**, 150–156 (2013)
4. Yang, L.W., Wang, X.F.: Leaf image recognition using fourier transform based on ordered sequence. Springer LNCS. **7389**, 393–400 (2012)
5. Wang, Q-P., Du, J-X., Zhai, C-M.: Recognition of leaf image based on ring projection wavelet fractal feature. In: International Journal Innovative Computing, Information and Control, pp. 240–246 (2012)
6. Candès, E., Donoho, D.: Curvelets—a Surprisingly Effective Nonadaptive Representation for Objects with Edges, pp. 1–10. Vanderbilt University Press, Nashville (2000)
7. Hu, M.K.: Visual pattern recognition by moment invariants. IRE Trans. Inf. Theory **8**(2), 179–187 (1962)
8. Hong, L., Wan, Y., Jain, A.K.: Fingerprint image enhancement: algorithm and performance evaluation. IEEE Trans. PAMI **20**(8), 777–789 (1998)
9. Flavia Dataset: http://sourceforge.net/projects/flavia/files/Leaf%20Image%20Dataset/
10. Kadir, A., Nogroho, L.E., Susanto, A., Santosa, P.I.: Neural network application on foliage plant identification. Int. J. Comput. Appl. **29**, 15–22 (2011)
11. Wang, X., Liang, J., Guo, F.: Feature extraction algorithm based on dual-scale decomposition and local binary descriptors for plant leaf recognition. Elsevier Digital Image Process. **34**, 101–107 (2014)
12. Caglayan, A., Oguzhan, G., Can, A.B.: A plant recognition approach using shape and color features in leaf images, vol. 8157, pp. 161–170 Springer LNCS (2013)

An Effective Prediction of Position Analysis of Industrial Robot Using Fuzzy Logic Approach

P. Kesaba, B.B. Choudhury and M.K. Muni

Abstract Industrial robots have been extensively used by many industries as well as organizations for different applications. This paper introduces some qualitative parameters to find out the best predictive value as per the comparison to experimental value. In the fuzzy-based method, the weight of each criterion and the rating of each alternative are described by using different membership functions and linguistic terms. By using four techniques triangular, trapezoidal, Gaussian and Hybrid membership functions, the effective prediction of robot angle with their position is determined in accordance to space of Robot. This paper compares four techniques and found that the Hybrid membership function is the effective one for determination of effective prediction measurement as it shows a good agreement with the experimental result. By taking help of this paper the user can easily access to any point of the workspace locations. It is very effective one by the fuzzy logic systems to analyze the work space.

Keywords Six axis industrial robot · Fuzzy logic · Triangular · Trapezoidal · Gaussian · Hybrid membership function

1 Introduction

Due to high labor costs and precision in repetitive works many industries have to rely on production automation to keep their competitive advantage. One of the most flexible and powerful automation technologies available today is industrial robotics. The selection of an industrial robot is a significant problem to the design engineer

P. Kesaba (✉) · B.B. Choudhury · M.K. Muni
Department of Mechanical Engineering, IGIT Sarang, Dhenkanal, India
e-mail: pkesaba@gmail.com

B.B. Choudhury
e-mail: bbcigit@gmail.com

M.K. Muni
e-mail: manoj1986nitr@gmail.com

A. Nagar et al. (eds.), *Proceedings of 3rd International Conference on Advanced Computing, Networking and Informatics*, Smart Innovation, Systems and Technologies 43, DOI 10.1007/978-81-322-2538-6_6

and depends on the task to be performed. Equipped with the right tool and operated in optimum path, standardized Industrial robots can perform numerous production tasks. Application of industrial robots are motivated by various technical and economical reasons like increase the quality of finished products, reduce the waste, increase degree of uniformity of quality, increase degree of operating safety etc. The control of a robot is very difficult in varying operational conditions like manufacturing environment. For these industrial applications the best alternative approach would be fuzzy logic. Fuzzy set theory provides a systematic calculus linguistically, and it performs numerical computation by using linguistic labels based on membership functions. Fuzzy logic controller gives better results and can be used where conventional controller is not suitable. It is evident that no detail mathematical description is required for fuzzy logic algorithm. Fuzzy logic algorithms can be best suited for non-accurate, subjective and uncertain source of information. It translates qualitative and imprecise linguistic statements into precise computer statements.

2 Literature Review

A fast path planning method by optimization of a path graph for both efficiency and accuracy is given by Hwang et al. [1]. The authors had proposed a path graph optimization technique employing a compact mesh representation. Gasparetto et al. [2] described a new method for smooth trajectory planning of robot manipulators. Bhalerao et al. [3] proposed an efficient parallel dynamics algorithm for simulation of large articulated robotic systems. The method presented in this paper relies on the Divide-and-Conquer-Algorithm (DCA) and the Articulated Body Algorithm (ABA). Srikanth et al. [4] addresses kinematic analysis of 3 D.O.F of serial robot for industrial applications. The study of motion is divided into kinematics and dynamics. A multi-agent approach based on fuzzy logic for a robot manipulator is given by Kazar et al. [5] proposed the method of modeling and control of a manipulator arm using fuzzy logic that can help the robot to join a well defined goal. The simulation results show the effectiveness of the approach. Abd et al. [6] presented a paper to solve a multi-objective problem in robotic flexible assembly cells. The methodology proposed by the authors is based on three main steps: (1) scheduling of the RFACs using different common rules, (2) normalization of the scheduling outcomes, and (3) selection of the optimal scheduling rules, using a fuzzy inference system. Ahmad et al. [7] presents a paper on investigations into the development of hybrid intelligent control schemes for the trajectory tracking and vibration control of a flexible joint manipulator using composite fuzzy logic control. The results are presented in time and frequency domains. The performances like input tracking capability, level of vibration reduction and time response specifications examined and a comparative assessment of the control techniques is presented and discussed. The simulation results show the effectiveness of the approach. Das and Parhi [8] have developed an methodical study on fuzzy system using some input parameters to the fuzzy membership functions for forecast of crack and relative crack depth.

3 Methods of Solution

3.1 Forward Kinematics Equation for Six Axis Industrial Robot

The Aristo robot has six DOF with all rotary type joints. The first three DOF are located in the arm which allows determining the robot position and the following three DOF are located in the end effectors to provide orientation. The present work deals with forward kinematics analysis of the robot where the position and orientation of the end effector are derived from the given joint angles and link parameters. The homogeneous transformation matrix for the 6R robot is as follows:

$$T = \begin{bmatrix} n_x & o_x & a_x & p_x \\ n_y & o_y & a_y & p_y \\ n_z & o_z & a_z & p_z \\ 0 & 0 & 0 & 1 \end{bmatrix} \quad (1)$$

where

p_x $C_1\,[d_6(C_{23}C_4S_5 + S_{23}C_5) + S_{23}d_4 + a_3C_{23} + a_2C_2] - S_1(d_6S_4S_5 + d_2)$

p_y $S_1\,[d_6(C_{23}C_4S_5 + S_{23}C_5) + C_{23}d_4 + a_3C_{23} + a_2C_2] - C_1(d_6S_4S_5 + d_2)$

p_z $d_6\,(C_{23}C_5 - S_{23}C_4S_5) + C_{23}d_4 - a_3S_{23} - a_2S_2)$

where $C_i = \cos\theta_i, S_i = \sin\theta_i$ and $C_{iJ} = \cos(\theta_i + \theta_J), S_{iJ} = \sin(\theta_i + \theta_J)\theta_i$ = Joint angle, d_i = Link Length

The input variables are six joint angles (A1–A6) within the given specified limits and outputs are the position of the end-effector (world coordinates X, Y and Z). The software used for the simulation of the robot is Aristo Version 1.4 presented in Fig. 1. Fifty seven numbers of input variables in terms of joint angles are taken experimentally and the output in terms of locations "X, Y and Z" are also obtained experimentally using forward Kinematics.

3.2 Fuzzy Logic System

Fuzzy logic is an extension of classical logic and uses fuzzy sets instead of classical sets. Fuzzy logic is a logic approximate reasoning which may be viewed as a simplification and addition of multi valued logic. Like classical set theory is used to develop classical logic, similarly fuzzy set theory is required to create fuzzy logic. Figure 2 outlines a simple architecture for a fuzzy logic controller.

The fuzzy controller has been developed where there are 6 inputs and 3 outputs parameter. The natural linguistic representation for the inputs is as follows

Industrial robot input capacity = "Input angle is θ_1 is A1", "Input angle is θ_2 is A2", "Input angle is θ_3 is A3", "Input angle is θ_4 is A4", "Input angle is θ_5 is A5" and "Input angle is θ_6 is A6",

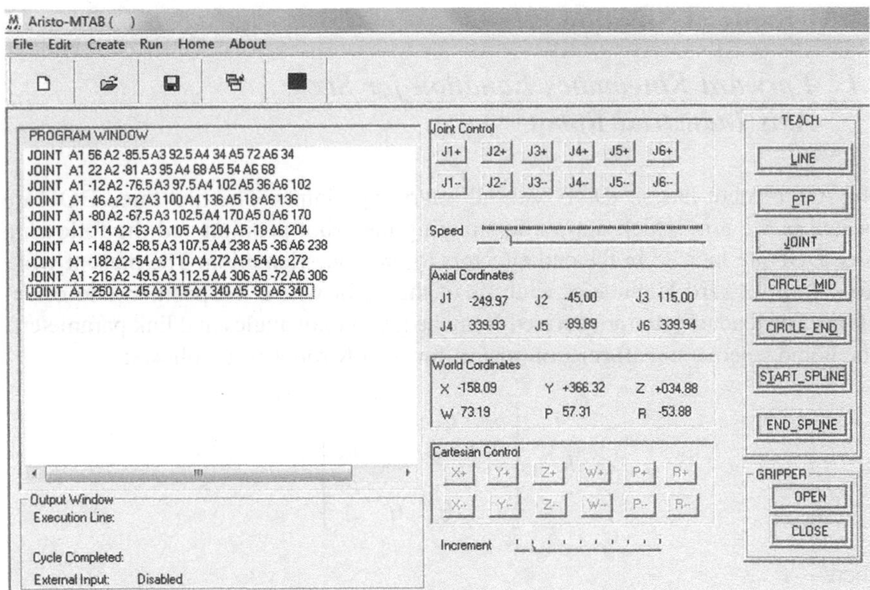

Fig. 1 Coordinates of robots using aristo simulation

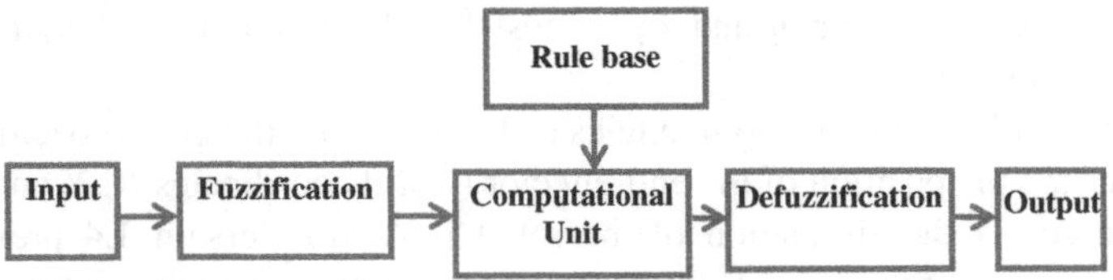

Fig. 2 Fuzzy controller architecture

Industrial robot output capacity = "Output Location is X is O1", "Output Location is Y is O2" and "Output Location is Z is O3",

The natural linguistic term used for the output is

Industrial robot Location = "OutputX= O1" and "OutputY= O2" and "OutputZ= O3" Based on the above fuzzy subset the fuzzy rules are defined in a general form as follows:

If A1 is $(A1)_i$ and A2 is $(A2)_j$ and A3 is $(A3)_k$ and A4 is $(A4)_l$ and A5 is $(A5)_m$ and A6 is $(A6)_n$ then O1 is $(O1)_{ijklmn}$ and O2 is $(O2)_{ijklmn}$ and O3 is $(O3)_{ijklmn}$

where i = 1 to 9; Because of "Input angles A1 to A6" are having nine membership functions. From the above expression, one set of rules are written as:

If A1 is (A1)i then Output (O1) is (O1)i. The Linguistic Terms Used for Fuzzy Membership Functions for angle A1 is given in Table 1. Similarly for other angles the linguistic terms are used accordingly.

The capacity of Robots is expressed in terms of four membership functions like triangular, trapezoidal, Gaussian and Hybrid as input to the fuzzy controller.

Table 1 Linguistic terms used for fuzzy membership functions

Membership functions	Terms	Linguistic term descriptions
A1VL	$A1_1$	Very low value of angle for input angle "A1"
A1L	$A1_2$	Low value of angle for input angle "A1"
A1M	$A1_3$	Medium value of angle for input angle "A1"
A1G	$A1_4$	Good value of angle for input angle "A1"
A1VG	$A1_5$	Very Good value of angle for input angle "A1"
A1VVG	$A1_6$	Very very good value of angle for input angle "A1"
A1H	$A1_7$	High value of angle for input angle "A1"
A1VH	$A1_8$	Very high value of angle for input angle "A1"
A1VVH	$A1_9$	Very very high value of angle for input angle "A1"

The triangular membership function is collection of three points forming a triangle. These membership functions are simple and fast when compared to other membership functions. Gaussian membership functions are popular methods for specifying fuzzy sets and have the advantage of being smooth. The trapezoidal membership function has a flat top and the advantage of simplicity. Hybrid membership function is combination of triangular, Trapezoidal as well as Gaussian membership function. The outputs obtained from fuzzy controllers are analyzed and compared to obtain the best effective technique for Robot selection. There are fifty seven fuzzy rules developed to acquire the outputs and some of the fuzzy rules are given in Table 2. The output of hybrid membership functions consists of four triangles, three Gaussians and two trapezoidal functions. It is a combination of all the membership functions.

Table 2 Fuzzy rules used for fuzzy inference system

Sl. no.	Some of rules used in fuzzy controller
1	If (ANGLE1 is AIVL) and (ANGLE2 is A2VL) and (ANGLE3 is A3VL) and (ANGLE4 is A4VL) and (ANGLE5 is A5VL) and (ANGLE6 is A6VL) then (OUTPUTX is O1VG)(OUTPUTY is O2VVH)(OUTPUTZ is O3VVH) (1)
2	If (ANGLE1 is AIVL) and (ANGLE2 is A2VL) and (ANGLE3 is A3VL) and (ANGLE4 is A4VL) and (ANGLE5 is A5VL) and (ANGLE6 is A6VL) then (OUTPUTX is O1H)(OUTPUTY is O2VVH)(OUTPUTZ is O3VH) (1)
3	If (ANGLE1 is AIL) and (ANGLE2 is A2L) and (ANGLE3 is A3L) and (ANGLE4 is A4L) and (ANGLE5 is A5L) and (ANGLE6 is A6L) then (OUTPUTX is O1VVH) (OUTPUTY is O2VH)(OUTPUTZ is O3VH) (1)
4	If (ANGLE1 is AIM) and (ANGLE2 is A2 M) and (ANGLE3 is A3 M) and (ANGLE4 is A4 M) and (ANGLE5 is A5 M) and (ANGLE6 is A6 M) then (OUTPUTX is O1VVH)(OUTPUTY is O2G)(OUTPUTZ is O3H) (1)
5	If (ANGLE1 is AIG) and (ANGLE2 is A2G) and (ANGLE3 is A3G) and (ANGLE4 is A4G) and (ANGLE5 is A5G) and (ANGLE6 is A6G) then (OUTPUTX is O1VH) (OUTPUTY is O2L)(OUTPUTZ is O3VVG) (1)

4 Results and Discussions

With the help of experimental setup fifty seven nos. of results are taken as given in Table 3 and in Fig. 1 and also fifty seven nos. of Fuzzy rules are developed. The different closeness values are calculated from the different fuzzy membership functions presented in Table 2. The six input variables and three outputs with Hybrid Membership Function are given in Table 4 and also in Fig. 3. The fuzzy controllers developed here for prediction of closeness with the experimental value. The predicted result from fuzzy controllers for closeness of Robots is compared with the experimental results and shows a very good agreement. It has been observed that the result of Fuzzy controller using Hybrid membership function shows more accurate result in comparison to other three controllers. So from this it can be presumed that the developed fuzzy controller along with the technique can be used as a robust tool for selection of Robots. Figure 4 shows the graphical representation of Input variables ("A1–A6") and Output (Location "O1–O3") with Hybrid Membership Function for Capacity of the Robot. All the results of four membership functions are developed and output values X, Y and Z of these membership functions are compared and presented in Figs. 5, 6 and 7 respectively.

Table 3 Experimental results taken from six axis industrial robot

	θ_1	θ_2	θ_3	θ_4	θ_5	θ_6	X	Y	Z
1	81.50	−88.88	90.63	8.50	85.50	8.50	83.70	298.10	350.00
2	64.50	−86.63	91.88	25.50	76.50	25.50	98.20	301.40	369.30
3	56.00	−85.50	92.50	34.00	72.00	34.00	97.65	300.12	357.98
4	47.50	−84.38	93.13	42.50	67.50	42.50	159.50	311.04	350.08
5	39.00	−83.25	93.75	51.00	63.00	51.00	210.70	280.90	340.60
6	30.50	−82.13	94.38	59.50	58.50	59.50	383.05	241.60	340.40
7	13.50	−79.88	95.63	76.50	49.50	76.50	389.10	152.60	328.70
8	5.00	−78.75	96.25	85.00	45.00	85.00	390.50	101.05	321.70
9	−54.50	−70.88	100.63	144.50	13.50	144.50	290.40	−351.23	233.39
10	−63.00	−69.75	101.25	153.00	9.00	153.00	224.89	−430.14	216.37
11	−71.50	−68.63	101.88	161.50	4.50	161.50	155.10	−456.45	198.95
12	−80.00	−67.50	102.50	170.00	0.00	170.00	130.00	−360.00	190.00

Table 4 Results from hybrid method

	θ_1	θ_2	θ_3	θ_4	θ_5	θ_6	X			Y			Z		
							EXP	HYB	% ERR	EXP	HYB	% ERR	EXP	HYB	% ERR
1	81.50	−88.88	90.63	8.50	85.50	8.50	83.70	83.30	0.48	298.10	295.00	1.04	350.00	340.00	2.86
2	64.50	−86.63	91.88	25.50	76.50	25.50	98.20	91.40	6.92	301.40	295.00	2.12	369.30	340.00	7.93
3	56.00	−85.50	92.50	34.00	72.00	34.00	97.65	96.90	0.77	300.12	295.00	1.71	357.98	340.00	5.02
4	47.50	−84.38	93.13	42.50	67.50	42.50	159.50	154.00	3.45	311.04	290.00	6.76	350.08	340.00	2.88
5	39.00	−83.25	93.75	51.00	63.00	51.00	210.70	202.00	4.13	280.90	252.00	10.29	340.60	340.00	0.18
6	30.50	−82.13	94.38	59.50	58.50	59.50	383.05	355.00	7.32	241.60	234.00	3.15	340.40	332.00	2.47
7	13.50	−79.88	95.63	76.50	49.50	76.50	389.10	383.00	1.57	152.60	141.00	7.60	328.70	340.00	−3.44
8	5.00	−78.75	96.25	85.00	45.00	85.00	390.50	383.00	1.92	101.05	110.00	−8.86	321.70	297.00	7.68
9	−54.50	−70.88	100.63	144.50	13.50	144.50	290.40	276.00	4.96	−351.23	−310.00	11.74	233.39	240.00	−2.83
10	−63.00	−69.75	101.25	153.00	9.00	153.00	224.89	219.00	2.62	−430.14	−404.00	6.08	216.37	214.00	1.10
11	−71.50	−68.63	101.88	161.50	4.50	161.50	155.10	144.00	7.16	−456.45	−449.00	1.63	198.95	188.00	5.50
12	−80.00	−67.50	102.50	170.00	0.00	170.00	130.00	116.00	10.77	−360.00	−386.00	−7.22	190.00	178.00	6.32

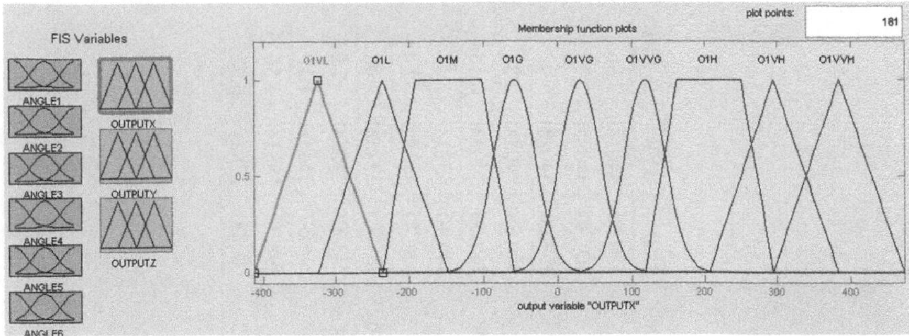

Fig. 3 Output (location "O1") with hybrid membership function for capacity of the robot

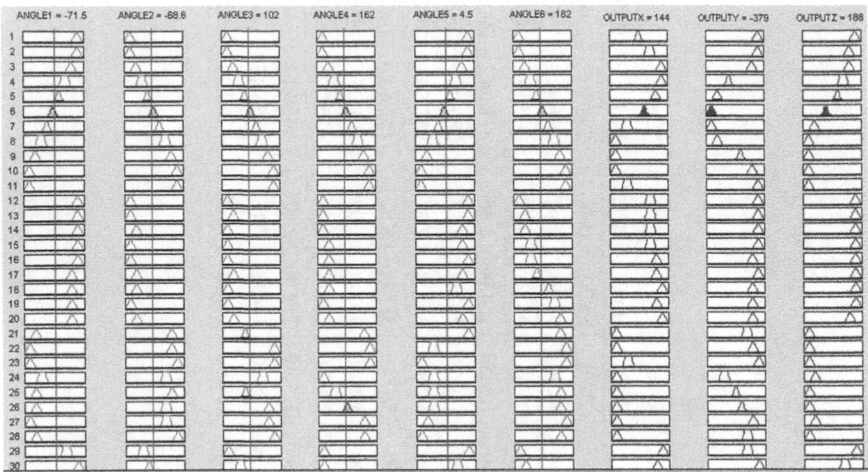

Fig. 4 Graphical representation of Input variables ("A1–A6") and output (location "O1–O3") with hybrid membership function for capacity of the robot

Fig. 5 Comparison of output value (location "X") of all four membership function with experimental value

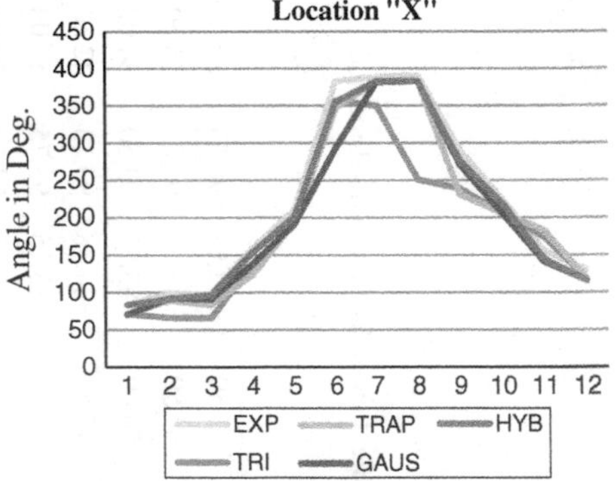

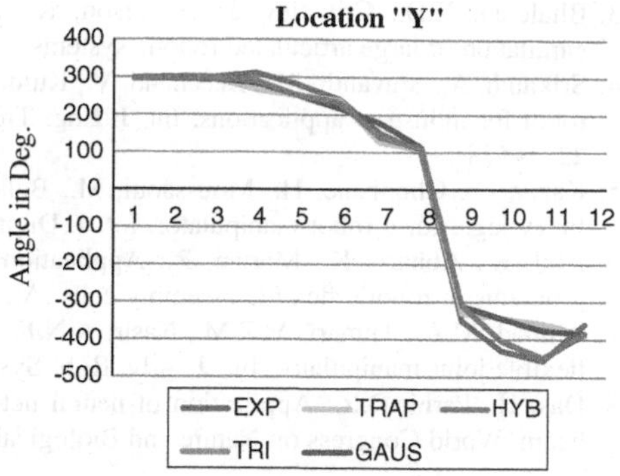

Fig. 6 Comparison of output value (location "Y") of all four membership function with experimental value

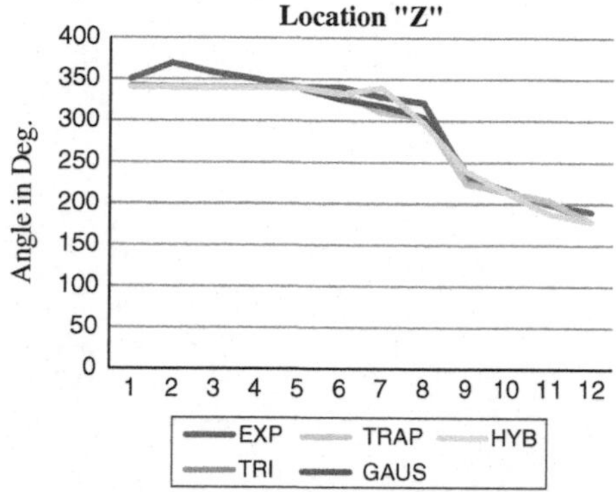

Fig. 7 Comparison of output value (Location "Z") of all four membership function with experimental value)

5 Conclusion

The results obtained from all four membership function viz. triangular, Gaussian, trapezoidal and hybrid of the fuzzy logic controller (FLC) are validated with experimental results of six axis industrial robot. From the above results it is concluded that the result obtained from hybrid membership function is very close to the experimental results. So within a specified capacity of the robot it is easy to predict the required position of the end-effector by using Fuzzy logic method. For future works the authors suggest that the technique can be applied for more DOF using Fuzzy logic and other soft computing techniques like GA, ACO etc.

References

1. Hwang, J.Y., Kim, J.S., Lim, S.S., Park, K.H.: A fast path planning by path graph optimization. IEEE Trans. Syst. Man Cybern. Part A Syst. Humans **33**(1), 121–128 (2003)
2. Gasparetto, A., Zanotto, V.: A new method for smooth trajectory planning of robot manipulators. Elsevier Mech. Mach. Theory **42**, 455–471 (2007)

3. Bhalerao, K.D., Critchley, J., Anderson, K.: An efficient parallel dynamics algorithm for simulation of large articulated robotic systems. Elsevier Mech. Mach. Theory **53**, 86–98 (2012)
4. Srikanth, A., Sravanth, M., Sreechand, V., Kumar, K.K.: Kinematic analysis of 3 DOF of serial robot for industrial applications. Int. J. Eng. Trends Technol. **4**(4), 1000–1004 (2013). ISSN: 2231–5381
5. Kazar, O., Ghodbane, H., Moussaoui, M., Belkacemi, A.: A multi-agent approach based on fuzzy logic for a robot manipulator. Int. J. Digital Content Technol. Appl. **3**(3), 86–90 (2009)
6. Abd, K., Abhary, K., Marian, R.: Application of fuzzy logic to multi-objective scheduling problems in robotic flexible assembly cells. Autom. Control Intell. Syst. **1**(3), 34–41 (2013)
7. Ahmad, M.A., Tumari, M.Z.M., Nasir, A.N.K.: Composite fuzzy logic control approach to a flexible joint manipulator. Int. J. Adv. Rob. Syst. **10**(58), 1–9 (2013)
8. Das, H., Parhi, D.R.: Application of neural network for fault diagnosis of cracked cantilever beam. World Congress on Nature and Biologically Inspired Computing, pp. 1303–1308 (2009)

Pitch Angle Controlling of Wind Turbine System Using Proportional-Integral/Fuzzy Logic Controller

Rupendra Kumar Pachauri, Harish Kumar, Ankit Gupta and Yogesh K. Chauhan

Abstract Blade pitch angle and tip speed ratio (TSR) control approaches are very important in Wind Turbine (WT) to achieve a constant output torque for stability of WT operation. Therefore, accurate and effective control techniques are investigated and designed to supply constant output torque to the Permanent Magnet Synchronous Generator (PMSG). Furthermore, two controlling approaches, i.e. Proportional-Integral (PI) and Fuzzy Logic (FL) based blade pitch angle and TSR control are implemented, and their performance is investigated in terms of torque stability and response time. The complete system is modeled in MATLAB/Simulink environment. The performance of the system with these control techniques are investigated under variable wind speed conditions. The performance of both the systems is found satisfactory, but system with FLC shows better performance as compared to PI controller.

Keywords Fuzzy logic controller · Blade pitch angle control · Permanent magnet synchronous generator · Tip speed ratio · Wind turbine

R.K. Pachauri (✉) · H. Kumar · A. Gupta · Y.K. Chauhan
Electrical Engineering Department, School of Engineering, Gautam Buddha University, Greater Noida 201308, India
e-mail: rupendra@gbu.ac.in

H. Kumar
e-mail: harishgbu@gmail.com

A. Gupta
e-mail: guptaankit299@gmail.com

Y.K. Chauhan
e-mail: yogesh@gbu.ac.in

A. Nagar et al. (eds.), *Proceedings of 3rd International Conference on Advanced Computing, Networking and Informatics*, Smart Innovation, Systems and Technologies 43, DOI 10.1007/978-81-322-2538-6_7

1 Introduction

The role of electrical energy is very important for developing a country. The shortage of fossil fuels e.g. petrol, diesel, coal and gas etc. is experienced globally due to exponential increase in the rate of energy consumption. It is forced to explore more sustainable energy resources [1]. In this context, the WT and photovoltaic systems are commonly used in various applications e.g. water pumping irrigation and rural electrification etc. Wind power technology has experienced a notable escalation, and progressing from last three decade. Today, WT is found as one of the best growing RE source [1].

The authors have formulated a mathematical model of WT and the mechanical torque produced by the WT is obtained in [2]. The predefined pitch angle WT model has been used for power generation with permanent magnet synchronous generator (PMSG). The authors [3] analysed the dynamic model of PMSG based on Wind Energy Conversion System (WECS). In [4], the authors proposed a wind turbine generator system (WTGS) connected with a variable speed turbine generator. Apart from the generator, the authors also investigated that WTGS contains of three parts i.e. wind turbine, drive train and PMSG. WT can be categorized into two kinds based on the axis in which it rotates namely horizontal axis wind turbines (HAWT) and vertical axis WTs (VAWT). A WT with a blade pitch angle control using the fuzzy logic (FL) to obtain the maximum output power is designed [5]. The authors have also shown that implementation of FL based pitch angle control to the WT is most suitable for the low level wind speed regions.

With the motivation of above literature review, the research aspect of this paper is to investigate and comparative analysis of PI and FL controller based blade pitch angle and TSR control for torque optimization of WT system.

2 System Description

The complete system comprises three major parts (a) Wind Turbine (b) Permanent magnet synchronous generator (c) Blade pitch angle and Tip speed ratio controlling schemes (i) Proportional-integral controller (ii) Fuzzy logic controller. The complete system is shown in Fig. 1.

The paper is outlined as follows. In Sect. 3, the WT modeling is presented. In Sect. 4, the mathematical modeling of PMSG is reported. In Sect. 5, the blade pitch angle and tip speed ratio control schemes are discussed in detail. Finally, the results are discussed in Sect. 6 and Sect. 7 concludes the paper.

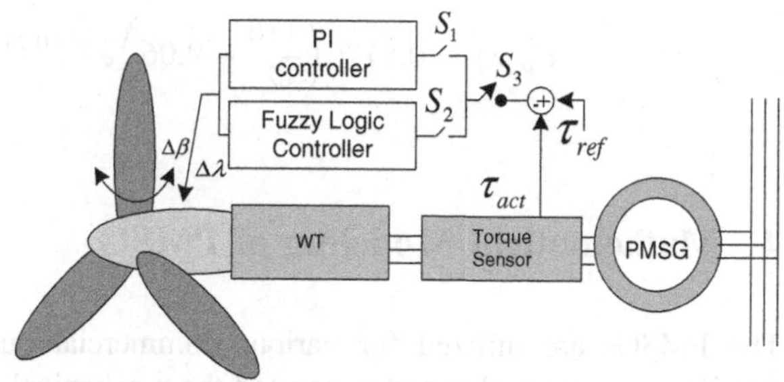

Fig. 1 Schematic diagram of blade pitch angle control of WT-PMSG system

3 Mathematical Modeling of WT

The WT converts the kinetic energy of wind potential into mechanical energy. A WT can function at a steady state speed with changeable blade pitch angle and TSR that allows it to produce the constant power at 50 Hz, which is required for connection with the grid [5]. The power available in the wind is equal to the amount of energy produce per second wind power (P_{wind}), power (P_m) and torque (τ_m) equations of a WT are given by Eq. (1) as,

$$P_{wind} = \frac{1}{2}A\rho V_w^3, \quad P_m = 0.5\rho A C_p V_w^3, \quad \tau_m = 0.5\rho A C_p V_w^2 \frac{R}{G\lambda} \tag{1}$$

where, A is the swept area of rotor blades, ρ is the air density, V_w is wind speed, C_p is coefficient of performance, R is rotor radius, G is gear ratio and λ is TSR. The power coefficient characteristic is highly non-linear in nature and reflects the aerodynamic behaviour of WT. The C_p is expressed in Eq. (2) as,

$$C_p(\lambda, \beta) = C_1\left(\frac{C_2}{\lambda_i} - C_3\beta - C_4\right)e^{\frac{-C_5}{\lambda_i}} + C_6\lambda, \quad \frac{1}{\lambda_i} = \frac{1}{\lambda + 0.08\beta} - \frac{0.035}{\beta^3 + 1} \tag{2}$$

where, C_1, $C_2 \ldots C_6$ are coefficients, β is blade pitch angle and λ_i is initial TSR. The TSR (λ) is expressed in Eq. (3) as the ratio between the speed of the tips of the blades of a WT and the wind speed.

$$\lambda = \frac{V_{TIP}}{V_W} = \omega \cdot R/V_w \tag{3}$$

where, ω is the rotor angular velocity (rad/s). The coefficient of performance C_p, is expressed as the fraction of energy extracted by WT of the total energy that would have followed through the area swept by the rotor. The coefficient of power is expressed in Eq. (4) [6] as,

$$C_p(\lambda) = 0.5176\left(\frac{116}{\lambda} - 9.06\right)e^{\frac{-21}{\lambda}+0.735} + 0.0068\lambda \tag{4}$$

4 Mathematical Modeling of PMSG

The PMSGs are utilized for various commercial purposes in wide range. The PMSGs are commonly used to convert the mechanical power output of turbines i.e. steam turbines, wind turbines etc. into electrical power for the intended system. The WT and the generator rotate in synchronism with the same shaft without gear assembly. The mathematical model of the PMSG in the state space equation form is given by Eq. (5) [6] as,

$$\frac{di_q}{dt} = \frac{(-R_s i_q + \omega_e(L_{qs} + L_{ls})i_d + u_q)}{L_{qs} + L_{ls}}, \quad \frac{di_d}{dt} = \frac{(-R_s i_d + \omega_e(L_{qs} + L_{ls})i_q + u_d)}{L_{ds} + L_{ls}} \tag{5}$$

where, R is the stator resistance, L_q and L_d are the inductances of the generator along d and q axis and L_{ls} is the leakage inductance of the generator. The electrical speed (ω_e) (rad/s) of the generator and electromagnetic torque equation of PMSG are expressed in Eq. (6) as,

$$\omega_e = p\omega_g, \quad \tau_{em} = 1.5p((L_{ds} - L_{ls})i_d i_q + i_q \psi_f) \tag{6}$$

where, i_d and i_q are the currents along d and q axis, ψ_f is the permanent magnetic flux, p is the number of pole pairs and τ_{em} is the electromagnetic torque of the generator.

5 Blade Pitch Angle and TSR Control Strategies

For controlling the torque using blade pitch angle and TSR, following two controllers namely PI and FLC are considered as,

5.1 Proportional—Integral Controller

The PI controller is a control which has feedback mechanism which is commonly used in industrial control operations [7]. The optimum combination of proportional and integral gain increases the speed of the response and also to minimize the steady state error. The PI controller equation is expressed in Eq. (7) as,

$$u(t) = K_p e(t) + K_i \int e(t)dt \tag{7}$$

where, K_p and K_i are proportional and integral gain, and used to control the response of the system. The input of the PI controller is the error between τ_{ref} and τ_{actual} and the output is change in blade pitch angle or TSR [8, 9].

5.2 *Fuzzy Logic Controller*

Recently, FLCs are introduced for blade pitch angle control for WT. These controllers are robust and advantageous as their design procedure does not require exact model information. Their major goal is to implement human knowledge/information in the form of a program [10, 11]. The main parts of a FLC are fuzzification, rule base, inference and defuzzification as shown in Fig. 2,

Both the inputs i.e. error (e) and change of error (ce) are shown in Eq. (8) as,

$$e(k) = \tau_{ref}(k) - \tau_{actual}(k),\ ce(k) = e(k) - e(k-1) \tag{8}$$

The rule base for FLC are shown in Table 1 as,

For the FL controller the output is the change in pitch angle or TSR. This generated pitch angle targets the WT to produce the optimal torque. The membership functions for both inputs and output are shown in Fig. 3a–c as,

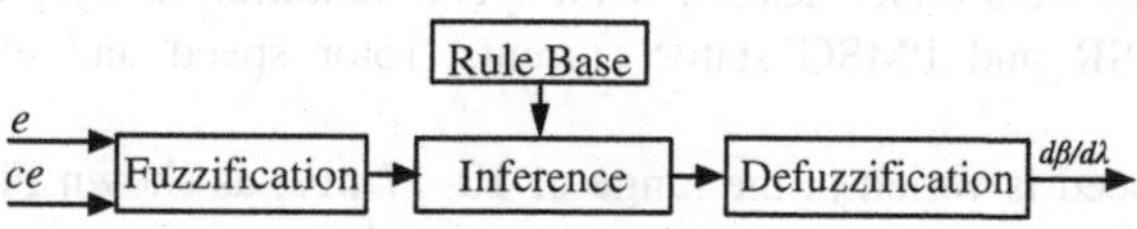

Fig. 2 Block diagram of FLC

Table 1 Rule base for FLC

e/ce	NB	NM	NS	ZE	PS	PM	PB
NB	NB	NB	NM	NM	NS	NS	ZE
NM	NB	NM	NM	NS	NS	ZE	PS
NS	NB	NM	NM	NS	ZE	PM	PM
ZE	NM	NS	NS	ZE	PM	PM	PB
PS	NS	NS	ZE	PS	PS	PM	PM
PM	NS	ZE	PS	PS	PM	PM	PB
PB	ZE	PS	PS	PM	PM	PB	PB

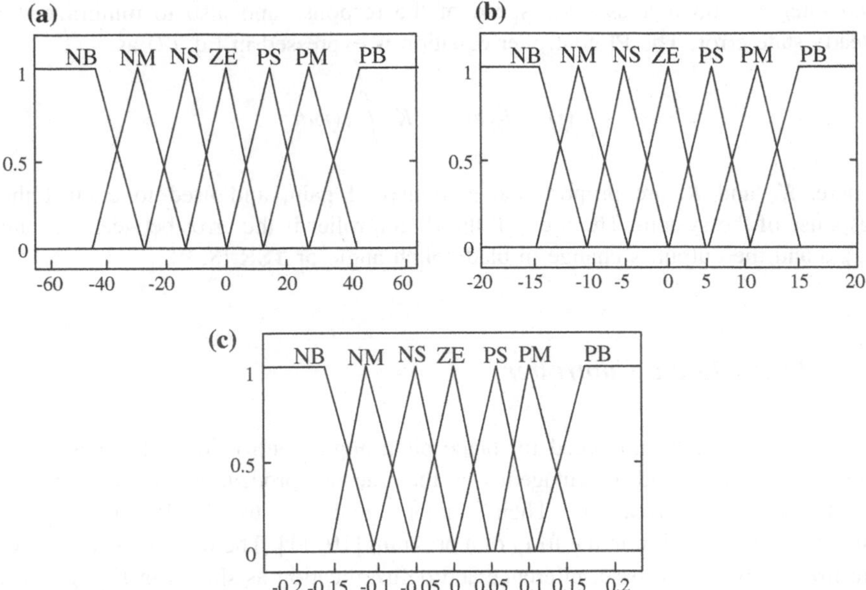

Fig. 3 Membership functions of fuzzy logic controller. **a** Input variable “e”. **b** Input variable “ce”. **c** Output variable “dβ/dλ”

6 Results and Discussion

The operating parameters of WT and PMSG are given in Appendix. The performance of blade pitch angle and TSR control systems is analyzed for both PI and FLC based controllers under desired wind speed condition using pitch angle, turbine torque, TSR and PMSG stator currents, rotor speed and electromagnetic torque.

The wind speed is varied in the range of 10–17 m/s, as shown Fig. 4. Both the cases of step increase and decrease in wind speed from reference point are considered. This pattern of wind speed variation is applied to both the systems with PI

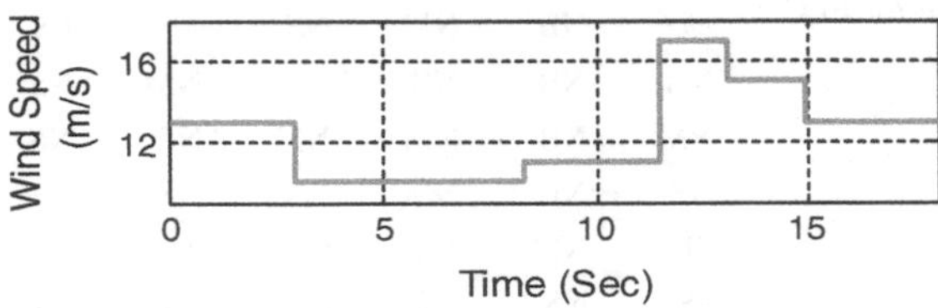

Fig. 4 Variation of wind speed

and FLC for blade pitch angle and TSR controlling. The transient response of blade pitch angle and TSR controlling using PI and FL controllers are shown in Figs. 5 and 6.

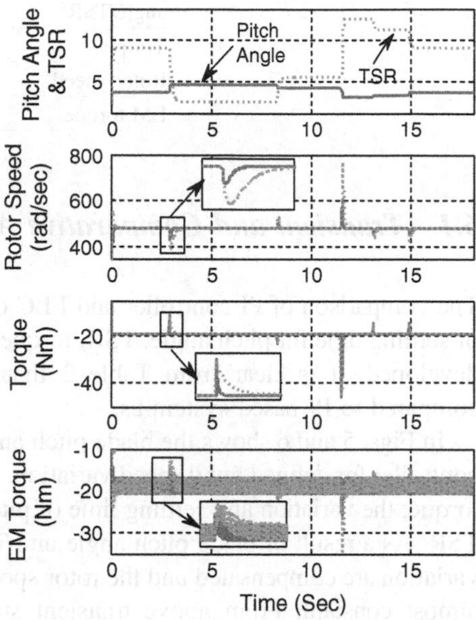

Fig. 5 Transient response of pitch angle and TSR controlling using PI controller

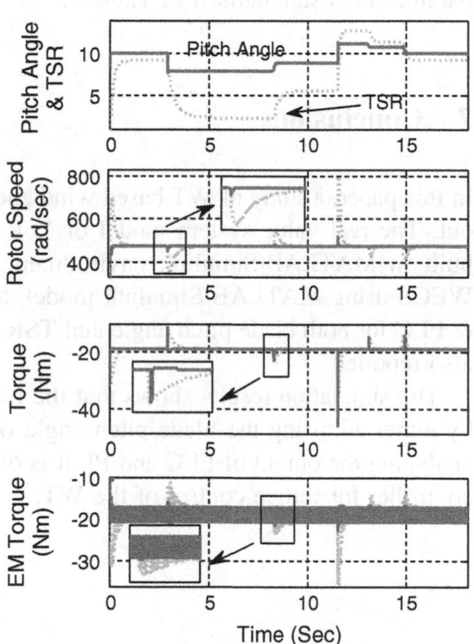

Fig. 6 Transient response of pitch angle and TSR controlling using FL controller

Table 2 Comparison of PI and FLC based system for blade pitch angle and TSR control

Parameters	Settling time (s)			
	PI controller		FL controller	
	Pitch angle	TSR	Pitch angle	TSR
Pitch angle/TSR	1.17	0.689	0.039	0.03
Torque	0.95	0.225	0.059	0.02
Rotor speed	0.94	0.319	0.289	0.09
EM torque	0.87	0.369	0.269	0.06

6.1 Transient and Comparative Analysis

The comparison of PI controller and FLC operated system is obtained on the basis of settling time for pitch angle, TSR, torque, rotor speed and electromagnetic torque developed. It is clear from Table 2 that FLC based system settles quickly as compared to PI based system as,

In Figs. 5 and 6 shows the blade pitch angle and TSR control by using PI and FL controller for defined wind speed variation. It is observed that for obtaining constant torque, the variation and settling time of pitch angle is less compared to variation in TSR. As a result of blade pitch angle and TSR variation, the effects of wind speed variation are compensated and the rotor speed, WT torque and EM torque are found almost constant. From above transient study, it is observed that the FL based controller has less settling time than PI controller. The comparative study of these parameters is summarised in Table 2.

7 Conclusions

In this paper, a study of WT based wind energy conversion system has been carried out. The real value system model of WT along with the system components are built in MATLAB/Simulink environment. With the simulation of the complete WECS using MATLAB/Simulink model, all the results have been obtained for PI & FLC for both blade pitch angle and TSR control and a their comparative study is also reported.

The simulation results shows that the reference torque is conveniently achieved by either adjusting the blade pitch angle or by TSR control using PI & FLC. By analyzing the output of FLC and PI, it is observed that FLC is better than PI based controller for torque control of the WT.

Appendix

Number of blades = 3, Blade radius (R) = 4.5 m, Gear ratio (G) = 40, Air density (ρ) = 1.2 kg/m^3, Wind speed (v_w) = 10–17 m/s, Power Coefficient (C_p) = 0.45, C_1 = 0.5176, C_2 = 116, C_3 = 4, C_4 = 5, C_5 = 21, C_6 = 0.0068, Stator Phase Resistance (R_s) = 0.4250 Ω, Inductance (L_d, L_q) = 0.0084, 0.0084 H, Inertia (J) = 0.001469 kg m^{2,} Friction factor (F) = 0.0003035 N m s, Pole pairs = 4, K_p = 0.02, K_i = 3.0.

References

1. Pachauri, R.K., Chauhan, Y.K.: Assessment of wind energy technology potential in Indian context. Int. J. Renew. Energy Res. **2**, 774–780 (2012)
2. Abdullah, M.A., et al.: Review of maximum power point tracking algorithms for wind energy systems. Renew. Sustain. Energy Rev. **16**, 3220–3227 (2012)
3. Luo, C., Banakar, H., Shen, B., Ooi, B.T.: Strategies to smooth wind power fluctuations of wind turbine generator. IEEE Trans. Energy Convers. **22**, 341–349 (2007)
4. Badoni, P., Prakash, S.B.: Modeling and simulation of 2 MW PMSG wind energy conversion systems. J. Electr. Electro. Eng. **9**, 53–58 (2014)
5. Pachauri, R.K., Chauhan, Y.K.: Mechanical control methods in wind turbine operations for power generation. J. Autom. Control Eng. **2**, 214–220 (2014)
6. Verma, J., et al.: Performance analysis and simulation of wind energy conversion system connected with grid. Int. J. Recent Technol. Eng. **2,** 33–38 (2014)
7. Kanellos, F.D., Hatziargyriou, N.D.: Control of variable speed wind turbines in isolated mode of operation. IEEE Trans. Energy Convers. **2**, 535–543 (2009)
8. Senjyu, T., Sakamoto, R., Urasaki, N., Funabashi, T., Fujita, H., Sekine, H.: Output power leveling of wind turbine generator for all operating regions by pitch angle control. IEEE Trans. Energy Convers. **21**, 467–475 (2006)
9. Uehara, A., et al.: A coordinated control method to smooth wind power fluctuations of a PMSG-based WECS. IEEE Trans. Energy Convers. **26**, 550–558 (2011)
10. Musyafa, A., et al.: Pitch angle control of variable low rated speed wind turbine using fuzzy logic controller. Int. J. Eng. Technol. **10**, 22–25 (2010)
11. Goyal, S., et al.: Power regulation of a wind turbine using adaptive fuzzy-PID pitch angle controller. Int. J. Recent Technol. Eng. **2,** 128–132 (2013)

Recognition of Repetition and Prolongation in Stuttered Speech Using ANN

P.S. Savin, Pravin B. Ramteke and Shashidhar G. Koolagudi

Abstract This paper mainly focuses on repetition and prolongation detection in stuttered speech signal. The acoustic and pitch related features like Mel-frequency cepstral coefficients (MFCCs), formants, pitch, zero crossing rate (ZCR) and Energy are used to test the effectiveness in recognizing repetitions and prolongations in stammered speech. Artificial Neural Networks (ANN) are used as classifier. The results are evaluated using combination of different features. The results show that the ANN classifier trained using MFCC features achieves an average accuracy of 87.39 % for repetition and prolongation recognition.

Keywords ANN · Energy · Formants · MFCCs · Pitch · Zero crossing rate

1 Introduction

Stuttering is one of the serious problems in speech pathology. It is a speech disorder in which flow of speech is disrupted by involuntary repetitions and prolongation of syllables, words or phrases and pausing. Different types of stuttering events are interjections, phrase repetitions, word repetitions, syllable repetitions and prolongations [1, 2]. The identification and evaluation of these disfluencies have many medical applications such as computing the severity of stammering, identification of problems causing stuttering (psychological and articulatory), therapy to reduce

---

P.S. Savin (✉) · P.B. Ramteke · S.G. Koolagudi
National Institute of Technology Karnataka, Surathkal 575025, Karnataka, India
e-mail: savinps@gmail.com

P.B. Ramteke
e-mail: ramteke0001@gmail.com

S.G. Koolagudi
e-mail: koolagudi@nitk.ac.in

A. Nagar et al. (eds.), *Proceedings of 3rd International Conference on Advanced Computing, Networking and Informatics*, Smart Innovation, Systems and Technologies 43, DOI 10.1007/978-81-322-2538-6_8

the stuttering and so on. Speech Language Pathologists (SLPs) evaluates stuttering by manually counting the number of disfluent speech units and types of disfluencies. This process of evaluation is time intensive and requires a careful attention of expert. Also the results of evaluation vary from expert to expert. Hence it is inconsistent and highly prone to errors [3]. To overcome these difficulties, there is a need of automatic assessment of stuttering disfluencies. It help SLPs in assessing the stuttered events and improve the inter-judge agreements between the evaluation of these events.

The repetition and prolongation are the two prominent stuttered events observed in patient compared to the other types of disfluencies [4]. The repetition events have similar acoustic features and prolongation of a phoneme has different features than its normal pronunciation, therefore it is possible to automatically diagnose these stuttering patterns. This paper mainly focuses on recognition of repetitions and prolongations.

Rest of the paper is organized as follows. The research carried out in this area is reviewed in Sect. 2. Methodology of the proposed approach is discussed in Sect. 3. Results are analyzed in Sect. 4. Section 5 concludes the paper and highlight the future scope of the work.

2 Literature Review

A considerable amount of research work has been done in recognition of different types of stuttering events in a speech signal. This section describes the previous works that have considered different features, the various datasets and classifiers or approaches, used for the recognition of stuttered events.

The MFCCs are claimed to be robust in recognition of human voice as it exactly maps the human auditory response [5]. Hence, most of the researches have considered MFCCs as a baseline features for the recognition of stuttered events with different classifiers [6–8]. An ideal approximation to the vocal tract spectral envelope is provided by the all-pole model of the Linear prediction coefficients (LPC) while it is applied to the analysis of speech signals which leads to a moderate estimation of source-vocal separation. Therefore, LPC and Linear prediction cepstral coefficients (LPCCs) are employed for the repetition and prolongation recognition [9, 10].

The dataset is a crucial part of the stuttering recognition system. The system trained using proper dataset may achieve better recognition accuracy. Many researches have used the dataset made available by University College London Archive of Stuttered Speech (UCLASS) for the analysis [9, 11]. It includes one sample each from 2 female and 8 male speakers ranged between 11 and 20 years, covers a broad range of stuttering rate. In [6], the speech data of 600 stuttering syllables is collected from 50 students in Beijing Speech Training Centre. The age range is from 17 to 42 years. A speech data of 7 males and 3 females collected from students of Faculty of Electrical Engg, UTM Skudai used in [12]. The word "Sembilan" (nine) is recorded for 20 samples of data for the normal speech and 15 samples of data for the artificially stuttered speech. In [13], the database consists recording from 15 clients around the age group of 25 where each person's speech consists of 150 words from standard English passage.

Various classifiers have been used to test the effectiveness of the approaches in recognizing stuttered events. The Support vector machine (SVM) attempts to obtain a good separating hyper-plane between two classes in the higher dimensional space. Hence, SVM is used for the repetition and prolongation recognition [8]. Hidden Markov Models (HMM) are used for recognition of any kind of pattern. The stuttered speech also consists of patterns for repetition, prolongation, etc. Hence, in [10] HMMs are used for identification of prolongations of fricatives. In [9], Linear discriminant analysis (LDA) and k-nearest neighbor (k-NN) algorithms used for classification of repetitions and prolongations. Both these classifiers try to find a linear transformation that maximizes class separability in reduced dimensional space.

From the literature survey, it is observed that many of the researches have been focused on stuttering event detection using MFCCs, LPC and LPCCs features. In this work an attempt is made to recognize repetitions and prolongations using different acoustic and pitch related features like formants, pitch, ZCR and energy with ANN as a classifier.

3 Methodology

In this work, the process of stuttered speech analysis is divided into four stages: segmentation, feature extraction, classification and output shown in Fig. 1. The following subsections explain them in detail.

3.1 Segmentation

In this approach, dataset is collected from audio recording of stuttering treatment in Hindi language. The four speakers of 25–40 years have repetition and prolongation disfluencies with the permission of the subject and concerned doctors, their speech has been recorded with a specific text. The speech signal is manually segmented into repetition, prolongation and normal speech units. Total 78 segments are obtained, out of which 32 are repetition, 18 are prolongation and 28 are normal speech units.

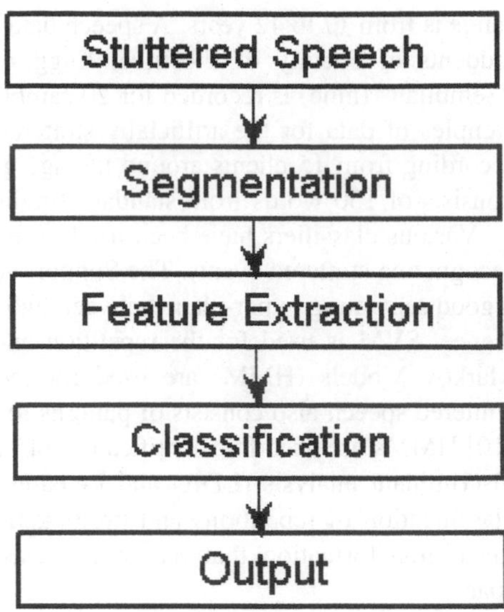

Fig. 1 Schematic block diagram of recognition of repetitions and prolongations

3.2 Feature Extraction

In this work, spectral and pitch related features are extracted for characterization of repetitions and prolongations. This subsection explains the process of feature extraction in brief.

The frequency bands of Mel frequency cepstrum follow human perception pattern i.e. frequency bands are spaced linearly up to 1 kHz and logarithmic beyond it [5]. Hence, MFCC features are considered to be robust for the speech recognition tasks. The articulation mechanism of the vocal tract is same for a particular repetition units and hence MFCC features are expected to be similar for those events. In this work, 13 MFCC features are considered to develop a baseline system.

Short time energy is a significant characteristic of speech processing [14]. The energy of the repeated speech unit is identical and it is unstable for prolongation, hence the energy may be considered for the recognition of repetitions and prolongations. Formants represents the resonance of the vocal tract. The position of formant frequencies and their corresponding magnitudes may differ depending on the phonemes as articulation mechanism of vocal tract for each phoneme is unique [15]. Repetition events have the same vocal tract articulation, hence may have similar formants. In this approach, 4 formants are considered for the experimentation. Pitch is an important attribute of voiced speech, refers to the fundamental frequency F_0 of vocal fold's vibration. Variation in pitch detected by cepstral analysis comprises of reduction in the spectrum band for cepstral analysis and estimation of the vocal tone frequency [16]. This leads to a noticeable change in cepstral maximum in pitch. There may be a significant change in pitch when stuttered events occurs in a speech, hence pitch can be considered as a feature for the repetition and prolongation recognition. Four statistical pitch parameters minimum pitch, maximum pitch, average pitch, standard deviation are considered as the features from each segment. The zero crossing rate is the rate at which the signal changes from positive to negative or back [14]. ZCR value for repeated phonemes will be almost same, hence it may distinguish repeated events.

3.3 *Classification*

A Multilayer Feed forward network using back propagation algorithm is used for speech pattern classification because of their discrimination and input-output mapping ability [17]. It consists of an input layer, one hidden layer and an output layer as shown in Fig. 2. In this approach, feature vector is considered as an input for the neural network hence size of feature vector decides the number of neurons in the input layer. The number of output neurons is equal to the number of output classes. The number of neurons in hidden layers is decided using Oja rule [17]. The Oja rule is given by, $H = \frac{\mathrm{T}}{5(\mathrm{N}+\mathrm{M})}$, where H is number of hidden layer neurons, N is the size of the input layer, M is the size of the output layer and T is the training set size. As the number of words used for training an ANN increases the number of hidden layer neurons also has to be increased.

4 Results and Discussion

In this section, the results achieved using the combination of different features are discussed. The speech database consists of 32 repetition, 18 prolongation and 28 normal speech units segmented at phoneme and word level from the speech recordings. The features are extracted from these speech units using the frame size of 25 ms with a shift of 10 ms. Different combinations of features have been considered to form the feature vectors which act as input to ANN. In this work, 75 % of the segments are used for training and 25 % is considered for testing. The number of input, output, and hidden layer neurons are chosen based on the

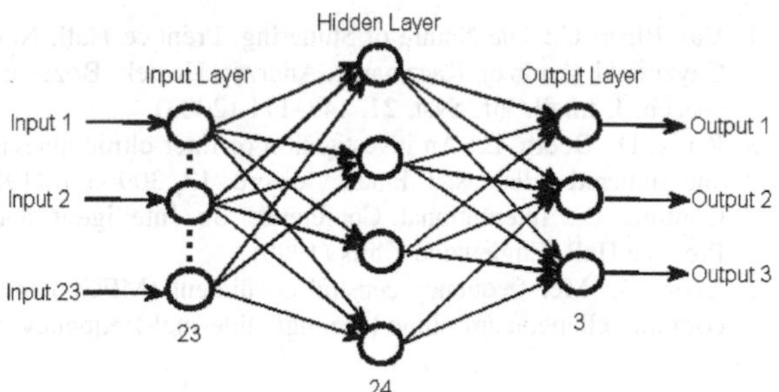

Fig. 2 Block diagram of artificial neural network

Table 1 Repetition and prolongation detection result using ANN

Features	Repetition	Prolongation	Normal	Average
MFCC	92.64	96.12	76.31	88.29
Formant + Pitch	93.38	96.31	70.31	86.69
MFCC + ZCR + energy	92.81	96.33	74.33	87.82
MFCC + Formant	90.89	96.71	74.56	87.34
Formant + Pitch + Energy	94.52	95.75	68.72	84.31
MFCC + Pitch + Formants + ZCR + Energy	91.18	96.82	72.87	87.39

number of input, output and number of speech segments used for training the network.

The classifiers (ANNs) are trained using different combination of features. 13 MFCCs, 4 Formants (F0, F1, F2, F3), 4 Pitch Parameters (Maximum Pitch, Minimum Pitch, Mean and Variance), ZCR and Energy are considered as a features for the experimentation. Table 1 shows the recognition accuracy achieved for each class using combination of these features. The MFCC features alone achieve better average accuracy of 88.29 % compared to the other combination of features. The combination of formants, pitch and energy achieves better accuracy of 94.52 % for repetition events. Prolongation events are classified by combination of MFCC and formants with the an accuracy of 96.71 %.

5 Summary and Conclusion

In this approach, the performance of ANN classifier in identifying the repetition and prolongation in the stuttered speech event is evaluated using combination of different features. Repeated and prolonged patterns are characterized in a better way with out MFCC features. Further, the proposed methodology can be extended to detect the interjection, and other speech disfluencies in different regional languages.

References

1. Van Riper, C.: The Nature of Stuttering. Prentice Hall, New Jersey (1971)
2. Czyzewski, Andrzej, Kaczmarek, Andrzej, Kostek, Bozena: Intelligent processing of stuttered speech. J. Intell. Inf. Syst. **21**, 143–171 (2003)
3. Kully, D., Boerg, E.: An investigation of inter-clinic agreement in the identification of fluent and stuttered syllables. J. Fluency disord. **13**, 309–318 (1988)
4. Conture, E.: International Conference on Intelligent and Advanced Systems, 2nd edn. Prentice-Hall, Englewood Cliffs (1990)
5. Lyons, J.: Mel frequency cepstral coefficient (MFCC) tutorial. http://practicalcryptography.com/miscellaneous/machine-learning/guide-mel-frequency-cepstral-coefficients-mfccs/

6. Zhang, J., Dong, B., Yan, Y.: A computer-assist algorithm to detect repetitive stuttering automatically. In: International Conference on Asian Language Processing, pp. 249–252 (2013)
7. Sin Chee, L., Chia Ai, O., Hariharan, M.: MFCC based recognition of repetition and prolongations in stuttered speech using artificial k-nn and lda. In: IEEE Student Conference on Research and Development, pp. 146–149 (2009)
8. Ravikumar, K.M., Rajagopal, R., Nagaraj, H.C.: An approach for objective assessment of stuttered speech using MFCC features. ICGST Int. J. Digital Signal Process. **9**, 19–24 (2009)
9. Chia Ai, O., Hariharan, M., Yaacob, S., Sin Chee, L.: Classification of speech dysfluencies with MFCC and LPCC features. J. Med. Syst. **39,** 2157–2165 (2012)
10. Wisniewski, M., Kuniszyk, J.W., Smolka, E., Suszynski, W.: Automatic detection of disorders in a continuous speech with the hidden markov models approach. Comput. Recogn. Syst. **2** (45), 445–453 (2007)
11. Sin Chee, L., Chia Ai, O., Hariharan, M., Yaacob, S.: Automatic detection of prolongations and repetitions using LPCC. In; 2009 International Conference Technical Postgraduates (TECHPOS) (2009)
12. Tan, T.S., Liboh, H., Ariff, A.K., Ting, C.M., Salleh, H.: Application of malay speech technology in malay speech therapy assistance tools. Int. Conf. Intell. Adv. Syst. **48**, 330–334 (2007)
13. Ravikumar, K.M., Balakrishna Reddy, Rajagopal, R., Nagaraj, H.C.: Automatic detection of syllable repetition in read speech for objective assessment of stuttered disfluencies. Proce. World Acad. Sci. **2**, 220–223 (2008)
14. Rabiner, L., Juang, B., Yegnanarayana, B.: Fundamentals of Speech Recognition. Pearson, India (2010)
15. Welling, L., Ney, H.: Formant estimation for speech recognition. IEEE Trans. Speech Audio Process. **6**, 36–48 (1998)
16. IIT Guwahati.: Estimation of pitch from speech signal. http://iitg.vlab.co.in/
17. Gevaert, W., Tsenov, G., Mladenov, V.: Neural networks used for speech recognition. J. Autom. Control **2**, 732–735 (2010)

Part II
Computational Intelligence: Algorithms, Applications and Future Directions

Primary Health Centre ATM: PHCATM See Healthcare Differently

Priyabrata Sundaray

Abstract PHCATM is a product that can accurately and efficiently find relief for common symptoms by interactively querying the user and providing real time suggestions. In addition, it has the ability to connect with the nearest medical officer with live chat whenever demanded for online health check-up. The user can discuss problems with the medical officer, and be involved in deciding the best care solution. Finally, the PHCATM module can dispense the prescribed medications and print out prescriptions written electronically. Too many people in India die due to lack of diagnosis in first place, availability of medicine in time or in affordable cost as second.

Keywords Primary health centre · OTC drugs · Chronic diseases · Vaccination · Online health check-up

1 Introduction

The ever increasing imbalance in economic as well as gender divergence are the key challenges to the health status of a nation in spite of several health orientated policies mandated by the government. More than three quarters of medical infrastructure, work force and other resources oriented towards health are connected with urban population which constitutes a mere 27 % of total population [1].

Mother India has manifested herself as the most inhabited nation in the world where around 73 % of population live in rural areas. The medical setups in Indian rural areas are mediocre. The system lacks in providing any sort of proper organizational framework in order to dispense quality health services for the 73 % of rural Indian residents [2]. Moreover while richer section of the society in urban areas can secure standard healthcare facilities, there is scarcely any proper medical

P. Sundaray (✉)
Department of Electrical & Electronics Engineering, SRM University, Chennai 603203, India
e-mail: pbsundaray@gmail.com

A. Nagar et al. (eds.), *Proceedings of 3rd International Conference on Advanced Computing, Networking and Informatics*, Smart Innovation, Systems and Technologies 43, DOI 10.1007/978-81-322-2538-6_9

care services available to poor slum community. This scenario vehemently express the hindrances encountered by the Indian villages in terms of health care and medical supports which they are liable to earn.

Morbidity pattern especially in rural areas is influenced by communicable diseases such as protozoan infections, chicken pox, whooping cough, pneumonia and tuberculosis or other respiratory infections & waterborne diseases like diarrhoea, amebiasis, hepatitis infection, typhoid & worm infestations [3]. The health condition of Indians, is becoming serious, especially for those living at the rural areas. This is evident from the life expectancy data which is only 63 years, the infant mortality rate being 80/1000 live births and maternal mortality rate being 438/1,00,000 live births [4].

Approximately 70 % of total deaths, where transmissible diseases are the cause of 92 % of deaths, is suffered by 20 % of the poorest population of the nation living at the countryside [5].

It is due the fact that

- The number of doctors assigned per every village is very less and they do not have the proper equipment to work in emergency.
- Being very far from the city they have shortage of medicines needed in much time.
- Due to lack in proper health check-up, people in villages are prone to many contagious diseases [6].

Most of the deaths in villages, which could have been prevented, are caused by contagious infections, parasitic & respiratory viruses. Infectious viral disease has more fatal effect in rural areas, having its effect on 40 % rural population as compared to 23.5 % urban [5]. It is extremely unfortunate that while the fatality level is twice more at rural as compared to that at urban areas, the people of village sides are not being provided with access to proper healthcare systems, as the system and infrastructure were constructed to serve the urban section of the society [7].

Main factors of health problems:

- Lack of knowledge regarding filtering and boiling water before drinking.
- Minimal sanitation (No pit latrines).
- Lack of awareness about vaccination.
- Unavailability of medicines in time of urgency.
- Illiteracy.

Today the perceived need of the villagers for easier access to medicine is an urgent need. The need to train traditional midwives in hygienic delivery, to make local health workers more aware of diseases and to educate the community on sanitation and hygiene, including the harmful effects of the unhygienic conditions is also necessary (Fig. 1). The exponential rise of population suffering from diseases is shown in Fig. 1.

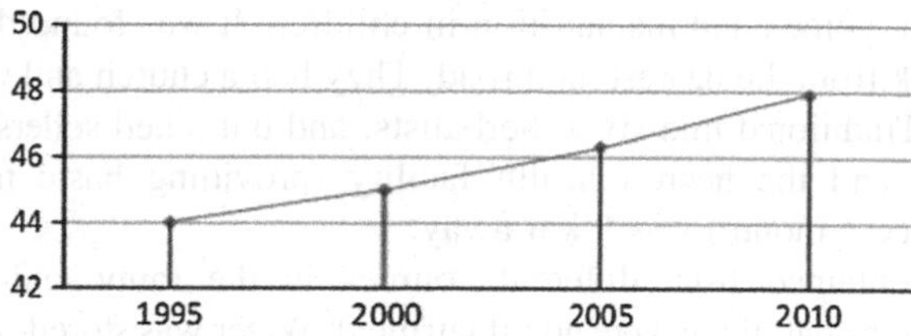

Fig. 1 Percentage of population suffering from diseases

1.1 Present Scenario of Primary Health Centres

Primary health centres (PHCs) started getting established after 1952. 1 PHC needs to be operational contributing 20,000 people in tribal or hilly region and 30,000 in case of plain region [8]. But the case is not so. Major problems faced by PHCs being:

1. Serving large population as PHCs cater to the requirements of more than 60 % of the population.
2. Lack in adequacy of staff and motivation among them to work for rural and tribal people.
3. Shortage of medicines or hindrance in dispensing those on time to the needy, has attributed to gross underutilization of the centers.
4. There is no scope for the involvement of the village people.

Thus the centerpiece idea of PHC, for a decentralized people based integrated service, based on curative, preventive and promotive measures has been totally undermined [9]. The concept being a person from the village community is to be selected & to be provided with the basic essential training, so that even the village people can cope up with emerging health problems effectively and more efficiently [10, 11].

1.2 Background

A survey of the environment, life-style, and health status, knowledge, attitudes and practices in the village of Chengalpattu was carried out prior to find the problems related to the people living there. In addition to the perceived needs of the villagers for a school, easier access to medicine and the study identified the need to train traditional midwives in hygienic delivery, to make local health workers more aware of diseases and to educate the community on sanitation and hygiene, including the harmful effects of the unhygienic conditions.

A health profile was obtained from a combined population of around 200. The profile was drawn from observation of village facilities, informal interviews with villagers, structured interviews with the village head, and a simple observation to detect anemia in women and malnutrition in children. It was found that the village is a few km walk from the nearest mud road. They had a church and weekly market but no school. Traditional midwives, herbalists, and untrained sellers of medicines were available, and the nearest health facility (providing basic medicines and vaccinations once a month) was 5 km away.

Water was obtained from different sources in the rainy and wet seasons. Sanitation was minimal (there were no pit latrines). Water was stored satisfactorily in 80 %, and the cooking areas were kept clean in 50 %. During the survey, more than 75 % of the population complained of illness (abdominal pain, scabies, respiratory infection, bloody stool, muscle strain, fever and insects bites) since the last two weeks. Fever in children was very common. 70 % of those who were ill sought no treatment. All of the adults interviewed were illiterate. 15 % of the women were anemic, and 10 % had received prenatal care at a hospital 15 km away. Only 10 % of the children had received one or more vaccinations, 30 % were malnourished.

2 PHCATM—A Medicine Vending Machine

PHCATM is much similar to the existing ATM machine but instead of the facility of transacting money it makes available the facilities such as medication, vaccination reminders and service on demand online check-up etc.

Features of PHCATM:

- Under a common account of a family, the members will be provided with free first aid.
- Medicines will be provided to the patient at the vending machine installed.
- Reminders about vaccinations due till date will be provided to each individual visiting the PHCATM.
- Even the people can avail the online check-up facility at their comfort in their village at regular intervals.
- The service will be available to the people under the “Unique ID” card which the Government aims at providing to each and every person so that the database of every patient can be stored.
- This database can be used for analyzing the situation in rural areas of the nation and allocating budget in the field of health and sanitation. In case of a patient suffering from the same disease again, the required drugs and doses can be provided in a quick and effective manner.
- Upon integration with Global Database, the individual’s medical history could be maintained and available as well as accessible anywhere and anytime for future reference.
- The machine would be able to check legitimate usage which includes having a limit on the amount of medications per person. Over-dispensing medication could mean illegitimate use which is dangerous to the user and creates a liability for the nation (Fig. 2 and Table 1).

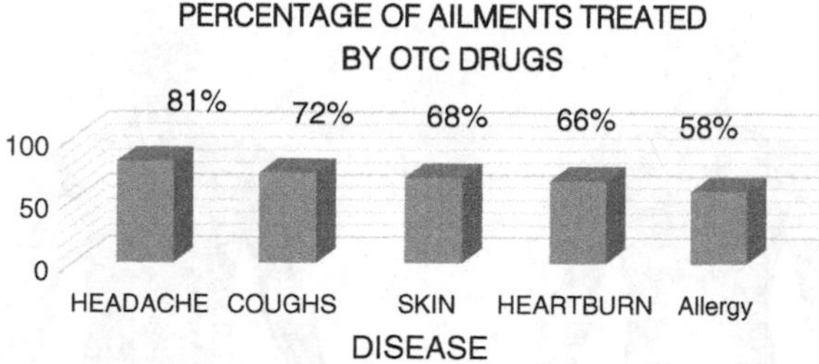

Fig. 2 Percentage of ailments most often treated with over-the-counter products

Table 1 List of most common over the counter drugs

Pain relievers	Antihistamines	Decongestants	Cough medicines
a. Non-steroidal anti-inflammatory drugs such as Aspirin, ibuprofen, naproxen	**Benadryl**	**Sudafed**	a. Antitussives—cough suppressants
b. Acetaminophen—relieves pain, reduces fever			b. Expectorants—Clears mucus

- The percentage of some common diseases being treated with OTC drugs is shown in Fig. 2.
- The most commonly used OTC drugs is listed in Table 1.

2.1 *Functioning of PHCATM*

In case of medicine, once the embedded system decides the proper medicines for the symptoms shown, **Inventory Controller** sends the information to PHCATM authorizing the machine to dispense the medicine. Once the prescribed medicines are dispensed, the processor prints a **receipt** for future assistance. The device would also be instrumented with voice controlled instructions in order to direct the patients on what to do next (Fig. 3).

Every PHCATM would be connected through server to the nearest medical officer in case a person ♂ ♀ demands to avail the online check-up facility. The medical officer would see to the situation and thus can solve the problem. In case there is an urgent need to appear before a doctor, the officer prescribes the patient to the nearest health center to have the treatment being prescribed by the officer.

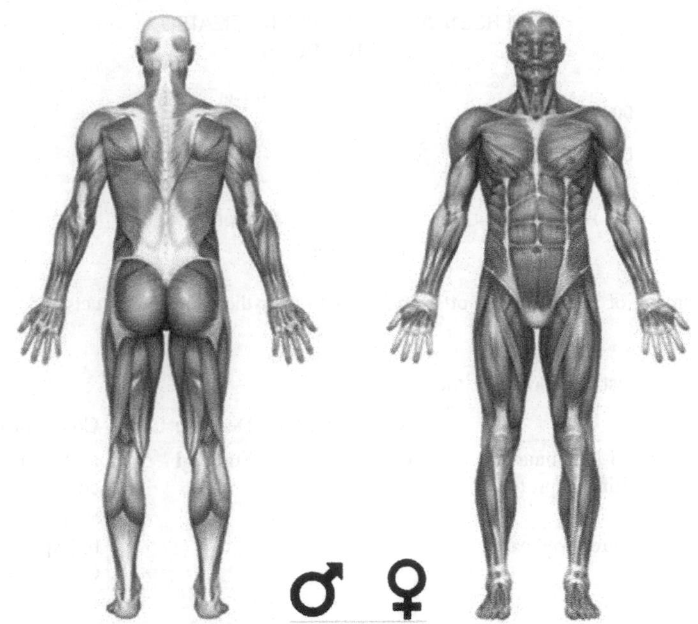

Fig. 3 Checkup graphical user interface

2.2 Accessibility of PHCATM

The machine has an inbuilt graphical user interface which comes as a handy for a patient to cope up with the digital environment, doesn't matter whether he/she is literate or not. **Problem of illiteracy of users can even be overcome by use of properly designed GUI**.

PHCATM is even accessible to blind and visually impaired people since the keypads at PHCATMs are equipped with **Braille.**

PHCATM can be incorporated with **Complete Blood Count Machine (CBC).** Which will result in helping Medical Officers to judge the illness better through accurate blood reports and benefit patients.

3 Technical Specifications

PHCATM is a simple data terminal consisting of input as well as output devices connected to it which can be accessed through UID (Aadhar Card in INDIA) which communicates through, a host processor, either through a leased line or a dial-up connection.

3.1 Inventory Controller

Controlling the inventory of drugs is critical to functioning of machine. GSM link would enable inventory controller to keep tab on inventory at regular intervals and report it to the control rooms. That way, replenishment is easy and the tracking too.

3.2 Input

Card reader It captures the account details of the Aadhar card (UID card) holder and provides back the past information. The host processor utilises this information to send the request to the nearest Primary Health Center or to the nearest medical officer.

Keypad The keypad helps the patient to let the host processor acknowledge which service is required (First aid, Vaccination, Online health checkup etc.).

GSM Link GSM link would enable inventory controller to keep tab on inventory at regular intervals and report it to the control rooms.

Camera Camera is important to identify symptoms of the patient or block in case of previous offenders/black-listed user.

Microphone In case of an illiterate or disabled user, keypad and display may not be of any use. Using microphone to analyze oral commands would be important.

3.3 Output

Speaker The speaker helps the patient with sound response when any key is pressed.

Display screen It helps the patient through every step of the medication process.

Receipt printer The receipt printer furnishes the dispensed medicine list along with a paper receipt of the services rendered along with the prescription.

Medicine dispenser With the help of inventory controller proper required medicines are dispensed.

COMPONENTS:

- Hardware
- Software

HARDWARE:

- CPU
- Magnetic/Chip reader
- Display
- Crypto processor
- Record Printer
- Vault

SOFTWARE:

- Front end
  - OS2, Linux, Unix, MS Windows
  - Firmware (coded in C ++/Java)
- Back end

  DBMS-Oracle, MS SQL Server etc.

4 Conclusion

The PHCATM vending machine is the product which will cater to the requirements of the rural population, since more than 60 % of patients visiting hospitals require pathological intervention before treatment. In order to facilitate people with first aid, medicines and vaccine reminders, PHCATM can be installed in every village.

Every PHCATM would have centers as their referral point for the village people. Which indicates there should be 1 PHCATM for each village.

- Once this machine is installed in every village, people can rely on it even in the time when there is no other option available to them since the service provided would be 24 × 7.
- Since the rural people can't afford to travel long distances to avail medical facilities, so they stick to the traditional ways of getting cured. But once this system is introduced it can save them from travelling such long distances and even their time would be saved.
- At the same time they can avail much better facility at the comfort of their homes.

This will surely change the medical facility available presently and will improve the health conditions prevailing in the rural parts of the world.

References

1. Balasubramaniam, K.: Structural adjustment programs and privatization of health. LINK (Newslett. Asian Community Health Action Netw.) **14**, 2 (1996)
2. Outcome Evaluation of the Mitanin Program.: A Critical Assessment of the Nation's Largest Ongoing Community Health Activist Program. SHRC, Chattisgarh (2004)
3. Park, K.: Communicable diseases. In: Banot, B. (ed) Park's Text Book of Preventive and Social Medicine, pp. 172–175, 16th edn. Banarsidas Bhanot, Jabalpur (2000)
4. WHO. The World Health Report: Conquering Suffering, Enriching Humanity, p. 1997. World Health Organisation, Geneva (1997)

5. Pneumonia and diarrhoea. United Nations Children's Fund (UNICEF) June 2012, Pneumonia and diarrhoea disproportionately affect the poorest, p. 7
6. CURRENT HEALTH SCENARIO IN RURAL INDIA
7. Child deaths due to pneumonia and diarrhoea are concentrated in the poorest regions and in mostly poor and populous countries in these regions. United Nations Children's Fund (UNICEF) June 2012, table 1.2, p. 9
8. Kumar, S.: Public-Private Partnership in Primary Healthcare
9. Towards sustainability in village health care by RUBY NORMA ELIASON, International Development, William Carey International University, Pasadena, CA, USA
10. Rasmuson, M., et al. (eds.): Community-based approaches to child health: BASICS experiences to date. BASICS Project, Arlington (1998)
11. Quinones, IdPL.: Community Health Workers: Profile and Training Process in Colombia. Pan American Health Organization and World Health Organization, Washington, DC (1999)

Hybrid Model with Fusion Approach to Enhance the Efficiency of Keystroke Dynamics Authentication

Ramu Thanganayagam and Arivoli Thangadurai

Abstract We propose in this paper a novel technique to enhance the performance of keystroke dynamic authentication using hybrid model with four fusion approach. Firstly, extract keystroke features from our database. Then generate template from extracted features, which is compact form of keystroke feature data. Hybrid model based on combination of Gaussian probability density function (GPDF) and Support Vector Machine (SVM) will convert test features into scores. At last, applied four fusion rules on hybrid model to fusing GPDF and SVM scores to improve the final result. Experimental results show that the performance of the proposed hybrid model can bring obvious improvement with error rate of 1.612 %.

Keywords Hybrid model · Biometric · Keystroke dynamic authentication and fusion approach

1 Introduction

Traditional authentication system using passwords, personal cards and PIN-numbers can easily be breached when a card is stolen or password is compromised. Furthermore, difficult passwords may be hard to remember by a legitimate user and simple passwords are easy to guess by an impostor. The use of biometrics offers an alternative means of identification which helps avoid the problems associated with conventional methods. Nowadays, losses due to identity theft is an issue of growing concern, especially considering the increased data exposure caused by some services on the Internet. In view of this scenario, there is a

R. Thanganayagam (✉)
Department of ECE, Kalasalingam University, Krishnankoil, India
e-mail: loginworld34@hotmail.com

A. Thangadurai
Department of ECE, Vickram College of Engineering, Arasanoor, India
e-mail: t.arivoli@gmail.com

A. Nagar et al. (eds.), *Proceedings of 3rd International Conference on Advanced Computing, Networking and Informatics*, Smart Innovation, Systems and Technologies 43, DOI 10.1007/978-81-322-2538-6_10

need for enhanced authentication mechanisms, such as by the use of biometrics. In security area, biometrics tries to recognize users by physiological or behavioral features of the person. Among the current biometric technologies, keystroke dynamics is a promising alternative due to several factors [1, 2]. First, it usually does not need any additional cost with hardware, as a common keyboard is enough to acquire keystroke data. Second, keystroke dynamics recognition may be performed in background, while the user is typing an e-mail or entering a password. Consequently, day-to-day tasks of users are not disturbed, which may contribute to a better acceptability of this technology. Keystroke dynamic is a type of behavioral biometrics based on the users typing rhythm, which is unique for different people. Keystroke timing patterns are captured without users knowledge based on the keystroke events gathered while users typing on a keyboard.

1.1 Motivation and Contribution

Password based authentication is not secure due to several drawbacks [3]:

1. Someone stolen the password
2. Brute force attack (try all possible combination of start with one digit, two digit passwords and so on)
3. Dictionary attack (try with list of password in the dictionary instead of all possible combination)
4. Password guessed (if someone look over while type or note it down on paper if password is difficult to remember)
5. User shared password to others
6. Someone hacked the password.

To overcome the above drawbacks, introduced keystroke dynamic is an additional parameter to secure password authentication. Keystroke dynamic analyze the users way of typing on keyboard that is typing pattern and it measures the time interval between each events of user holding the key and switchover between the key (one key to another key). Individual keystroke pattern or features are different, so it is maintain consistency and uniqueness. We study different types of keystroke features and analyze the performance of individual and combination of features. We propose a Hybrid model with different fusion approach to combine the scores from Gaussian probability density function (GPDF) and support vector machine (SVM) with combination of feature data. The following contributions has been done:

- Created keystroke database of 100 users
- Extract four types of keystroke features and analyze the performance
- Keystroke feature data transform to scores using hybrid model
- The efficient combination of four fusion approach
- Evaluate Equal error rate (EER).

1.2 Organization

The rest of paper is organized in the following way: Sect. 2 discussed the review of related research in the field while Sect. 3 presents the proposed methodology, data collection, feature extraction and template generation. Section 4 presents our proposed hybrid model and Sect. 5 presents proposed fusion approach. The experimental results analyzed and discussed in Sect. 6. Finally, Sect. 7 provides conclusions.

2 Related Works

Recent years, researchers were focusing on the collection of number of users keystroke biometric data for accurate evaluation on user database benchmark, decreasing the evaluated result errors or improve accuracy and concentrating keystroke latency as feature data. Hosseinzadeh et al. [4] conducted experiment on keystroke authentication using Gaussian Mixture Model. Training and testing data were collected from 8 users who typed their full name consecutively ten times No complexity pattern involved in the name characters due to smaller character length, so it can be easily replicated. Experiment was utilized only two extracted keystroke features and Expectation Maximization algorithm used to train the Gaussian Mixture Model. Then, Log-likelihood test platform was performed to identify the probability of closest data of testing and training data to confirm genuine user authentication. Overall experiment result is 2.4 % FRR and 2.1 % FRR, this error rate is high and also the number of users tested is not enough to conclude the final results obtained. Sang and Shen et al. [5] authors were implemented SVM classifier in the keystroke dynamics. Keystroke data collected from ten users. Experiment was performed on one class SVM which is simulating genuine data and two-class SVM is used to separate genuine and imposters' data. The results are reliable but significant weakness is only ten samples were collected. In [6] authors have implemented Hidden Markov Models as classifier in keystroke recognition. Twenty people were enrolled to this experiment with their password ten times in four different sessions. However, lower length of eight digit password implemented. A total of 800 samples collected which is enough for the experiment. The final result of EER is 3.6 % which is considerably higher error rate. Guven et al. [7] proposed new classifier for keystroke authentication. New classifier is similar to neural network structure. Keystroke raw data was collected from sixteen users, then extracted the keystroke latency (successive key press or down). Experiment was conducted on similar to neural network structure to calculate the weights using statistical method. Statistical method include mean and standard deviation of the keystroke latencies. User test sample latency value was compared to standard deviation of reference latency, result is genuine if test latency fall two times within the standard deviation reference latency, then assume whole string being considered as valid. Due to assumption, experiment produced result with high error rate of FRR of 17 % and FAR of 26 % which was poor performance of keystroke authentication. Azevedo et al. [8] developed hybrid system based on the combination of stochastic optimization algorithm (Genetic algorithm) and support vector machine (SVM) and particle swarm optimization. Hybrid system select the keystroke features from enrolled users. Experiment was implemented on SVM uses a Genetic algorithm and particle swarm optimization. First test was conducted on SVM with Genetic algorithm (evolutionary algorithm) for feature selection with minimum error rate of 5.18 % when FAR of 0.43 % and FRR of 4.75 %. Second test was carried on particle swarm optimization with global acceleration of 1.5 gave a minimum total error of 2.21 % with error rate of 0.41 % FRR and 2.07 % FAR. This paper, proposed hybrid model with different fusion approach which is merge the scores produced by the GPDF and SVM. This approach is able to considerably improve the overall result.

3 Proposed Methodology

We introduce hybrid model in this research, which are the combination of two matching function namely the Gaussian Probability Density Function (GPDF) and Support vector machine (SVM). Figure 1 shows proposed hybrid model with two keystroke features combined and applied fusion rules against the combinations performed on SVM and GPDF. Hybrid model is developed to calculate the score between the test user templates against the reference user template (stored in database). Then, fusion applied for both SVM and GPDF matching scores. The function of fusion is fusing both matching scores and then fused output score is compared with a predefined threshold before making a final decision. The decision should be accepted if fusion output score is greater than threshold value or else rejected the user authentication. In this paper, four fusion rules are studied.

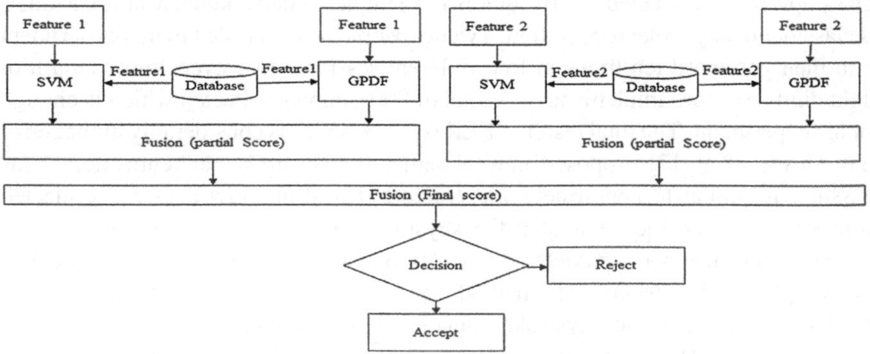

Fig. 1 Proposed hybrid model with two keystroke features

3.1 Data Collection

Captured keystroke biometric data with the help of GREYC Keystroke software developed at GREYC Laboratory [9]. This software is downloadable from the following address www.ecole.ensicaen.fr/~rosenber/keystroke.html. 100 users' data was collected with an interval of 6 months apart. These users are university academic and administrative staffs. Initially, each user is allowed to choose their choice of username and password during the enrolment process. Next, same users have to continuously type fixed line of text "credential evaluation" for fifteen times. So we collect fifteen samples of each user and total of 1500 (100 user * 15) samples are stored in the database.

3.2 Feature Extraction and Template Generation

When user type a character on keyboard, two types of events occurred namely, key press (P) and key release (R). Based on occurrence of specific events, we can extract keystroke feature data. Four types of keystroke features could be generated [10] as shown Fig. 2. Extracted four features: (Press-to-Release $PR = R_1 - P_1$) the time between a key being pressed until the key being released, (Press-to-Press $PP = P_2 - P_1$) time between two successive keys being pressed, (Release-to-Press $RP = P_2 - R_1$) time between a key being released to the next key being pressed and (Release-to-Release $RR = R_2 - R_1$) the time between two successive keys being released. RP feature value may occur negative due to next key being pressed before

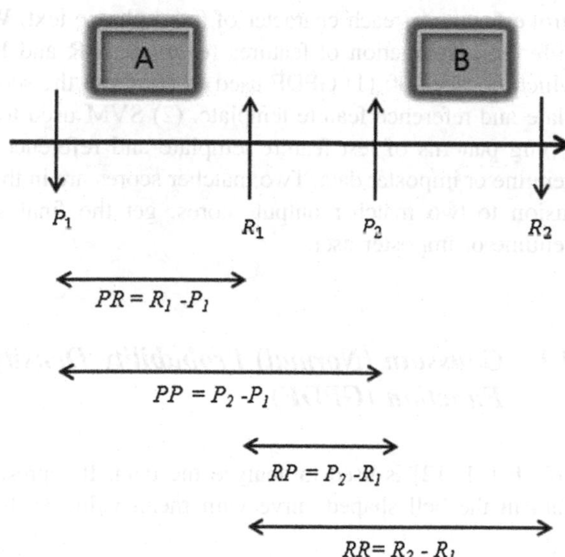

Fig. 2 Four keystroke features extracted from phrase "AB"

previous key being released. Template generation is the compact form of four keystroke features, which is extracted from raw keystroke data. Each user keystroke data could be converted to one template which consists four types of features of their mean and standard deviation of keystroke feature of each character of fixed phrase text. User templates are stored in database and could be retrieved for authentication purpose. The formula of mean (μ) and standard deviation (σ) as below,

$$\mu = \frac{1}{T} * \sum_{j=1}^{T} t_j \quad and \; \sigma = \sqrt{\frac{\sum_{j=1}^{T} (t_j - \mu)^2}{T}} \tag{1}$$

where T represents the number of training samples and t_j denotes the value of each keystroke feature. The user is required to continuously type fixed phrase text "credential evaluation" for 15 times which yields fifteen samples from each user. Testing purpose, randomly divide fifteen samples of each user into five and ten whereas five samples are converted to templates serves as reference for future authentication and ten samples reserved for testing purpose. Generated five templates are stored in the database for comparison while test user authentication.

4 Hybrid Model

Hybrid model have two matchers namely, GPDF and SVM. Test and reference feature data will be converted into individual template. Each template consists four types of features (PR, PP, RP, RR) of their mean and standard deviation of keystroke feature of each character of fixed phrase text. We use both template of two different combination of features (example: PR and RP) fetch into hybrid model which consists of: (1) GPDF used to compute the score between test feature template and reference feature template. (2) SVM used to compare the scores of each typing patterns of test feature template and reference feature template to identify genuine or imposter data. Two matcher scores are in the range of 0 to 1. Then apply fusion to two matcher output scores, get the final score which will decide the genuine or imposter user.

4.1 Gaussian (Normal) Probability Density Function (GPDF)

GPDF [11, 12] is used to analyze the data. It represents the normally distributed data in the bell shaped curve with mean value is the centroid and variance is a measure of dispersion of data around mean. This paper, GPDF is used to calculate the matching score between user test feature data template and reference feature data template. Matching score between the ranges of 0 to 1. GPDF modified form as below

$$Score_{GPDF} = \sum_{i=1}^{N} \exp\left[-\frac{(n-\mu)^2}{2\sigma^2}\right] \tag{2}$$

where $Score_{GPDF}$ represents the GPDF matching score, n denotes the test keystroke feature of a particular character, μ and σ^2 denotes the mean and variance of each character from reference feature data, respectively. Calculate the matching score between two templates of test and reference data by apply variance and mean of reference data and test data into the Eq. (2). Eventually decide the final result of matching score if closer to 1, then reference feature data template and test feature data template are similar. Now the test was conducted for one feature of the template, the matching score named as sub score. Same experiment should be performed for all four features of the template, eventually final score has been calculated with average of all sub scores.

4.2 Support Vector Machine (SVM)

SVM is to compare the scores of each typing patterns of training data samples and test data samples for identifying authorized and unauthorized user. We have done experiment on linear version of SVM which maps the input user data into a high dimensional feature space through linear kernel. Detail description of SVM can be found in [13]. SVM is the determination of the optimal hyper plane which will optimally separate the two classes of genuine and imposter of input user dataset. Based on linear kernel function, SVM maps the input user dataset samples in a high-dimensional feature space and then separate the dataset from the origin with a maximum margin. SVM algorithm function *f* is defined as the region that majority of data from input dataset which contained in one pattern (genuine) as +1, and data outside this region is −1 (imposter), function *f*(*x*) as below

$$f(x) = \sum_{i}^{N} \alpha_i y_i K(x_i, x) + b \tag{3}$$

where N represents the size of training data and x_i denotes the supporting vector. $K(x_i, x)$ is the kernel function representing the inner product between x_i and x in feature space. To maximize the margin that is distance between the nearest point of the training set and the hyper plane is called optimization problem. It could be solved can be stated as

$$\max_{\alpha_i \geq 0} \sum_i \alpha_i - \frac{1}{2} \sum_{j,k} \alpha_j \alpha_k y_j y_k K(x_j, x_k), \quad 0 \leq \alpha_i \leq C \ for \ \forall_i \quad and \ \sum_i \alpha_i y_i = 0 \quad (4)$$

where C denotes the penalization coefficient of data points on hyper plane. Based on C value to set the width of margin between data points in the middle of the hyper plane. In order to maximize the performance, we have set the values for the parameter C = 128.

5 Fusion Approach

Four fusion rules were applied in the hybrid model fusion approach. Table 1 shows formula of four fusion rules namely sum, weighted sum, product and maximum. Hybrid model have two matcher of SVM and GPDF whose output range is 0 to 1, so score normalization is not necessary before fetching to next process of fusion. Fusion helps to combine the significant information of SVM and GPDF, so it could increase the overall performance. Reason to employ fusion method to improve the performance significantly. In this research, we propose fusion to hybrid model that is fusing between SVM and GPDF scores to produce a final score. At last, fusion score will decide the user is genuine or imposter.

6 Experimental Results and Discussions

6.1 Experimental Setup

Our experiments were performed with fixed phrase text of users keystroke data. Collected the raw data of 15 samples from 100 users. Extracted four different keystroke features (PR, PP, RP, and RR) from 100 users raw data. Then, template could be generated based on extracted four keystroke features with calculated their mean and standard deviation for each and every character of the user typed. Each user data consists of four different templates. Testing purpose, randomly divide fifteen samples of each user into five and ten whereas five samples are converted to templates named as training samples and ten samples reserved for testing purpose

Table 1 Various fusion rules

Fusion rule	Formula
Sum	$Score_{SUM} = \frac{Score_{SVM} + Score_{GPDF}}{2}$
Weighted sum	$\mathrm{Score}_{\mathrm{Wsum}} = \mathrm{W}_1\mathrm{Score}_{\mathrm{SVM}} + \mathrm{W}_2\mathrm{Score}_{\mathrm{GPDF}}$
Product	$Score_{product} = \frac{Score_{SVM} Score_{GPDF}}{2}$
Max	$Score_{Max} = MAX(Score_{SVM}, Score_{GPDF})$

named as testing samples. Experiment was carried on all sets of users' templates with five training samples versus ten testing samples. Error rates could be calculated as false rejection rate and false acceptance rate. The false rejection rate (FRR) is the ratio of genuine user rejected and the total number of user samples attempted. The false acceptance rate (FAR) is the ratio of approved imposters as genuine users and the total number of user samples attempted. The experiment was carried by comparing the test data sample score against threshold value within the range of 0 to 1 (interval of 0.01), calculate FAR and FRR. Repeated the experiment with increase the interval each time 0.01, calculated the FAR, FRR values. After tabulation of FAR and FRR values, equal error rate (EER) is calculated when FAR is near to FRR value. Tested fifteen different combinations with five training data and ten testing data. For each combination of sample data, we have obtained final results could be the average of EER. Experimental results discussed for the next section could be described with the average value of EER, FAR and FRR.

7 Results

7.1 Keystroke Features Without Fusion

This approach, each keystroke features (PR, PP, RP, and RR) have been applied on matcher SVM and GPDF without using any fusion approaches. Observing the performance of four keystroke features on SVM and GPDF in Fig. 3 we notice that feature PR is obtained better result compared to other keystroke features (PP, RP, RR). Observing the EER % of four keystroke features without fusion in Table 2 we notice that PR feature lead the best result of 3.8214 EER % while using SVM. SVM results of all four keystroke features are better than GPDF results. Apart from that we performed this experiment with 100 user samples even though the result remains consistent.

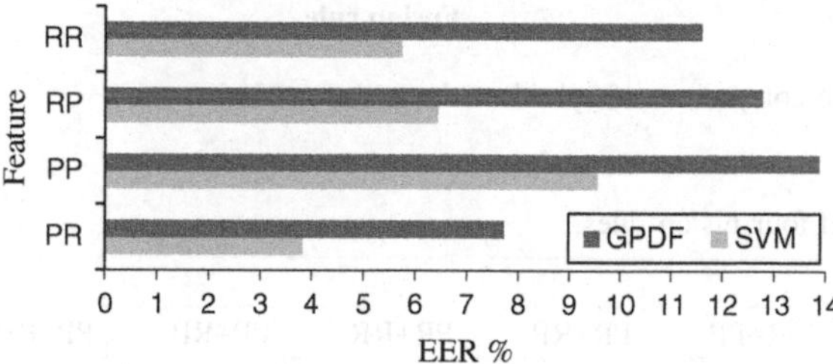

Fig. 3 Performance comparison of four keystroke features without fusion

Table 2 EER % of four features on SVM and GPDF without fusion

Method	PR	PP	RP	RR
SVM	3.8214	9.5531	6.4372	5.7315
GPDF	7.7199	13.875	12.776	11.602

7.2 Hybrid Model with Fusion Approach

We proposed, hybrid model using combination of two keystroke features with four fusion approach. Initially, two keystroke features have been applied on two matcher SVM and GPDF, output scores of each matcher named partial score. Then partial score fused with four fusion rules namely sum, weighted sum, product and maximum. Each fusion rules were applied separately to partial score and analyzed the output score shown in Fig. 4. This testing was repeated with all possible combination of two keystroke features and also repeated with different fusion rules. Eventually fused output score is compared with a predefined threshold before making a final decision. The final decision should be accepted if the score is greater than threshold value or else rejected. Experiment results shown in Table 3 we noticed that PR+RP feature combination with weighted sum fusion rule produced the best result of 1.612 % EER among other feature combination. PR feature combination with any other three features provided better results compared to without PR of remaining feature combination. Noticing that the performance improvement on fusion approach than the without fusion approach on individual feature. Observed the better results of two combination features than single keystroke feature used in without fusion

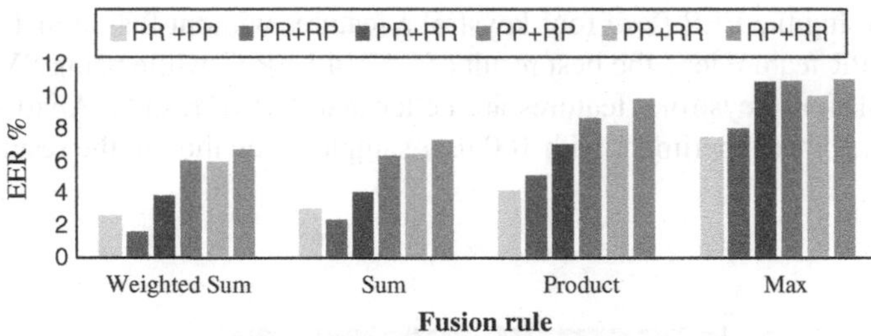

Fig. 4 Performance comparison of hybrid model with four fusion rules

Table 3 EER % of four fusion rules

EER %						
Fusion rule	PR+PP	PR+RP	PR+RR	PP+RP	PP+RR	RP+RR
Weighted sum	2.57	1.612	3.842	6.043	5.961	6.714
Sum	3.015	2.36	4.074	6.341	6.432	7.315
Product	4.124	5.112	7.031	8.64	8.184	9.842
Max	7.443	8.004	10.921	10.944	10.03	11.054

approach. Among the combination of features used in hybrid model with various fusion rule, weighted sum rule shown better results than the other fusion rule. Among the four fusion rules, weighted sum rule results are better and worst results obtained on max rule. Analyzed the major performance difference between weighted sum and max rule, scenario of max rule final output is the maximum probability output between two matchers that is best among the two matcher scores and discard the even small point difference score value of one matcher but weighted sum rule utilize both matcher scores with weighted value. Therefore, weighted sum rule have imposter acceptance is very less, so overall performance is high. Best equal error rate obtained among all four fusion rule is weighted sum rule at 1.612 %. At the experimental stage, weighted sum fusion rule was tested with bias weight of $W_{1ScoreSVM}$ and $\mathrm{W_{1ScoreGPDF}}$ in the range starting from 0 with step size 0.1 till 1 value, observed the best result obtained at bias value of $\mathrm{W_{1ScoreSVM}} = 0.73$ and $\mathrm{W_{2ScoreGPDF}} = 0.27$. Weighted sum rule performs better than sum and product rule due to the setting of bias weight. Overall observation, two keystroke feature combination enhance the performance using weighted fusion rules on hybrid model than individual features used on without fusion method.

8 Conclusions

In light of the current need for enhanced authentication mechanisms, keystroke dynamics shows as a promising alternative. We discussed a promising method for the performance enhancement of keystroke dynamic authentication using hybrid model with four fusion approach. We showed the two keystroke features using hybrid model with four fusion approach to improve the efficiency of a keystroke dynamic authentication system. We described that hybrid model by fusing the scores from two matchers of SVM and GPDF, the result can be improved significantly than using them individually. We showed in our experiment that using two keystroke features combination is able to provide best result than individual keystroke features. The experimental results showed that proposed hybrid model with weighted sum rule using two keystroke feature combination is able to obtain better result of 1.612 % of EER, due to fusion approach helps to increase the performance.

References

1. Hosseinzadeh, D., Krishnan, S.: Gaussian mixture modeling of keystroke patterns for biometric applications. IEEE Trans. Syst. Man Cybern. Part C Appl. Rev. **38**(6), 816–826 (2008)
2. Peacock, A., Ke, X., Wilkerson, M.: Typing patterns: A key to user identification. IEEE Secur. Priv. **2**(5), 40–47 (2004)
3. Sasse, M., Brostoff, S., Weirich, D.: Transforming the 'weakest link' a human/computer interaction approach to usable and effective security. BT Technol. J. **19**(3), 122–131 (2001)

4. Hosseinzadeh, D., Krishnan, S., Khademi, A.: Keystroke identification based on Gaussian mixture models. In: Proceedings IEEE International Conference on Acoustics, Speech and Signal Processing (ICASSP), vol. 3, pp. 1144–1147 (2006)
5. Sang, Y., Shen, H., Fan, P.: Novel impostors' detection in keystroke dynamics by support vector machine. In: Parallel and Distributed Computing: Applications and Technologies, pp. 666–669. Springer (2005)
6. Rodrigues, R.N., Yared, G.F.G.: Biometric access control through numerical keyboards based on keystroke dynamics. In: Zhang, D., Jain, A.K. (eds.) International Conference of Biometrics (ICB 2006), LNCS 3832, pp. 640–646 (2005)
7. Guven, O., Akyokus, S., Uysal, M., Guven, A.: Enhanced password authentication through keystroke typing characteristics. In: Proceedings of 25th IASTED international multi-conference: artificial intelligence and applications, Innsbruck, Austria, pp. 317–322 (2007)
8. Azevedo, G., Cavalcanti, G., Filho, E.C.: Hybrid solution for the feature selection in personal identification problems through keystroke dynamics. In: International Joint Conference on Neural Networks, pp. 1947–1952 (2007)
9. Giot, R., El-Abed, M., Rosenberger, C.: GREYC keystroke: a benchmark for keystroke dynamics biometric systems. In: IEEE 3rd International Conference on Biometrics, pp. 1–6 (2009)
10. Ramu, T., Arivoli, T.: A Framework of secure biometric based online exam authentication: an alternative to traditional exam. IJSER **4**(11), 52–60 (2013)
11. Archambeau, C., Valle, M., Assenza, A., Verleysen, M.: Assessment of probability density estimation methods: Parzen window and finite gaussian mixtures. ISCAS, pp. 3245–3248 (2006)
12. Parzen, E.: On estimation of a probability density function and mode. Ann. Math. Stat. **33**, 1065–1076 (1962)
13. Tax, D.M.J., Duin, R.P.W.: Support vector data description. Mach. Learn. **54**, 45–66 (2004)

Multi-class Twin Support Vector Machine for Pattern Classification

Divya Tomar and Sonali Agarwal

Abstract In this paper, we propose a novel algorithm for multi-class classification, called as Multi-class Twin Support Vector Machine (MTWSVM) which is an extension of the binary Twin Support Vector Machine (TWSVM). MTWSVM is based on "one-against-one" strategy in which the patterns of each class are trained with the patterns of another class. To speed up the training phase, optimization problems are solved by Successive Over Relaxation (SOR) technique. The experiment is performed on eight benchmark datasets and the performance of the proposed approach is compared with the existing multi-class approaches based on Support Vector Machines and Twin Support Vector Machines.

Keywords Twin support vector machine · Multi-class twin support vector machine · Successive over relaxation · Pattern classification

1 Introduction

Support Vector Machine (SVM) is a binary classifier which separates the patterns of two classes by generating a maximum margin hyper-plane [1–3]. For this purpose, it finds solution for a complex Quadratic Programming Problem (QPP). SVM provides global solution by constructing unique hyper-plane to separate the patterns of different classes rather than local boundaries as compared to other existing classification methods. Since SVM follows the Structural Risk Minimization (SRM) principle, so it reduces the occurrence of risk during the training phase as well as enhances its generalization capability. SVM has shown better performance as compared to the other existing machine learning approaches due to which it is

D. Tomar (✉) · S. Agarwal
Indian Institute of Information Technology, Allahabad 211 012, India
e-mail: divyatomar26@gmail.com

S. Agarwal
e-mail: sonali@iiita.ac.in

A. Nagar et al. (eds.), *Proceedings of 3rd International Conference on Advanced Computing, Networking and Informatics*, Smart Innovation, Systems and Technologies 43, DOI 10.1007/978-81-322-2538-6_11

one of the most widely used classification method that has applications in many fields ranging from text categorization, speech recognition, defect prediction, disease detection, intrusion detection, bankruptcy prediction, emotion detection, face identification, time series forecasting and etc. [4–19]. It is also efficiently extended to multi-class problems domain [20–24]. But one of the main issues with the conventional SVM is its high computational complexity. To reduce the computational complexity of SVM, Jayadeva et al. proposed Twin Support Vector Machine (TWSVM), a supervised machine learning approach for binary classification [25]. TWSVM solves two SVM-type Quadratic Programming Problems (QPPs), each of which are smaller than the QPP in a traditional SVM.

TWSVM is suitable for only binary classification while most of the real world applications such as speaker recognition, disease detection, image classification, face detection, semantic analysis etc. demand for a multi-classifier. So, in this paper we proposed a multi-class classifier, MTWSVM which is obtained by extending the formulation of binary TWSVM on the basis of "one-against-one" strategy. For M-class classification, MTSVM solves M (M-1)-QPPs and constructs M (M-1) hyper-planes, (M-1) planes for each class. Thus, it generates M (M-1) binary TWSVM classifiers where each classifier is trained with the patterns of another class. The class is assigned to the test pattern on the basis of "max-win" voting strategy i.e., the class with maximum vote is assigned to the pattern. The vote is given to a class on the basis of distance of a pattern from its corresponding hyper-plane. If a pattern lies closer to a class as compared to another class, then the vote is given to it. In this paper, we analyze and compare the performance of the proposed approach with other existing approaches.

The paper is organized as follows. Section 2 provides the brief overview of TWSVM. Section 3 presents the formulation of the proposed approach for both linear and non-linear cases. Section 4 discusses the experimental results and performance comparison of the proposed approach with the existing approaches. Finally, concluding remarks are given in Sect. 5.

2 Twin Support Vector Machine

TWSVM is a binary classifier that performs the classification task by constructing two non-parallel hyper-planes rather than single hyper-plane as in SVM. Consider a training dataset $T = \{(x_1, y_1), (x_2, y_2), \ldots, (x_l, y_l)\}$ for two-class classification problem, where $x_i \in R^n, i = 1, 2, \ldots, l$ represents input training samples or patterns and $y_i \in \{+1, -1\}$ refers to the corresponding class label. Let each class comprises l_1 and l_2 patterns. Consider two matrices $A_{+1} \in R^{l_1 \times n}$ and $A_{-1} \in R^{l_2 \times n}$ contain the patterns of class +1 and class −1 correspondingly. TWSVM determines following two non-parallel hyper-planes

$$x^T w_{+1} + b_{+1} = 0 \text{ and } x^T w_{-1} + b_{-1} = 0 \quad (1)$$

by solving two QPPs as follows:

$$\min (w_{+1}, b_{+1}, \xi) \frac{1}{2} \|A_{+1} w_{+1} + e_{+1} b_{+1}\|^2 + c_{+1} e_{-1}^T \xi$$

$$\text{s.t.} - (A_{-1} w_{+1} + e_{-1} b_{+1}) + \xi \geq e_{-1}, \xi \geq 0 \quad (2)$$

$$\min (w_{-1}, b_{-1}, \eta) \frac{1}{2} \|A_{-1} w_{-1} + e_{-1} b_{-1}\|^2 + c_{-1} e_{+1}^T \eta$$

$$\text{s.t.} (A_{+1} w_{-1} + e_{+1} b_{-1}) + \eta \geq e_{+1}, \eta \geq 0 \quad (3)$$

where $w_{+1}, w_{-1} \in R^n$ are normal vectors to their respective hyper-plane, b_{+1} and b_{-1} are bias terms. $e_{+1} \in R^{l_1}$ and $e_{-1} \in R^{l_2}$ are the vectors of 1's of suitable dimension, $c_{+1} > 0$ and $c_{-1} > 0$ are penalty parameters and $\xi \in R^{l_2}$ and $\eta \in R^{l_1}$ are slack variables due to class −1 and class +1 correspondingly. TWSVM constructs hyper-plane in such a way that each plane lies closest to the patterns of one class and as far as possible from the patterns of another class.

It assigns the class to a new pattern according to its distance from each hyper-plane. The pattern is assigned into a class depends on which of the two planes is nearest to it. TWSVM predicts the class according to the following decision function:

$$f(x) = \arg \min_{i=+1,-1} \frac{|w_i \cdot x + b_i|}{\|w_i\|} \quad (4)$$

where | . | is the absolute value. TWSVM also works well for the classification of non-linearly separable data samples or patterns [25]. For this purpose, it determines two non-parallel kernel generated surfaces.

In, TWSVM the patterns of one class provide constraints to another class and vice versa, hence, it solves two smaller size QPPs unlike SVM which solved a single QPP. For ‘l’ data samples or patterns, the computational complexity of SVM is $O(l^3)$. Suppose data samples are distributed equally in each class. So, each class comprises $l/2$ data samples. Thus the computational complexity of TWSVM classifier is $O(2 \times (l/2)^3$ which is four times faster as compared to the conventional SVM.

3 Multi-class Twin Support Vector Machine

The formulation of Multi-class Twin Support Vector Machine (MTWSVM) is based on "one-against-one" strategy in which patterns of one class is trained with the patterns of another class. For this purpose, it solves M (M-1) QPPs and generates M (M-1) binary classifiers, where M is the number of classes present in the dataset. MTWSVM determines (M-1) non-parallel hyper-planes for every class. Suppose the training dataset contains *l* patterns. The format of training dataset for multi-class is given below:

$$T = \{(x_1, y_1), (x_2, y_2)\ldots, (x_l, y_l)\} \tag{5}$$

where $x_i, i = 1, \ldots, l$ represents input pattern or data samples in n-dimensional real space R and $y_i \in \{1, \ldots, M\}$ represents the class label corresponding to each pattern. Consider the training of the patterns of ith class with the patterns of jth class. The proposed classifier assumes the patterns of ith class with positive class labels and the patterns of jth class with negative class labels and vice versa. Suppose the matrices $A_i \in R^{l_i \times n}$ and $A_j \in R^{l_j \times n}$ represent the patterns of ith and jth class correspondingly. MTWSVM works well for both linear and non-linear type of examples and its formulation for both cases are obtained as:

3.1 Linear MTWSVM

Consider ith class and jth class, when both classes are trained with each other, the linear MTWSVM classifier solves following pair of QPPs:

$$\text{MTWSVM(i)} \quad \min(w_{ij}, b_{ij}, \xi_{ij}) \frac{1}{2} \left\| A_i w_{ij} + e_{i1} b_{ij} \right\|^2 + c_i e_{j1}^T \xi_{ij}$$

$$\text{s.t.} - \left(A_j w_{ij} + e_{j1} b_{ij}\right) + \xi_{ij} \geq e_{j1}, \xi_{ij} \geq 0 \tag{6}$$

$$\text{MTWSVM(j)} \quad \min(w_{ji}, b_{ji}, \xi_{ji}) \frac{1}{2} \left\| A_j w_{ji} + e_{j1} b_{ji} \right\|^2 + c_j e_{i1}^T \xi_{ji}$$

$$\text{s.t.} \left(A_i w_{ji} + e_{i1} b_{ji}\right) + \xi_{ji} \geq e_{i1}, \xi_{ji} \geq 0 \tag{7}$$

Equations (6) and (7) determine non-parallel planes for each class as:

$$f_{ij} = (w_{ij}.x) + b_{ij} = 0 \text{ and } f_{ji} = (w_{ji} \cdot x) + b_{ji} = 0 \tag{8}$$

where $w_{ij}, w_{ji} \in R^n$ are normal vectors to the hyper-plane f_{ij} and f_{ji} correspondingly and $b_{ij}, b_{ji} \in R$ represent bias terms. $e_{i1} \in R^{l_i}$ and $e_{j1} \in R^{l_j}$ are two vectors with all values as 1. $c_i > 0$ and $c_j > 0$ correspond to penalty parameters and $\xi_{ij} \in R^{l_j}$ and

$\xi_{ji} \in R^{l_i}$ are slack variables. The first term of Eq. (6) or (7) determines the sum of squared distances of the patterns of ith class or jth class from its corresponding hyper-plane f_{ij} or f_{ji}. The constraint requires that the patterns of jth class or ith class to be at a distance 1 from the hyper-plane f_{ij} or f_{ji} with soft margin. Minimization of the first term of Eq. (6) or (7) keeps the corresponding hyper-plane in the close proximity of the ith class or jth class. The second term of the Eq. (6) or (7) is attempting to minimize misclassification error due to the patterns of jth class or ith class. Thus the patterns of each class lie in the close proximity of their corresponding hyper-plane and as far as possible from other planes. Lagrangian corresponding to Eq. (6):

$$\begin{aligned} L(w_{ij}, b_{ij}, \xi_{ij}, \alpha_i, \beta_i) = \frac{1}{2}\left\|A_i w_{ij} + e_{i1} b_{ij}\right\|^2 + c_i e_{j1}^T \xi_{ij} \\ - \alpha_i^T\left(-\left(A_j w_{ij} + e_{j1} b_{ij}\right) + \xi_{ij} - e_{j1}\right) - \beta_i^T \xi_{ij} \end{aligned} \tag{9}$$

where $\alpha_i \in R^{l_j}$ and $\beta_i \in R^{l_j}$ are non-negative lagrangian multipliers. The Karush-Kuhn-Tucker (KKT) necessary and sufficient optimality conditions are obtained as follows:

$$A_i^T\left(A_i w_{ij} + e_{i1} b_{ij}\right) + A_j^T \alpha_i = 0 \tag{10}$$

$$e_{i1}^T\left(A_i w_{ij} + e_{i1} b_{ij}\right) + e_{j1}^T \alpha_i = 0 \tag{11}$$

$$c_i e_{j1} - \alpha_i - \beta_i = 0 \tag{12}$$

$$-\left(A_j w_{ij} + e_{j1} b_{ij}\right) + \xi_{ij} \geq e_{j1}, \xi_{ij} \geq 0 \tag{13}$$

$$\alpha_i^T\left(\left(A_j w_{ij} + e_{j1} b_{ij}\right) - \xi_{ij} + e_{j1}\right) = 0, \beta_i^T \xi_{ij} = 0 \tag{14}$$

$$\alpha_i \geq 0, \beta_i \geq 0 \tag{15}$$

Since $\beta_i \geq 0$, from Eq. (12):

$$0 \leq \alpha_i \leq c_i \tag{16}$$

Equations (10) and (11) lead to:

$$\begin{bmatrix} A_i^T \\ e_{i1}^T \end{bmatrix} [A_i e_{i1}] \begin{bmatrix} w_{ij} \\ b_{ij} \end{bmatrix} + \begin{bmatrix} A_j^T \\ e_{j1}^T \end{bmatrix} \alpha_i = 0 \tag{17}$$

Define $B_i = [A_i e_{i1}]$ and $B_j = [A_j e_{j1}]$, $u_{ij} = \begin{bmatrix} w_{ij} \\ b_{ij} \end{bmatrix}$. With these notations, Eq. (17) may be reformulated as:

$$B_i^T B_i u_{ij} + B_j^T \alpha_i = 0 \text{ or } u_{ij} = -(B_i^T B_i)^{-1} B_j^T \alpha_i \tag{18}$$

The inverse of $B_i^T B_i$ is essential for the solution of Eq. (18). Although it is always positive semi-definite but in some situations it may be ill-conditioned. To solve such condition, a regularization term ϵI may be added to $B_i^T B_i$. Here, $\epsilon > 0$ and I is an identity matrix of appropriate dimensions. With regularization term, Eq. (18) is modified as:

$$u_{ij} = -(B_i^T B_i + \epsilon I)^{-1} B_j^T \alpha_i \tag{19}$$

The dual of MTWSVM(i) is given by:

$$\text{DMTWSVM(i)} \max_{\alpha_i} e_{j1}^T \alpha_i - \frac{1}{2} \alpha_i^T B_j (B_i^T B_i)^{-1} B_j^T \alpha_i \text{ s.t. } 0 \le \alpha_i \le c_i \tag{20}$$

Similarly, the dual of MTWSVM(j) is obtained as:

$$\text{DMTWSVM(j)} \quad \max_{\alpha_j} e_{i1}^T \alpha_j - \frac{1}{2} \alpha_j^T B_i (B_j^T B_j)^{-1} B_i^T \alpha_j \text{ s.t. } 0 \le \alpha_j \le c_j \tag{21}$$

$$\text{and } u_{ji} = (B_j^T B_j)^{-1} B_i^T \alpha_j \tag{22}$$

Equations (18) and (22) determine the hyper-plane parameters. In this way, we can determine the hyper-planes for each class. MTWSVM seeks M (M-1) hyper-planes for M-class classification problem. MTWSVM classifier measures the distance of a test pattern from each hyper-plane and assigns class to it by using "Max-wins voting strategy" according to which the class with maximum vote wins i.e., the final class of the test pattern. The vote is given to a class according to the perpendicular distance of the pattern from the hyper-plane. For example, the perpendicular distances of a test pattern 'x' from f_{ij} and f_{ji} planes are measured as:

$$d_{ij} = \frac{|w_{ij} \cdot x + b_{ij}|}{\|w_{ij}\|} \text{ and } d_{ji} = \frac{|w_{ji} \cdot x + b_{ji}|}{\|w_{ji}\|} \tag{23}$$

where | . | is the absolute value. If $d_{ij} < d_{ji}$, i.e., the distance of the pattern from ith class is less as compared to the distance from jth class, then the vote for the ith class is added by one, otherwise the vote is given to jth class. Thus, for a given pattern, we calculate the vote of each class and the class with maximum number of vote is assigned to it. The algorithm of linear MTWSVM is defined as:

Algorithm

Training Phase

For i=1 to M, j=i+1 to M

If i^{th} class is trained with j^{th} class then Define two matrices B_i and B_j as:

$B_i = [A_i\ e_{i1}]$ and $B_j = [A_j e_{j1}]$, *where matrices A_i and A_j include the patterns of ith and j^{th} classes.*

Select penalty parameter $c_i, c_j > 0$.

Solve the QPPs (20) and (21) and obtains α_i and α_j.

Determine u_{ij} and u_{ji} and constructs hyper-planes f_{ij} and f_{ji}.

Testing Phase

Training phase generates M (M-1) hyper-planes.

For i=1 to M, j=i+1 to M

For a test data point, calculates its distance from each hyper-plane according to equation (23).

Initialize the vote of each class. *Vote(i)=0 and Vote(j)=0.*

Compare the distances d_{ij} and d_{ji}.

If ($d_{ij} < d_{ji}$)

Then Vote(i)= Vote(i)+1

ElseVote(j)= Vote(j)+1

End If.

Assign test data point to a class with maximum vote.

Figure 1 shows the geometric representation of the proposed classifier for linearly-separable patterns in $\boldsymbol{R}^2$. Here, we consider three classes. The patterns of each class are denoted by different shapes. The proposed MTWSVM classifier seeks total 6 hyper-planes, two ($3-1 = 2$) planes for each class.

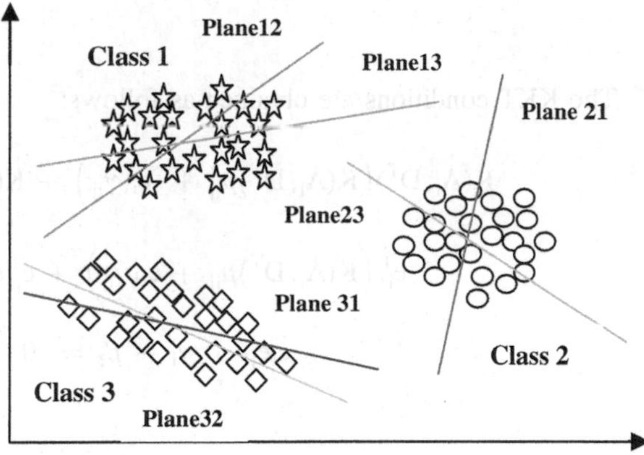

Fig. 1 Linear MTWSVM classifier

3.2 Non-Linear MTWSVM

Sometime it is not possible to separate the classes with linear boundaries. So, a classifier must be able to work efficiently for both linear and non-linear type of examples. The formulation of non-linear MTWSVM is obtained with the help of kernel trick, in which the patterns are mapped to higher dimensional space so that their separation could become easier. For this purpose, following kernel-generated surfaces are considered instead of planes:

$$K\left(x,D^T\right)\mu_{ij} + \gamma_{ij} = 0 \text{ and } K\left(x,D^T\right)\mu_{ji} + \gamma_{ji} = 0 \text{ where } i,\ j = 1,\ldots,M \tag{24}$$

where K is any suitable kernel function and $D = [A_i A_j \ldots A_M]^T$. For ith class and jth class, when both class are trained with each other, the non-linear MTWSVM classifier solves following pair of QPPs:

$$\text{KMTWSVM(i)} \quad \min\left(\mu_{ij}, \gamma_{ij}, \xi_{ij}\right) \frac{1}{2}\left\| K(A_i,D^T)\mu_{ij} + e_{i1}\gamma_{ij}\right\|^2 + c_i e_{j1}^T \xi_{ij}$$

$$\text{s.t.} - \left(K(A_j,D^T)\mu_{ij} + e_{j1}\gamma_{ij}\right) + \xi_{ij} \geq e_{j1}, \xi_{ij} \geq 0 \tag{25}$$

$$\text{KMTWSVM(j)} \quad \min\left(\mu_{ji}, \gamma_{ji}, \xi_{ji}\right) \frac{1}{2}\left\| K(A_j,D^T)\mu_{ji} + e_{j1}\gamma_{ji}\right\|^2 + c_j e_{i1}^T \xi_{ji}$$

$$\text{s.t.}\left(K(A_i,D^T)\mu_{ji} + e_{i1}\gamma_{ji}\right) + \xi_{ji} \geq e_{i1}, \xi_{ji} \geq 0 \tag{26}$$

Lagrangian corresponding to Eq. (25):

$$\begin{aligned} L\left(\mu_{ij}, \gamma_{ij}, \xi_{ij}, \alpha_i, \beta_i\right) = & \frac{1}{2}\left\| K(A_i,D^T)\mu_{ij} + e_{i1}\gamma_{ij}\right\|^2 \\ & + c_i e_{j1}^T \xi_{ij} - \alpha_i^T\left(-\left(K(A_j,D^T)\mu_{ij} + e_{j1}\gamma_{ij}\right) + \xi_{ij} - e_{j1}\right) - \beta_i^T \xi_{ij} \end{aligned} \tag{27}$$

The KKT conditions are obtained as follows:

$$K(A_i, D^T)\left(K(A_i, D^T)\mu_{ij} + e_{i1}\gamma_{ij}\right) + K(A_j, D^T)\alpha_i = 0 \tag{28}$$

$$e_{i1}^T\left(K(A_i, D^T)\mu_{ij} + e_{i1}\gamma_{ij}\right) + e_{j1}^T\alpha_i = 0 \tag{29}$$

$$c_i e_{j1} - \alpha_i - \beta_i = 0 \tag{30}$$

$$-\left(\mathrm{K}(\mathrm{A_j}, \mathrm{D^T})\mu_{\mathrm{ij}} + \mathrm{e_{j1}}\gamma_{\mathrm{ij}}\right) + \xi_{ij} \geq \mathrm{e}_{j1}, \xi_{ij} \geq 0 \tag{31}$$

$$\alpha_{\mathrm{i}}^{\mathrm{T}}\left(\left(\mathrm{K}(\mathrm{A_j},\mathrm{D^T})\mu_{\mathrm{ij}} + \mathrm{e_{j1}}\gamma_{\mathrm{ij}}\right) - \xi_{\mathrm{ij}} + \mathrm{e_{j1}}\right) = 0, \beta_{\mathrm{i}}^{\mathrm{T}}\xi_{\mathrm{ij}} = 0 \tag{32}$$

$$\alpha_{\mathrm{i}} \geq 0, \beta_{\mathrm{i}} \geq 0 \tag{33}$$

Equations (28) and (29) lead to:

$$\begin{bmatrix} \mathrm{K}(\mathrm{A_i},\mathrm{D^T}) \\ \mathrm{e_{i1}^T} \end{bmatrix} [\mathrm{K}(\mathrm{A_i}, \mathrm{D^T})\mathrm{e_{i1}}] \begin{bmatrix} \mu_{\mathrm{ij}} \\ \gamma_{\mathrm{ij}} \end{bmatrix} + \begin{bmatrix} \mathrm{K}(\mathrm{A_j}, \mathrm{D^T}) \\ \mathrm{e_{j1}^T} \end{bmatrix} \alpha_{\mathrm{i}} = 0 \tag{34}$$

Let $\mathrm{H_i} = [\mathrm{K}(\mathrm{A_i}, \mathrm{D^T})\mathrm{e_{i1}}]$ and $\mathrm{H_j} = [\mathrm{K}(\mathrm{A_j}, \mathrm{D^T})\mathrm{e_{j1}}]$. Then Eq. (34) can be reformulated as:

$$\mathrm{H_i^T H_i} \begin{bmatrix} \mu_{\mathrm{ij}} \\ \gamma_{\mathrm{ij}} \end{bmatrix} + \mathrm{H_j^T}\alpha_{\mathrm{i}} = 0 \text{ i.e., } \begin{bmatrix} \mu_{\mathrm{ij}} \\ \gamma_{\mathrm{ij}} \end{bmatrix} = -(\mathrm{H_i^T H_i})^{-1}\mathrm{H_j^T}\alpha_{\mathrm{i}} \tag{35}$$

The dual of KMTWSVM(i) is given by:

$$\text{DKMTWSVM(i)} \quad \max_{\alpha_i} e_{j1}^T\alpha_i - \frac{1}{2}\alpha_i^T H_j (H_i^T H_i)^{-1} H_j^T \alpha_i \text{ s.t. } 0 \leq \alpha_i \leq c_i \tag{36}$$

Similarly, the dual of KMTWSVM(j) is obtained as:

$$\text{DKMTWSVM(j)} \quad \max_{\alpha_j} e_{i1}^T\alpha_j - \frac{1}{2}\alpha_j^T H_i (H_j^T H_j)^{-1} H_i^T \alpha_j \text{ s.t. } 0 \leq \alpha_j \leq c_j \tag{37}$$

$$\text{and} \begin{bmatrix} \mu_{\mathrm{ji}} \\ \gamma_{\mathrm{ji}} \end{bmatrix} = (\mathrm{H_j^T H_j})^{-1}\mathrm{H_i^T}\alpha_{\mathrm{j}} \tag{38}$$

The perpendicular distances of a test pattern 'x' from kernel surfaces are calculated as:

$$d_{ij} = \frac{\left|\mu_{\mathrm{ij}}.\mathrm{K}(\mathrm{x}, D^{\mathrm{T}}) + \gamma_{\mathrm{ij}}\right|}{\left\|\mu_{\mathrm{ij}}\right\|} \text{ and } d_{ji} = \frac{\left|\mu_{\mathrm{ji}}.\mathrm{K}(\mathrm{x}, D^{\mathrm{T}}) + \gamma_{\mathrm{ji}}\right|}{\left\|\mu_{\mathrm{ji}}\right\|} \tag{39}$$

Non-linear MTWSVM also assigns class on the basis of max-wins voting strategy.

3.3 Successive Over Relaxation (SOR) for MTWSVM

MTWSVM contains M (M-1) QPPs which need to be solved. QPP (20) can be reformulated as follows [26, 27]:

$$\max_{\alpha_i} e_{j1}^T \alpha_i - \frac{1}{2} \alpha_i^T G_i \alpha_i \text{ s.t. } 0 \le \alpha_i \le c_i \tag{40}$$

where $G_i = B_j (B_i^T B_i)^{-1} B_j^T$. The SOR algorithm is given below:

Choose parameter $t_i \in (0, 2)$ and start with any initial value $\alpha_i^0 \in R^{l_i}$. Having α_i^k, compute α_i^{k+1} according to the following formula:

$$\alpha_i^{k+1} = (\alpha_i^k - t_i D_i^{-1} (G_i \alpha_i^k - e_{j1} + L_i (\alpha_i^{k+1} - \alpha_i^k)))_\sharp \tag{41}$$

where $L_i \in R^{l_i \times l_i}$ and $D_i \in R^{l_i \times l_i}$ are strictly the lower triangular matrix and diagonal matrix correspondingly. The process terminate if $\left\| \alpha_i^{k+1} - \alpha_i^k \right\|$ is less than some given tolerance else replace α_i^k by α_i^{k+1} and k by k + 1 and repeat the above process.

3.4 Computational Complexity

The proposed MTWSVM classifier solves M (M-1) QPPs and generates M (M-1) binary classifiers. Consider each class contains approximately $l/_M$ patterns. In "one-versus-one" MTWSVM classifier the patterns of each class provide constraints to another class and each binary classifier contains $l/_M$ constraints. Therefore the computational complexity of linear MTWSVM is $M(M-1)(\frac{l}{M})^3 = \frac{l^3(M-1)}{M^2}$. Computational complexity of "one-versus-one" SVM is $O\left(\frac{4(M-1)l^3}{M^2}\right)$. The ratio of computational complexity of "one-versus-one" SVM and MTWSVM is four i.e., our proposed approach is four times faster than that of "one-versus-one" SVM which also proves the basic concept of traditional TWSVM which is four times faster than that of binary SVM.

4 Experimental Results

In order to prove the validity of the proposed approach, we performed the experiment on eight benchmark datasets such as Iris, Wine, Thyroid, Glass, Ecoli, PageBlock, Shuttle and PenBased using 10-fold cross validation method. All these datasets are taken from KEEL Data Repository [28]. The performance of the proposed MTWSVM classifier is compared with the existing multi-classification

Table 1 Performance comparison on eight benchmark datasets (predictive accuracy in percentage (%))

Datasets	OVA SVM		Multi SVM		MBSVM		Twin KSVC		MTWSVM	
	Lin	Non-lin	Lin	non-lin	Lin	Non-lin	Lin	Non-lin	Lin	Non-lin
Iris (150 × 4×3)	95.67	97.33	95.72	97.33	96.12	98.00	94.24	98.13	**97.73**	**98.38**
Wine (178 × 13 × 3)	97.35	**99.43**	95.83	98.71	**97.86**	98.24	97.09	97.75	97.21	97.33
Thyroid (215 × 5×3)	94.61	97.98	92.17	96.11	97.03	98.95	94.50	98.59	**100**	**100**
Glass (214 × 13 × 6)	70.44	72.49	71.18	72.06	75.31	69.52	70.89	63.21	**80.42**	**85.23**
Ecoli (327 × 7×5)	75.01	85.56	74.64	83.35	74.77	86.36	71.62	82.01	**85.35**	**88.42**
Page block (548 × 10 × 5)	76.74	80.29	73.84	79.67	80.28	85.43	79.09	82.39	**89.43**	**90.58**
Shuttle (2175 × 9×5)	73.82	76.08	72.39	75.86	82.07	85.8	79.23	82.48	**82.38**	**86.27**
PenBased (1100 × 16 × 10)	82.00	86.71	81.73	86.27	**87.79**	**88.89**	82.18	88.56	79.95	84.53

approaches based on SVM and TWSVM such as Multi-SVM, One-against-One SVM (OVA SVM), Multiple Birth Support Vector Machine (MBSVM) [29], Twin - KSVC [30]. All classifiers are implemented by using matlabR2012a on Windows 7 PC with Inter Core i-7 processor (3.4 GHz) with 12-GB RAM. For non-linear cases, this research work used Gaussian Kernel function $K(x_i, x_j) = \exp\left(-\frac{\|x_i - x_j\|^2}{2\sigma^2}\right)$. The selection of parameters also affects the performance of the proposed MTWSVM. Parameters such as c_i, c_j and sigma (σ) are obtained by using Grid Search approach from the range- c_i, $c_j \in \{10^{-8}, \ldots, 10^{3}\}$, sigma $\in \{2^{-5}, \ldots, 2^{10}\}$. Accuracy, also referred as correct classification rate, is obtained by taking the average of 10-times testing accuracies.

Table 1 presents the comparison of predictive accuracy of linear and non-linear MTWSVM with the other existing approaches. In Table 1, ". × . × ." indicates the size of the dataset in following order-total number of patterns, attributes and total number of classes. The better predictive accuracies are indicated by bold values in the Table 1. From the table, it is observed that the proposed MTWSVM classifier obtains better predictive accuracy for Iris, Thyroid, Glass, Ecoli, Pageblock and Shuttle datasets. For other two datasets it also shows comparable performance.

5 Conclusion

In this research work, we proposed a novel multi-classifier, named as Multi-class Twin Support Vector Machine (MTWSVM), which is obtained by extending the formulation of binary TWSVM. MTWSVM utilizes "one-versus-one" concept according to which the patterns of each class are trained with the patterns of another class. The training procedure of MTWSVM is speed up by utilizing SOR technique for solving dual problems. The experimental results prove the validity of the proposed MTWSVM classifier. The results show that the MTWSVM obtains better predictive accuracy for six datasets like Iris, Thyroid, Glass, Shuttle, Pageblock and Ecoli datasets while for other datasets it also shows comparable performance. The computational complexity of MTWSVM is also analyzed which is four times faster than that of "one-versus-one" SVM and thus follows the basic concept of TWSVM. The future work is to optimize the parameters selection procedure using some optimization approaches such as Genetic Algorithm, Ant Colony Optimization, Particle Swarm Optimization etc. We also want to explore the real world application of the proposed MTWSVM classifier. This classifier generates large number of hyper-planes which increases with the classes thus the implication of having higher number of planes can also be explored in future work.

References

1. Cortes, C., Vapnik, V.: Support-vector networks. Machine learning **20**(3), 273–297 (1995)
2. Vapnik, V.: The nature of statistical learning theory. Springer (2000)
3. Cristianini, N., Shawe-Taylor, J.: An introduction to support vector machines and other kernel-based learning methods. Cambridge university press (2000)
4. Wang, T.Y., Chiang, H.M.: One-against-one fuzzy support vector machine classifier: an approach to text categorization. Expert Syst. Appl. **36**(6), 10030–10034 (2009)
5. Wang, T.Y., Chiang, H.M.: Fuzzy support vector machine for multi-class text categorization. Inf. Process. Manage. **43**(4), 914–929 (2007)
6. Ganapathiraju, A., Hamaker, J.E., Picone, J.: Applications of support vector machines to speech recognition. Sig. Process. IEEE Trans. **52**(8), 2348–2355 (2004)
7. Manikandan, J., Venkataramani, B.: Design of a real time automatic speech recognition system using modified one against All SVM classifier. Microprocess. Microsyst. **35**(6), 568–578 (2011)
8. Elish, K.O., Elish, M.O.: Predicting defect-prone software modules using support vector machines. J. Syst. Softw. **81**(5), 649–660 (2008)
9. Can, H., Jianchun, X., Ruide, Z., Juelong, L., Qiliang, Y., Liqiang, X.: A new model for software defect prediction using particle swarm optimization and support vector machine. In 25th Control and Decision Conference (CCDC), Chinese, pp. 4106–4110 (2013)
10. Chen, H.L., Yang, B., Liu, J., Liu, D.Y.: A support vector machine classifier with rough set-based feature selection for breast cancer diagnosis. Expert Syst. Appl. **38**(7), 9014–9022 (2011)
11. Übeyli, E.D.: Multiclass support vector machines for diagnosis of erythemato-squamous diseases. Expert Syst. Appl. **35**(4), 1733–1740 (2008)
12. Sweilam, N.H., Tharwat, A.A., Abdel Moniem, N.K.: Support vector machine for diagnosis cancer disease: a comparative study. Egypt. Inform. J. **11**(2), 81–92 (2010)
13. Li, Y., Xia, J., Zhang, S., Yan, J., Ai, X., Dai, K.: An efficient intrusion detection system based on support vector machines and gradually feature removal method. Expert Syst. Appl. **39**(1), 424–430 (2012)
14. Kuang, F., Xu, W., Zhang, S.: A novel hybrid KPCA and SVM with GA model for intrusion detection. Appl. Soft Comput. **18**, 178–184 (2014)
15. Shin, K.S., Lee, T.S., Kim, H.J.: An application of support vector machines in bankruptcy prediction model. Expert Syst. Appl. **28**(1), 127–135 (2005)
16. Min, S.H., Lee, J., Han, I.: Hybrid genetic algorithms and support vector machines for bankruptcy prediction. Expert Syst. Appl. **31**(3), 652–660 (2006)
17. Han, B.J., Ho, S., Dannenberg, R.B., Hwang, E.: SMERS: Music emotion recognition using support vector regression (2009)
18. Guo, G., Li, S.Z., Chan, K.L.: Support vector machines for face recognition. Image Vis. Comput. **19**(9), 631–638 (2001)
19. Kim, K.J.: Financial time series forecasting using support vector machines. Neurocomputing **55**(1), 307–319 (2003)
20. Hsu, C.W., Lin, C.J.: A comparison of methods for multiclass support vector machines. Neural Netw. IEEE Trans. **13**(2), 415–425 (2002)
21. Kerebel, U.H.G.: Pairwise Classification and Support Vector Machines: In Advances in Kernel Methods, pp. 255–268. MIT Press, Cambridge (1999)
22. Weston, J., Watkins, C.: Multi-class support vector machines. Technical Report CSD-TR-98-04, Department of Computer Science, Royal Holloway, University of London, May (1998)
23. Lee, Y., Lin, Y., Wahba, G.: Multicategory support vector machines. In Proceedings of the 33rd Symposium on the Interface (2001)
24. Platt, J.C., Cristianini, N., Shawe-Taylor, J.: Large Margin Dags For Multiclass Classification. In NIPS, vol. 12, 547–553 (1999)

25. Jayadeva, K., Khemchandani, R., Chandra, S.: Twin support vector machines for pattern classification. IEEE Trans. Pattern Anal. Mach. Intell. **29**(5), 905–910 (2007)
26. Shao, Y.H., Zhang, C.H., Wang, X.B., Deng, N.Y.: Improvements on twin support vector machines. IEEE Trans. on Neural Netw. **22**(6), 962–968 (2011)
27. Mangasarian, O.L., Musicant, D.R.: Successive overrelaxation for support vector machines. IEEE Trans. Neural Netw. **10**(5), 1032–1037 (1999)
28. http://sci2s.ugr.es/keel/imbalanced.php. Accessed 20 May 2014
29. Yang, Z.X., Shao, Y.H., Zhang, X.S.: Multiple birth support vector machine for multi-class classification. Neural Comput. Appl. **22**(1), 153–161 (2013)
30. Xu, Y., Guo, R., Wang, L.: A twin multi-class classification support vector machine. Cognitive Comput. **5**(4), 580–588 (2013)

Multi-label Classifier for Emotion Recognition from Music

Divya Tomar and Sonali Agarwal

Abstract Music is one of the important medium to express the emotions such as anger, happy, sad, amazed, quiet etc. In this paper, we consider the task of emotion recognition from music as a multi-label classification task because a piece of music may have more than one emotion at the same time. This research work proposes the Binary Relevance (BR) based Least Squares Twin Support Vector Machine (LSTSVM) multi-label classifier for emotion recognition from music. The performance of the proposed classifier is compared with the eight existing multi-label learning methods using fourteen evaluation measures in order to evaluate it from different point of views. The experimental result suggests that the proposed multi-label classifier based emotion recognition system is more efficient and gives satisfactory outcomes over the other existing multi-label classification approaches.

Keywords Multi-label classification · Emotion recognition · Binary relevance · Least squares twin support vector machine

1 Introduction

Music plays an important role in everyone's life. According to the quotes of famous German philosopher Friedrich Nietzsche, "without music, life would be a mistake". Researchers are taking interest in automatically analyzing the emotional contents of music and several recent developments in the field of Music Information Retrieval system have been reported [1–7]. Emotion can be conveyed through music. The recognition of emotion from music plays significant role in the improvement of information retrieval of music as music database is growing in size. Retrieval of

D. Tomar (✉) · S. Agarwal
Indian Institute of Information Technology, Allahabad 211 012, India
e-mail: divyatomar26@gmail.com

S. Agarwal
e-mail: sonali@iiita.ac.in

A. Nagar et al. (eds.), *Proceedings of 3rd International Conference on Advanced Computing, Networking and Informatics*, Smart Innovation, Systems and Technologies 43, DOI 10.1007/978-81-322-2538-6_12

music on the basis of emotion is useful for various applications such as music recommendation systems [8]; content based searching, music therapy, song selection in hand portable devices [9], in various TV and radio programs etc. Lots of work has been done in the field of analyzing and recognizing emotional contents of music. But mostly researchers consider this task as a single label classification task [3–5, 10]. While a piece of music may contains several different emotions at the same time. So, the task of emotion recognition from music is a multi-label classification task. Hence, the objective of this research work is to focus on multi-label classification methods and to develop an emotion recognition system using music database.

The goal of multi-label classification is to obtain a model that assigns a set of class labels to each object or data samples unlike multi-class classification in which the classifier predicts single class [11–13]. In recent years, multi-label learning has gained popularity in the research community which results the development of a variety of multi-label learning algorithms. In multi-label learning, a classifier learns from a number of data samples or instances, where each data sample can be associated with multiple classes and so after be able to predict the possible class labels for a new data sample. Multi-label classification is different from single label classification problem where each data sample belongs to only one class label from a set of disjoint class labels $\mathcal{L}$ Single label classification problem can be recognized as a binary classification for $|\mathcal{L}| = 2$ or multi-class classification for $|\mathcal{L}| > 2$. The need of multi-label classification emerges from the various real world problems such as in the text categorization a text may belong to different categories, in the medical diagnosis where a patient may have diabetes and cancer at the same time, in image and email classification etc.

There are two ways to handle multi-label classification task which we will discuss in details in the second section. In this research work, we proposed Binary Relevance based Least Squares Twin Support Vector Machine (BR-LSTSVM) classifier in which the multi-label problem is divided into several single-label classification problems. The reason for which we have used Binary Relevance method will be discussed in Sect. 3. From the literature survey, it is found that Support Vector Machine (SVM) has shown better performance as compared to the other existing classifiers [14–19]. But the problem with this is its high computational complexity [20, 21]. To handle this problem, Jayadeva et al. proposed a novel classifier Twin Support Vector Machine (TWSVM) which is four times faster than that of conventional SVM [21]. TWSVM is not only better in terms of speed but also shows better performance over SVM. But again, TWSVM requires the optimization of two Quadratic Programming Problems (QPPs). In order to utilize the better speed of TWSVM, Kumar et al. proposed a novel binary classifier named as LSTSVM which is the least squares variant of TWSVM [22]. In LSTSVM, two complex QPPs are transformed into two linear equations which are easy to solve. So, in this paper we used LSTSVM as a base classifier because it has several advantages such as better generalization ability, faster computational speed and easier implementation. Therefore, this paper has adopted Binary Relevance based LSTSVM multi-label classifier for the emotion recognition from music.

The paper is organized as follows: Section 2 discusses the literature survey on emotion detection system using music data and various multi-label classification methods. Section 3 gives the detail description of the proposed approach. Dataset description, performance evaluation parameters and results are discussed in Sect. 4. Conclusion is drawn in Sect. 5.

2 Literature Survey

2.1 Emotion Recognition System

Feng et al. analyzed two musical features (tempo and articulation) and by using these features recognized four different emotions such as happiness, fear, anger and sadness [23]. Lie et al. recognized emotion from music acoustic data [24]. They used music rhythm and timbre feature to represent stress dimension and intensity feature for energy dimension of Thayer model. Yang et al. developed a Music Emotion Recognition System by utilizing the idea of content based retrieval [3]. For this purpose, they used regression approach to recognize emotion contents from music. They predicted the Arousal Valence values of each music sample which became a point in the arousal-valence plane so that users can efficiently retrieve the songs by specifying a point in that plane. Han et al. developed a music emotion recognition system (SMERS) by using Support Vector regression [5]. They focused on predicting the arousal and valence of an audio content of a song. They extracted seven different musical features such as rhythm, pitch, tonality, harmonics, temp, key and loudness and recognized eleven emotions for example- angry, excited, relaxed, sleepy, pleased, bored, sad, nervous, calm, happy and peaceful based on these features. In another research work, Yang et al. considered both lyrics and audio features and recognized emotion from music by using SVM [25]. They divided the music samples into several frames and extracted both textual audio features of it. Trohidis et al. used the "Tellegen-Watson-Clark" model of mood and developed emotion detection system as a multi-label classification task [7]. Authors compared four multi-label classification approaches and among which random k-labelsets (RAkEL) performed well. Tzacheva et al. used Thayer's model to represent different emotions [4]. They assumed the task of emotion detection as a single-label classification task and compared the performance of Bayesian Neural Network and J48 Decision Tree to detect four emotions.

2.2 Multi-label Classification

Multi-label classification problem is mainly divided into two categories: Problem transformation methods and algorithm adaptation methods.

2.2.1 Problem Transformation Method

In the problem transformation method, the multi-label classification problem is transformed into a set of single label problems. This approach is independent from the algorithm and any existing classification technique can be applied to multi-label classification problems. Various problem transformation methods are available in the literature and used by the researchers for transferring multi-label classification problem into single label problems [11–13, 26, 27].

- ***Binary Relevance(BR)***: This is one of the most popular approach of problem transformation method in which multi label dataset is divided into k single label datasets (k = $|\mathcal{L}|$), each for one class label and a binary classifier is constructed for each label.Spyromitros et al. proposed a classifier for handling multi-label data called BRkNN which is the BR followed by kNN (k-Nearest Neighbor) [28]. If the computation cost of computing k nearest neighbors is 'C', then the computational cost ofBRkNNis |L| × C. This problem can be resolved by adopting single search for kNN but at the same time it does not consider the correlation among class labels and generate independent predictions for each class label. Classifier Chaining (CC) method which is closely related to BR method includes 'k' binary classifiers as in BR. This method is proposed by Read et al. in which k binary classifiers are linked along a chain in which ith classifier handles the classification problem associated with the class label 'i' [29].
- ***Ranking by Single Label***: In this method, the multi-label dataset is transformed into single label datasets by using several ways such as-ignorance of instances with multi-label, random selection of label, counts for each label and find maximum and minimum counts of labels and assign weight to each label. A single label classifier generates vote (Probability) and assigns rank to each class label [13, 30].
- ***Ranking by Pairwise Comparison (RPC)***: This method transforms the multi-label dataset into k(k–1)/2 binary label datasets where |k = $\mathcal{L}$| and requires the training of k(k–1)/2 binary classifiers, one for each pair of class labels. The binary label dataset for each pair of class labels is generated by taking those examples which associated with at least one of the class label but not both. For a new data sample, all the binary classifiers are invoked and ranking is assigned to each label by counting the vote obtained from each classifier [13, 30].
- ***Calibrated Label Ranking (CLR)***: This method is proposed by Furnkranz et al. which is an extension of previously discussed RPC approach [31]. In this method, an additional label c_0 (also known as calibration label) is added to the original multi label dataset which partitioned the labels into relevant and irrelevant class labels. It generates the ranking of the label as: $c_{i1} > c_{i2} > \cdots > c_{ij} > c_0 > c_{ij+1} > \cdots > c_{ik}$. Each data sample that is associated with a class label is treated as a positive data sample for that particular label and negative for the calibration label. Then binary classifier is trained with

these datasets in order to discriminate between the class labels and calibrated labels and for a new data sample, ranks are assigned to each label by counting the vote [11, 31]. A variant of CLR, named as Quick Weighted algorithm for Multi-label Learning (QWML) is proposed by Mencia et al. [32]. The voting strategy of QWML is different from the majority voting used by CLR. This approach focuses on classes with "low voting loss".

- ***Label Powerset (LP)***: This method considers each unique set of class labels in the original dataset as single class label and transformed the original multi-label dataset into single label datasets. So, the task becomes a single label classification problem and assigns the most probable class label to the new data sample. The random k-labelsets (RAkEL) approach is obtained by ensemble of LP classifiers where different random subset is used to train each LP classifier.

2.2.2 Algorithm Adaptation Method

Clare and King use C4.5 algorithm for multi-label classification problem with the modified entropy calculation [33]:

$$\text{Entropy} = -\sum_{i=1}^{N}(p(\lambda_i)\log p(\lambda_i) + q(\lambda_i)\log q(\lambda_i) \qquad (1)$$

where, $(p(\lambda_i)$ represents relative frequency of class λ_i and $q(\lambda_i) = 1 - p(\lambda_i)$. This modified entropy based C4.5 approach allows multiple class labels at the leaves.

- ***Boosting***: AdaBoost.MH and AdaBoost.MR are two extended version of basic AdaBoost algorithm for multi-label data classification. AdaBoost.MH takes examples in the form of example-label pairs and the weight of misclassified example-label is increased in each iteration. AdaBoost.MH minimizes the hamming loss. While, AdaBoost.MR finds the hypothesis and arranges the correct class labels on the basis of ranking. Ensemble of classifier chain (ECC) is a multi-label classification technique that is obtained by combining several CC classifiers $c_1, c_2, c_3, \ldots, c_n$. Each CC classifier $c_k$is trained with the random subset of multi label dataset and gives different multi label predictions. The result of each CC classifier is combined per label and each label obtains a number of votes [29].
- ***Lazy Learning***: Various algorithm adaptation approaches based on K-Nearest Neighbor are proposed by the researchers. The process of aggregation of class labels of a given data samples differs with each other. Multi-label k-Nearest Neighbor (ML-kNN) approach is a lazy learning approach used for the multi-label data classification. This algorithm utilizes maximum a posteriori (MAP) rule to predict the class labels by reasoning with the class labeling information embodied in the neighbors [34].

3 Proposed Algorithm

Consider a training dataset $D = (x_i, Y_i), 1 \leq i \leq n$ of 'n' data samples where $x_i \in \chi$ (data sample space) represents input data sample and $Y_i \in y$ (label vector space) represents class label. Let the training dataset consists 'm' features. The objective of the multi-label learning is to obtain a multi-label classifier that optimizes evaluation parameters. In this research work, we adopted Binary Relevance (BR) problem transformation method to develop multi-label classifier for emotion recognition from music. Although, BR does not handle the label dependency, yet still it has several advantages over other existing methods such as, it has linear complexity regarding the number of class labels and any binary classifier can be considered as a base learner [35]. Binary Relevance method divides the multi-label dataset into $|L|$ dataset each for one class. In this way, it divides the multi-label classification problems into $|L|$ single label classification problem. For each dataset $D_j, 1 \leq j \leq k$, (where $k = |L|$)considers the data samples of jth label as positive and others as negative. Binary Relevance predicts the class for a new data sample by combining the labels that are positively predicted by each classifier. As a base classifier, we used Least Squares Twin Support Vector Machine due to its better generalization ability and faster computational speed. LSTSVM is a binary classifier that constructs two hyper-planes, one for each class by solving following two linear equations:

$$\begin{aligned} &\min(\mathrm{w}_1, \mathrm{b}_1, \xi)\frac{1}{2} \parallel \mathrm{X}_1\mathrm{w}_1 + \mathrm{e}_1\mathrm{b}_1 \parallel^2 + \frac{\mathrm{c}_1}{2}\xi^{\mathrm{T}}\xi \\ &\mathrm{s.t.}-(\mathrm{X}_2\mathrm{w}_1 + \mathrm{e}_2\mathrm{b}_1) + \xi = \mathrm{e}_2 \end{aligned} \quad (2)$$

$$\begin{aligned} &\min(\mathrm{w}_2, \mathrm{b}_2, \eta)\frac{1}{2} \parallel \mathrm{X}_2\mathrm{w}_2 + \mathrm{e}_2\mathrm{b}_2 \parallel^2 + \frac{\mathrm{c}_2}{2}\eta^{\mathrm{T}}\eta \\ &\mathrm{s.t.}(\mathrm{X}_1\mathrm{w}_2 + \mathrm{e}_1\mathrm{b}_2) + \eta = \mathrm{e}_1 \end{aligned} \quad (3)$$

where X_1 and X_2 are two matrices contain the data samples of positive and negative class correspondingly. W_1 and W_2 are the normal vectors to the hyper-plane, b_1 and b_2 are bias terms, e_1 and e_2 are the two vectors of one's, c_1 and c_2 are positive penalty parameters and ξ and η are slack variables due to the negative and positive class respectively. Hyper-plane parameters are obtained by solving above two equations as:

$$\begin{bmatrix} \mathrm{w}_1 \\ \mathrm{b}_1 \end{bmatrix} = -\left(\mathrm{B}^{\mathrm{T}}\mathrm{B} + \frac{1}{\mathrm{c}_1}\mathrm{A}^{\mathrm{T}}\mathrm{A}\right)^{-1}\mathrm{B}^{\mathrm{T}}\mathrm{e}_2 \quad (4)$$

$$\begin{bmatrix} \mathrm{w}_2 \\ \mathrm{b}_2 \end{bmatrix} = \left(\mathrm{A}^{\mathrm{T}}\mathrm{A} + \frac{1}{\mathrm{c}_2}\mathrm{B}^{\mathrm{T}}\mathrm{B}\right)^{-1}\mathrm{A}^{\mathrm{T}}\mathrm{e}_1 \quad (5)$$

where, $A = [X_1 e_1]$ and $B = [X_2 e_2]$. These parameters generate hyper-planes by using following equation:

$$x^T w_1 + b_1 = 0 \text{ and } x^T w_2 + b_2 = 0 \tag{6}$$

The class label is assigned to a given data sample according to the following decision function:

$$f(x) = \arg \min_{i=+1,-1} \frac{|w_i \cdot x + b_i|}{\| w_i \|} \tag{7}$$

LSTSVM also gives promising results in the classification of non-linearly separable data samples with the help of kernel function. Non-linear LSTSVM classifier determines following kernel surfaces in high dimensional space:

$$\mathrm{Ker}(x^T, Z^T)\mu_1 + \gamma_1 = 0 \text{ and } \mathrm{Ker}(x^T, Z^T)\mu_2 + \gamma_2 = 0 \tag{8}$$

Here, 'Ker' refers to the kernel function and $Z = [X_1 X_2]^T$. Non-linear LSTSVM solves following two linear equations to separate the data samples of two classes:

$$\begin{gathered} \min(\mu_1, \gamma_1, \xi) \frac{1}{2} \| \mathrm{Ker}(X_1, Z^T)\mu_1 + e\gamma_1 \|^2 + \frac{c_1}{2} \xi^T \xi \\ \text{s.t.} -(\mathrm{Ker}(X_2, Z^T)\mu_1 + e\gamma_1) = e - \xi \end{gathered} \tag{9}$$

$$\begin{gathered} \min(\mu_2, \gamma_2, \xi) \frac{1}{2} \| \mathrm{Ker}(X_2, Z^T)\mu_2 + e\gamma_2 \|^2 + \frac{c_2}{2} \eta^T \eta \\ \text{s.t.} (\mathrm{Ker}(X_1, Z^T)\mu_2 + e\gamma_2) = e - \eta \end{gathered} \tag{10}$$

Equations (9) and (10) determine kernel surface parameters as:

$$\begin{bmatrix} \mu_1 \\ \gamma_1 \end{bmatrix} = -(Q^T Q + \frac{1}{c_1} P^T P)^{-1} Q^T e \tag{11}$$

$$\begin{bmatrix} \mu_2 \\ \gamma_2 \end{bmatrix} = (P^T P + \frac{1}{c_2} Q^T Q)^{-1} P^T e \tag{12}$$

where $P = [K(X_1, D^T)e] Q = [K(X_2, D^T)e]$. New data sample is classified according to the following formulation:

$$\mathrm{class}(j) = \mathrm{argmin}(j = 1, 2) \frac{|x^T \mu_j + \gamma_j|}{\mu_j} \tag{13}$$

In this paper, we used Gaussian Kernel function which is defined as:

$$K_G(x_i, x_j) = \exp\left(-\frac{\|x_i - x_j\|^2}{2\sigma^2}\right) \tag{14}$$

where x_i and x_j are two input vectors. In this way, we construct the binary classifier for each dataset and determine the class label for each data sample. The positively predicted class labels by each classifier are combined for a given data sample i.e., the data sample belongs to those class labels which are positively predicted by the classifier in each dataset for that particular data sample. For a new data sample, each binary classifier predicts the class label. Then the result of each classifier is combined for positive class labels.

4 Numerical Experiment

4.1 Dataset Description

The experiment is performed on Emotions dataset which is taken from Mulan's repository [36]. The domain of Emotions dataset is music and it contains 593 instances. Each music instance can be labeled with six different emotions such as L1 angry-aggressive (189 examples), L2 quiet-still (148 examples), L3 amazed-surprised (173 examples), L4 sad-lonely (168 examples), L5 happy-pleased (166 examples) and L6 relaxing-calm (264 examples).The dataset contains 72 features which are broadly falls into two main categories: rhythmic and timbre. It comprises 8 rhythmic features and 64 timbre features. Label cardinality of the multi-label dataset is the average number of class labels of the data samples. Label Density of the multi-label dataset is the average number of class labels of the data samples divided by total number of labels $|L|$. Label cardinality and density of emotions dataset is 1.869 and 0.311 respectively.

4.2 Performance Evaluation Parameters

The metrics require for the performance evaluation of a multi-label classifier are different than those used in conventional single-label multi-classifier. Let the set of class labels predicted by the proposed classifier ML-LSTSVM for a given data sample x_i be represented by P_i. This research work used two different types of evaluation measures-example based and label based. "The example based evaluation metrics measure the average differences of the actual and the predicted sets of labels over all data samples of the evaluation dataset". On the other hand,

Table 1 Performance evaluation measure

Example based metrics	Label based metrics
Accuracy $= \frac{1}{n}\sum_{i=1}^{n}\frac{\lvert Y_i \cap P_i\rvert}{\lvert Y_i \cup P_i\rvert}$	Macro Precison $= \frac{1}{k}\sum_{j=1}^{k}\frac{TP_j}{TP_j + FP_j}$
Precision $= \frac{1}{n}\sum_{i=1}^{n}\frac{\lvert Y_i \cap P_i\rvert}{\lvert P_i\rvert}$	Macro Recall $= \frac{1}{k}\sum_{j=1}^{k}\frac{TP_j}{TP_j + FN_j}$
Recall $= \frac{1}{n}\sum_{i=1}^{n}\frac{\lvert Y_i \cap P_i\rvert}{\lvert P_i\rvert}$	Macro Accuracy $= \frac{1}{k}\sum_{j=1}^{k}\frac{TP_j + TN_j}{TP_j + FP_j + TN_j + FN_j}$
$F1$ score $= \frac{1}{n}\sum_{i=1}^{n}\frac{2\times\lvert Y_i \cap P_i\rvert}{\lvert Y_i\rvert + \lvert P_i\rvert}$	Macro $F1 = \frac{1}{k}\sum_{j=1}^{k}\frac{2\times \text{Precison}_j \times \text{Recall}_j}{\text{Precison}_j + \text{Recall}_j}$
Subset Accuracy $= \frac{1}{n}\sum_{i=1}^{n} I(Y_i = P_i)$	Micro Precison $= \frac{\sum_{j=1}^{k} TP_j}{\sum_{j=1}^{k} TP_j + \sum_{j=1}^{k} FP_j}$
Hamming Loss $= \frac{1}{n}\sum_{i=1}^{n}\frac{\lvert Y_i \Delta P_i\rvert}{\lvert L\rvert}$	Micro Recall $= \frac{\sum_{j=1}^{k} TP_j}{\sum_{j=1}^{k} TP_j + \sum_{j=1}^{k} FN_j}$
	Micro Accuracy $= \frac{\sum_{j=1}^{k} TP_j + \sum_{j=1}^{k} TN_j}{\sum_{j=1}^{k} TP_j + \sum_{j=1}^{k} FN_j + \sum_{j=1}^{k} TN_j + \sum_{j=1}^{k} FP_j}$
	Micro $F1 = \frac{2\times \text{micro\_precision} \times \text{micro\_recall}}{\text{micro\_precision} + \text{micro\_recall}}$

"label-based evaluation metrics measure the predictive performance for each label separately and then average the performance over all labels". Example based performance evaluation measure includes six metrics (accuracy, precision, recall, subset accuracy, F1 score, hamming loss) and label-based evaluation measure includes 8 metrics (macro accuracy, macro precision, macro recall, macro-F1, micro accuracy, micro precision, micro recall, micro F1) as shown in Table 1. Here,TP_i, TN_i, FP_i, FN_i indicate the number of True Positive, True Negative, False Positive and False Negative data samples of ith class.

4.3 Results and Discussion

The performance of the proposed BR-LSTSVM based emotion recognition system is compared with nine existing multi-label classification approaches such as BR, CC, RAkEL, CLR, MLkNN, ML C4.5, QWML and ECC. In this study, Support Vector Machine is used as a base classifier in BR, CC, RAkEL, CLR, QWML and ECC. All these approaches are evaluated against 14 evaluation metrics for example Hamming Loss, Accuracy, Precision, Recall, Subset Accuracy, F1-Score, Micro Precision, Macro Precision, Micro Recall, Macro Recall, Micro Accuracy, Macro Accuracy, Micro F1 and Macro F1. The proposed multi-label classifier BR-LSTSVM and other existing multi-label classifiers used in this study are implemented in matlab on windows 7 with Intel Core i-7 processor (3.4 GHz) with 12-GB RAM. The experiment is performed by using 10-fold cross validation

Table 2 Performance comparison for emotions dataset

	BRSVM	RAkEL	CLR	ML-kNN	QWML	CC	ECC	ML C4.5	BR-LSTSVM
Hamming loss↓	0.1943	0.1849	0.1930	0.2616	0.2543	0.2561	0.2810	0.2472	**0.1703**
Accuracy↑	0.5185	0.5876	0.5271	0.3427	0.3732	0.3563	0.4325	0.5364	**0.6547**
Precision↑	0.6677	0.7071	0.6649	0.5184	0.5481	0.5518	0.5801	0.6062	**0.7738**
Recall↑	0.5938	0.6962	0.6142	0.3802	0.4292	0.3976	0.5334	0.7033	**0.7982**
Subset accuracy ↑	0.2759	**0.7009**	0.6378	0.4379	0.1492	0.1246	0.1688	0.2771	0.6033
F1-score↑	0.6278	0.3395	0.2830	0.1315	0.4813	0.4610	0.5562	0.6512	**0.7837**
Mirco precision↑	0.7351	0.7081	0.7270	0.6366	0.6804	0.6980	0.5793	0.6078	**0.7702**
Micro recall↑	0.5890	0.6925	0.6103	0.3803	0.4315	0.3931	0.5314	0.7127	**0.7581**
Micro F1↑	0.6526	0.6993	0.6622	0.4741	0.5286	0.5033	0.5542	0.6553	**0.7641**
Micro accuracy↑	0.7465	0.8241	**0.8529**	0.7540	0.5845	0.5237	0.5407	0.6325	0.7685
Macro precision↑	0.6877	0.7059	0.7036	0.4608	0.6602	0.5815	0.5310	0.6021	**0.7782**
Macro recall↑	0.5707	0.6765	0.5933	0.3471	0.3983	0.3645	0.5084	0.7023	**0.7581**
Macro F1↑	0.6001	0.6768	0.6212	0.3716	0.4584	0.4206	0.5006	0.6304	**0.7681**
Macro accuracy↑	0.7343	0.8115	**0.8374**	0.7185	0.2542	0.4129	0.5887	0.6518	0.8143

method and the parameters are selected by using Grid Search approach. Penalty parameter is selected from the set of $c_i \in \{10^{-8}, \ldots 10^{3}\}$ and Gaussian kernel parameter sigma is chosen from the set of $\sigma \in \{2^{-5}, \ldots 2^{10}\}$. Table 2 shows the performance comparison of existing multi-label classification approaches with the proposed classifier BR-LSTSVM for emotions dataset.

The downward symbol associated with the metrics indicates that the value of the corresponding metric should be as less as possible i.e., the multi-label classifier performs well if the value of Hamming Loss is low. The upward symbol associated with the metrics indicates that the value of the corresponding metric should be as high as possible. i.e., the multi-label classifier performs well if the value of accuracy, precision or any other associated metrics is high. The best results obtained by the multi-label learning approaches in each evaluation measure are indicated by bold value. From the Table 2, it is clear that the BR-LSTSVM based emotion recognition system has achieved better performance as compared to the other existing approaches for 11 evaluation metrics. BR-LSTSVM based music emotion recognition system performs well on Hamming Loss, Accuracy, Precision, Recall, F1-score, Micro Precision, Micro Recall, Micro F1, Macro Precision, Macro Recall and Macro F1. The performance with respect to other evaluation parameters is also comparable with other existing approaches.

5 Conclusion

It is established through the literature survey that there are several emotion recognition systems based on music database exist but most of them considered the task of emotion recognition as a single label classification task. While a song or a piece of music may contains different emotions at the same time. Also, there were many researchers worked on multi-label music emotion recognition system but did not achieve better performance. So, in this paper we developed a multi-label classifier BR-LSTSVM based emotion recognition system. The proposed system achieves better performance in terms of eleven evaluation parameters as compared to the other existing multi-label approaches. In this research work, we only considered the musical features. In future it is interesting to analyze the textual features such as lyrics of the song with musical features for emotion recognition of a musical content.

References

1. Van de Laar, B.: Emotion detection in music, a survey. In: Twente Student Conference on IT. vol. 1, p. 700 (2006)
2. Yang, Y.H., Chen H.H., Machine recognition of music emotion: a review. ACM Trans. Intell. Syst. Technol. (TIST). **3**(3), 40 (2012)
3. Yang, Y.H., Lin, Y.C., Su, Y.F. Chen, H.H.: A regression approach to music emotion recognition. IEEE Trans. Audio Speech Lang. Process. **16**(2), 448–457 (2008)

4. Tzacheva, A.A., Schlingmann, D., Bell, K.J.: Automatic detection of emotions with music files. Int. J. Soc. Netw. Mining **1**(2), 129–140 (2012)
5. Han, B.J., Ho, S., Dannenberg, R.B., Hwang, E.: SMERS: Music emotion recognition using support vector regression (2009)
6. Li, T., Ogihara, M.: Detecting emotion in music. IISMIR, **3**, 239–240 (2003)
7. Trohidis, K., Tsoumakas, G., Kalliris, G., Vlahavas, I.P.: Multi-label classification of music into emotions. ISMIR **8**, 325–330 (2008)
8. Cai, R., Zhang, C., Wang, C., Zhang, L, Ma, W.Y.: Musicsense: contextual music recommendation using emotional allocation modeling. In: Proceedings of the 15th International Conference on Multimedia, 553–556 (2007)
9. Tolos, M., Tato, R., Kemp, T.: Mood-based navigation through large collections of musical data. In: Proceedings of the 2nd IEEE Consumer Communications and Networking Conference (CCNC 2005) pp. 71–75 (2005)
10. Rocha, B., Panda, R., Paiva, R.P.: Music Emotion Recognition: The Importance of Melodic Features. In: 5th International Workshop on Machine Learning and Music, Prague, Czech Republic (2013)
11. Tsoumakas, G., Katakis, I., Vlahavas, I.: Mining multi-label data: In Data mining and knowledge discovery handbook, pp. 667–685. Springer, US (2010)
12. Bi, W., Kwok, J.: Efficient Multi-label Classification with Many Labels. In: Proceedings of the 30th International Conference on Machine Learning (ICML-13), pp. 405–413 (2013)
13. Tsoumakas, G., Katakis, I.: Multi-label classification: an overview. Int. J. Data Warehouse. Min. (IJDWM) **3**(3), 1–13 (2007)
14. Cortes, C., Vapnik, V.: Support vector network. Mach. Learn. **20**(3), 273–297 (1995)
15. Vapnik, V.: The nature of statistical Learning, @ndedn. Springer, New York (1998)
16. Chen, D., Odobez, J.M.: Comparison of Support Vector Machine and Neural Network for Text Texture Verification. IDIAP, Switzerland (2002)
17. Tomar, D., Agarwal, S.: A survey on data mining approaches for healthcare. Int. J. Bio-Sci. Bio-Technol. **5**(5), 241–266 (2013)
18. Sweilam, H.N., Tharwat, A.A., Moniem, N.K.A.: Support vector machine for diagnosis cancer disease: a comparative study. Egypt. Inform. J. **11**(2), 81–92 (2010)
19. Agarwal, S., Pandey, G.N.: SVM based context awareness using body area sensor network for pervasive healthcare monitoring. In: Proceedings of the First International Conference on Intelligent Interactive Technologies and Multimedia, pp. 271–278 (2010)
20. Mangasarian, O.L., Wild, E.W.: Multisurface proximal support vector classification via generalized eigenvalues. IEEE Trans. Pattern Anal. Mach. Intell. **28**(1), 69–74 (2006)
21. Jayadeva, Khemchandani, R., Chandra, S.: Twin Support vector Machine for pattern classification. IEEE Trans. Pattern Anal. Mach. Intell. **29**(5), 905–910 (2007)
22. Kumar, M.A., Gopal, M.: Least squares twin support vector machines for pattern classification. Expert Syst. Appl. **36**, 7535–7543 (2009)
23. Feng, Y., Zhuang, Y., Pan, Y.: Music information retrieval by detecting mood via computational media aesthetics. In Proceedings of the IEEE/WIC International Conference on Web Intelligence, WI. pp. 235–241 (2003)
24. Lu, L., Liu, D., Zhang, H.J. Automatic mood detection and tracking of music audio signals. IEEE Trans. Audio, Speech, Lang. Process. **14**(1), 5–18 (2006)
25. Yang, Y.H., Lin, Y.C., Cheng, H.T., Liao, I.B., Ho, Y.C. Chen, H.H.: Toward multi-modal music emotion classification. In Advances in Multimedia Information Processing-PCM 2008, Springer Berlin Heidelberg, pp. 70–79 (2008)
26. Sorower, M.S.: A literature survey on algorithms for multi-label learning. Oregon State University, Corvallis (2010)
27. Zhang, M., Zhou, Z. A review on multi-label learning algorithms. 1–1 (2013)
28. Spyromitros, E., Tsoumakas, G. Vlahavas, I.: An empirical study of lazy multilabel classification algorithms. In Artificial Intelligence: Theories, Models and Applications, Springer Berlin Heidelberg, pp. 401–406 (2008)

29. Read, J., Pfahringer, B., Holmes, G., Frank, E.: Classifier chains for multi-label classification. Mach. Learn. **85**(3), 333–359 (2011)
30. Prajapati, P., Thakkar, A., Ganatra, A.A: Survey and current research challenges in multi-label classification methods. Int. J. Soft Comput. 2
31. Fürnkranz, J., Hüllermeier, E., Mencía, E.L., Brinker, K.: Multilabel classification via calibrated label ranking. Mach. Learn. **73**(2), 133–153 (2008)
32. Mencía, E.L., Park, S.H., Fürnkranz. J.: Efficient voting prediction for pairwise multilabel classification. Neurocomput. **73**(7), 1164–1176 (2010)
33. Clare, A., King, R.D.: Knowledge discovery in multi-label phenotype data: In principles of data mining and knowledge discovery, pp. 42–53. Springer, Berlin (2001)
34. Zhang, M.L., Zhou, Z.H.: ML-KNN: a lazy learning approach to multi-label learning. Pattern Recogn. **40**(7), 2038–2048 (2007)
35. Luaces, O., Díez, J., Barranquero, J., Coz, J.J.D., Bahamonde, A.: Binary relevance efficacy for multilabel classification. Prog. Artif. Intell. **1**(4), 303–313 (2012)
36. Emotions Dataset. http://mulan.sourceforge.net/datasets-mlc.html. Accessed 20 August 2014

Visibility Enhancement in a Foggy Road Along with Road Boundary Detection

Dibyasree Das, Kyamelia Roy, Samiran Basak and Sheli Sinha Chaudhury

Abstract Images and videos of outdoor scenes suffer from reduced clarity due to presence of fog/haze/mist, and thus it becomes difficult to drive in bad weather conditions. Several methods have already been proposed to improve the images acquired in foggy weather conditions. In this paper a novel method of dehazing using dark channel prior along with masking the sky regions has been proposed, the output has improved considerably due to clear visibility of separation of surrounding edges from the sky as well as reduced artifacts. Focus on road edge detection has also been emphasized on, in this work along with dehazing leading to prominent visibility in foggy conditions.

Keywords Dark channel prior · Edge detectors · Hough transform · RoadEdge detection

1 Introduction

Presence of haze degrades the outdoor images/videos to a great extent and thus has become a major problem for outdoor driving in bad weather conditions. Hampered visibility makes it difficult for the drivers to identify the road edges and get a clear view of the road while driving. Dehazing has become an important research topic in

---

D. Das (✉) · K. Roy · S. Basak · S. Sinha Chaudhury
Department of Electronics and Telecommunication Engineering,
Jadavpur University, Kolkata, India
e-mail: dibyasree.d@gmail.com

K. Roy
e-mail: kyamelia_rain@yahoo.co.in

S. Basak
e-mail: samiranpk@gmail.com

S. Sinha Chaudhury
e-mail: shelism@rediffmail.com

A. Nagar et al. (eds.), *Proceedings of 3rd International Conference on Advanced Computing, Networking and Informatics*, Smart Innovation, Systems and Technologies 43, DOI 10.1007/978-81-322-2538-6_13

many computer vision based applications such as video surveillance, remote sensing, object recognition, and tracking. The paper focuses on haze removal along with road edge detection from single image and video. Haze is caused due to scattering and absorption of light by tiny air particles known as aerosols in atmosphere before it reaches the camera [3]. The non-deflected scene light reaching the camera together with the light reflected from different direction forms the airlight [2]. This phenomenon fades the true color and contrast of the scene objects. Since haze is dependent on an unknown depth which cannot be measure accurately, dehazing an image completely becomes impossible but however improvement in visibility can be rendered by the various approaches of dehazing and visibility restoration. The various models that have been proposed till date includes, model proposed by Satherley and Oakley [14] and Tan and Oakley [13], assuming that the scene depths were known they formulated a physics-based technique to restore scene color and contrast without using predicted weather information. Narasimhan and Nayar [6] analyzed the color variation in scene objects under the effect of homogeneous haze based on a dichromatic atmospheric scattering model. They considered two images of the same scene taken at different time intervals. Scene contrast recovery using this model is somewhat ambiguous as the color of the haze and the scene points are almost same. Fattal [8] presented a method for estimating the transmission in hazy scenes taking into consideration that the medium transmission function and the surface shading are locally and statistically uncorrelated. The dark channel prior model by He et al. [7] aimed at dehazing a single image based on the outdoor haze free image information, a common drawback of the above two methods is their computational cost and time complexity.

The segmentation part when comes into play gives the idea of the road boundary detection or the lane detection. The existing methods are purely based on the Hough transform followed by some edge detection process. The idea proposed by [9] is vision-based road boundary detection. The optimal path is calculated by Dynamic Programming (DP) and then randomized Hough Transform is applied. In terms of ambiguity, the computational time is quite high because of the Dynamic Programming. In [10] a fast lane detection system is organized using Hough transform and with the 2D filter. Here, image binarization is done separately which is an extra step, in terms of complexity. In [11] edge detection for road boundary is proposed. The filters used are Prewitt and Sobel, which are not much efficient in detecting edges. The original images which are taken are in form of gray image and detected edges are not superimposed on the original one, which would have given a better assistance to the driving system.

The dark channel prior model is effectively used in this paper for real time application with reduced timing complexity. After dehazing, the artifacts in the resulting image present mostly in the sky pixels were removed by masking the sky portion from the image, which resulted in improved output. The proposed model has been extended to video application along with road edge detection using Hough transform in the first case and boundary detection of the edges in the second case and comparatively studied the output images of both the cases. The rest of the paper is organized as follows. In Sect. 2 the proposed method has been described.

Section 3 presents the experimental results on both image and video, a comparison with a few previous methods is also contained. Finally in Sect. 4 proposed model has been summarized.

2 Proposed Method

2.1 *Haze Model*

The two factors which are mainly responsible for formation of haze image are direct contrast attenuation and airlight [2]. Contrast of images are hugely affected due to haze, the attenuation is caused due to scene light passing through the atmosphere consisting aerosols where the light gets scattered in different directions or absorbed [3]. Airlight is caused by scattering of scene light after falling on an air particle which again gets re-scattered or refracted. This repeated scattering restricts it to reach the camera directly. The attenuation in contrast and color suffered by the image depends on the nature of the medium through which the scene light passes and the scene depth of the original image. So the optical model for the formation of hazy image as shown by author [2] can be described as:

$$\mathbf{I(x)} = \mathbf{J(x)}\,\mathbf{t(x)} + \mathbf{A}\,(\mathbf{1} - \mathbf{t(x)}). \tag{1}$$

where I(x) is the original hazy image intensity, x is the pixel index, J(x) is the scene radiance [3]. The haze removal algorithms aim to recover J from I. A is the global atmospheric light and t is the un-deflected scene transmission light that reaches the camera or the observer directly. The transmission t(x) denotes the field depth of the scene objects. In a homogeneous medium the transmission is expressed as shown in [7] as:

$$\mathbf{t(x)} = \mathbf{e}^{-\boldsymbol{\beta}\mathbf{d}}. \tag{2}$$

β is the scattering coefficient of the medium. The above equation indicates that the scene radiance is attenuated exponentially with the scene depth d.

2.1.1 Histogram Stretching for Contrast Enhancement

Most of the haze images suffer from degraded contrast of the scene objects. In this model histogram stretching was done before dehazing in order to shift the pixel values to fill the entire brightness range, resulting in high contrast. The RGB normalized image was converted to HSV, histogram stretching was also applied to the S and V channel before converting it back to RGB.

2.1.2 Dark Channel Prior Model

The dark channel prior model [7] proposed by He et al. [7] is capable of extracting the transmission map of the original haze effected image assuming the airlight is already known. According to this model in haze free outdoor images, the non-sky regions consist of minimum one color channel in which the intensity is very low at some pixels. Thus as stated by the authors in [7], the dark channel value of an image J can be formulated as:

$$\mathbf{J^{dark}(x) = min_{channel \in \{r,g,b\}}\left(min_{y \in \Omega(x)}\left(J^{channel}(y)\right)\right)}. \tag{3}$$

$J^{channel}$ represents the color channel of J and $\mathbf{\Omega}(x)$ is a patch around pixel x. The dark channel value of a haze free image generally tends to zero. This has been represented by the authors in [7] as:

$$\mathbf{J^{dark}(x) \rightarrow 0}. \tag{4}$$

The color of the most haze opaque pixels in I was taken as A. After estimating A, the transmission map t(x) was derived by the authors in [7] as:

$$\mathbf{t(x) = 1 - w\,min_{channel \in \Omega(x)}\left[min_{channel} I^{channel}(y)/A^{channel}\right]}. \tag{5}$$

where w is a constant parameter used to keep some negligible amount of haze in distant objects to make the images look real, the value of w varies from 0 to 1 [4]. Several techniques have tried to refine the transmission using laplacian matting, bilateral filtering, guided filters etc., however these steps though helped in providing better results, it takes a lot of time to process and thus has been avoided here. Finally from Eq. (1) the dehazed image can be recovered as:

$$J(x) = \frac{I(X) - A}{\mathbf{t(x)}} + A. \tag{6}$$

2.1.3 Masking Sky Patches for Removing Artifacts and Preserving the Haze Covered Surrounding Edges

It is observed that the resultant dehazed image obtained consist of a lot of artifacts present mostly in the sky patches, moreover the separation of edges of scene objects closer to the sky is not clear. Since sky is mostly blue the pixels for the sky were selected by picking values in the blue plane that are very high. The primary sky pixels were selected by keeping a threshold > 200 and a mask was applied to each color plane setting the mask pixels to a maximum value of 255. The resultant image obtained was visibly clearer and contained less artifacts. However the masking is only effective in daylight.

2.2 Edge Detection

Filters are used in the process of identifying image by locating the sharp edges which are discontinuous. The discontinuities are basically the changed pixel intensity values which give the boundary of the image. An image has a non-zero value for the gradient corresponding to many points. Among these points, all of them do not belong to an edge for some specific application. Thus, few methods are used which define or specify the existing edge points. Recurrently, detection is achieved by the thresholding criterion provided [10]. Image segmentation comprises edge detection as its first step, has fueled an exhaustive search for a good edge detection algorithm.

2.3 Edge Detection Methods

There are many edge detection techniques. Amongst them the most frequently used are:

(1) Roberts edge detection [12]
(2) Sobel edge detection [12]
(3) Prewitt edge detection [12]
(4) Canny edge detection [12]

A comparative study was performed using all the four edge detectors on the video frames as shown in Fig. 1. The Canny edge detector smoothes the image with the Gaussian filter. The edges are detected and linked, applying non-maxima suppression to the gradient magnitude. Evidently, it is noticeable in the Fig. 1 the result produced by the Canny edge detector is the best amongst the remaining detectors used here.

2.3.1 Road Boundary Detection

The separation and location of objects in images as well as in video is rendered with segmentation as an essential process in image processing. MATLAB offers tool for

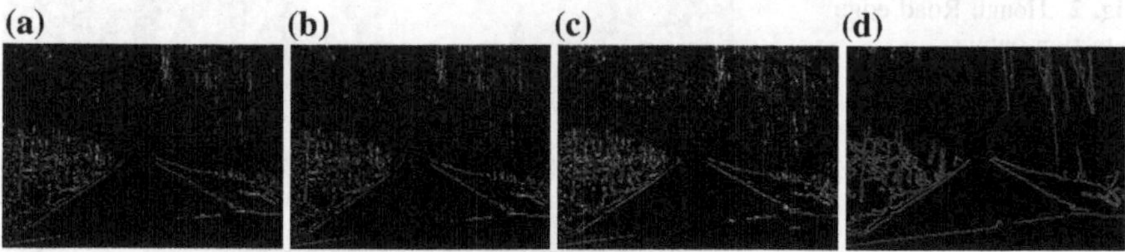

Fig. 1 Edge detection by **a** Sobel. **b** Prewitt. **c** Roberts. **d** Canny

image processing. MATLAB offers tool for image processing, however detecting a specific object or segmenting the specific edge is non-trivial. In this paper, two methods are used for detecting the road edges. The first method used is Hough transform and the second one is boundary detection technique with the MATLAB function.

(a) Hough Transform: In image processing applications and computer vision algorithms, Hough transform is used as a toolto extract image features. The classical Hough transform [12] identifies and detects the lines in an image. Detecting straight lines is the simplest case in Hough transform. The straight line is represented as y = mx + b, where m representing the slope and b is the intercept. The equation of the line represented in the normal form [12] as:

$$r = x\cos\ominus + y\sin\ominus \tag{7}$$

where, r = the geometric distance between the line and the origin

θ = the angle formed by the orthogonal vector directed towards the line in the first quadrant of the geometric plane.

The linear Hough transform algorithm is used to detect the presence of a line in an image. A two dimensional array is used along with an accumulator. Accumulator's dimension is same as the number of unknown parameters i.e. r and θ in pair (r, θ). The Hough transform algorithm applied at each pixel at (x, y) and its neighborhood seeks for enough evidence to judge the presence of a straight line. If the search result is positive then the parameters r and θ of the line are calculated and then the bin of the accumulator is looked for putting the values, thereby incrementing the value of the bin. The local maxima in the accumulator space is approximated by the bin's maximum value, which in term gives the lines to be extracted. Finally, the linear Hough transform [12] results in a 2D matrix containing the quantized angle. θ and quantized distance r. After, the Hough transform is performed on the video frame the resultant edge detected output images is shown by blue solid lines in Fig. 2. One of the edges of the road is detected very correctly but next to that many lines has been produced which do not show any edge. The wrongly detected edges have been highlighted by red circles. Another edge is present, which is though an edge not the road edge but a snow line. In Fig. 3 a plot of Hough peaks is shown which is a plot of r versus θ.

Fig. 2 Hough Road edge detection output

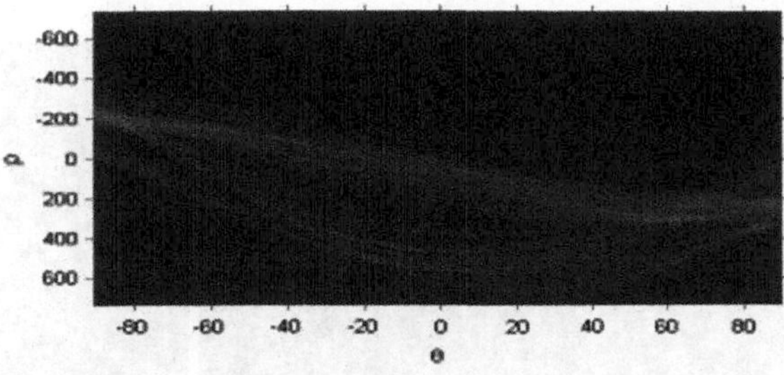

Fig. 3 Plot of r versus θ

(b) Tracing Boundary: Boundary tracing method gives better result in comparison to the previous technique. In the first method with Hough transform the edges of the road are detected but it segmented the strong edge of the road as well as any strong edge present in the video frame. The result obtained with Hough transform as shown in Fig. 4a gives a distinguishable mark in the road edge detection where there are bends present in the road lines. The road bends are not detected, as well as in the crossing points of the road, the edge line is extended through the path and gives a wrong detection. Intuitively, it makes sense to focus on the boundary detection method. The boundary detection method gives visibly good results as shown in Fig. 4b, segmenting the desired road edges, the bushes if present and more remarkably the bends of the roads. This method is also able to detect any obstacle or any other vehicle present in the road as highlighted with red box in Fig. 5b, thus, helps in avoiding accidents. The image to be traced has to be a binary image where non-zero pixel belongs to an object and the background is constituted with the 0-pixel. After the boundaries are detected they are superimposed on the original image and the output is obtained. The image to be processed has to be logical or numeric and it must be real with 2-D and non-sparse.

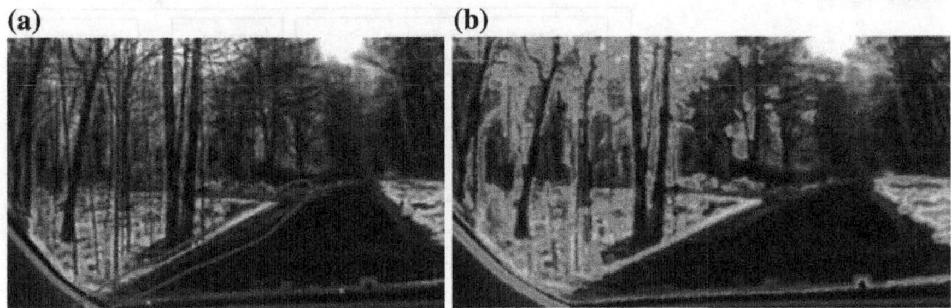

Fig. 4 Output after applying. **a** Hough transform and **b** Boundary detection methods

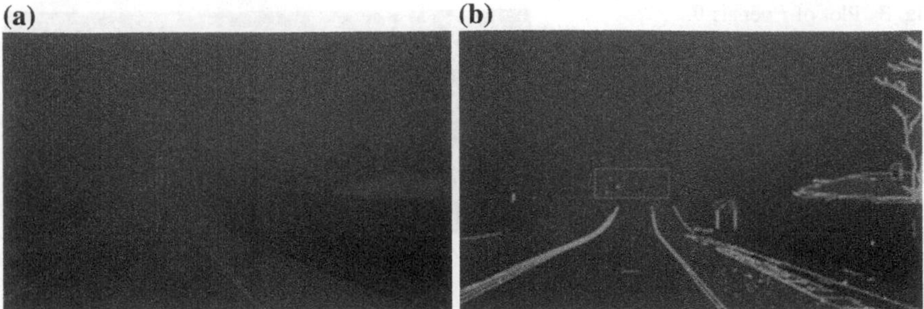

Fig. 5 Sample image showing vehicle as well as road boundary detection. **a** Original image. **b** Output image

2.4 Extension to Video Application

The proposed model can be easily extended to video application. For a video we have also included road boundary detection along with dehazing the frames for better driving during bad weather conditions. The video that is captured is separated into several frames. The segmentation technique is then applied on each of the frames. The segmentation procedure encompasses the Hough transform as one method and boundary detection as other, along with the edge detection. The method can be applied for real time video applications as well, as the proposed model gives satisfactory results within a suitable time period. The complete model has been presented with the help of a block diagram in Fig. 6.

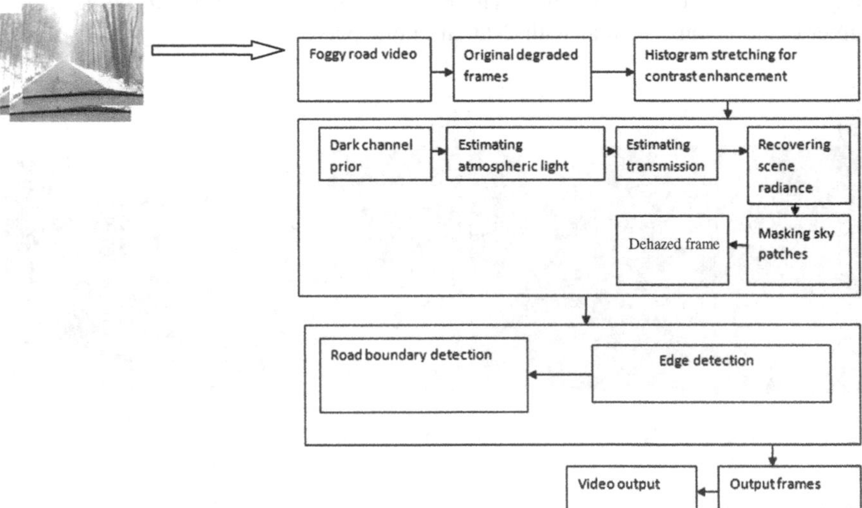

Fig. 6 Block diagram of proposed model

3 Experimental Results

The applied dehazing method has been shown step wise in Fig. 7 where (a) refers to the original hazy image, (b) shows the resultant after applying dark channel prior as discussed in Eq. (4), (c) shows the estimated transmission as discussed in Eq. (5), (d) shows the recovered scene radiance and finally in (e) we have shown the resultant dehazed image after masking.

A comparison of the dehazed outputs using techniques by different authors along with the proposed technique has been shown in Figs. 8 and 9.

The final results are shown in Fig. 10. The respective steps of road edge detection are performed and adjacently represented in Fig. 11.

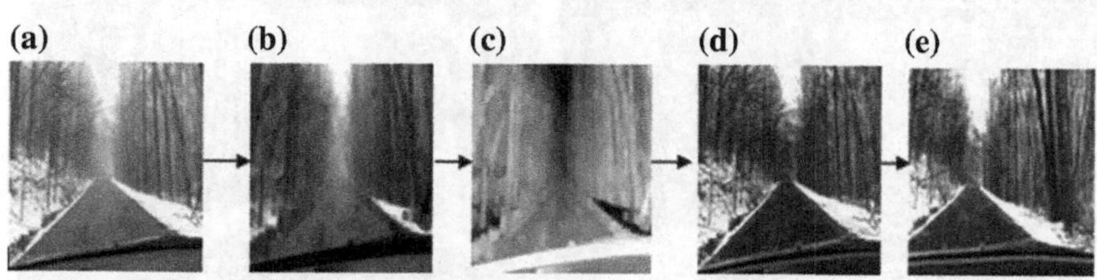

Fig. 7 Dehazing performed on a single frame of a video

Fig. 8 **a** Original hazy image, **b** dehazing by Fattal [1], **c** by He et al. [1] **d** by Kopf et al. [1] **e** by Tan [1] **f** proposed method

Fig. 9 **a** Original hazyimage **b** Dehazing by Tarel et al. **c** Proposed method output

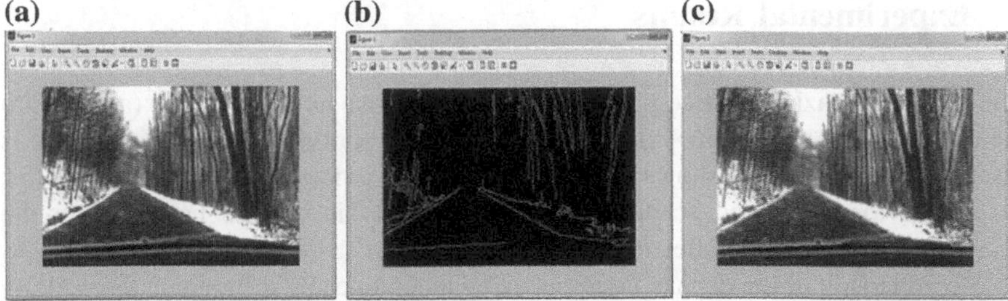

Fig. 10 **a** Original dehazed image. **b** Edge detector output. **c** Final output

Fig. 11 Screenshot of foggy video and final dehazed video output with road edge detection

Total time taken for processing each frame is around 8.3 s or less. On measuring the contrast to noise ratio a lot of improvement was observed on the dehazed frame, the ratio of the original frame is 59.3005 whereas that of the dehazed frame is 94.5843, thus the contrast of the resultant frame has significantly improved as obvious from the result.

4 Conclusion

In this paper the technique proposed is a novel method for dehazing video, frame by frame and applying road boundary detection for improving the visibility of road while driving in bad weather conditions. As the sky patches are unimportant for a driver, we have completely masked it, which has not only resulted in an enhanced image but has also reduced artifacts along the surrounding edges. The final output is comparable with the existing techniques and quite suitable for real time video processing as the time taken for processing each frame is very less. Moreover the added feature of road edge detection along with on-road vehicle recognition makes it a lot easier for a driver to drive in bad weather conditions. However in case of dense fog on applying the dark channel prior model the resultant image turned dark, though the road edge detection worked perfectly thus the dehazing was not satisfactory. Thus the future work of this model will stress on improving visibility in case of denser fog.

References

1. Tarel, J.-P., Hautiere, N.: Fast visibility restoration from a single color or gray level image, In: IEEE 12th international conference on Computer Vision, pp. 2201–2208 (2009)
2. Koschmieder, H.: Theorie der horizontalen sichtweite, Beitr.Phys.Freien Atm., **12**, 171–181 (1924)
3. Lv, X., Chen, W., Shen, I.F.: Real-time dehazing for image and video. In: 18th Pacific Conference on Computer Graphics and Applications (PG), pp. 62–69, Sept (2010)
4. Yeh, C.: Kang, Li-Wei, Lee, M., Lin, C.,: Haze Effect Removal from Image via Haze Density estimation in Optical Model. Opt. Express **21**(22), 27127–27141 (2013)
5. Yeh, C., Kang, L.-W., Lee, M., Lin, C.: Efficient image/video dehazing through haze density analysis based on pixel-based dark channel prior. In: International Conference on Information Security and Intelligent Control, August (2012)
6. Narasimhan, S.G., Nayar, S.K.: Contrast restoration of weather degraded images. IEEE Trans. Pattern Anal. Mach. Intell. **25**(6), 713–724 (2003)
7. He, K., Sun, J., Tang, X.: Single image haze removal using dark channel prior. In: IEEE Conference on Computer Vision and Pattern Recognition, pp. 1956–1963. Miami (2009)
8. R., Fattal: Single image dehazing. In: Proceeding ACM SIGGRAPH 2008 papers, Article No. 72, vol. 27, Issue 3, Aug 2008
9. Lin, H., Kim, H., Lin, C., Chua, L.O.: Road boundary detection based on the dynamic programming and the randomized hough transform. In: International symposium on Information Technology Convergence, pp. 63–67. IEEE (2007)
10. Joshy, N., Jose, D.: Improved detection and tracking of lane marking using hough transform. IJCSMC **3**(8), 507–513 (2014)
11. Routray, A., Mohanty, K.B.: A fast edge detection algorithm for road boundary extraction under nonuniform light condition. In: 10th International Conference on Information Technology, pp. 38–40. 17–20 Dec 2007
12. Gonzalez, R.C., Woods, R.: Digital Image Processing Book. Third Edition, Pearson Education India (2009)
13. Tan, K., Oakley, J.P,: Physics based approach to color image enhancement in poor visibility conditions. J. Optical Soc. Am. A **18**(10), 2460–2467 (2001)
14. Oakley, J.P, Satherley, B.L,: Improving image quality in poor visibility conditions using a physical model for degradation. In: IEEE Transactions Image Processing, vol. 7, Feb (1998)

Artificial Neural Network Based Prediction Techniques for Torch Current Deviation to Produce Defect-Free Welds in GTAW Using IR Thermography

N.M. Nandhitha

Abstract In recent years, on-line weld monitoring is the potential area of research. In this work, torch current deviation prediction systems are developed with Artificial Neural Networks to produce welds free from Lack of Penetration. Lack of penetration is deliberately introduced by varying the torch **current**. Thermographs are acquired during welding and hotspots are extracted using Euclidean Distance based segmentation and are quantitatively characterized using the second order central moments. Exemplars are then created with central moments as input parameters and deviation in torch current as the output parameter. Radial Basis Networks (RBN) and Generalized Regressive Neural Networks (GRNN) are then trained and tested to assess the suitability for torch current prediction. GRNN outperforms RBN in predicting the torch current deviation with 98.95 % accuracy.

Keywords GTAW · Lack of penetration · RBN · GRNN · Torch current deviation

1 Introduction

Welding is defined as the process of joining metals in industries. With the advent of automated welding, large number of weld pieces is produced within a short span of time. In spite of accurate parameter settings, defects do occur in welds. Hence these welds are sent to strict quality assessment before dispatched to the end users. Welds that do not satisfy the standards specified by American Society of Mechanical Engineers (ASME) are rejected. Rejection of weld pieces result in loss of time, money and manpower. Hence online weld monitoring is a potential area of research in recent years.

N.M. Nandhitha (✉)
Deparment of Electronics and Communications Engineering,
Sathyabama University, Chennai 600 119, India
e-mail: nandhi_n_m@yahoo.co.in

A. Nagar et al. (eds.), *Proceedings of 3rd International Conference on Advanced Computing, Networking and Informatics*, Smart Innovation, Systems and Technologies 43, DOI 10.1007/978-81-322-2538-6_14

On-line weld monitoring involves the following: Selection of a suitable Non Destructive Testing Technique (NDT) that uses sensors for capturing the defects in welds during welding, image segmentation techniques that accurately isolate the weld defects, appropriate features that exactly represent the defects and suitable non-linear systems for predicting the deviations in weld parameters. In this work, InfraRed Thermography is used as a NDT technique for on-line weld monitoring. It uses an IR camera that captures the heat patterns and maps it into thermographs. This choice is justified because of the following reasons: In Gas Tungsten Arc Welding (GTAW), the arc is formed between the metal electrode (Tungsten) and the weld plate. Heat produced by the arc melts the plate and forms a pool which in turn forms the weld. In nutshell, heat produced at the weld pool mostly determines the quality of the weld. Heat transferred into the weld pool is in turn dependent on the torch current, torch speed etc. Hence heat can be used to assess the weld quality.

This paper is organized as follows: Related work is described Sect. 2. Section 3 provides the proposed methodology. Results are discussed in Sect. 4. Conclusions and future directions are provided in Sect. 5.

2 Related Work

Considerable research is carried out in the area of weld pool monitoring using thermal cameras. Also suitable signal and image processing techniques for analyzing thermal signatures obtained from both active and passive thermography are also cited in the literature. Sreedhar et al. [1] inferred that thermal analysis of weld pool has distinct features for defective and defect free welds. Vasudevan et al. [2] developed a computer controlled GTA machine by using Infrared thermography for sensing the characteristics of the weld pool. Leksir et al. [3] proposed an on-line weld quality monitoring system for Submerged Arc Welding (SAM). In the proposed technique, weld plates are in motion whereas the welding equipment is stationary. Weld pool temperature is used as an indicator for weld quality assessment. Fuzzy logic is then used to assess the weld quality as poor or fair or good. Swiderski and Hlosta [4] used pulsed eddy current stimulated thermography for the assessment of joints' quality in metal sheets. It was concluded that thermal contrast is better than vibrothermography. Aitor Garcia De La Yedra et al. [5] inferred that Discrete Fourier Transform (DFT), Discrete Wavelet Transform (DWT) and Thermographic Signal Reconstruction (TSR) results in enhanced weld quality assessment from thermal images. In all these literatures, it is found that the research has not proceeded to predicting the deviations in the physical parameters responsible for the defect. Hence in this paper, torch current deviation prediction system is developed using Radial bases. In order to train the network, input features are obtained from the descriptors used for representing the hotspots in weld thermographs.

3 Proposed Methodology

Of the various defects that occur in GTAW, Lack of Penetration is a very serious defect and results in immediate rejection of the weld pieces [6]. According to American Society of Mechanical Engineers (ASME), Lack of penetration is not permitted except when are of shallow and very short lengths [7]. Hence in this work, on-line weld monitoring system for Lack of Penetration is proposed using industrial Infrared Thermography is proposed.

3.1 Image Acquisition and Preprocessing

Lack of Penetration is deliberately introduced during by reducing the torch current from its optimal value. All the other physical parameters are kept constant throughout the experiment. Thermal videos are obtained using IR camera (during welding) and are stored as ".avi" files. As the acquired thermographs are videos, initially frames are extracted from these files. Also as the first 100 thermographs do not depict welding, they are not considered for segmentation and further processing. From the manual interpretation of the thermographs, it is concluded that the size and shape of the hotspot varies with that of the torch current. Figure 1 shows the hotspot variation for thermographs acquired with torch current of 70, 80 and 90 A (90 A is the optimal current). In the pseudocolouring of thermographs, hotspot is represented with yellow color. From the manual interpretation, it is found that as the torch current increases to the optimal value, the size of the hotspot also increases.

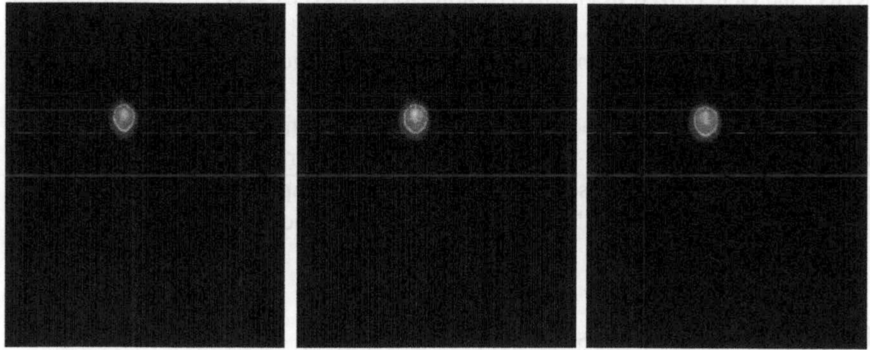

Fig. 1 Thermographs (frame 162) depicting the weld pool acquired with torch current of 70, 80 and 90 A

3.2 Image Segmentation for Feature Extraction

Of the various image segmentation techniques, Euclidean distance based image segmentation is used for extracting the hotspot. It is chosen as it does not involve the overhead of converting a pseudocolor thermograph into gray scale thermograph. Also the quantization error involved in color to gray scale conversion can be avoided. Euclidean Distance based segmentation accurately isolates the defect region (i.e. to the true size of the abnormality) and completely removes the undesirable regions from the weld thermographs. In Euclidean distance based segmentation, the Euclidean distance is calculated between the average intensities (Red, Green, Blue domains) of the hotspot and all the pixels (corresponding domains) in the thermograph. An output image is then obtained by retaining the pixels with Euclidean distance less than the threshold. Intensities of the other pixels are made as zero [8, 9]. Once the hotspot region is isolated, the hotspot is represented in terms of central moments. Central moments are chosen because they are shift and translation invariant. Exemplars are generated with frame number and the central moments as input parameters and torch current deviation as the output parameter. The next task is to develop a non-linear predicting system that predicts the deviation in the physical parameter responsible for the defect. As the output parameter is also available for training the network, supervised neural networks are chosen.

3.3 ANN Based Classifiers

Initially Back Propagation Network was used for current deviation prediction. In this paper, feasibility of other supervised learning algorithms is studied. Two most commonly used networks that use radial functions are Radial Basis Networks (RBN) and Generalized Regressive Neural Networks (GRNN) [10]. Performance of these networks is dependent on the choice of the spread functions. Spread function should neither be large nor be less. Larger spread function results in fast but abrupt convergence while smaller spread function results in slow but accurate convergence. In this work, accuracy of prediction is the major concern. As the neural network predictors are pre-trained with the exemplar set and only the weight updated neural network is used for prediction, time complexity is not a major issue. The spread function for RBN and GRNN is 0.0001 and 0.7 respectively.

4 Results and Discussion

A set of 11439 exemplars were created of which two different sets of 5719 exemplars are used for training and testing. Performance of the network is shown in Table 1. From the Table 1, it is found that both RBN and GRNN results in accurate

Table 1 Performance evaluation of RBN and GRNN based predictors for torch current deviation

Desired deviation in torch current	Predicted deviation in torch current	
	From RBN	From GRNN
0.4	0.4	0.4
0.35	0.35	0.35
0.25	0.25	0.25
0.25	0.25	0.25
0.2	0.2	0.2

prediction for the shown set of exemplars. However there are deviations between few sets of actual and desired values in both the classifiers. It is reflected in the calculation of accuracy. It is found that GRNN has an overall accuracy of 98.95 % while RBN has only 86.71 %.

5 Conclusion

In this paper, feasibility of RBN and GRNN for the prediction of torch current deviation is studied for on-line weld monitoring. The performance is compared to that BPN based predictor. It is found that GRNN results in a better accuracy than both RBN and BPN for the test dataset that resembles the trained dataset. However the accuracy of RBN and GRNN for a different set if input parameters are only 13.64 and 25 % respectively. In order to improve the performance of the network, it is necessary to consider the shape and Fourier descriptors of the hotspot.

References

1. Sreedhar, U., Krishnamurthy, C.V., Balasubramaniam, K., Raghupathy V.D., Ravisankar S.: Automatic defect identification using thermal image analysis for online weld quality monitoring. J. Mater. Process. Tech. **212**(7), 1557–1566 (2012)
2. Vasudevan, M., Chandrasekhar, N., Maduraimuthu, V., Bhaduri, A.K., Raj, B.: Real-time monitoring of weld pool during GTAW using infra-red thermography and analysis of infra red thermal images. Weld. World **55**(7–8), 83–89 (2012)
3. Leksir, Y.L.D., Bouhouche, S., Boucherit, M.S., Bast, J.: Submerged arc welding online quality evaluation using infrared thermography based fuzzy reasoning. In: 13th International Symposium on Nondestructive Characterization of Materials (2013)
4. Swiderski, W., Hlosta, P.: Pulsed eddy current thermography for defects detection in joints of metal sheets. In: 11th European Conference on Non-Destructive Testing (2014)
5. De La Yedra, A.G., Echeverria, A., Beizama, A., Fuente, R., Fernández, E.: Infrared thermography as an alternative to traditional weld inspection methods thanks to signal processing techniques. In: 11th European Conference on Non-Destructive Testing (2014)
6. Lancaster, J.: Handbook of Structural Welding, Processes, Materials and methods used in the Welding of Major Structures, pipelines and process plants, vol. 260. Abington Publishing, (1997)

7. Halmshaw, R.: Industrial Radiology, Theory and Practice, vol. 230. Chapman & Hall Publications, London (1995)
8. Gonzalez, R.C., Woods, R.E.: Digital Image Processing. Prentice Hall, New Delhi (2005)
9. Selvarasu, N., Nachiappan, A., Nandhitha, N.M.: Abnormality detection from medical thermographs in human using Euclidean distance based color image segmentation. In: Proceedings of 2010 International Conference on Signal Acquisition and Processing, pp. 73–75, (2010)
10. Freeman, J.A., Skapura, D.M.: Neural Networks Algorithms. Applications and Programming Techniques, Pearson Education (1997)

A Study on Cloud Based SOA Suite for Electronic Healthcare Records Integration

Sreekanth Rallapalli and R.R. Gondkar

Abstract In order to exchange healthcare information reliability, security and cost-effectiveness are three important factors where healthcare industry has to focus upon. Versatile platforms for enterprise-wide information sharing are needed for payers and providers. To provide quality care to patient's accurate information through networks should be required for clinicians and integrated information for the business operation is needed for administrators. Information access from various systems like innovation, performance improvement, demand, monetary and many such other systems is required from both the sides of organization. These organizations must share data externally from application data, medical records of all patients, medicinal data, chemistry reports and symptomatic information from various mediator bodies and strictly following the policies and procedures for storing, modifying, disseminating the electronic health records (EHR). In this era most of the Health care industries are moving to cloud services for processing and storing the healthcare data. There is a need to build cloud based SOA suite for EHR integration. This SOA suite will help healthcare organizations with widespread integrated capabilities and a fused middleware platform. In this paper we study a cloud based SOA suite for EHR integration by empowering the transparent, extensible, protected methods for distributing real and secured information only for intended recipients.

Keywords EHR · SOA · Healthcare · Networks · Cloud

S. Rallapalli (✉)
R&D Centre, Bharathiyar University, Coimbatore, India
e-mail: rsreekanth1@yahoo.com

R.R. Gondkar
AIT, Bangalore, India
e-mail: rrgondkar@gmail.com

A. Nagar et al. (eds.), *Proceedings of 3rd International Conference on Advanced Computing, Networking and Informatics*, Smart Innovation, Systems and Technologies 43, DOI 10.1007/978-81-322-2538-6_15

1 Introduction

One of the challenging tasks in the healthcare organizations is to exchange the data within the internal and external partners and also with government agencies [1, 2]. Creating unique or custom-built results for every business challenge is implemented by most of organization due to lack of effective strategies. This type of applications is leading to heterogeneous environment and involves too much cost to operate. As per Gartner a leading survey organization large and mid size healthcare organizations will have to spend additionally on integrating the healthcare applications. The complexity in exchanging the data in health care organizations is all about the stringent regulations like HIPAA, PHIN, and NHIN [3]. These regulations say how the data is formatted, archived and exchanged throughout. Assistance is required for health care organizations to implement EHR systems that improve the health care data transaction and implement guidelines to assure the confidentiality and safety. The benefits of implementing EHR includes in improving the quality and convenience for patient care. It also helps to improve the accuracy of diagnoses and health outcomes. EHR minimizes the medical errors and increase the security and efficiency of health care administration [4]. It also minimizes the difficulty in integrating backend information systems.

1.1 Cloud Computing

Service providers who offer various services in a single place to meet the expectations of the customer use Cloud computing as a platform [5]. Cloud computing provides convenient way of using the services but there are security [6] concerns like authenticity and confidentiality of the data [7].

1.2 Service Oriented Architecture (SOA)

Any architecture deployed on cloud should provide reusable services, platform independent services from loosely-integrated suite which is possible with Service Oriented Architecture (SOA) [5, 8]. SOA supports business integration where in it is possible to integrate EHR data of healthcare industries.

2 Middleware Platform

A Cloud has many advantages. Its resources are held in house if it is private cloud. Security, reliability and networks can be provided as per the customer requirement. It is necessary to provide middleware (Fig. 1) and its functions have to be expanded

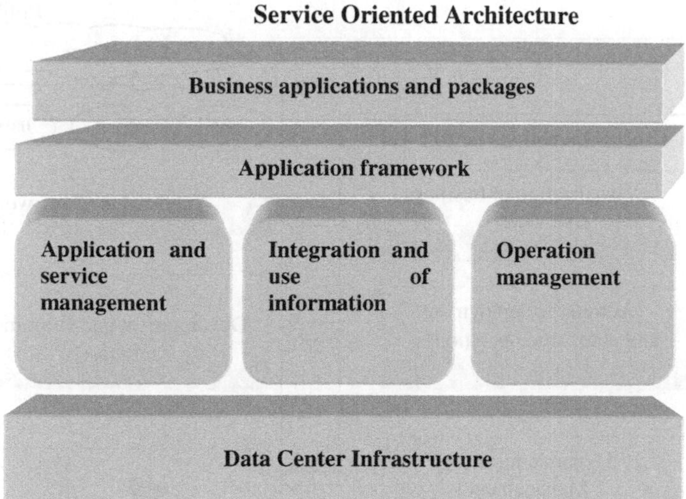

Fig. 1 Middleware platform

in each of the infrastructure and application layers on cloud. The cloud SOA suite for health care integration is a component of middleware environment intended to evaluate data points, link applications, and fulfill various challenges of the today data based organizations which are highly controlled [9]. The healthcare lifecycle is streamlined starting from initial stage of initiation and till reporting.

In order to share internal and external information between the health care organizations and other providers like insurance carriers, a middleware tool can provide an interface for the same [10, 11]. In order to adhere the standards set by the healthcare industry such as HL7 and information templates like National Information Exchange Model (NIEM) which can easily reply to the citizens and meet all the requirements set by the government.

3 Middleware Architecture for Cloud

The middleware architecture for cloud is shown in Fig. 2. Application servers which form the hardware on which the applications and software packages run. Execution platform is provided by this middleware [9]. Roles of cloud operations middleware, which manages and control the layers shown in the Fig. 2, are explained below.

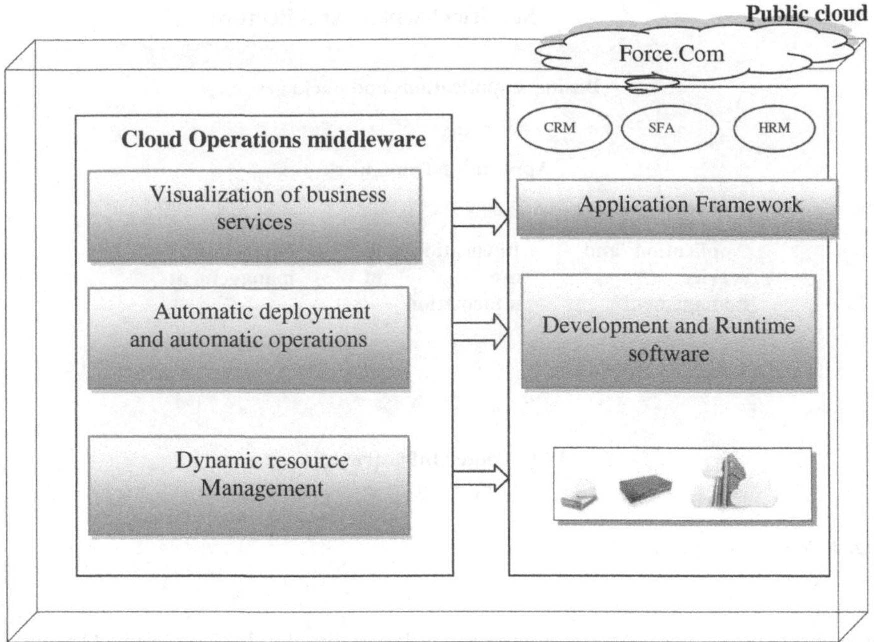

Fig. 2 Middleware architecture for cloud

3.1 Dynamic Resource Management

Data center and infrastructure regions use this middleware for managing the hardware resources which consists of servers, storage and networks.

3.2 Automatic Deployment and Automatic Operation

Operation management region use this middleware for automating the deployment, setting and operations of the software stack.

3.3 Visualization of Business Services

Application and service management region use this middleware for visualizing individual business service units.

3.4 Development and Execution Software

Application and service management region, information integration and utilization region use this middleware.

4 Cloud SOA Suite for Electronic Healthcare Integration

In this section Cloud SOA suite for EHR Integration can be studied and analyzed for its optimal usage. Messaging and data collaboration are the two important aspects of healthcare IT operations. SOA Suite for healthcare integration[1] simplifies these aspects and its goal can be accomplished by the infrastructure for combined applications and organization information policies. Simplified dashboards, advanced monitoring tools are added with this domain-specific application. In order to track the files and other information without much difficulty dashboards can help throughout the lifecycle for administrators and clinicians in healthcare organizations. In order to manage business applications reusable components along with SOA Suite, a middleware application helps organizations in planning and deployment. By enforcing the standards such as HL7 and X12N will facilitate the association between organization health systems and by enforcing certain standards for messaging. Any Service oriented architecture suite should be supported by data exchange standards. Cloud based SOA suite also follows the same protocol standards such as TCP/IP and MLLP, and AS2. All these standards provide the regulations for the internet data transfer.

All the stakeholders involved in Health care organization can utilize the capabilities provided by the cloud SOA suite for healthcare. The information can also be exchanged via various devices which have the capability to connect to the cloud services. In order to share information and promote quality at low cost the suite connects the information systems on the cloud. In a cloud environment a general-purpose middleware will simplify the exchange of information.

Cloud based SOA suite can be scalable in line with the property of cloud scalable feature which can spread across multiple business needs. This feature allows to import and export the information for any purpose of visits to the healthcare organizations and in case of emergencies. Figure 3 shows the cloud SOA suite for electronic health records integration.

The SOA architecture provided to the healthcare organizations will have better interaction with other healthcare providers in terms of exchange of clinical information. Healthcare organizations can also cut down the costs.

Cloud SOA for healthcare organizations can have many benefits [12].

[1]Oracle SOA suite for healthcare integration connecting clinical and administrative process with SOA.

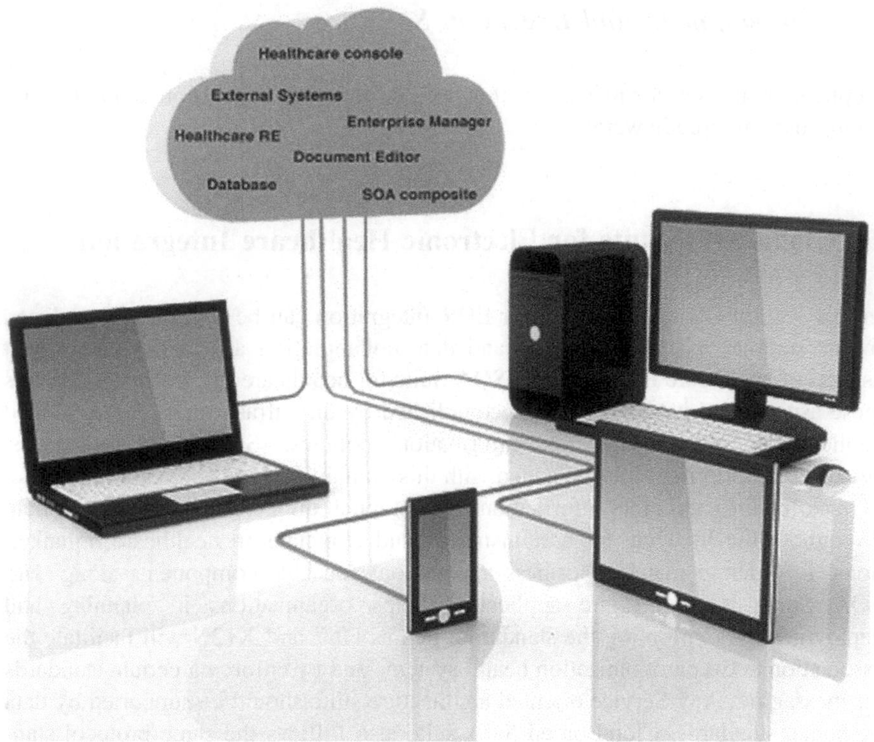

Fig. 3 Cloud SOA suite for electronic health records integration

4.1 Platform Integration

To integrate different platforms across cloud computing it is necessary that we use XML messages between different services. By implementing the XML data the integration of platforms such as PHP and Java can be achieved. So External systems component in cloud SOA can be used to get messages from different platforms.

4.2 Service Combination

SOA contains different parts like Healthcare console, Document editor, Database, Healthcare RE and Enterprise manager which are well defined and function independently to provide various services.

4.3 *Reusability*

Services provided by SOA can be reused at large without major changes or modifications.

4.4 *Security*

Cloud SOA provides various security measures while exchanging the health care information among various users. By providing the authentication like username passwords and encrypting the information exchange between the client data and server data security among the healthcare providers can be achieved. Security features provided by cloud SOA can ensure that the data is received for only the intended recipients.

5 Conclusion

A reliable SOA infrastructure is required for SOA suite in order to integrate the Healthcare records. The applications can share the information constantly to leverage essential business processes for healthcare integration. Cloud computing will ensure to provide scalable infrastructure like hardware, software and any healthcare applications. SOA make sure that it delivers the software as a service to all healthcare providers. By implementing cloud SOA healthcare organizations can minimize the security risk involved in exchange of the information. In this paper middleware platform and middleware architecture for cloud is studied. Also the SOA suite required for Health industry is studied. Finally the Cloud based SOA Suite for healthcare integration is proposed using the existing SOA suite for Electronic health records integration.

References

1. Andrei, T.: Cloud computing challenges and related security issues. http://www.cse.wustl.edu/~jain/cse571-09/ftp/cloud/
2. Yu, W.D., Jothiram, V.: Security in wireless mobile technology for healthcare systems. In: Proceedings of the 9th IEEE International Conference on e-Health Networking, Applications and Services, pp. 308–311. Taipei, 19–22 June 2007
3. Buecker, A., Ashley, P., Borrett, M., Lu, M. Muppidi, S., Readshaw, N.: Understanding SOA security, design and implementation. http://www.redbooks.ibm.com/redbooks/pdfs/sg247310.pdf (2007, November). Accessed 24 March 2010
4. Axel, B., Paul, A., Martin, B., Ming, L., Sridhar, P., Neil, R.: Understanding SOA security design and implementation. http://www.redbooks.ibm.com/redbooks/pdfs/sg247310.pdf

5. Yu, W.D., Joshi, B., Chandola, P.: A service modeling approach to service requirements in SOA and cloud computing—using a u-healthcare system case. In: 2011 IEEE international conference on e-health networking, applications and services
6. Brodkins, J.: Gartner: seven cloud-computing security risks. www.informworld.com (2009). 2 July 2009
7. Marcia, S.: Security challenges with cloud computing services. http://searchsecurity.techtarget.com/news/article/0,289142,sid14_gci1368905,00.htm (2010). Accessed March 2010
8. Yu, W.D., Ramani, A.: Design and implementation of a personal mobile medical assistant. J. Inf. Technol. Healthc. **4**(2), 92–102 (2006)
9. Nagakura, H.: Middleware for creating private clouds. Fujitsu Sci. Tech. J. **47**(3), 263–269
10. Arsanjani, A., Ghosh, S., Allam, A., Abdollah, T., Ganapathy, S., Holley, K.: SOMA: a method for developing service-oriented solutions. http://www.research.ibm.com/journal/sj/473/arsanjani.html (2008, August 6). Accessed 24 March 2010
11. Azure—Part 1—Introduction http://geekswithblogs.net/shaunxu/archive/2010/02/24/azure—part-1–introduction.aspx
12. Mulholland, A., Daniels, R., Hall, T.: The Cloud and SOA: Creating Architecture for Today and for the Future: HP, 2008

Hap: Protecting the Apache Hadoop Clusters with Hadoop Authentication Process Using Kerberos

V. Valliyappan and Parminder Singh

Abstract Hadoop is a disseminated framework that gives an appropriated file system and MapReduce group employment handling on vast bunches utilizing merchandise servers. Despite the fact that Hadoop is utilized on private groups behind an association's firewalls, Hadoop is regularly given as an issue multi-inhabitant administration and is utilized to store delicate information; as an issue, solid validation and approval is important to ensure private information. Adding security to Hadoop is testing in light of the fact that all the co operations don't take after the fantastic client server design: the record framework is parceled and disseminated obliging approval checks at different focuses; a submitted bunch employment is executed at a later time on hubs not the same as the hub on which the customer verified and submitted the employment; occupation assignments from distinctive clients are executed on the same register hub; In this paper to address these difficulties, the base Kerberos confirmation system is supplemented by assignment and ability like access tokens and the idea of trust for optional administrations.

Keywords MapReduce · HDFS · ACL · GPS · Kerberos · RPC · Lucene · Nutch · Oozie · Setuid

1 Introduction

While the open source schema has empowered the foot shaped impression of Hadoop to coherently stretch, endeavor associations face arrangement and administration challenges with enormous information. Hadoop's center determinations are

V. Valliyappan (✉) · P. Singh
School of Computer Engineering, Lovely Professional University,
Jalandhar, Punjab 144411, India
e-mail: valliyappan.vaiyapuri@outlook.com

P. Singh
e-mail: parminder.16479@lpu.co.in

A. Nagar et al. (eds.), *Proceedings of 3rd International Conference on Advanced Computing, Networking and Informatics*, Smart Innovation, Systems and Technologies 43, DOI 10.1007/978-81-322-2538-6_16

as of now being created by the Apache group and, up to this point, don't sufficiently address venture necessities for hearty security, arrangement authorization, and administrative agreeability. While Hadoop may have its difficulties, its approach, which takes into account the appropriated preparing of extensive information sets crosswise over groups of machines, speaks to the fate of big business registering [1]. Hadoop and comparative NoSQL information stores empower any association, huge or little, to gather, oversee and investigate huge information sets, yet these beginning innovations were not outlined on account of extensive security. The reaction from information security sellers, who give answers for conventional organized databases, has been to alter their current off the-rack items to secure the bunch [2] environment. However well-meaning these autonomous methodologies may be, each one fails to offer a complete furthermore centered security answer for Hadoop. Just another approach that addresses the one of a kind structural planning of appropriated registering can meet the security prerequisites of the venture server farm and the Hadoop group environment [1]. This paper audits the security holes that exist in all open source Hadoop appropriations while investigating and assessing the unique ways to Hadoop security being taken by Hadoop dissemination and information security sellers. At last, a robust pathway for securing disseminated registering situations in the undertaking is given.

1.1 Enormous Data Shows a New Security Challenge

Enormous information starts from various sources including sensors used to assemble atmosphere data, presents on online networking locales, advanced pictures and features, buy exchange records, and cell GPS signs, to name a couple. On account of distributed computing and the socialization of the Internet, petabytes of unstructured information are made every day online and much of this data has an inherent business esteem on the off chance that it can be caught and dissected. Case in point, versatile correspondences organizations gather information from cell towers; oil and gas organizations gather information from refinery sensors and seismic investigation; electric force utilities gather information from force plants and appropriation frameworks [1, 3]. Organizations gather extensive measures of client created information from prospects and clients including charge card numbers, standardized savings numbers, information on purchasing propensities and examples of use. The flood of enormous information and the need to move this data all through an association has made a huge new focus for programmers and different cybercriminals. This information, which was at one time unusable by associations is currently exceptionally profitable, is liable to security laws also agreeability regulations, and must be ensured (Fig. 1).

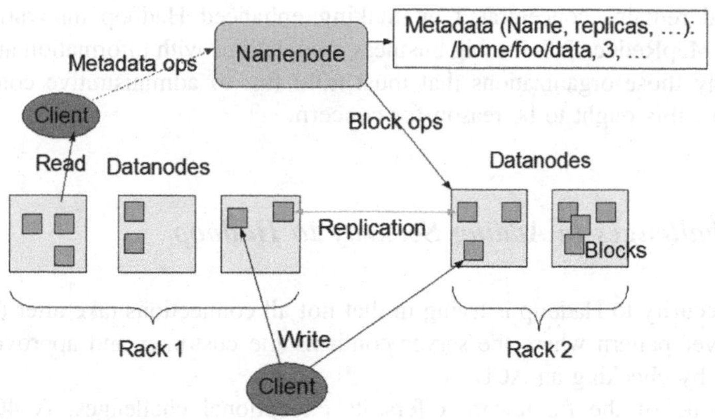

Fig. 1 HDFS architecture

1.2 Hadoop Is not a Secure Method

Hadoop, in the same way as other open source advances, for example, UNIX and TCP/IP, was not made on account of security. Hadoop developed from other open-source Apache undertakings, coordinated at building open source web crawlers. Hadoop was a twist off sub-task of Apache Lucene [4] and Nutch [5] ventures, which utilized a MapReduce office and a dispersed record framework with no implicit security. Hadoop is additionally the open-source rendition of the Google MapReduce structure, and no security was planned into the product as the information being put away (open Urls) was not subject to protection regulation [3]. The open source Hadoop group underpins some security offers through the current execution of Kerberos, the utilization of firewalls, and essential HDFS authorizations. Kerberos is not a required necessity for a Hadoop bunch, making it conceivable to run whole groups without sending any security. Kerberos is additionally hard to introduce and design on the group, and to incorporate with Active Directory (AD) and Lightweight Directory Access Protocol, (LDAP) administrations. This makes security dangerous to convey, what's more therefore obliges the reception of even the most fundamental security capacities for clients of Hadoop [3]. Venture associations have been subjected to the dangers connected with information security breaks throughout recent decades, and expect that any new engineering that is embraced by IT and introduced in the datacenter will meet a base set of security necessities. Endeavors need the same security capacities for huge information as that set up for "non-huge information" data frameworks, including arrangements that address client confirmation and access control, strategy authorization and administration, also information covering and encryption. Numerous associations require these enormous information shields so as to keep up administrative agreeability with HIPAA, HITECH, SOX, PCI/DSS, and other security and protection commands. To date, the open source group has not tended to these security holes, and remains concentrated on making enhanced Hadoop innovations, for example, MapReduce 2.0. For big business associations with information at danger, particularly those organizations that must hold fast to administrative consistence commands, this ought to be reason for concern.

1.3 Challenges in Adding Security to Hadoop

Adding security to Hadoop is trying in that not all connections take after the usual client-server pattern where the server confirms the customer and approves every operation by checking an ACL.

The scale of the framework offers its exceptional challenges. A 4000 hub common bunch [2] is relied upon to serve in excess of 100,000 simultaneous assignments; we expect the bunch size and the quantity of undertakings to increment about whether. Normally accessible Kerberos [6] servers won't be capable to handle the scale of such a variety of errands confirming straightforwardly. The record framework is parceled and dispersed: the Name-node forms the introductory open/make record operations and approves by checking document ACLs. The customer gets to the information straightforwardly from the Data-nodes which don't have any ACLs and subsequently have no chance to get of approving gets to. A submitted cluster occupation is executed at a later time on hubs not quite the same as the hub on which the client confirmed and submitted the employment. Henceforth the client's confirmation at the employment accommodation machine needs to be engendered for later utilization at the point when the occupation is really executed (note the client has typically detached from the framework when the occupation really gets executed). This spread qualifications raises issues of trust and offers opportunities for security infringement. The errands of an occupation need to safely get data, for example, assignment parameters, transitional yield of errands, assignment status, and so on. Transitional Guide yield is not put away in HDFS; it is put away on the neighborhood circle of each one register hub and is gotten to by means of a HTTP-based convention served by the Task-trackers. Undertakings from diverse inhabitants can be executed on the same machine. This imparted environment offers opportunities for security infringement by means of the APIs of the neighborhood working arrangement of the figure hub: access to the halfway yield of different occupants, access to simultaneous undertakings of other occupations and access to the HDFSs neighborhood piece stockpiling on the hub. Clients can get to the framework through helper administration, for example, Oozie, our work process framework [3]. This raises new issues. Case in point, ought to the clients qualifications be passed through these optional administrations or ought to these administrations be trusted by whatever is left of the Hadoop framework?

Some of these issues are special to frameworks like Hadoop. Others, as individual issues, happen in different frameworks. Where conceivable we have utilized standard approaches and consolidated them with new arrangements. The general set of arrangements and how they collaborate, to our insight are genuinely special and educational to an architect of complex dispersed frameworks.

2 Secure and Non-secure Deployments of Hadoop

Hadoop is conveyed in numerous distinctive associations; not every one of them oblige an exceedingly secure arrangement. Despite the fact that, to date, just Yahoo! has sent Secure Hadoop bunches, numerous associations are wanting to change over their surroundings to a safe one. Henceforth Hadoop needs the choices for being arranged secure (with solid validation) or non-secure. The non-secure setup depends on customer side libraries to send the customers side accreditations as decided from the customer side working framework as a component of the convention; while not secure, this arrangement sufficient for numerous organization that depend on physical security. Approval checks through ACLs and document authorizations are still performed against the customer supplied user id. A safe arrangement obliges that one designs a Kerberos [6] server; this paper depicted the instruments and arrangements for such a protected arrangement.

The Physical Environment—Every Hadoop bunch at Yahoo! is autonomously overseen and joined with the more extensive venture system. The security arrangement of our association directs that each one bunch has doors through which occupations can be submitted or from which HDFS can be gotten to; the portals firewall every Hadoop group. Note this is a specific approach of our association also not a limitation of Hadoop: Hadoop itself permits access to HDFS or MapReduce from any customer that can achieve it through the system. Every hub in the group is physically secure and is stacked with the Hadoop programming by framework directors; clients don't have immediate access to the hubs and can't introduce any product on these hubs. Clients can however login on a group's door hubs [3, 6]. A client can't turn into a super user (root) on any of the group hubs. Further, a client can't join a non-group hub, (for example, client workstation) to the group system and snoop on the system. In spite of the fact that HDFS and MapReduce groups are structurally separate, the groups are normally arranged to superimpose to permit processing to be performed near its information.

3 HDFS

Correspondence between the customer and the HDFS administration is made out of two parts:

i. A RPC association from the customer to the Name-node to, say, open or make a document. The RPC association can be verified through Kerberos or through an appointment token. On the off chance that the application is running on a machine where the client has logged into Kerberos then Kerberos verification is sufficient. Assignment token is required just when a right to gain entrance is needed as a feature of MapReduce employment. In the wake of checking consents on the document way, Name-node returns block ids, square areas and piece access tokens [7, 8].

ii. A streaming attachment association is utilized to peruse on the other hand compose the piece of a document at a Data-node. A data-node requires the customer to supply a block access token produced by the Name-node for the square being gotten to. Interchanges between the Name-node and Data-nodes is by means of RPC and shared Kerberos verification is performed. Figure 2 demonstrates the correspondence ways and the system used to verify these ways.

The Lecture Notes in Computer Science volumes are sent to ISI for inclusion in their Science Citation Index Expanded.

Tokens as Supplementary Mechanisms—A client submitting a vocation confirms with the Job Tracker utilizing Kerberos. The occupation is executed later, perhaps after the client has separated from the framework. How would we spread the qualifications? There are a few choices: have the client pass the secret key to the occupation tracker pass the Kerberos accreditations (TGT or administration ticket for the Name-node) Use an uncommon appointment token passing the secret key is obviously inadmissible. We decide to utilize an uncommon appointment token rather than passing the Kerberos accreditations (reasons clarified beneath). After introductory validation to Name-node utilizing Kerberos accreditations, a customer acquires an appointment token, which is given to a vocation for resulting validation to Name-node. The token is actually a mystery key imparted between the customer and Name-node and should be ensured when disregarded frail channels. Any individual who gets it can imitate as the client on Name-node. Note that a customer can only obtain new designation tokens by validating utilizing Kerberos [6, 7].

The configuration of designation token is:

```
Token-id = {owner-id, renewer-id, issue date, max date,
sequence number}
```

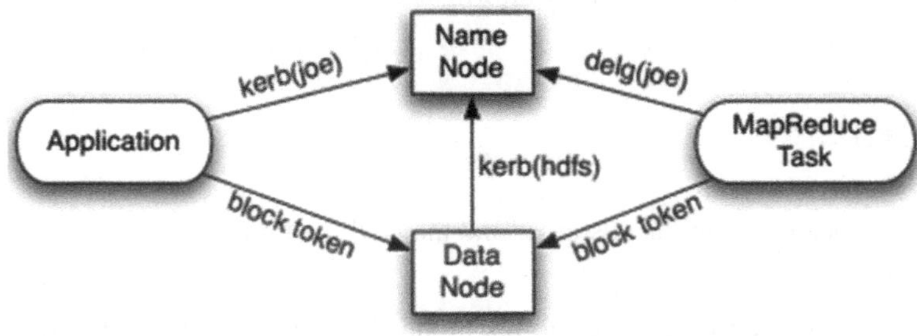

Fig. 2 HDFS authentication

At the point when a customer gets an appointment token from Name-node, it tags a renewer that can restore or drop the token. Of course, assignment tokens are legitimate for 1 day from when they are issued and may be recharged up to a greatest of 7 days. Since MapReduce occupations may last more than the legitimacy of the assignment token, the Job-tracker is determined as the renewer. This permits the Job-tracker to recharge the tokens connected with a vocation once a day until the employment finishes. At the point when the employment finishes, the Job Tracker demands the Name-node to cross out the work's appointment token. Recharging an appointment token does not change the token, it simply redesigns its lapse time on the Name-node, and the old assignment token proceeds to work in the MapReduce errands [9].

4 MapReduce

A MapReduce employment includes the accompanying stages (as portrayed in Fig. 3). A customer interfaces with the Job-tracker to demand an occupation id and a HDFS way to compose the employment definition documents. The Map-reduce library code composes the points of interest of the employment into the assigned HDFS arranging index and gains the important HDFS assignment tokens [10]. The majority of the employment records are noticeable just to the client, yet rely on upon HDFS security. The Job-tracker gets the employment and makes an arbitrary mystery, which is known as the employment token. The work token is utilized to verify the work's errands to the MapReduce skeleton. Employments are broken down into assignments, each of speaks to a segment of the work that needs to be

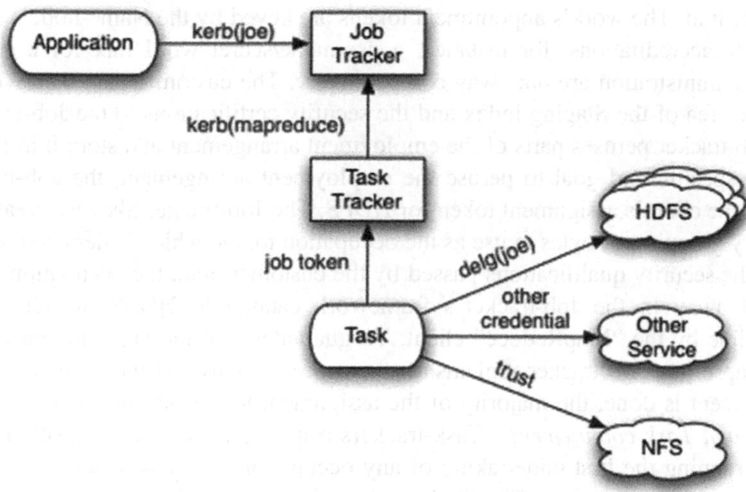

Fig. 3 MapReduce authentication

done. Since assignments (or the machine that they run on) may come up short, there can be various endeavors for each one errand. At the point when an assignment endeavor is allotted to a particular Task tracker, the Task tracker makes a secure environment for it. Since errands from distinctive clients may run on the same figure hub, we have decided to utilize the host working framework's security surroundings and run the undertaking as the client. This empowers utilization of the neighborhood document framework and working framework for disengagement between clients in the group. The tokens for the employment are put away in the neighborhood record framework and set in the errand's surroundings such that the assignment process what's more any sub-courses of action will utilize the work's tokens. Each one running undertaking reports status to its Task tracker and decrease assignments bring guide yield from different Task-trackers. These gets to are validated utilizing the employment token. At the point when the occupation finishes (or falls flat), the greater part of the HDFS designation tokens connected with the employment are renounced [8, 9].

4.1 Job Submission

A customer submitting a vocation or checking the status of an occupation verifies with the Job-tracker utilizing Kerberos over RPC. For occupation accommodation, the customer composes the work setup, the employment classes, the data parts, also the Meta data about the info parts into an index, called the Job Staging registry, in their home index. This registry is secured as perused, compose, and execute exclusively by the client. Employments (by means of its assignments) may get to a few distinctive HDFS and different administrations [10]. Consequently, the occupation needs to bundle the security qualifications in a manner that an undertaking can later find. The work's appointment tokens are keyed by the Name-node's URL. Alternate accreditations, for instance, a username/secret word mix for a certain HTTP administration are put away comparatively. The customer then uses RPC to pass the area of the Staging index and the security certifications to the Job-tracker. The Job-tracker peruses parts of the employment arrangement and store it in RAM. With a specific end goal to peruse the employment arrangement, the Job-tracker utilizes the client's assignment token for HDFS. The Job-tracker likewise creates an arbitrary grouping of bytes to use as the occupation token, which is depicted in area [11]. The security qualifications passed by the customer, and, the occupation token are put away in the Job-tracker's framework catalog in HDFS, which is just intelligible by the "MapReduce" client. To guarantee that the appointment tokens don't lapse, the Job tracker replenishes them occasionally. At the point when the employment is done, the majority of the assignment tokens are negated.

Job and Task confinement—Task-trackers runs errands of the occupations, and before running the first undertaking of any occupation, it needs to set up a secure environment for the same. The Task-tracker dispatches a little setuid program (a moderately little program written in C) to make the occupation registry, and make the holder of that employment index the occupation manager client. A setuid system is utilized since root benefits are obliged to allocate responsibility for index to some other client. The setuid program additionally duplicates the qualifications document from the Job-tracker's framework index into this registry [8, 11, 9]. The setuid program then switches again to the employment holder client, and does the rest of the confinement work. That incorporates—duplicating of the employment documents (arrangement and the classes) from the employment holder's staging index. It utilizes the client's appointment token for HDFS from the certifications document for these. As a component of the undertaking dispatch, an errand catalog is made every assignment inside the occupation index. This index stores the transitional information of the undertaking (for instance, the guide yield records).

The gathering possession at work catalog and the undertaking registries is situated to the "MapReduce" client. The bunch (cluster) [2] possession is situated along these lines so that the Task tracker can serve things like the guide yields to Reducers later. Note that just the employment holder client and the Task tracker can read the substance of the occupation registry.

Task—The undertaking runs as the client who submitted the occupation. Since the capacity to change client ids is constrained to root the setuid system is utilized here as well. It dispatches the assignment's JVM as the right occupation manager client. The setuid program additionally handles murdering the JVM if the errand is slaughtered. Running with this client id guarantees that one client's employment cannot send working framework signals to either the Task-tracker or other client's undertakings. It likewise guarantees that nearby document authorizations are sufficient to keep data private [8].

Job Token—At the point when the occupation is submitted, the Job-tracker makes a mystery key that is just utilized by the undertakings of the work when distinguishing themselves to the structure. As said prior, this token is put away as a component of the certifications record in the Job-tracker's framework index on HDFS. This token is utilized for the RPC through DIGEST-Md5 when the Task speaks with the Task-tracker to demands undertakings or report status. Furthermore, this token is utilized by Pipes undertakings, which run as sub-courses of action of the MapReduce undertakings. Utilizing this imparted mystery, the tyke and guardian can guarantee that they both have the mystery [8].

Shuffle—At the point when a guide undertaking completes, its yield is given to the Task-tracker that dealt with the guide undertaking. Each one diminish in that employment will contact the Task-tracker and get its area of the yield by means of HTTP. The system needs to guarantee that different clients may not get the guide yields. The decrease undertaking will figure the HMAC-Sha1 of the asked for URL and the current timestamp and utilizing the occupation token as the mystery. This HMAC-Sha1 will be sent alongside the appeal furthermore the Task-tracker will just serve the appeal if the HMAC-Sha1 is the right one for that URL and the timestamp is inside the keep going N minutes. To guarantee that the Task-tracker hasn't been supplanted with a Trojan, the reaction header will incorporate a HMAC-Sha1 created from the asking HMAC-Sha1 and secured utilizing the employment token. The rearrange in the decrease can check that the reaction came from the Task-tracker that it at first reached. The preference of utilizing HMAC-Sha1 over DIGEST-Md5 for the verification of the mix is that it dodges a roundtrip between the server and customer. This is an imperative thought since there are numerous mix associations, each of which is exchanging a little measure of information.

5 Experimental Result

RPC Hadoop customers get to most Hadoop administrations through Hadoop's RPC library. In shaky renditions of Hadoop, the client's login name is resolved from the customer OS and sent crosswise over as a feature of the association setup and are not confirmed; this is unstable on the grounds that a learned customer that comprehend the convention can substitute any client id. For validated bunches, all RPC's interface utilizing Simple Authentication and Security Layer (SASL). SASL arranges a sub-convention to utilize and Hadoop will help either utilizing Kerberos (by means of GSSAPI) or DIGEST-Md5 [8].

Most applications run on the entryways use Kerberos tickets, while errands in MapReduce occupations use tokens. The playing point of utilizing SASL is that utilizing another confirmation plan with Hadoop RPC would just oblige actualizing a SASL interface and adjusting a tiny bit of the paste in the RPC code. The backed systems are:

The customer stacks any Kerberos tickets that are in the client's ticket reserve. MapReduce additionally makes a token reserve that is stacked by the assignment. At the point when the application makes a RPC association, it utilizes a token, in the event that a proper one is accessible. Else, it employments the Kerberos qualifications. Every RPC convention characterizes the kind(s) of token it will acknowledge. On the customer side, all tokens comprise of a parallel identifier, a double secret key, the sort of the token (appointment, piece get to, or work), and the specific administration for this token (the particular Job-tracker on the other hand Name-node). The token identifier is the serialization of a token identifier protest on the server that is particular to that sort of token [8, 9]. The watchword is created by the server utilizing HMAC-Sha1 on the token identifier and a 20 byte mystery key from Java's secure random class. The mystery keys are moved intermittently by the server what's more, if essential, put away as a component of the server's relentless state for utilization if the server is restarted.

6 Conclusion

We were shrewd to supplement Kerberos with tokens as Kerberos servers would not have scaled to many a large number of simultaneous Hadoop errands. Amid the outline of appointment tokens, we had considered the option of augmenting a

Kerberos execution to consolidate instruments so that a Kerberos administration ticket could be designated and recharged in the manner required for Hadoop. It was a decent choice to not do that as it would have been trying to have an adjusted Kerberos server embraced for non-Hadoop use in our association. Further it would have upset the more extensive appropriation of secure Hadoop in the business.

References

1. Inukollu, V.N., Arsi, S., Ravuri, S.R.: Security issues associated with big data in cloud computing. Int. J. Netw. Secur. Its Appl. (IJNSA) **6**(3) May 2014
2. Kumari, M., Tyagi, S.: A Three layered security model for data management in Hadoop environment. Int. J. Adv. Res. Comput. Sci. Soft. Eng. **4**(6) June 2014
3. Priya P. Sharma, Chandrakant P. Navdeti: Securing big data hadoop: a review of security issues, threats and solution. Int. J. Comput. Sci. Inf. Technol. (IJCSIT) **5**(2), 2126–2131 (2014)
4. Apache Lucene: http://lucene.apache.org/core/4_10_2/demo/overview-summary.html#overview_description
5. Apache Nutch: https://wiki.apache.org/nutch/FrontPage#What_is_Apache_Nutch.3F
6. Baliello, C., Basso, A., Giusto, C.D.: Kerberos Protocol: An Overview, Distributed Systems, Fall (2002)
7. Gaikwad, R.L., Dakhane, D.M., Pardhi, R.L.: Network Security Enhancement in Hadoop Clusters. Int. J. Appl. Innov. Eng. Manage. (IJAIEM) **2**(3) March 2013
8. Das, D., Radia, S.: Adding Security to Apache Hadoop. Hortonworks, IBM
9. Securing your enterprise Hadoop ecosystem by cloudera
10. Bardiya, P.D., Gulhane, R.A., Karde P.P.: Data Security using Hadoop on Cloud Computing. Int. J. Comput. Sci. Mob. Comput. **3**(4), 802–809 April 2014
11. Donnelly, P., Bui, P., Thain, D.: Attaching cloud storage to a campus grid using parrot, chirp, and Hadoop. Computer Science and Engineering University of Notre Dame

Image Analysis for Efficient Surface Defect Detection of Orange Fruits

Thendral Ravi and Suhasini Ambalavanan

Abstract This work portrays a novel approach for the improvement of a real-time computerized vision based model for automatic orange fruit peel defect detection. In this paper at first, different filtering methods and wavelet based method has been used to denoise the given input image and performs their comparative study. Based on this study, the wavelet based approach is used for smoothening of the images together with removing the higher energy regions in an image for better defect detection as well as makes the defects more retrievable. Finally, orange fruit skin color defects are identified by using RGB and HSI color spaces. The experimental test results indicate that the designed algorithm is scalable, computationally effective and robust for identification of orange fruit surface defects.

Keywords Image processing · Color spaces · Machine vision · Background segmentation · Defect detection

1 Introduction

India is the largest producer of fruits and vegetables, which contributes the nation's growth by increasing the export potential. Diverse agro-climate ensures availability of all variety fresh fruits in India. It ranks second place in the world with fruit production rate of 44.04 million tonnes, from an area of 6.1 million hectares. This implies 10 % of the Indian production rate mainly depends on the fruit cultivation. Numerous types of fruits are cultivated in India, such as banana, mango, orange, apple, guava, papaya, pineapple, pomegranate and grapes are the major ones. Out of

T. Ravi (✉) · S. Ambalavanan
Department of Computer Science and Engineering, Annamalai University, Chidambaram, Tamil Nadu 608 002, India
e-mail: thendralamutha@gmail.com

S. Ambalavanan
e-mail: suha_babu@yahoo.com

A. Nagar et al. (eds.), *Proceedings of 3rd International Conference on Advanced Computing, Networking and Informatics*, Smart Innovation, Systems and Technologies 43, DOI 10.1007/978-81-322-2538-6_17

all the fruits, orange is dominant because India is one of the top three orange producers in the world.

Oranges are cultivated, later they are shifted to the packaging plant for testing various quality parameters which helps to decide their grade and price. The visual appearance of orange is one of the prominent considerations while grading, also helps consumers to select the better quality oranges. The criteria for evaluating the external appearance of orange, contain a good shape, a visually appetizing look, and uniform color allocation on the surface. Out of all the factors, the visibility of surface defects is the most prominent consideration for deciding grade and price of the fresh fruits, because the quality of fruits is identified by consumers using good visual appearance along with the presence of non-defect surface area on the fruit.

Fruit quality check by human beings mainly depends on the physical condition, knowledge, and mood of the human involved in the grading work. Furthermore, the manual inspection can be inefficient and very time consuming, particularly when dealing with large productivity. Defect detection using manual analysis of an object is not a reliable approach because of human errors. For this purpose, packinghouses need more sophisticated systems that are highly effective in an automated visual inspection system for detecting skin defects. Several studies have been performed in order to find the defects and their connection with the quality parameters of fresh fruits such as olives [1], peaches [2], oranges [3], potatoes [4], bell peppers [5], stonefruit [6], pistachio [7], dry dates [8], sweet cherry [9], mushroom [10] and apples [11–13].

The color space is used to represent the pixel color of an image with three color channels. Generally RGB color space is used in digital images and computers. This color space depending on three primary colors such as red, green, and blue channels. Usually objects are described by different color spaces. In some cases a simple color ratio can give useful information than the other complex techniques. For example, RGB ratios [14] were used to classify four kinds of pomegranate arils. They used two image segmentation methods. One method based on the average values of RGB color coordinates and another method based on the simple threshold value of R\G ratio. Test results showed that threshold of R\G ratio could be valuable in classification of the pomegranate arils with 90 % accuracy.

Citrus fruit external defects are identified and compared by using five different color space transformation methods [15]. The test results show that the highest discriminatory power obtained by using only the HSI color space. RGB color space is transformed into HSI color space. Fuji apples are classified into four color categories using both RGB and HSI color spaces [16]. From HSI color space separate the ‘H’ color component [17] to classify starfruits into four maturity stages. Lab color space is also used to classify the fruits. Simple algorithm based on ‘a’ color channel was used to classify strawberry fruits into three color categories [18].

Following on from this, our work proposes (1) to restore the maximum information from the acquired noisy images for succeeding process such as segmentation; (2) to select an efficient color channels with highest discriminatory power for detecting the surface defects of oranges.

2 Materials and Methods

To validate the proposed defect detection algorithm, 40 orange fruit images were randomly selected from the internet. These images were then transferred to the computer and all proposed algorithms were developed in the MATLAB environment using wavelet and image processing toolbox version 7.0.

2.1 Preprocessing Operation

An image, is known as an accumulation of information along with the occurrence of noises during capturing and transferring the data. Presence of noises degrades the quality of the image, so the information related to an image gets to be lost or damage. It must be prominent to restore the maximum information from the acquired noisy images. Filtering is a technique for enhancing the image. In this paper, six different image filtering methods are compared using the root mean square error (RMSE) and peak signal to noise ratio (PSNR). If the value of RMSE is low and the value of PSNR is high, then the applied denoising method is better.

$$\mathrm{MSE} = \frac{1}{m \times n} \sum_{x=0}^{m-1} \sum_{y=0}^{n-1} [\mathrm{im}(x,y) - f(x,y)]^2$$

where the original image is im(*x*, *y*), reconstructed image is *f*(*x*, *y*) and $n \times n$ is the picture size.

$$\mathrm{PSNR} = 10\log 10\left(\mathrm{MAX}^2 | \mathrm{MSE}\right)$$

The peak value of the pixels within an image is represented as MAX. In the 8-bit pixel format image, MAX value is represented by 255.

A statistical measurement of input and denoised input image of one orange fruit is reported in Table 1. From the statistical measurement, DWT has the high PSNR value and low RMSE value. The experimental evaluation of our proposed 2D-DWT

Table 1 RMSE, PSNR value between input and denoised input image of orange

Filtered method	RMSE	PSNR (dB)
Mean	0.05	25.57
Gaussian	0.02	33.99
Median	0.05	26.51
Bilateral	0.02	33.21
Weiner	0.05	26.05
DWT	0.01	35.18

decomposition and reconstruction shows that it removes noise significantly and more effectively than the other denoising methods. This filtering output serves as input to succeeding processes. The fundamental objective of the image preprocessing is to improve the image data quality by eliminating undesired distortions (noise removal) and improving the required features for further processing.

2.2 Background Segmentation

After preprocessing step, the subsequent step is distinguish the fruit from the image background. To achieve this separation, we accomplished color based segmentation. To be able to design a more accurate method to accomplish this separation, we select the HSI color space because it provides more efficient segmentation than various other color spaces, resulting in a clear difference between fruit and background colors. In the HSI color space saturation is the amount of gray (0–100 %) in the color. This means that by analyzing this region separation between fruit and background is straightforward. The following functions permit the transformation of RGB to HSI space

$$H = \cos^{-1}\left\{\frac{\frac{1}{2}[(R-G)+(R-B)]}{[(R-G)^2+(R-B)(G-B)]^{\frac{1}{2}}}\right\}$$

$$S = 1 - \frac{3}{R+G+B}[\min(R,G,B)]$$

$$I = \frac{1}{3}(R+G+B)$$

Input image transformed into the HSI color model and separate the ‘S’ plane to recognize the ripen fruits. After this separation, ‘S’ plane image is transformed into a binary image in which the background areas are represented as black color and the fruit areas are represented as white color with the pixel value of ‘0’ and ‘1’ respectively. After this binary conversion a closing operation was carried out, to close small gaps in the object on the image and to smooth the edges.

This resulting binary image was changed over to the same type of the input image for removal of the background region. For this implementation this binary mask image was individually multiplied by red, green and blue channels of the original input image. The color image was restored by composition of red, green and blue channels obtained from the past step. The resultant image shows the recognized fruit regions only, from which pixels belong to the background are zero (black), and those corresponding to the fruit have their original skin color. Figure 1 shows the key steps involved in the background removal and defects detection algorithm based on the proposed method.

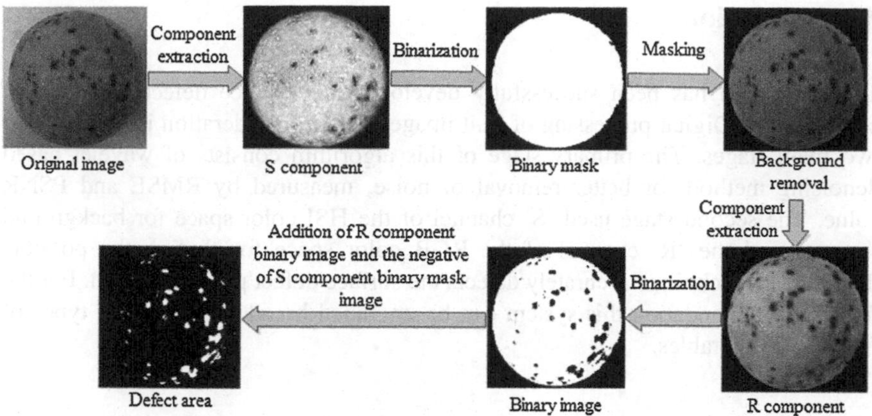

Fig. 1 Steps involved in the defects detection algorithm

2.3 *Defect Detection Algorithm*

The background removal RGB image serves as input to this algorithm. In this work, the red (R) channel image was preferred to build the defect detection algorithm because the contrast between the defect region and the normal region in the fruit image was maximized, thus allowing for a clear distinction of these defect spots from the rest of the fruit areas. This R channel image was transformed into the binary image to determine the characteristics from which to evaluate possible defects on the fruit. These pixels are visible after combining the binary image of the red component and the negative of the S component binary image. The results of our proposed surface defect detection algorithm are shown in Fig. 2. In this diagram the given input RGB images are shown in the first row, segmented binary defect regions are shown in the second row.

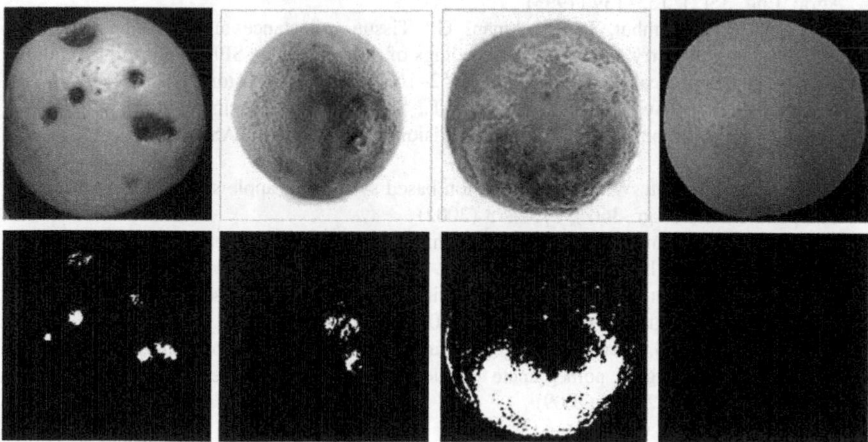

Fig. 2 Example of surface defect detection. Input RGB images (*top*), Binary defect area applying proposed algorithm (*bottom*)

3 Conclusion

This algorithm has been successfully developed for surface defect detection of orange fruits. Digital processing of fruit images under consideration is divided into two main stages. The primary stage of this algorithm consists of wavelet based denoising method for better removal of noise, measured by RMSE and PSNR value. The second stage used 'S' channel of the HSI color space for background removal and the 'R' channel of the RGB color space for finding the possible defects. This technique accurately detects the surface defect part of the fruit. For the future recommendation this system can be upgraded by adding the multi types of fruits and vegetables.

References

1. Diaz, R., Gil, L., Serrano, C., Blasco, M., Molto, E., Blasco, J.: Comparison of three algorithms in the classification of table olives by means of computer vision. J. Food Eng. **61** (1), 101–107 (2004)
2. Miller, B.K., Delwiche, M.J.: Peach defect detection with machine vision. Trans. ASAE **34**(6), 2588–2597 (1991)
3. Cerruto, E., Failla, S., Schillaci, G.: Identification of blemishes on oranges. In: International Conference on Agricultural Engineering, AgEng 96, Madrid, EurAgEng Paper No. 96F-017 (1996)
4. Muir, A.Y., Porteous, R.L., Wastie, R.L.: Experiments in the detection of incipient diseases in potato tubers by optical methods. J. Agr. Eng. Res. **27**(2), 131–138 (1982)
5. Shearer, S.A., Payne, F.A.: Color and defect sorting of bell peppers using machine vision. Trans. ASAE **33**(6), 1245–1250 (1990)
6. Singh, N., Delwiche, M.J.: Machine vision methods for defect sorting stonefruit. Trans. ASAE **37**(6), 1989–1997 (1994)
7. Pearson, T.: Machine vision system for automated detection of stained pistachio nuts. LWT—Food Sci. Tech. **29**(3), 203–209 (1996)
8. Wulfsohn, D., Sarig, Y., Algazi, R.V.: Defect sorting of dry dates by image analysis. Can. Agric. Eng. **35**(2), 133–139 (1993)
9. Guyer, D., Uthaisombut, P., Stockman, G.: Tissue reflectance and machine vision for automated sweet cherry sorting. In: Proceedings of the conference SPIE, optics in agriculture, forestry, and biological processing II, pp. 152–165, vol. 2907, Boston (1996)
10. Heinemann, P.H., Hughes, R., Morrow, C.T., Sommer, H.J., Beelman, R.B., Wuest, P.J.: Grading of mushrooms using a machine vision system. Trans. ASAE **37**(5), 1671–1677 (1994)
11. Li, Q., Wang, M., Gu, W.: Computer vision based system for apple surface defect detection. Comput. Electron. Agr. **36**(2), 215–223 (2002)
12. Wen, Z., Tao, Y.: Building a rule-based machine vision system for defect inspection on apple sorting and packing lines. Expert Syst. Appl. **16**, 307–313 (1999)
13. Leemans, V., Destain, M.F.: A real-time grading method of apples based on features extracted from defects. J. Food Eng. **61**(1), 83–89 (2004)
14. Blasco, J., Cubero, S., Gómez Sanchis, J., Mira, P., Moltó, E.: Development of a machine for the automatic sorting of pomegranate (Punica granatum) arils based on computer vision. J. Food Eng. **90**(1), 27–34 (2009)

15. Blasco, J., Aleixos, N., Gómez, J., Moltó, E.: Citrus sorting by identification of the most common defects using multispectral computer vision. J. Food Eng. **83**(3), 384–393 (2007)
16. Xiaobo, Z., Jiewen, Z., Yanxiao, L.: Apple color grading based on organization feature parameters. Pattern Recogn. Lett. **28**, 2046–2053 (2007)
17. Abdullah, M.Z., Mohamad Saleh, J., Fathinul Syahir, A.S., Mohd Azemi, B.M.N.: Discrimination and classification of fresh-cut starfruits (Averrhoa carambola L.) using automated machine vision system. J. Food Eng. **76**(4), 506–523 (2006)
18. Liming, X., Yanchao, Z.: Automated strawberry grading system based on image processing. Comput. Electron. Agr. **71**(S1), S32–S39 (2010)

Segregation of Rare Items Association

Dipti Rana, Rupa Mehta, Prateek Somkunwar, Naresh Mistry and Mukesh Raghuwanshi

Abstract Nowadays there are many applications including rare itemsets. Here, this paper is concentrating Associations of rare itemsets as association rule mining is considered as one of the most important data mining techniques utilized in the area of market basket data analysis, stock data analysis for frequent items mining. Also it is applied for rare itemsets mining in applications like intrusion detection, medical science, etc. as they have special characteristic like appearing for less number of times. This paper is categorizing them according to the usages of different basic approach, storage structure, mining of items, number of database scans and threshold(s) used, proposing the approach to segregate the rare items from the study of the number of research works done in this area and analyzed the result.

Keywords Association rules mining · Frequent itemsets mining · Rare itemsets mining · Clustering

---

D. Rana (✉) · R. Mehta · P. Somkunwar · N. Mistry
Sardar Vallabhbhai National Institute of Technology, Surat, Gujarat, India
e-mail: dpr@coed.svnit.ac.in

R. Mehta
e-mail: rgm@coed.svnit.ac.in

P. Somkunwar
e-mail: prateeksomkunwar9@gmail.com

N. Mistry
e-mail: njm@ced.svnit.ac.in

M. Raghuwanshi
Yeshwantrao Chavan College of Engineering, Nagpur, India
e-mail: m_raghuwanshi@rediffmail.com

A. Nagar et al. (eds.), *Proceedings of 3rd International Conference on Advanced Computing, Networking and Informatics*, Smart Innovation, Systems and Technologies 43, DOI 10.1007/978-81-322-2538-6_18

1 Introduction

Data mining is nowadays utilized in many applications area as it is providing number of techniques to mine different types of knowledge like association, grouping of labelled data and unlabeled data. In contrast to standard statistical methods, data mining techniques has major advantage of that it mines interesting information without demanding a priori hypotheses [1]. One of the major data mining techniques used is association rule mining which derives the relation between attribute values. This technique individually and successfully applied in many application domains such as biology, finance, marketing, etc. But, here this research is to utilize the major feature of this technique to generate rare associations among the data which may have different features and where relation exists among the features as well as the among the feature values.

In view of this, the researchers who have interest in frequent association rule mining can refer [2, 3]. The next Sect. 2 is describing the problem statement and accordingly Sect. 3 is summarizing the rare association rule mining approaches. Section 4 is proposing the work and the future work to extend from this analyzation.

2 Problem Formulation

Weather data have number of different attributes or features having time stamp information. From the study of weather data, it is found that most of the weather events have frequent occurrence while some weather event association is infrequent. For example, normal increment/decrement in temperature is frequent weather event while the temperature raise and fall to the peak value of temperature series is infrequent or rare event. There is a need to mine rare weather event from the weather data. In view of this, the next section is describing the various approaches of rare association rule mining which already have took place and to have the scope to continue further in this area.

3 Association Rules Mining

Association Rules Mining (ARM) is an important analysis topic within the knowledge discovery space. For a powerful association rule two measures support and confidence are widely used [1, 4]. Support of an itemset is defined as the percentage of datum consists of that itemset. While confidence of rule is percentage of datum which consists of both antecedent and consequent to the datum which has antecedent.

The main goal of most of ARM approaches is to find rules which satisfy the given minimum support and minimum confidence. The ARM problem can be solved in two phases, initially find all the frequent patterns and then generate the association rule from them. The key component that creates association rule mining sensible is the minimum support threshold which prunes the search space to limit the amount of rules generated.

Mining frequent pattern is not of interest always. In several applications rare pattern is more interesting like in retailing business, customers buy luxury goods rarely but they yield more profit than the low price good which are bought frequently. These infrequent items are called rare items. These rare items can generate more profit than frequent items. The problem of mining rules with low support and high confidence is named Rare Association Mining.

Initially the algorithm for ARM is used for RARM with low support. But it generates many too many meaningless frequent patterns and that they will overload the decision makers, who may find it difficult to know the patterns generated by data processing algorithms. Other application of RARM is in the area of medical science, in intrusion detection system etc.

From the literature survey found different categorized work done in the area of RARM based on Apriori, FP-tree and an evolutionary approach as discussed here.

Apriori Based Approach

Based on the downward closure property following approaches have worked out:

- MS Apriori: In the Mutliple Support Apriori (MS Apriori) algorithm author has proposed that the use of single minimum support is unable to determine the nature of different items and therefore different support Minimum Item Support (MIS) for each individual item is defined [5]. More support for frequent item and less support for the rare item prevent to pruning of rare items and help to find frequent rules as well rare rules. The support of a rule is minimum MIS value out of items present in that rule. MIS values for item

$$\text{MIS(item)} = \text{M(item) if M(item)} > \text{LS, otherwise LS} \quad (1)$$

$$\text{M(item)} = \beta * \text{f(i)},\ 0 < \beta < = 1 \quad (2)$$

where f(i) = actual frequency of item i and LS is the Least Support value which must be satisfied.

All items are in increasing order of their MIS values. Then the first item with lowest MIS value which has actual support more than its MIS value is chosen to prune the remaining itemset on the basis of that MIS value. Length 1 itemset list is generating by adding all items to the list which has support more than the MIS value of first selected item. This is important because an itemset which is not frequent may become frequent by adding an item to it. From the Length 1 list it generates Length 2 list by trying combinations. For the list of length more than 2 say k, join any two element of list of length k-1 which have k-2 item same. This is the candidate list of length k. for any itemset this list if it's all subset of length k-1 is not found in the list of length k-1, it is removed from the list. That is how pruning of list is done. Drawback of MSApriori is that the MIS value of each item depends on the user defined value of β which is hard to determine the proper value.

- Relative Support Apriori Algorithm (RSAA): The approach adopts two supports one for rare items and another for frequent items and defines the relative support as critical value [6]. Relative Support (RSup) of itemset i_1, i_2, i_3, … can be given as

$$\begin{aligned}\mathrm{RSup}(i_1, i_2, i_3, \ldots) = {} & \max(\mathrm{sup}(i_1, i_2, i_3\}\ldots)/\mathrm{sup}(i_1),\ \ \mathrm{sup}(i_1, i_2, i_3\ldots)/\mathrm{sup}(i_2),\\ & \mathrm{sup}(i_1, i_2, i_3\ldots)/\mathrm{sup}(i_3), \ldots)\end{aligned} \tag{3}$$

where, i_1, i_2, i_3, … are items and sup(x) is support of item x. RSup is always between 0 and 1. Two lists are generated each time, one of rare itemset and another by combination of rare and frequent itemset. Pruning is done as in [4] and also uses relative support threshold. High value of relative support indicates the high co-occurrence.
- Apriori Inverse: The maximum supports threshold and minimum confidence threshold [7] are used to find the rare item in this method. If a support of a rule is below maximum support and confidence is above the minimum confidence, these rules are referring as sporadic rules. Apriori Inverse is able to find all the sporadic rules. The superset of rare item is always rare is the inverse property. This algorithm determines only the sporadic rules using one minimum support threshold to avoid noise and one maximum support threshold value to find rare items. The sporadic rules have the property that they fall below user define maximum support but they fall above the minimum confidence value. The disadvantage of this approach is it is unable to find all rare itemset. It is faster for finding sporadic rules.
- Apriori for Rare Association Rule Mining (AfRARM): The main idea of this algorithm is to traverse the dataset in top down manner [8] opposite to Apriori which is bottom up approach. The process first finds all rare itemset of largest length then in the next level finds all its subset and checks whether they satisfy the rare support or not. The subset of rare item may be rare. This process continues till all the length 1 rare itemset is found. These rare patterns are used to generate rules. If rare items are less in the database this approach is efficient.
- RARITY: RARITY algorithm [9] uses the same property as AfRARM [8]. It starts with the largest itemset and move downwards to the itemset of smaller length. Initially it starts form the itemset of largest length and at each level the length of itemset reduces by 1. However for implementation it uses candidate list, veto list and rare list. Initially all the largest itemset are in the candidate list, if itemset is found rare it is moved to the rare list otherwise to the veto list. Veto list contains frequent itemlist. In the next level appropriate subset of rare itemsets are the new candidates. For each candidate if it is found in the veto list then it is discarded, if it is not found in veto list the support of candidate is calculated, if it is rare moved to the rare list otherwise to the veto list. This process continues until we get the itemset of length one. It is more efficient than AfRARM [8] because of use of veto list, it uses less number of database scans.

- Improved Multiple Support Apriori Algorithm (IMSApriori): In this method the novel notion of Support Difference (SD) is proposed to determine the minimum support of each item [10]. SD refers to the appropriate deviation of an item from its frequency (or support) in order that an itemset involving that items are often thought of as a frequent itemset which is referred by equation, SD = $\lambda(1 - \alpha)$ where λ represents the parameter like mean, median, mode and α represents the maximum support threshold ranging between 0 to 1. Minimum Item Support (MIS) is determined by MIS (item) = Support (item) − SD if support (item) – SD > Least Support otherwise MIS(item) is set to the least support. Further IMSApriori uses the same approach used in [3] approach for rare rule generation.
- NBD-Apriori-FR: NBD-Apriori-FR uses the same downward closure property and bottom-up approach of Apriori algorithm [11]. It takes database D and minsup as inputs and produces both rare and frequent itemsets as outputs. Initially for first level it generates three list one of rare itemsets, second of frequent itemset and third is the zero list for the items which has zero support. After that for each level it generate three lists first list of frequent itemsets which has support above the threshold second list of rare itemsets which has support less than threshold and third list contains the itemsets yield by combining the frequent items and rare items. A zero list is also maintained for the itemset which has zero support. If any subset of itemset is found in zero list, the itemset is moved to the zero list. Before database scan zero list is searched. Finally rare rule can be generated from the first and third list. This algorithm generates all the rare rules.

Tree Based Approaches

The best approach of frequent association rule mining is based on tree which reduces the number of database scan.

- CFP-Growth: In this approach a new data structure MIS-tree is proposed based on FP-tree [12] in which each item has different support as in [3] together with their MIS value. Initially the MIS-tree is generated without generating separate header table. The transaction items are sorted and inserted in the MIS-tree on the basis of descending order of MIS value. And at the end of database scan, items which have support less than the Minimum of MIS value are deleted and a compact tree is generated. For generation of compact tree children of deleted node is merged to the parent to delete node. To extract rule form this compact tree the conditional pattern base tree is constructed for each item and based on MIS value of that item rule is extracted. This structure provides ease for tune the MIS value.
- IPD based Approach: The method is utilizing a novel notion of Item to Pattern Difference (IPD) to filter the uninteresting pattern [13]. IPD is defined as difference of maximum support of individual element in the pattern and support of a pattern. Maximum Item to Pattern Difference (MIPD) is used defined values set as a threshold. For each item different support is calculated as in [10]. The tree

construction is same as in MIS-tree [12]. To mine the compact MIS tree use the conditional pattern base chooses each frequent length 1 pattern in the compact MIS tree as the suffix-pattern. For this suffix-pattern construct its conditional pattern bases. From the conditional pattern bases, construct MIS tree, called conditional MIS tree, with all those prefix-sub paths that have satisfied the MIS value of the suffix-pattern and MIPD. Finally, recursive mining on conditional MIS-tree results in generating all frequent patterns.

- RP-tree: The RP-tree approach is to mine a subset of rare association rules using a tree structure which is similar to FP-tree [14]. In the first scan support of each item is calculated. Rare items are those support is less than given support threshold. In the next scan RP-tree is generated using transaction which has at least one rare item. The order of items in each transaction during insertion is according to the item frequency of the original database. The resultant RP-Tree consists of rare itemsets only. This tree is used for rule generations.

RARM Using Evolutionary Algorithm

The evolutionary approach is also utilized for rare association rule mining which is based on different concepts other than Apriori and Tree approaches.

- Rare-G3PARM: The algorithm Rare Grammar Guided Genetic Programming for Association Rule Mining (Rare-G3PARM) starts by generating a set of new individual conformant to the specified grammar [16]. The algorithm extends the Grammar Guided Genetic Programming for Association Rule Ming (G3PARM) approach which is used to mine frequent patterns [15]. The context free grammar is used for each individual and encoded in a tree shape through the application of production rules. Rare-G3PARM starts by generating a set of new individual conformant to the specified grammar. In order to obtain new individuals, the algorithm selects individuals from the general population and the pool to act as parents and a genetic operator is applied over them immediately afterwards with a certain probability. These new individuals are evaluated.
- The elite population or pool is empty for the first generation, otherwise it comprises the n most reliable individuals obtained along the evolutionary process, and the population is combined to form a new set. Then, this new set is ranked by their fitness, so only the best ones are selected until the new population is completed. The update procedure is carried out ranking by confidence the new set of individuals this ranking serving to select the best n individuals from the new set for the updating process.
- Only those individuals having a fitness value greater than zero, a confidence value greater than the minimum-confidence threshold, and a lift value greater than unity are considered prompting the discovery of infrequent, reliable and interesting association rules. An important feature of Rare-G3PARM is the use of the lift measure, which represents the interest of a given association rule. Traditionally, ARM proposals make use of a support and confidence framework,

including G3PARM, attempting to discover rules which have support and confidence values are greater than given thresholds.

$$\text{Lift}(X \rightarrow Y) = \text{Confidence}(X \rightarrow Y)/\text{Support}\ (Y) \tag{4}$$

- A new Genetic Operator which modifies the highest support condition of a rule to obtain a new condition having a lower support value, has been implemented. Notice that the lower the support values of the conditions, the lower the support value of the entire rule.

4 Summary

Here, it is summarized that, from the study of these all methods, that from one and half decade works are going on in this area and still it is going on. Here, we have studied different approaches and divided into three categories like Apriori, FP-Growth and based on Evolutionary algorithm.

In approach based on Apriori which is using candidate generation, in all the approaches considering this approach and requires database scan up to the large itemset size and mining of direction is either top-down or bottom-up. And the major threshold used is minimum support and in some cases used maximum support and other related support measures.

In FP-Growth based approach which is mainly based on the usage of storage structure to mine the information faster with less number of database scans. And accordingly all the methods used the variation of FP-Tree like structures and require equal or less number of database scan than the FP-Tree approach which showed the improvement in the approaches with complexity at the structure level.

Evolutionary algorithm is also applied to mine the association for rare items, which is very novel for association rule mining which depends upon the characteristics of data and utilization function.

5 Proposed Work

The rare items are having special characteristic like appearing for less number of times. From the literature survey it is found that numbers of approaches are utilized to discover the rare associations. The approaches discussed here are complex and takes more execution time and more number of database scans.

And majorly, when want to mine rare association of item, one can think about the partitioning of data by utilization of other data mining technique before mining of rare association.

Herewith, proposed an approach by utilizing another data mining technique clustering for rare association rule mining. Clustering is the technique which groups the data according to the data characteristics where one group data are different than the other group. Up to the knowledge of the author this approach is yet not utilized by any researcher.

The approach is proposing to apply basic clustering on the dataset, with the cluster thresholds like the number of instances then repeatedly apply the clustering technique to minimize the cluster size with less intra cluster distance and more inter cluster distance. The process is repeated until; the approach is generating clusters of rare items having the good accuracy parameter as discussed in the next section and then applies association rule mining concepts only for those clusters.

Here, as the association rules mining will be applied only for the clusters which are small in size, that indicates rarely occurring data and thus minimizing the overall association rule mining tasks.

6 Experiment Analysis and Future Work

The experiment is performed on weather data of Surat from the year 2007 to 2012, containing different parameters like temperature, humidity, precipitation, etc. Accuracy of clusters is measured using 3 parameters like average intra cluster distance, average inter cluster distance and inter/intra cluster ratio. Here, the results are shown up to the clusters only, not for associations.

Experiment is considered using K-means Clustering of weka. After preprocessing of data, and after the number of experiments where varied the number of clusters to achieve the higher inter/intra ratio.

The results indicate that 30 clusters are achieved with 2.18 inter/intra ratio. From which 15 clusters are having good density and 15 clusters are having rare density. The Tabel 1 shows the achieved result, indicating that both the intra cluster distance and inter cluster distance are large enough.

But, from the result it is also analyzed that the intra cluster is 3.34 which is also quite high. Also, after analyzation of clustered records, it is found that for this type of weather data, the generated clusters are not having typical pattern which does not make all of them different from each other.

For better clusters, intra cluster distance parameter requires less value and inter cluster distance parameter requires higher value. Moreover, the number of clusters

Table 1 Experiment result for clusters

Parameters	K-means clustering
Number of cluster	30
Average intra cluster distance	3.34
Average inter cluster distance	7.30
Parameter (inter/intra)	2.18

are even large enough, here equals in both the cases of frequent and rare. Thus, it is still required to have less number of clusters with higher inter/intra cluster ratio and uniqueness in clusters items associations.

Acknowledgments This research work is carried out under the research project grant for SVNIT Assistant Professors' bearing circular number: Dean(R&C)/1503/2013-14.

References

1. Han, J., Kamber: Data mining concepts and techniques. In: Morgan Kaufmann Publishers, March 2006
2. Patel, M.R., Rana, D.P., Mehta, R.G.: FApriori: A modified Apriori algorithm based on checkpoint, In: IEEE International Conference on Information Systems and Computer Networks (ISCON), pp. 50–53 (2013)
3. Rana, D.P., Mistry N.J., Raghuwanshi, M.M.: Memory cutback for FP-tree approach. In: Int. J. Comput. Appl. (IJCA) **87**(3), (2014)
4. Agrawal, R., Srikant, R.: Fast algorithms for mining association rules. In: 20th International Conference on Very Large Databases, Santiago, Sept (1994)
5. Liu, B., Hsu, W., Ma, Y.: Mining association rules with multiple minimum supports. In: SIGKDD, pp. 337–341. ACM Press, New York (1999)
6. Yun, H., Ha, D., Hwang, B., Ryu, K.: Mining association rules on significant rare data using relative support. In: Journal of Systems and Software, pp. 181–191. Elsevier (2003)
7. Koh, Y., Rountree, N.: Finding sporadic rules using apriori-inverse. PAKDD, Hanoi, Vietnam, Springer LNCS **3518**, 97–106 (2005)
8. Adda, M., Wu, L., Feng, Y.: Rare itemset mining. In: Sixth International Conference on Machine Learning and Applications. IEEE (2007)
9. Troiano, L., Scibelli, G., Birtolo, C.: A fast algorithm for mining rare itemset. In: Ninth International Conference on Intelligent Systems Design and Applications. IEEE (2009)
10. Kiran, R.U., Reddy, P.K.: An improved multiple minimum support based approach to mine rare association rules. In: IEEE Symposium on Computational Intelligence and Data Mining, pp. 340–347 (2009)
11. Hoque, N., Nath, B., Bhattacharyya, D.K.: A new approach on rare association rule mining. Int. J. Comput. Appl. **53**(3) (2012)
12. Hu, Y., Chen, Y.: Mining association rules with multiple minimum supports: a new mining algorithm and a support tuning mechanism. Elsevier (2004)
13. Kiran, R.U., Reddy, P.K.: Mining rare association rules in the datasets with widely varying items' frequencies. Springer, Berlin Heidelberg, LNCS **5981**, 49–62 (2010)
14. Tsang, S., Sing Koh, Y., Dobbie, G.: Finding interesting rare association rules using rare pattern tree. In: TLDKS VIII, vol. 7790, pp. 157–173. Springer, Berlin, LNCS (2013)
15. Luna, J.M., Romero, J.R., Ventura, S.: On the adaptability of G3PARM to the extraction of rare association rules. Knowl. Inf. Syst. **38** 391–418 (2013) (Springer, London)
16. Hoai, R.I., Whigham, N.X., Shan, P.A., O'neill, Y., McKay, M.: Grammar-based genetic programming: a survey. Genet. Program. Evolvable Mach. **11**(3–4), 365–396 (2011) (Springer)

Local Gabor Wavelet-Based Feature Extraction and Evaluation

T. Malathi and M.K. Bhuyan

Abstract Feature extraction is an essential step in many image processing and computer vision applications. It is quite desirable that the extracted features can effectively represent an image. Furthermore, the dominant information visually perceived by human beings should be efficiently represented by the extracted features. Over the last few decades, different algorithms are proposed to address the major issues of image representations by the efficient features. Gabor wavelet is one of the most widely used filters for image feature extraction. Existing Gabor wavelet-based feature extraction methodologies unnecessarily use both the real and the imaginary coefficients, which are subsequently processed by dimensionality reduction techniques such as PCA, LDA etc. This procedure ultimately affects the overall performance of the algorithm in terms of memory requirement and the computational complexity. To address this particular issue, we proposed a local image feature extraction method by using a Gabor wavelet. In our method, an image is divided into overlapping image blocks, and subsequently each of the image blocks are separately filtered out by Gabor wavelet. Finally, the extracted coefficients are concatenated to get the proposed local feature vector. The efficacy and effectiveness of the proposed feature extraction method is evaluated using the estimation of mean square error (MSE), peak signal-to-noise ratio (PSNR), and the correlation coefficient (CC) by reconstructing the original image using the extracted features, and compared it with the original input image. All these performance evaluation measures clearly show that real coefficients of the Gabor filter alone can effectively represent an image as compared to the methods which utilize either the imaginary coefficients or the both. The major novelty of our method lies on our claim—capability of the real coefficients of a Gabor filter for image representation.

Keywords Image feature extraction · Gabor filter · Image reconstruction

---

T. Malathi (✉) · M.K. Bhuyan
Department of Electronics and Electrical Engineering, Indian Institute of Technology, Guwahati 781039, India
e-mail: malathi@iitg.ernet.in

M.K. Bhuyan
e-mail: mkb@iitg.ernet.in

A. Nagar et al. (eds.), *Proceedings of 3rd International Conference on Advanced Computing, Networking and Informatics*, Smart Innovation, Systems and Technologies 43, DOI 10.1007/978-81-322-2538-6_19

1 Introduction and Related Works

Feature extraction is a crucial step which greatly affects the performance of the intended applications, and it is an open research problem of Computer Vision. Depending on the application, nature of the feature to be extracted varies. Feature extraction can be performed in two ways; namely local and global methods. Global methods represent the whole image with a single feature vector, whereas in local methods, a single image can have feature descriptors at multiple feature points [1]. These feature descriptors describe the local image neighborhood around the computed interest points. The features described by both global and local methods provide different information about the image as the support (neighborhood) used to extract the intended features is different. Global methods are compact representation of an image, and the feature vector of an image represented by a global method corresponds to a single point in higher dimensional space making it suitable for analysis. As global methods represent the entire image with a single feature, it assumes that the image contains only a single object and the method fails when the image has more than one object. This drawback is overcome by local methods.

Feature extraction using Gabor wavelets has many important applications in Computer Vision. Although the Gabor wavelets are used to extract features for various applications, but the procedure of feature extraction varies from one application to another. One important application is face recognition. In face recognition, Gabor wavelet-based feature extraction is implemented in two ways—Analytical methods and Holistic methods. In analytical methods, Gabor features are extracted from a pre-defined feature point, whereas Gabor features are used to represent the entire image in the holistic methods [2].

In facial expression recognition only a single frequency is selected or the interval between the neighboring frequencies is increased to reduce the dimensionality of Gabor feature for each of the orientations [3]. Dimensionality can be further reduced by using principal component analysis (PCA), and linear discriminant analysis (LDA). Most of the widely available Gabor-based feature extraction methods use magnitude information, but the method proposed in [4] only uses Gabor phase information for facial expression recognition. In [5], a spatially maximum occurrence model (SMOM) for facial expression recognition (FER) is proposed. Similarity comparisons between the images are performed by elastic shape-texture matching (ESTM) algorithm which is based on shape and texture information. Texture information in the high-frequency spectrum is obtained by applying Gabor wavelets on the edge images.

Traditional Gabor filter fails in rotation invariant texture analysis as the sinusoidal grating varies in one direction. In [6], a circular Gabor filter in which the sinusoid varies in all orientation is proposed.

Zhang and Ma used Gabor features for medical image retrieval [7]. The energy from the output of each Gabor filter is given as input to fuzzy sets, and features are extracted. To improve the retrieval rate, mean and standard deviation of the filter output is used instead of energy.

Chen and Liu proposed a face coding algorithm using Gabor wavelet network (GWN) [8]. GWN combines the local-feature and template-based characteristics. Feature extracted by GWN is converted to bit strings, and given them as input to genetic algorithm for coding.

Gabor wavelet is also used for determining stereo correspondence. Phase information of Gabor wavelet at different scales is used to find the corresponding matching pixels along the same scanline [9].

All the existing Gabor wavelet-based feature extraction methods either use Gabor magnitude or phase as features, which requires the computation of both real and imaginary coefficients. Subsequently PCA, LDA or statistical measures are used to reduce the dimensionality of the feature. Additionally, Gabor features are extracted at specific orientation in order to reduce the dimension of the feature. All these dimensionality reduction techniques affect the performance of an algorithm. Hence, the use of both real and the imaginary components of Gabor wavelets unnecessarily create a burden in terms of memory requirement and the computational complexity. To address this specific issue, we proposed a new local feature extraction method only using real coefficients of Gabor filter in spatial domain. It is experimentally verified that the real coefficients can produce significantly good results for most of the applications. The organization of the paper is as follows: Sect. 2 describes the proposed feature extraction method, Sect. 3 shows the experimental results, and Sect. 4 concludes the paper.

2 Proposed Local Gabor Feature Extraction Method

Feature extraction plays a crucial role in computer vision and image processing applications such as face recognition, facial expression recognition, texture classification and segmentation, image retrieval, face reconstruction, fingerprint recognition, iris recognition and stereo correspondence. Gabor wavelet is a widely used feature extraction tool in these areas. The motivation behind using Gabor wavelet Eq. (1) for feature extraction is as follows [10]:

- Simple cells in the visual cortex of mammalian brains can be best modeled by Gabor function.
- Gabor wavelet is a bandpass filter, and it is an optimal conjoint representation of images in both space and frequency domain that occupies the minimum area.
- The orientation and scale tunable property of Gabor wavelet helps in detecting edge and bars which aids in texture feature extraction [11].
- 2D Gabor wavelet has good spatial localization, orientation and frequency selectivity property.
- Image perception by human visual system is similar to image analysis by Gabor function.

Gabor function is Gaussian modulated complex Eq. (2) sinusoids which is given by [12]

$$g(x,y) = \frac{1}{\sqrt{2\pi}} e^{-\frac{1}{8}(4x^2+y^2)} \left[e^{i\kappa x} - e^{-\frac{\kappa^2}{2}} \right] \tag{1}$$

Gabor wavelet is referred as a class of self-similar functions generated by the process of orientation and the scaling of the 2D Gabor function which is given by

$$\begin{aligned} g_{mn}(x,y) &= a^{-m} g(x_a, y_a),\ a > 1 \\ x_a &= a^{-m}(x\cos\theta + y\sin\theta) \text{ and} \\ y_a &= a^{-m}(-x\sin\theta + y\cos\theta) \end{aligned} \tag{2}$$

where, $\theta = \frac{n\pi}{k}$, m and n are two integers and k is the total number of orientations.

To find the feature vector for a pixel of interest, a small image patch around the pixel is considered and convolved with Gabor wavelet with different orientations and scaling. The obtained filter outputs are concatenated to obtain the desired feature vector.

Figure 1 shows the block diagram of the proposed feature extraction method. Let us consider an image I of size $P \times Q$. In order to find the feature vector for the pixel $I(i,j)$, a certain neighborhood $N(i,j)$ of size $u \times v$ is considered, where (i,j) is the pixel coordinates. This patch is convolved with the Gabor filter kernel g_{mn} for different orientations and scaling. The features are then extracted by concatenating the obtained coefficients given by

$$\begin{aligned} F(i,j) &= \text{concat}(\gamma_{mn}(i,j)) \\ \gamma_{mn}(i,j) &= N(i,j) * g_{mn} \end{aligned} \tag{3}$$

where, "*concat*" is the concatenation operator and "$*$" is the convolution operator. This procedure is repeated for all the pixels of the image. Since Gabor filter is a complex filter, we extracted three different features: only real, only imaginary, and both real and imaginary (magnitude) coefficients. Figure 2 shows the output (magnitude, real and imaginary part) of Gabor wavelet for teddy image.

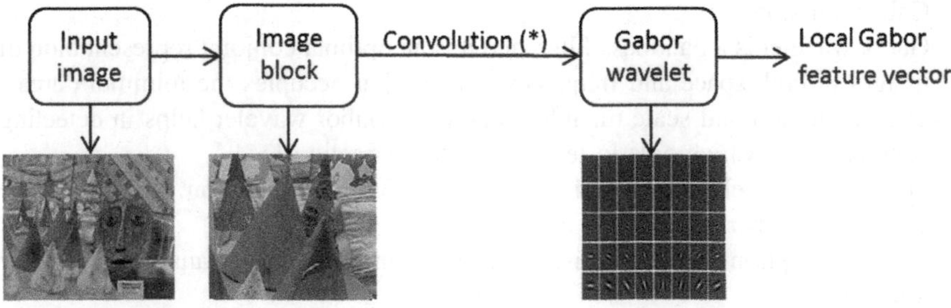

Fig. 1 Block diagram of the proposed local Gabor feature extraction method

Fig. 2 Output of Gabor wavelet filtered image. **a** Input teddy image, **b** both the real and the imaginary coefficients (magnitude), **c** real part and **d** imaginary part

In general, any function can be reconstructed by the linear superposition of its bases weighted by the wavelets. 2D Gabor wavelets are nonorthonormal bases *i.e.*, they are nonorthogonal wavelets. The function can be approximately reconstructed when the family of wavelets is orthonormal basis [12]. This is achieved when $0 < A \leq B < 2$, and the function is reconstructed using the inversion formula given by

$$f = \frac{2}{A+B} \sum_{mn} \langle g_{mn}, f \rangle g_{mn} \tag{4}$$

where, $(A+B)/2$ is a measure of the redundancy of the frame, B/A is a measure of the tightness of the frame. When $A = B$, the frame is called a tight frame and the reconstruction by linear superposition using the above inversion formula is exact.

Reconstruction of the teddy image using the above inversion formula Eq. (4) is shown in Fig. 3. Similar to feature extraction, reconstruction of the original image is also performed by using only real coefficients, only the imaginary coefficients, and the magnitude. The reconstructed image using the extracted local features are compared with the original image, and the performance of the reconstruction is evaluated by using the metrics Mean Square Error (MSE), Peak Signal-to-Noise ratio (PSNR), and cross correlation (CC).

Extraction of Gabor (magnitude and phase) features is performed by convolving an image with the Gabor filter. Since Gabor wavelet is complex, feature extraction

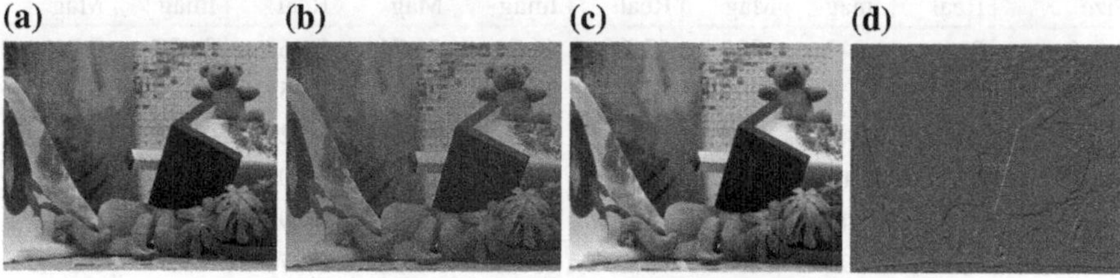

Fig. 3 Reconstructed teddy image. **a** Input teddy image, **b** magnitude of the output, **c** real part and **d** imaginary part

is done in two steps: (i) convolution of an image with real Gabor filter to obtain real coefficients, (ii) convolving an image with the imaginary Gabor filter to obtain the imaginary coefficients. In general, the computational complexity of Gabor filter is $O(PQh^2)$, where $P \times Q$ is the size of input image, $h \times h$ is the Gabor kernel size. So, the computational complexity is significantly reduced in our method as we proposed to use only the real coefficients. Apparently, the memory requirement for storing of all the coefficients of Gabor filter is also reduced. This is because of the fact that both the real and imaginary coefficients have to be computed and stored in order to compute the magnitude and phase information. On the other hand, the proposed method requires half of the memory since it uses only the real coefficients. Hence, the applications which demand fast computation of features, and also the implementation with a reduced system memory, the proposed method can be effectively employed.

3 Experimental Results

The performance of the proposed feature extraction method is evaluated on the Middlebury stereo images [13, 14]. The feature vectors are extracted using the Eq. (2).

The original image is reconstructed from the real and imaginary feature vectors using the Eq. (4). The performance of the proposed feature extraction method is evaluated by comparing the reconstructed image R with the original input image. The metrics used to evaluate the performance are MSE, PSNR and CC. The performance of the proposed method is evaluated for different window size, number of orientations and scales. For all these cases, these three parameters are obtained. The performance of the proposed method is compared with the holistic Gabor features [2]. Tables 1, 2, 3, 4 and 5 shows the MSE, PSNR and CC for different window sizes, number of orientations and number of scales. Parameters used to obtain our

Table 1 Comparison of MSE, PSNR and CC of the proposed method for different window sizes

Window	MSE (proposed)			PSNR (proposed)			CC (proposed)		
Size	Real	Imag	Mag	Real	Imag	Mag	Real	Imag	Mag
3×3	211	642	223	57.58	46.72	56.76	0.968	0.927	0.968
5×5	297	1108	391	54.03	41.17	51.38	0.957	0.799	0.957
7×7	464	1541	682	49.64	38.52	45.68	0.927	0.771	0.927
9×9	623	1718	817	46.77	36.96	43.92	0.901	0.724	0.901
11×11	681	1345	739	45.02	38.40	44.45	0.891	0.742	0.891
13×13	955	2391	1009	43.20	33.65	41.98	0.871	0.574	0.871
15×15	953	2818	1106	43.33	32.18	41.30	0.872	0.491	0.872

Table 2 Comparison of MSE and PSNR of the proposed method with the holistic approach for different number of orientations

No. of orientations	MSE (proposed)			MSE (H)	PSNR (proposed)			PSNR (H)
	Real	Imag	Mag	Mag	Real	Imag	Mag	Mag
1	211	1222	320	1366	57.58	40.98	53.96	39.30
2	211	642	223	1046	57.58	46.72	56.76	43.57
3	211	643	232	547	57.54	46.71	56.39	48.47
4	211	642	231	524	57.54	46.72	56.42	49.03
5	211	642	232	556	57.54	46.72	56.41	48.41
6	211	642	232	557	57.54	46.72	56.41	48.41
7	211	642	232	556	57.54	46.72	56.41	48.42
8	211	642	232	556	57.54	46.72	56.41	48.42

Table 3 Comparison of Correlation coefficients (CC) of the proposed method with the holistic approach for different numbers of scaling

No. of orientations	CC (proposed)			CC (H)
	Real	Imag	Mag	Mag
1	0.968	0.897	0.968	0.800
2	0.968	0.927	0.968	0.902
3	0.968	0.927	0.968	0.971
4	0.968	0.927	0.968	0.973
5	0.968	0.927	0.968	0.971
6	0.968	0.927	0.968	0.971
7	0.968	0.927	0.968	0.971
8	0.968	0.927	0.968	0.971

Table 4 Comparison of MSE and PSNR of the proposed method with the holistic approach for different numbers of scaling

No. of scaling	MSE (proposed)			MSE (H)	PSNR (Proposed)			PSNR (H)
	Real	Imag	Mag	Mag	Real	Imag	Mag	Mag
1	211	642	223	398	57.58	46.72	56.76	53.17
2	211	642	223	1046	57.58	46.72	56.76	43.57

Table 5 Comparison of Correlation coefficients (CC) of the proposed method with the holistic approach for different numbers of scaling

No. of scaling	CC (proposed)			CC (H)
	Real	Imag	Mag	Mag
1	0.968	0.927	0.968	0.987
2	0.968	0.927	0.968	0.902

experimental results are as follows: Number of scales—1 and 2, window sizes —$3 \times 3, 5 \times 5, 7 \times 7, 9 \times 9, 11 \times 11, 13 \times 13$ and 15×15, number of orientations —1, 2, 3, 4, 5, 6, 7 and 8. Apparently, low SNR, high PSNR and CC show better performance. In all these cases, we can see that the performance of the image reconstructed by using only the real coefficients outperforms the reconstruction using the imaginary coefficients and the magnitude of the coefficients. Additionally, reconstruction only by real coefficients is significantly better than the reconstruction by the magnitude of the coefficients of the holistic Gabor features. This is because, the receptive field profiles of human visual system can be best modeled by the product of a Gaussian and either a cosine or sine function [15]. Even receptive fields are bar detectors, while the odd receptive fields are edge detectors [16]. In following tables, H corresponds to holistic Gabor features, and the *real*, *imag* and *mag* denote the real, imaginary and magnitude information respectively.

4 Conclusion

Feature extraction is a vital step in many Computer Vision applications. Due to some important characteristics of Gabor wavelet, it is one of the widely used filters for feature extraction. In this paper, we proposed a local Gabor feature extraction method. To evaluate the performance of the proposed method, the original input image is reconstructed with the help of the extracted feature vector, and subsequently, the performance of our method is evaluated using the metrics such as MSE, PSNR and CC. Experimental results show that the proposed feature extraction method can efficiently represent an image. Additionally, we found that the reconstruction of an image only using the real coefficients outperforms the other two reconstruction methods *i.e.,* reconstruction only by imaginary coefficients, and reconstruction by using both the real and the imaginary coefficients (magnitude). The proposed method clearly has two important attributes—reduced computational complexity and the less memory requirement. So, the proposed method can be effectively used for the applications of image representation by a Gabor filter, which need less computation and also less memory.

References

1. Lisin, D.A., Mattar, M.A., Blaschko, M.B., Benfield, M.C., Learned-miller, E.G.: Combining local and global features for object class recognition. In: Proceedings of the IEEE International Conference on Computer Vision and Pattern Recognition, pp. 47–55 (2005)
2. Shen, L., Bai, L.: A review on Gabor wavelets for face recognition. Pattern Anal. Appl. **9**, 273–292 (2006)
3. Deng, H.-B., Jin, L.-W., Zhen, L.-X., Huang, J.-C.: A new facial expression recognition method based on local Gabor filter bank and PCA plus LDA. Int. J. Inform. Tech. **11**, 86–96 (2005)

4. Dongcheng, S., Fang, C., Guangyi, D.: Facial expression recognition based on Gabor wavelet phase features. In: Proceedings of the International Conference on Image and Graphics, pp. 520–523 (2013)
5. Xie, X., Lam, K.-M.: Facial expression recognition based on shape and texture. Pattern Recogn. **42**, 1003–1011 (2009)
6. Zhang, J., Tan. T., Ma, L.: Invariant texture segmentation for circular Gabor filters. In: Proceedings of the IEEE International Conference on Pattern Recognition, pp. 201–204 (2002)
7. Zhang, G., Ma, Z-M.: Texture feature extraction and description using Gabor wavelet in content-based medical image retrieval. In: Proceedings of the International Conference on Wavelet Analysis and Pattern Recognition, pp. 2–4 (2007)
8. Chen, B., Liu, Z-Q.: A novel face coding scheme based on Gabor wavelet networks and genetic algorithm. In: Proceedings of the International Conference on Wavelet Analysis and Pattern Recognition, pp. 1489–1492 (2007)
9. Sanger, T.: Stereo disparity computation using Gabor filters. Biol. Cybern. **54**, 405–418 (1988)
10. Lee, T.S.: Image representation using 2D Gabor wavelets. IEEE Trans. Pattern Anal. Mach. Intell. **18**, 959–971 (1996)
11. Daugman, J.G.: Complete discrete 2D Gabor transforms by neural networks for image analysis and compression. IEEE Trans. Acoust. Speech Signal Process. **36**, 1169–1179 (1988)
12. Bhagavathy, S., Tesic, J., Manjunath, B.S.: On the Rayleigh nature of Gabor filter outputs. In: Proceedings of the IEEE International Conference on Image Processing, pp. 745–748 (2003)
13. Scharstein, D., Szeliski, R.: A taxonomy and evaluation of dense two-frame stereo correspondence algorithms. Int. J. Comput. Vision **47**, 7–42 (2002)
14. Scharstein, D., Szeliski, R.: High-accuracy stereo depth maps using structured light. In: Proceedings of the IEEE International Conference on Computer Vision and Pattern Recognition, pp. 195–202 (2003)
15. Pollen, D. A., Ronner, S. F.: Visual cortical neurons as localized spatial frequency filters. IEEE Trans. Syst., Man, Cybern. **13**, 907–916 (1983))
16. Navarro, R., Tabernero, A.: Gaussian wavelet transform: two alternative fast implementations for images. Multidimension. Syst. Signal Process. **2**, 421–436 (1991)

Human Interaction Recognition Using Improved Spatio-Temporal Features

M. Sivarathinabala and S. Abirami

Abstract Human Interaction Recognition (HIR) plays a major role in building intelligent video surveillance systems. In this paper, a new interaction recognition mechanism has been proposed to recognize the activity/interaction of the person with improved spatio-temporal feature extraction techniques robust against occlusion. In order to identify the interaction between two persons, tracking is necessary step to track the movement of the person. Next to tracking, local spatio temporal interest points have been detected using corner detector and the motion of the each corner points have been analysed using optical flow. Feature descriptor provides the motion information and the location of the body parts where the motion is exhibited in the blobs. Action has been predicted from the pose information and the temporal information from the optical flow. Hierarchical SVM (H-SVM) has been used to recognize the interaction and Occlusion of blobs gets determined based on the intersection of the region lying in that path. Performance of this system has been tested over different data sets and results seem to be promising.

Keywords Video surveillance · Blob tracking · Spatio temporal features · Interaction recognition

1 Introduction

Video Surveillance is one of the major research fields in video analytics and this has been mainly used for security purpose. Human Interaction Recognition becomes a key step towards understanding the human behavior with respect to the scenes.

M. Sivarathinabala (✉) · S. Abirami
Department of Information Science and Technology, College of Engineering, Anna University, Chennai, India
e-mail: sivarathinabala@gmail.com

S. Abirami
e-mail: abirami_mr@yahoo.com

A. Nagar et al. (eds.), *Proceedings of 3rd International Conference on Advanced Computing, Networking and Informatics*, Smart Innovation, Systems and Technologies 43, DOI 10.1007/978-81-322-2538-6_20

Activity/Interaction recognition from the surveillance videos has been considered as challenging task. The situations such as background clutter, occlusion and illumination changes may cause difficulty in recognizing the activity of the person. Recognizing human interactions can be considered as the extension of single person action. In literatures [1–6], Action refers to the single person movement composed of multiple gestures such as arm/leg motion and torso motion, Activity refers to the combination of single or multiple people movement. Interaction may also refer to the activity that happens between two persons.

In videos, Action can be represented using global features as well as local features. Global features such as features considered from the entire image frame and local features have been considered from the local portion from the image frame. In literatures, Activity recognition approach relies on global feature representation or fusing all the local features or by computing the histogram. These approaches limit the performance of activity recognition under occlusion. To recognize interaction, body parts location and its movement during occlusion is still an open challenge. Thus we are motivated to propose a new middle level features to identify the activity/interaction even under occlusion.

Our Contribution lies in threefold: a new interaction recognition approach has been introduced to recognize/identify the activity of the person whenever there is crossover also. In feature extraction phase, Middle level features have been extracted to analyze the spatial and temporal relationships between two persons. The Hierarchical SVM classifier has been used to classify the interactions between two persons.

2 Related Works

Major works in the field of video analytics have been devoted in the object tracking and activity recognition phase and they are addressed in this section. Tracking the particular person/object under illumination conditions, occlusion and dynamic environments is still a challenging research. In general, Occlusion [2] can be classified into three categories: self occlusion, partial occlusion and total occlusion. In the previous work [7], to handle self and partial occlusion problems, a combination of blob tracking method and particle filter approach, has been employed by using Contour as shape representation and color as feature for tracking. In addition to this, blob splitting and merging approach has been attempted to identify occlusion.

Activity/Interaction of the tracked person has to be identified in order to increase the security in the environment. Human Interaction Recognition is crucial phase to understand from the nature of the moving persons. Arash Vahdat et al. [8] modelled activity with a sequence of key poses performed by the actors. Spatial arrangements between the actors are included in the model, for temporal ordering of the key poses. Chen et al. [9] proposed an automated reasoning based hierarchical framework for human activity recognition. This approach constructs a hierarchical structure for representing the composite activity. This structure is then transformed into logical formulas and rules, based on which the resolution based automated reasoning is applied to recognize the composite activity. There comes uncertainty in temporal and spatial relationship. This problem can be solved using mid level spatio-temporal features. Patron-Perez et al. [10] develop a per-person descriptor that uses head orientation and the local spatial and temporal context in a neighbourhood of each detected person. They also employed structured learning to capture spatial relationships between interacting individuals.

In literatures, many attempts have done to improve the interaction recognition rate using new feature extraction techniques and using new learning methods robustness to several complex situations such as background clutter, illumination changes and occlusion conditions. Here in this work, we have been provided solution to recognize the interaction between two persons under various situations.

3 Human Interaction Recognition

In this research, a special attempt has been made to identify the interaction between two persons from the tracked blobs and improved spatio-temporal features. Middle-level features [11], which connect local features and global features, are apparently suitable to represent complex activities. Our approach relies on corner detector and HOG descriptor to describe the pose of the person. Temporal features have been analyzed using optical flow for every corner points and in each of the body part. Pose information for head, arm, torso and leg has been obtained separately and clustered, in addition with the temporal features provided by the optical flow bins and then the activity/Interaction between two persons have been recognized. Semantics has been added with actions to recognize interactions without any confusion. Hierarchical SVM classifier has been used to classify the interaction from pose and activity classifiers. In this framework, spatio-temporal relationship has been maintained by constructing 5 bins in the optical flow descriptor [12, 13]. Real time human interaction recognition framework has been shown in Fig. 1.

4 Middle Level Features

Video Sequence has been represented by middle level features and their spatio-temporal relationships. In videos, local spatial temporal feature extraction has been widely used that involves interest point detection and feature description. From the tracked persons, local spatio-temporal features have been extracted directly to provide representation with respect to spatio-temporal shifts and scales. The spatio-temporal features can provide information about multiple motions in the scene. Feature detector usually selects the spatio-temporal locations and scales in the video. In this work, Harris corner detector [5] has been used to detect the spatio-temporal

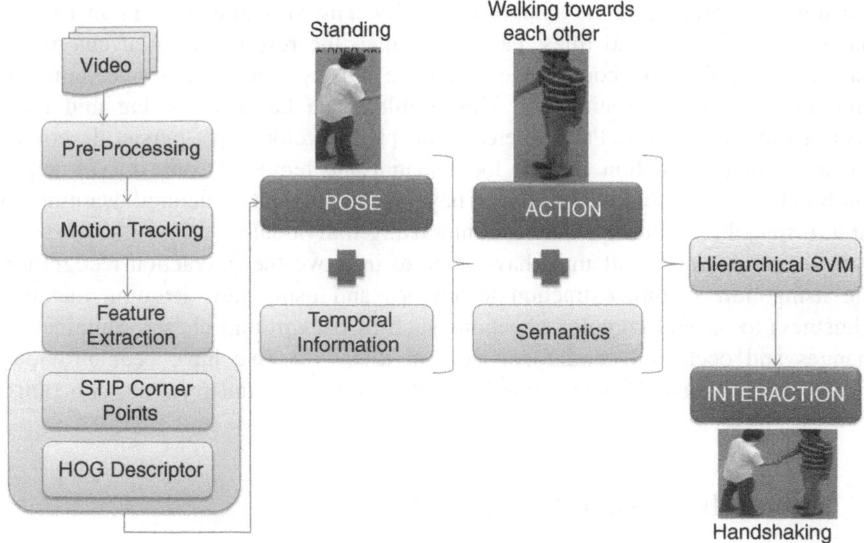

Fig. 1 Framework for human interaction recognition

points. Harris Corner detector, a simple detector that detects the interest points in the frame. Let the corner interest points be $p_1, p_2, p_3, \ldots, p_n$ described by the 2D co-ordinate. A densely sampled key point has been extracted that are stable in low resolution videos, clutter and also in fast motion and the motion has been analyzed from each corner points using optical flow algorithm. The distance between corresponding corner points of both the blobs makes separate spatial information. $d_t = [d[p_i] - d[p_j]]$ where i and j represents both the blobs, d_t represents the distance between the corresponding corner points in each frame at time t.

In general, detected interest points can be described using a suitable descriptor. Histogram of oriented gradients [10] that captures the spatial location of the body part and that is encoded by relative positions of HOG within the detector window. Motion information has been added with HOG i.e. a good feature combining this with the appearance (or) shape information makes a strong cue to represent the feature. Optical flow provides differential flow that will gives information about limb body relative motions. Temporal relations have been maintained throughout the video by analyzing optical flow from Hog descriptor. Optical flow motions has been differentiated into five different bins such as left, right, up, down and no motion.

Pose prediction is the first step in the interaction recognition and it can be done using HOG and SVM models. Posture has been estimated separately for the body parts such as head, torso, arm and leg. HOG descriptor builds a body part model pre trains the body parts into head, torso, arm and leg and SVM classification has been performed. Each forms a separate cluster and distance between the neighborhood points has been analyzed. $H_p = \{H_{1i}, H_{1j}, H_{2i}, H_{2j}, \ldots, H_{ni}, H_{nj}\}$; $A_p = \{A_{1i}, A_{1j},$

A_{2i}, A_{2j}, …, A_{ni}, A_{nj}}; T_p = {T_{1i}, T_{1j}, T_{2i}, T_{2j}, …, T_{ni}, T_{nj}}; L_p= {L_{1i}, L_{1j}, L_{2i}, L_{2j}, …, L_{ni}, L_{nj}} Where H_p, A_p, T_p and L_p represents the Head, Arm, Torso and Leg posture. $d_{Hp} = [d[H_{1i}]_t - d[H_{1i}]_{t+1}]$ where t and t + 1 represents the current frame and next frame. Similarly distance has been calculated and pose has been predicted. Next to the pose, actions have been analyzed from the spatial and temporal informations. Five classes of the actions such as walking, running, boxing, touching and hand waving that are trained and the actions have been classified.

5 Interaction Modeling

In each frame we have a set of human body detections $X = [x \ldots x_M]$. Each detection $xi = [l_x, l_y, s, q, v]$, has information about its corner location (l_x, l_y), scale (s), discrete body part orientation (q), and v represents SVM classification scores. Associated with each frame is a label $Y = [y_1 \ldots y_{k}, y_c]$. This label is formed by a class label y_i and K for each detection (where K is the number of interaction classes, with 0 representing the no-interaction class) and a configuration label y_c that serves as an index for one of the valid pairings of detections. (i, j) indicates that detection i is interacting with detection j and the 0 index means there is no interaction. The match between an input X and a labeling Y has been measured by the following cost function:

$$S(X, Y) = \sum\nolimits_{i}^{m} \alpha_{yiqi} v_{yi} + \sum \alpha_{yiqi} + \sum\nolimits_{(i,j)\in P_{yc}} (\delta_{ij}\beta_{yi}q_i + \delta_{ji}\beta_{yj}q_j) \quad (1)$$

where v_{yi} is the SVM classification score for class y_i of detection i, P_{yc} is the set of valid pairs defined by configuration index y_c, d_{ij} and d_{ji} are indicator vectors codifying the relative location of detection j with respect to detection i . y_iq_i are scalar weighting and bias parameters that measure the confidence that we have in the SVM score of class y_i when the discrete orientation of the body part is q_i. b_{yiqi} is a vector that weights each spatial configuration given a class label and discrete head orientation. Once the weights are learnt, we can find the label that maximizes the cost function by exhaustive search, which is possible given the small number of interaction classes and number of people in each frame. Interaction modeling has been shown in the Fig. 2a, b the handshake interaction model has been given.

The person 1 on the left side walks person 2 on the right, and they shaking their hands then rapidly depart. The poses such as arm stretch and arm stay and action such as walking has been correctly classified. From the optical flow bins, the directions has been identified such as the person moving right or left direction. A semantic description for the handshaking action has been shown in Fig. 2b. Along with the semantic descriptions, Interactions has been classified using hierarchical SVM. Another example is of kicking interaction has been shown in Fig. 3. Here in this case, leg posture has been considered. The poses such as leg stretch and leg stay and action such as walking has been correctly classified. From the optical flow bins, the directions has been identified such as the person moving right or left direction.

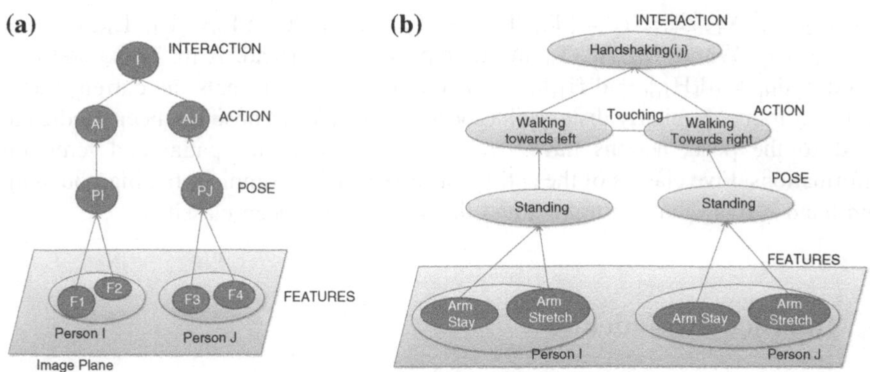

Fig. 2 **a** Interaction modeling and **b** handshaking interaction model

Kicking Interaction

Person 1 <Walk> Touching <Walk> Time

Leg Stretch Leg stay Leg Stretch

Person 2 <Walk> Touching <Walk> Time

Arm stay Arm stay Arm Stay

Fig. 3 Semantic descriptions of kicking interaction

Hierarchical Support Vector Machine (H-SVM) has been used to classify the interaction between two persons. To perform human interaction recognition, we fuse the features at two levels, (1) the output of the pose classifier and activity classifier has been concatenated and given as the input to classifier, and (2) the classifiers for the two sources are trained separately and classifier combination is performed subsequently to generate final result. To combine the classifier outputs of both the spatial and temporal features, classifier outputs have been interpreted as probability measure.

6 Results and Discussion

The implementation of this object tracking system has been done using MATLAB (Version2013a). MATLAB is a high performance language for technical computing. The input videos are taken from UT interaction dataset [14] and BIT interaction dataset [15]. The proposed algorithm have been applied and tested over many

Fig. 4 Sample frames from UT and BIT interaction datasets

(a)

(b)

Fig. 5 **a** Handshaking—interaction. **b** Pushing—interaction

different test cases and two of the scenarios have been shown here. The sample frames from the datasets has been shown in Fig. 4.

Figure 5a, b represents the handshaking and pushing interaction. HS represents Hand Shake and HF represents High Five interaction. Table 1a, b shows the confusion matrix for the UT and BIT Interaction datasets respectively. It is evident from the table that, the seven interactions classes has been trained and tested using hierarchical SVM. Our method using midlevel features obtained the accuracy of 90.1 % in UT Interaction dataset (set1) and 88.9 % in BIT Interaction dataset. Our approach has been compared with the existing methods used in interaction modelling shown in Table 2.

Table 1 Confusion matrix of UT interaction dataset and BIT interaction respectively

a					b				
	HS	Hug	Push	Punch		HF	Kick	Push	Pat
HS	1	X	X	X	HF	0.94	X	0.06	X
Hug	X	1	X	X	Kick	X	0.95	0.02	0.03
Push	X	X	0.98	0.02	Push	X	0.06	0.92	0.02
Punch	X	X	0.05	0.95	Pat	0.02	0.01	0.08	0.89

Table 2 Comparison with other existing methods

Previous works	Dataset considered	Accuracy (%)
BOW [16]	UT interaction (set1)	58.20
Visual co-occurrence [17]	UT interaction (set1)	40.63
Our method	UT interaction (set1)	90.1
	BIT interaction	88.9

7 Conclusion

In this research, an automated interaction recognition system has been developed using new spatio-temporal features which are called as mid level features. The features have been considered from the tracked blob in the spatial and also in temporal domain. We have been integrated spatio-temporal relationship between every consecutive frame in the video sequence. The activities of the each person have been identified and the activities/Interaction that happened between two persons has been recognized through the midlevel features and high level semantic descriptions. The intersecting regions between the potential detects the occlusion states. The proposed algorithm has been tested over BIT interaction dataset and UT interaction dataset. This system has the ability to recognize the interaction of the person even if there is a person/object crossover also. In future, this system could be extended along with the detection of heavy occlusion and multiple objects tracking too.

References

1. Yilmaz, A., Javed, O., Shah, M.: Object tracking :a survey. ACM Comput. Surv. **38**(4) (2006) (Article no. 13)
2. Ryoo, Agarwal, J.K.: Human activity analysis: a survey. ACM Comput. Surv. **43**(3), 16:1–16:43 (2011)
3. Dollar, P., Rabaud, V., Cottrell, G., Belongie, S.: Behavior recognition via sparse spatio temporal features. In: 2nd Joint IEEE International Workshop on Visual Surveillance and Performance Evaluation of Tracking and Surveillance, pp. 65–72 (2005)
4. Gowsikhaa, D., Manjunath, Abirami, S.: Suspicious human activity detection from surveillance videos. Int. J. Internet Distrib. Comput. Syst. **2**(2), 141–149 (2012)
5. Gowshikaa, D., Abirami, S., Baskaran, R.: Automated human behaviour analysis from surveillance videos: a survey. Artif. Intell. Rev. (1046). doi:10.1007/s2-012-9341-3(2012)
6. Gowsikhaa, D., Abirami, S., Baskaran, R.: Construction of image ontology using low level features for image retrieval. In: Proceedings of the International Conference on Computer Communication and Informatics, pp. 129–134 (2012)
7. Sivarathinabala, M., Abirami, S.: Motion tracking of humans under occlusion using blobs. Advanced Computing, Networking and Informatics, vol 1. Smart Innovation, Systems and Technologies, vol. 27 , pp. 251–258 (2014)

8. Vahdat, A., Gao, B., Ranjbar, M., Mori, G.: A discriminative key pose sequence model for recognizing human interactions. In: Proceedings of the IEEE International Conference on Computer Vision (2011)
9. Chen, S., Liu, J., Wang, H.: A hierarchical human activity recognition framework based on automated reasoning. In: The Proceedings of IEEE International Conference on Systems, Man, and Cybernetics, pp. 3495–3499 (2013)
10. Patron-Perez, A., Marszalek, M., Zisserman, A., Reid, I.: High five: recognising human interactions in TV shows. In: British Machine Vision Conference (2010)
11. Bruhn, A., weickert, J.: Lucas/Kanade meets Horn/Schunck: combining local and global optic flow methods. Int. J. Comput. Vision **61**(3), 211–231 (2005)
12. Jain, M., Jégou, H., Bouthemy, P.: Better exploiting motion for better action recognition. In: The Proceedings of IEEE International Conference on Computer Vision and Pattern Recognition, pp. 2555–2562 (2013)
13. Huang, K., Wang, S., Tan, T., Maybank, S.: Human behaviour analysis based on new motion descriptor. In: IEEE Transactions on Circuits and Systems for Video Technology (2009)
14. Ryoo, M.S., Aggarwal, J.K: UT Interaction Dataset, ICPR contest on Semantic Description of Human Activities (SDHA) (2010)
15. Yu Kong and Yunde Jia and Yun Fu, "Learning Human Interaction by Interactive Phrases", Book title,European Conference on Computer Vision,pp.300–313, vol.7572,2012
16. Schuldt, C., Laptev, I., Caputo, B.: Recognizing human actions: local SVM approach. In: The Proceedings of ICPR, Cambridge (2004)
17. Nour el houda Slimani, K., benezeth, Y., Souami, F.: Human interaction recognition based on the co-occurence of visual words. In: The Proceedings of IEEE Conference on Computer Vision and Pattern Recognition (CVPR) Workshops, pp. 455–460 (2014)

Common Coupled Fixed Point Results in Fuzzy Metric Spaces Using JCLR Property

Vishal Gupta, Ashima Kanwar and Naveen Gulati

Abstract The present study is devoted to use the notion of joint common limit (shortly, *JCLR* property) for coupled maps and utilize this concept to prove common coupled fixed point theorem on fuzzy metric space. In this paper, we also prove some fixed point theorems using the concept *CLR* property, E.A property and integral type contractive condition in fuzzy metric space. Illustrative examples supporting main results have been given.

Keywords Fixed point · Fuzzy metric space · Weakly compatible mappings · E.A property · *CLR* property · *JCLR* property

1 Introduction

The notion of a fuzzy set stems from the observation made by Zadeh [1] that "more often than not, the classes of objects encountered in the real physical world do not have precisely defined criteria of membership". In this paper, we are considering the fuzzy metric space in the sense of Kramosil and Michalek [2]. Following this concept, the notion of continuous t-norm is pertains to Georage and Veeramani [3].

2010 Mathematics Subject Classification: 54H25, 47H10, 47S40, 54A40.

V. Gupta (✉) · A. Kanwar
Maharishi Markandeshwar University, Mullana, Ambala 133207, Haryana, India
e-mail: vishal.gmn@gmail.com

A. Kanwar
e-mail: kanwar.ashima87@gmail.com

N. Gulati
S.D College, Ambala, Haryana, India
e-mail: dr.naveengulati31@gmail.com

A. Nagar et al. (eds.), *Proceedings of 3rd International Conference on Advanced Computing, Networking and Informatics*, Smart Innovation, Systems and Technologies 43, DOI 10.1007/978-81-322-2538-6_21

Bhaskar and Lakshmikantham [4], Lakshmikantham and Ćirić [5] discussed the mixed monotone mappings and gave some coupled fixed point theorems, which can be used to discuss the existence and uniqueness of solution for a periodic boundary value problem.

Definition 1.1 [4] An element $(x, y) \in X \times X$ is called a coupled fixed point of the mapping $P : X \times X \to X$ if $P(x, y) = x$, $P(y, x) = y$.

Definition 1.2 [5] An element $(x, y) \in X \times X$ is called a coupled coincidence point of the mappings $P : X \times X \to X$ and $a : X \to X$ if $P(x, y) = a(x)$, $P(y, x) = a(y)$.

Definition 1.3 [5] An element $(x, y) \in X \times X$ is called a common coupled fixed point of the mappings $P : X \times X \to X$ and $a : X \to X$ if

$$x = P(x, y) = a(x),\ y = P(y, x) = a(y).$$

Definition 1.4 [5] An element $x \in X$ is called a common fixed point of the mappings $P : X \times X \to X$ and $a : X \to X$ if $x = P(x, x) = a(x)$.

Definition 1.5 [6] The mappings $P : X \times X \to X$ and $a : X \to X$ are called weakly compatible if $P(x, y) = a(x)$, $P(y, x) = a(y)$ implies that
$aP(x, y) = aP((x), a(y))$, $aP(y, x) = aP((y), a(x))$ for all $x, y \in X$.

Aamri and Moutawakil [7] generalized the concept of non compatibility by defining E.A. property for self mappings. Sintunavarat and Kumam [8] defined the notion of common limit in the range property (or *CLR*) property in fuzzy metric spaces.

Definition 1.6 Let $(X, M, *)$ be fuzzy metric space. The mappings $P : X \times X \to X$ and $a : X \to X$ are said to satisfy E.A property if there exist sequences $\{x_n\}$ and $\{y_n\}$ in X, such that $\lim_{n\to\infty} P(x_n, y_n) = \lim_{n\to\infty} a(x_n) = x$, $\lim_{n\to\infty} P(y_n, x_n) = \lim_{n\to\infty} a(y_n) = y$ for some x, y in X.

Definition 1.7 Let $(X, M, *)$ be fuzzy metric space. The mappings $P : X \times X \to X$ and $a : X \to X$ are said to satisfy $CLR(a)$ property if there exist sequences $\{x_n\}$ and $\{y_n\}$ in X, such that $\lim_{n\to\infty} P(x_n, y_n) = \lim_{n\to\infty} a(x_n) = a(s), \lim_{n\to\infty} P(y_n, x_n) = \lim_{n\to\infty} a(y_n) = a(w)$, for some s, w in X.

Remark 1.8 Let the class Φ of all mappings $\phi : [0, 1] \to [0, 1]$ satisfying the properties such that ϕ is continuous and non-decreasing on $[0, 1]$ and $\phi(t) > t$ for all $t \in (0, 1)$. We note that $\phi \in \Phi$, then $\phi(1) = 1$ and $\phi(t) \geq t$ for all $t \in [0, 1]$.

2 Main Result

First, we explain the concept of the joint common limit (shortly, *JCLR* property) for coupled maps as follows:

Let $P, Q : X \times X \to X$ and $a, b : X \to X$ be four mappings. The pairs (P, a) and (Q, b) are satisfy the joint common limit in the range of a and b property if there exists sequences $\{x_n\}$, $\{y_n\}$, $\{u_n\}$ and $\{v_n\}$ such that

$$\lim_{n\to\infty} P(x_n, y_n) = \lim_{n\to\infty} a(x_n) = a(l) = \lim_{n\to\infty} Q(u_n, v_n) = \lim_{n\to\infty} b(u_n) = b(l) \text{ and}$$
$$\lim_{n\to\infty} P(y_n, x_n) = \lim_{n\to\infty} a(y_n) = a(m) = \lim_{n\to\infty} Q(v_n, u_n) = \lim_{n\to\infty} b(v_n) = b(m), \tag{C1}$$

for some l, m in X.

Now, we are ready to prove our first main result.

Theorem 2.1 *Let* $(X, M, *)$ *be FM-space, where* $*$ *is a continuous t-norm such that* $a * b = \min\{a, b\}$. *Let* $P, Q : X \times X \to X$ *and* $a, b : X \to X$ *be mappings such that the pairs* (P, a) *and* (Q, b) *are satisfy JCLR*(ab) *property and also satisfying following conditions: for all* $x, y \in X$, $t > 0, k \in (0, 1)$ *and* $\phi \in \Phi$.

$$M(P(x,y), Q(u,v), kt) \geq \phi \left\{ \begin{array}{l} M(P(x,y), b(u), t) * M(Q(u,v), a(x), t) \\ * \, M(P(x,y), a(x), t) * M(Q(u,v), b(u), t) \\ * \, M(a(x), b(u), t) \end{array} \right\}, \tag{C2}$$

Then, the pairs (P, a) *and* (Q, b) *have common coupled coincident point. Moreover, if* (P, a) *and* (Q, b) *are weakly compatible, then* (P, a) *and* (Q, b) *have a unique common fixed point in* X.

Proof The pairs (P, a) and (Q, b) are satisfy the joint common limit in the range of a and b property it satisfies the condition (C1) which is discussed above.

The proof is divided into five steps.

Step I: From condition (C1) and letting $n \to \infty$, we have

$$M(P(x_n, y_n), Q(l, m), kt) \geq \phi \left\{ \begin{array}{l} M(P(x_n, y_n), b(l), t) * M(Q(l, m), a(x_n), t) \\ * \, M(P(x_n, y_n), a(x_n), t) * M(Q(l, m), b(l), t) \\ * \, M(a(x_n), b(l), t) \end{array} \right\};$$

we have $Q(l, m) = b(l)$. In similar way,

$$Q(m, l) = b(m). \tag{1}$$

Step II: By using (C2) and considering $n \to \infty$ and conditions (C1), (1), we have

$$M(P(l,m), Q(u_n,v_n), kt) \geq \phi\left\{\begin{array}{l} M(P(l,m), b(u_n), t) * M(Q(u_n,v_n), a(l), t)* \\ M(P(l,m), a(l), t) * M(Q(u_n,v_n), b(u_n), t) \\ * M(a(x), b(u_n), t) \end{array}\right\},$$

$$\left.\begin{array}{r} P(l,m) = a(l) = Q(l,m) = b(l) \\ \text{Similarly}, P(m,l) = a(m) = Q(m,l) = b(m) \end{array}\right\}. \tag{2}$$

Hence we conclude that the pairs (P,a) and (Q,b) have common coupled coincident point l,m in X.

Now, we assume $P(l,m) = a(l) = Q(l,m) = b(l) = r_1$ and

$$P(m,l) = a(m) = Q(m,l) = b(m) = r_2, \text{where } r_1, r_2 \in X. \tag{3}$$

Since the pairs (P,a) and (Q,b) are weakly compatible, we get that

$$a(P(l,m)) = P(a(l), a(m)) \& a(P(m,l)) = P(a(m), a(l)).$$
$$b(Q(l,m)) = Q(b(l), b(m)) \& b(Q(m,l)) = P(b(m), b(l)).$$

From (3), we have

$$a(r_1) = P(r_1,r_2) \& a(r_2) = P(r_2,r_1),\ b(r_1) = Q(r_1,r_2) \& b(r_2) = Q(r_2,r_1). \tag{4}$$

Step III: We shall prove that $r_1 = P(r_1,r_2) \& r_2 = P(r_2,r_1)$.

For this, using (C2) and (2), (3), (4), we get

$$M(P(r_1,r_2), Q(l,m), kt) \geq \phi\left\{\begin{array}{l} M(P(r_1,r_2), b(l), t) * M(Q(l,m), a(r_1), t) * M(P(r_1,r_2), a(r_1), t) \\ * M(Q(l,m), b(l), t) * M(a(r_1), b(l), t) \end{array}\right\}.$$

This implies

$$r_1 = P(r_1,r_2) = a(r_1) \text{ and } r_2 = P(r_2,r_1) = a(r_2). \tag{5}$$

Step IV: Now, we shall show that $r_1 = Q(r_1,r_2) \& r_2 = Q(r_2,r_1)$.

Again using (C2) and (3), (4), (5),

$$M(P(l,m), Q(r_1,r_2), kt) \geq \phi\left\{\begin{array}{l} M(P(l,m), b(r_1), t) * M(Q(r_1,r_2), a(l), t) * M(P(l,m), a(l), t) \\ * M(Q(r_1,r_2), b(r_1), t) * M(a(x), b(r_1), t) \end{array}\right\},$$

we get $r_1 = P(r_1,r_2) = a(r_1) = Q(r_1,r_2) = b(r_1)$.

Similarly, we can show that

$$r_2 = P(r_2, r_1) = a(r_2) = Q(r_2, r_1) = b(r_2). \tag{6}$$

Step V: We shall assert that (P, a) and (Q, b) have common fixed point in X.For this, we shall prove that $r_1 = r_2$. Let suppose that $r_1 \neq r_2$. so,

$$\begin{aligned} M(r_1, r_2, t) &= M(P(r_1, r_2), Q(r_2, r_1), t) \\ &\geq \phi \left\{ \begin{array}{l} M(P(r_1, r_2), b(r_2), t) * M(Q(r_2, r_1), a(r_1), t) * M(P(r_1, r_2), a(r_1), t) \\ * \ M(Q(r_2, r_1), b(r_2), t) * M(a(r_1), b(r_2), t) \end{array} \right\}, \end{aligned}$$

this is contradiction to our supposition.

This implies $r_1 = r_2$. Thus, we proved that $r_1 = P(r_1, r_1) = a(r_1) = Q(r_1, r_1) = b(r_1)$. □

So, we conclude that the pairs (P, a) and (Q, b) have common fixed point in X. The uniqueness of the fixed point can be easily proved in the same way as above by using condition (C2). This completes the proof of Theorem 2.1.

Using the above theorem, we now state next theorem in which a pair of mappings satisfies *CLR* property.

Theorem 2.2 Let $(X, M, *)$ *be FM-space, where* $*$ *is a continuous t-norm such that* $a * b = \min\{a, b\}$. *Let* $P : X \times X \to X$ *and* $a : X \to X$ *be mappings such that the pair* (P, a) *is satisfies CLR*(a) *property and satisfying following condition:*

$$M(P(x, y), P(u, v), kt) \geq \phi \left\{ \begin{array}{l} M(P(x, y), a(u), t) * M(P(u, v), a(x), t) \\ * \ M(P(x, y), a(x), t) * M(P(u, v), a(u), t) \\ * \, M(a(x), a(u), t) \end{array} \right\}, \tag{C3}$$

for all $x, y \in X$, $k \in (0, 1)$, $t > 0$ *and* $\phi \in \Phi$.

Then, the pair (P, a) *has common coupled coincident point. Moreover, if the pair* (P, a) *is weakly compatible, then the pair* (P, a) *has a unique common fixed point in* X.

Proof By considering $P = Q$ and $a = b$ in Theorem (2.1) and the pair (P, a) is satisfies $CLR(a)$ property then there exist sequences $\{x_n\}$, $\{y_n\}$ in X such that $\lim_{n\to\infty} P(x_n, y_n) = \lim_{n\to\infty} a(x_n) = a(l)$ and $\lim_{n\to\infty} P(y_n, x_n) = \lim_{n\to\infty} a(y_n) = a(m)$, for some l, m in X. So, in the same way as in Theorem (2.1), we get the result. □

Now, we give some examples to support main results.

Example 2.3 Let $X = (0, 1]$, $a * b = \min(a, b)$, $M(x, y, t) = \frac{t}{t+|x-y|}$,

for all $x, y \in X$, $t > 0$. Then $(X, M, *)$ is a fuzzy metric space. Let P, Q : $X \times X \to X$ and a, $b : X \to X$ defined as

$$P(x, y) = \frac{x+y}{2},\ a(x) = x,\ Q(x, y) = x - y + 1,\ b(x) = \begin{cases} 1, & x = 1 \\ 0, & x \in (0, 1) \end{cases}.$$

Now consider the sequences $\{x_n\} = 1 + \frac{1}{n}$, $\{y_n\} = 1 - \frac{1}{n}$, $\{u_n\} = \frac{1}{n}$, $\{v_n\} = -\frac{1}{n}$ and $\phi(t) = \sqrt{t}$.

$$\lim_{n\to\infty} P(x_n, y_n) = \lim_{n\to\infty} a(x_n) = \lim_{n\to\infty} Q(u_n, v_n) = \lim_{n\to\infty} b(u_n) = 1 = a(1) = b(1),$$
$$\lim_{n\to\infty} P(y_n, x_n) = \lim_{n\to\infty} a(y_n) = \lim_{n\to\infty} Q(v_n, x_n) = \lim_{n\to\infty} b(v_n) = 1 = a(1) = b(1).$$

These imply the pairs (P, a) and (Q, b) are satisfying $JCLR(ab)$ property.

Also, $a(P(1, 1)) = P(a(1), a(1)) = 1$, $b(Q(1), Q(1)) = Q(b(1), b(1)) = 1$. Then (P, a) and (Q, b) are weakly compatible. Then all conditions in Theorem 2.1 are satisfied .So, we concluded that $x = 1$ is the unique fixed point of P, Q, a, b.

Example 2.4 Let $(X, M, *)$ be FM-space, where $X = [0, 1]$, $*$ is a continuous t-norm such that $a * b = \min\{a, b\}, M(x, y, t) = \frac{t}{t+|x-y|}$. Let $P : X \times X \to X$ and $a : X \to X$ are mappings defined as $P(x, y) = x + y$, $a(x) = x$, $\forall x, y \in X$.Now consider the sequences $\{x_n\} = \frac{1}{n}, \{y_n\} = -\frac{1}{n}$ and $\phi(t) = \sqrt{t}, t > 0$.

$$\lim_{n\to\infty} P(x_n, y_n) = \lim_{n\to\infty} a(x_n) = 0 = a(0),\ \lim_{n\to\infty} P(y_n, x_n) = \lim_{n\to\infty} a(y_n) = 0 = a(0).$$

This implies the pair (P, a) is satisfies $CLR(a)$ property. Also, the pair (P, a) is weakly compatible. All the conditions of Theorem 2.2 are satisfied. Thus the pair (P, a) has $x = 0$ unique common fixed point in X.

Theorem 2.5 *Let $(X, M, *)$ be FM-space, where $*$ is a continuous t-norm such that $a * b = \min\{a, b\}$. Let $P : X \times X \to X$ and $a : X \to X$ be a mapping such that the pair (P, a) is satisfies the E.A property and the range of $a(X)$ is closed subspace of X satisfying condition* (C3) *of Theorem 2.2. Then, the pair (P, a) has common coupled coincident point. Moreover, if the pair (P, a) is weakly compatible, then the pair (P, a) has a unique common fixed point in X.*

Proof Since the pair (P, a) is satisfies the E.A property then there exist sequences $\{x_n\}$, $\{y_n\}$ in X such that $\lim_{n\to\infty} P(x_n, y_n) = \lim_{n\to\infty} a(x_n) = u$ and $\lim_{n\to\infty} P(y_n, x_n) = \lim_{n\to\infty} a(y_n) = v$, for some u, v in X. □

Here, the range of $a(X)$ is closed subspace of X. By using the property closed subspace of X that there exist l, m in X such that $u = a(l)$, $v = a(m)$. Therefore, the pair (P, a) is satisfies $CLR(a)$ property. So, from Theorem (2.2), we get result immediately.

Branciari-Integral contractive type condition [9] result is very valuable in fixed point theory on fuzzy metric space. In next result, contractive condition of integral type is used which is the extension of main result using *JCLR* property.

Theorem 2.6 *Let $(X, M, *)$ be FM-space, where $*$ is a continuous t-norm such that $a * b = \min\{a, b\}$.Let $P, Q : X \times X \to X$ and $a, b : X \to X$ be mappings such that the pairs (P, a) and (Q, b) are satisfy $JCLR(ab)$ property and also satisfying this condition $\int_0^s \chi(t)dt \geq \phi\left(\int_0^w \chi(t)dt\right)$, where $\chi(t)$ is Lebesgue- integrable function, $k \in (0, 1)$, $t > 0$, $\phi \in \Phi$ and $s = M(P(x, y), Q(u, v), kt)$,*

$$w = M(P(x, y), b(u), t) * M(Q(u, v), a(x), t) * M(P(x, y), a(x), t) \\ * M(Q(u, v), b(u), t) * M(a(x), b(u), t)$$

Then, the pairs (P, a) and (Q, b) have common coupled coincident point. Moreover, if the pairs (P, a) and (Q, b) are weakly compatible, then (P, a) and (Q, b) have a unique common fixed point in X.

Proof By assuming $\chi(t) = 1$, we obtained the desired result with help of Theorem (2.1). □

References

1. Zadeh, L.A.: Fuzzy Sets. Inform. Control, **8**, 338–353 (1965)
2. Kramosil, I., Michalek, J.: Fuzzy metric and Statistical metric spaces, Kybernetica, **11**, 326–334 (1995)
3. George, A., Veeramani, P.: On some results in fuzzy metric spaces. Fuzzy Sets Syst. **64**, 336–344 (1994)
4. Bhaskar, T.G., Lakshmikantham, V.: Fixed point theorems in partially ordered metric spaces and applications. Nonlinear Anal. TMA. **65**, 1379–1393 (2006)
5. Lakshmikantham, V., Ćirić, LjB: Coupled fixed point theorems for nonlinear contractions in partially ordered metric space. Nonlinear Anal. TMA. **70**, 4341–4349 (2009)
6. Abbas, M., Ali Khan, M., Radenovic, S.: Common coupled fixed point theorems in cone metric spaces for w-compatible mappings. Appl. Math. Comput. **217**, 195–202 (2010)
7. Aamri, M., Moutawakil, D.E.I.: Some new common fixed point theorems under strict Contractive conditions. J. Math. Anal. Appl. **270**, 181–188 (2002)
8. Sintunavarat, W., Kuman, P.: Common fixed point theorems for a pair of weakly compatible mappings in fuzzy metric spaces, J. Appl. Math. Article ID: 637958 (2011)
9. Branciari, A.: A fixed point theorem for mappings satisfying a general contractive condition of integral type. Int. J. Math. Sci. **29**, 531–536 (2000)

10. Ćirić, L.j.: Common fixed point theorems for a family of non-self mappings in convex metric spaces. Nonlinear Anal. **71**, 1662–1669 (2009)
11. Deng, Z.: Fuzzy pseudometric spaces. J. Math. Anal. Appl. **86**, 74–95 (1982)
12. Gupta, V., Kanwar, A.: Fixed point theorem in fuzzy metric spaces satisfying E.A. property, Indian. J. Sci. Technol. **5**, 3767–3769 (2012)
13. Hu, X.-Q.: Common coupled fixed point theorems for contractive mappings in fuzzy metric spaces, Fixed Point Theor. Appl. Article ID 363716 (2011)
14. Mishra, S.N., Sharma, N., Singh, S.L.: Common fixed points of maps on fuzzy metric spaces. Int. J. Math. Sci. **17**, 253–258 (1994)
15. Schweizer, B., Sklar, A.: Statistical metric spaces, Pacific J. Math. **10,** 314–334 (1960)
16. Kumar, S., Singh, B., Gupta, V., Kang, S.M.: Some Common Fixed Point Theorems for weakly compatible mappings in fuzzy metric spaces. Int. J. Math. Anal. **8**, 571–583 (2014)

Speech Based Arithmetic Calculator Using Mel-Frequency Cepstral Coefficients and Gaussian Mixture Models

Moula Husain, S.M. Meena and Manjunath K. Gonal

Abstract In recent years, speech based computer interaction has become the most challenging and demanding application in the field of human computer interaction. Speech based Human computer interaction offers a more natural way to interact with computers and does not require special training. In this paper, we have made an attempt to build a human computer interaction system by developing speech based arithmetic calculator using Mel-Frequency Cepstral Coefficients and Gaussian Mixture Models. The system receives arithmetic expression in the form of isolated speech command words. Acoustic features such as Mel-Frequency Cepstral Coefficients features are extracted from the these speech commands. Mel-Frequency Cepstral features are used to train Gaussian mixture model. The model created after iterative training is used to predict input speech command either as a digit or an operator. After successful recognition of operators and digits, arithmetic expression will be evaluated and result of expression will be converted into an audio wave. Our system is tested with a speech database consisting of single digit numbers (0–9) and 5 basic arithmetic operators ($+, -, \times, /$ and $\%$). The recognition accuracy of the system is around 86 %. Our speech based HCI system can provide a great benefit of interacting with machines through multiple modalities. Also it supports in providing assistance to visually impaired and physically challenged people.

Keywords MFCC · GMM · EM algorithm

M. Husain (✉) · S.M. Meena · M.K. Gonal
B.V.B College of Engineering and Technology, Vidyanagar, Hubli 580031, Karnataka, India
e-mail: moulahusain@bvb.edu

S.M. Meena
e-mail: msm@bvb.edu

M.K. Gonal
e-mail: mkgonal@bvb.edu

A. Nagar et al. (eds.), *Proceedings of 3rd International Conference on Advanced Computing, Networking and Informatics*, Smart Innovation, Systems and Technologies 43, DOI 10.1007/978-81-322-2538-6_22

1 Introduction

In recent years, many of the emerging technologies are focusing towards improving the interaction between man and the machine. Speech is the most prominent modality which can offer a more natural and convenient way to interact with computers. Training a human being to understand and speak a spoken language is much easier compared to training computers, and make them to understand and speak our language. Automatic speech recognition (ASR) [1] technology is used to train machines which will map speech words into machine understandable form. Recognition of speech involves two major steps: audio feature extraction and, speech recognition using pattern recognition and machine learning algorithms. Generally, speech signal features will have great amount of internal variations. Such variations are caused by difference in age, accent, pronunciation, gender, environmental conditions and physiological aspects of the speaker. This inherent variability nature of speech signal demands for an efficient speech feature extraction and recognition system.

In this work, we implement a system which receives arithmetic expression in the form of speech commands. We extract most popularly used audio features, Mel-scaled Frequency Cepstral Coefficients (MFCC) from input speech commands. These features are used to train Gaussian Mixture Models (GMM). The model will be built by training GMM using Expectation Maximization (EM) algorithm. The model created using GMM can be used for recognizing speech commands and converting it into machine readable form. Once after recognizing digits and operators, it will compute the expression and store the result in the form of text. At the end, result of arithmetic expression present in the form of text is converted back to audio signal. The system has been tested with speech database of English language. Figure 1 shows the complete flow of feature extraction, recognition and evaluation of arithmetic expression.

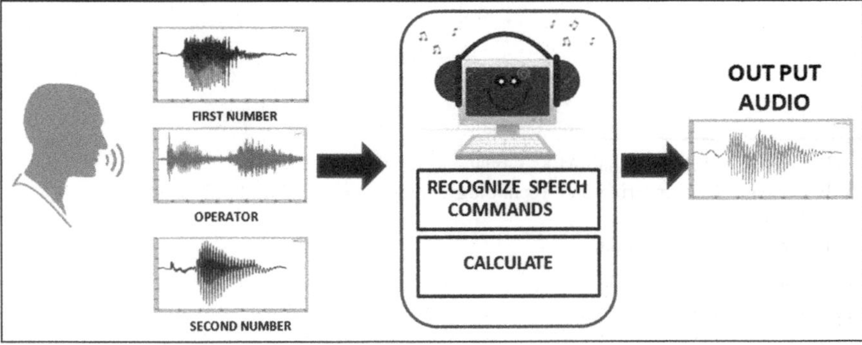

Fig. 1 Speech command based arithmetic calculator receiving input as speech command and generating result of computation in the form of speech

Generally, speech and speaker recognition systems will have two main stages: Front-end processing and Recognition using speech features. The most commonly used feature descriptors in the field of speech recognition are Linear prediction Coefficients (LPC) [2], Perceptual Linear Prediction (PLP) [3] and Mel Frequency Cepstral Coefficients (MFCC) [4]. In LPC feature extraction technique, repetitive nature of speech is exploited by predicting current sample as the linear combination of past m samples. The predicted sample $S_p(n)$ is given by

$$S_p(n) = -\sum_{i=0}^{m} a_i S(m-i) \tag{1}$$

S(m) is the short term spectrum of speech and a_i are coefficients of linear prediction. PLP also works based on short term spectrum of speech similar to LPC but uses some psychophysical based transformations. In this technique, a set of filter-banks are used to transform frequency spectrum in linear scale to a Bark scale given by

$$f_{bark} = \log\left(\frac{f_{lin}}{600} + \sqrt{1 + \left(\frac{f_{lin}}{600}\right)^2}\right) \tag{2}$$

where f_{bark} and f_{lin} are bark frequency and linear scale frequencies respectively. Such transformation of frequency scales will provide better representation of vocal tract or speech envelope. MFCC feature is almost similar to PLP feature but transforms frequencies from linear scale to Mel scale given by

$$f_{mel} = 2595 \times \log_{10}(1 + f_{lin}/700) \tag{3}$$

where f_{mel} represents mel-scale frequency after transformation and f_{lin} represents linear frequencies. In [4], Davis and Mermelstein have proved the advantages and noise robustness characteristics of MFCC over other speech features.

After extracting relevant features from speech signals we need to use a classifier for recognizing audio inputs. Hidden Markov Models (HMM) [5] and Gaussian Mixture Models (GMM) [6] are the two dominant statistical tools used in the speech and speaker recognition applications. HMMs are popularly known for speech recognition applications over 3 decades. GMMs are largely used for text independent speaker recognition applications. GMMs are sometimes considered as a component of HMMs. In this proposed work, our system receives inputs (numbers and operators) in the form of isolated audio waves. As our input commands are non continuous, we have implemented recognition of speech commands using GMM classifier. An iterative procedure, E-step and M-step (EM algorithm [7–9]) is used to build the GMM model.

In the next section, we describe extraction of MFCC features, in third section we explain GMM as classifier and training by using EM algorithm, fourth section gives details of the experiments conducted and results obtained, and in the final section conclusion and future scope is discussed.

2 Acoustic Feature Extraction

Extraction of features from speech signals play a significant role in recognizing audio words. In this section, we describe steps involved in extracting acoustic features from the audio waves. Figure 2 shows the various steps for computing and generating MFCC features from speech commands.

Pre-emphasis is the first step in feature extraction. It reduces the dynamic range of speech spectrum by boosting signals at higher frequencies. Pre-emphasis of speech waves can be performed by applying first order difference equation on input signal IS(n):

$$O(n) = I(n) - \alpha I(n-1) \tag{4}$$

where O(n) is the output signal. α is the pre-emphasis coefficient and typically lies in the range from 0.9 to 1.0. Z transform of pre-emphasis equation is given by

$$Z(x) = 1 - \alpha \times x^{-1} \tag{5}$$

Speech signals generally vary with time and statistically non stationary in nature. In order to extract stable acoustic features from the speech signals, we segment speech into successive frames of length typically 30 ms. In this short duration, speech signal will be considered as reasonably stationary or quasi stationary. Accuracy of features extracted depends on the optimum length of the frame. If the length of frame

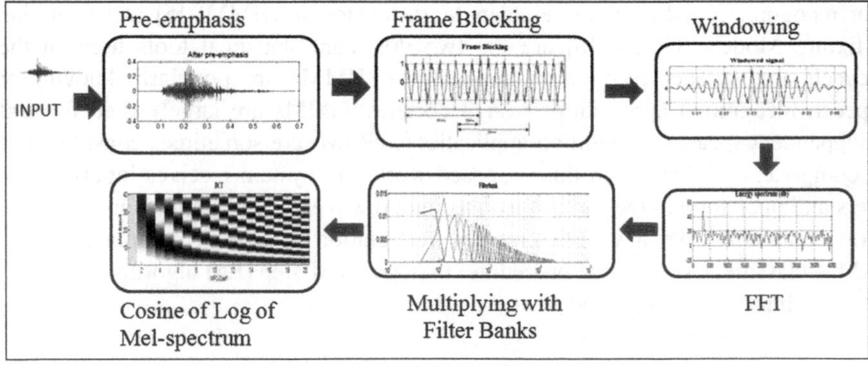

Fig. 2 MFCC feature extraction steps and related waveforms

is very small, we may loose temporal characteristics of the speech. The information loss due to quick transitions between successive frames can be controlled by overlapping successive frames with a duration of around 30 % of the frame length.

After framing of speech signal, each frame is multiplied with a window of duration equal to length of the frame. The window will be moved for every new frame by a duration equal to frame shift. The most simple window is rectangular window defined as

$$w(n) = \begin{cases} 1 & \forall 0 \leq \mathrm{n} \leq \mathrm{N}-1 \\ 0 & \text{otherwise} \end{cases} \tag{6}$$

But the drawback of rectangular window is, it generates side-lobes in the frequency domain because of abrupt transitions at the edge positions. These large side-lobes may cause spectral leakage where energy at a given frequency leaks into adjacent regions. This problem can be resolved by taking the window containing smooth decay at the edges. The most commonly used window for preventing spectral leakage by reducing the height and width of side-globes is Hamming window given by

$$w(n) = \begin{cases} 0.54 - 0.46 \cos\frac{(2\pi n)}{N} & \forall 0 \leq \mathrm{n} \leq \mathrm{N}-1 \\ 0 & \text{otherwise} \end{cases} \tag{7}$$

Approximately N samples of each frame indexed by n = 0 to N − 1 will be transformed from time domain to frequency domain using Discrete Fourier Transform (DFT) $F(k)$ given by

$$F(k) = \sum_{n=0}^{N-1} f(k) e^{\frac{-j2\pi nk}{N}} \tag{8}$$

where F(k) are Fourier coefficients and M is the length of FFT vector. The transformation from time domain signal to frequency spectrum gives sensible information for recognizing speech information. In practice, the DFT coefficients are calculated by using Fast Fourier Transform (FFT) algorithms which take time of $n \log n$ instead of n^2.

FFT of speech signal contains magnitude and phase components. Generally magnitude part of frequency response is useful for distinguishing speech signals. Phase component of frequency response may be useful for locating source of audio signal. Hence we neglect phase component of frequency spectrum. The squared magnitude frequency response of *i*th frame, by discarding phase component is given by

$$|F_i(k)|^2 = F_i(k) F_i^T(k) \tag{9}$$

According to psychophysical studies, human auditory perception follows a non linear frequency scale. This non-linear property can be achieved by multiplying power spectrum with a bank of k triangular filters. These triangular band pass filters are spaced uniformly over a mel-scale in such a way that it reflects non linear auditory behaviour of the human voice system. These mel-bank filters generate an array of filtered values, whose length is equal to the number of bank filters used. Also, mel-bank filters provide the benefit of reducing dimensions of the data.

The filter-bank energies are very much sensitive to sounds of very high and low loudness values. The sensitivity of these filter banks can be reduced by applying logarithmic function. Accuracy of speech recognition increases with the application of log function.

In the final step, we apply cosine function on log of mel-spectrum obtained in the previous step. The features obtained by applying cosine function on logarithm of Mel scaled spectrum is called MFCCs.

DCT coefficients obtained in the last step are useful for representing useful information required for speech recognition. The lower order DCT coefficients contain slowly varying vocal tract information (spectral envelope) where as higher order coefficients represent excitation information. In speech recognition as only spectral envelope information is required, we discard higher order coefficients and retain only lower order coefficients.

3 Gaussian Mixture Model for Recognition

Gaussian mixture models are parametric probabilistic models represented by the combination of finite set of multivariate normal distributions. Unlike K-means [7, 9, 10], GMM performs soft clustering and incorporates mixture weights, covariance and mean of the distributions. GMMs are generally used to model continuously varying information or biometric related features. Given a feature vector f, the mixture density for the speech sample 's' is given by

$$p(f|\lambda_s) = \sum_{j=1}^{K} w_j \mathcal{N}(f|M_j, C_j) \tag{10}$$

where $\mathcal{N}$ is a normal density function defined by

$$\mathcal{N}(f|M_j, C_j) = \frac{e^{-1/2(f-M_j)C_j^{-1}(f-M_j)^T}}{(2\pi)^{D/2}|C_j|^{1/2}} \tag{11}$$

M_j and C_j are mean and covariance vectors. w_j are the mixture weights and follow the relation $\sum_{j=1}^{K} w_j = 1$. In speech recognition each class of speech sample

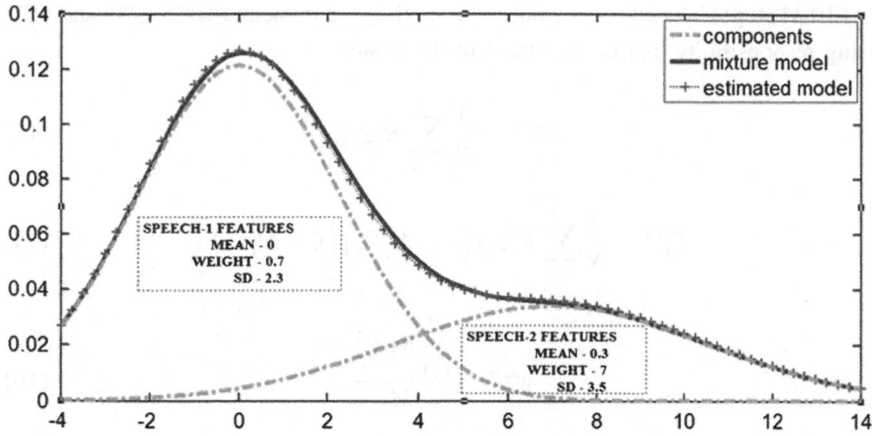

Fig. 3 Gaussian mixture model representation for two Gaussian components

(in our case numbers and operators) are represented by a unique GMM model. This GMM model λ_s can be used to predict the class of input sample. The GMM model λ_s is represented by a set of three parameters w_j, M_j and C_j. Figure 3 gives representation of speech signals and their distribution details. In speech recognition, given a feature vector of any sample 's', goal is to estimate the model λ_s parameters w_j, M_j and C_j such that its distribution matches with one of the training samples. The model parameters are estimated by using iterative method of Expectation and Maximization (EM) algorithm.

In EM algorithm of estimating model parameters, we first begin with an initial model λ_{cur}. Using EM algorithm we estimate GMM model λ_{new} parameters w_j, M_j and C_j such that $p(f|\lambda_{new}) \geq p(f|\lambda_{cur})$. The new model λ_{new} becomes the current model λ_{cur}. This procedure is repeated until it converges to an optimal point.

The EM algorithm estimates new model parameters as follows

(i) Initialize the mixture weights w_{cur}, means M_{cur} and covariances C_{cur}. Evaluate initial value of log likelihood given by

$$\ln p(f|\lambda_s) = \sum_{i=1}^{M} \ln \sum_{j=1}^{K} w_j \mathcal{N}(f_i|M_j, C_j) \tag{12}$$

(ii) **E step**: Calculate responsibility matrix that assign data points to the clusters.

$$R(z_{ij}) = \frac{w_j \mathcal{N}(f_i|M_j, C_j)}{\sum_{l=1}^{L} w_l \mathcal{N}(f_i|M_j, C_l)} \tag{13}$$

(iii) **M step**: Calculate new values of model λ_{new} parameters M_j^{new}, C_j^{new} and w_j^{new} using responsibility matrix calculated in the E-step.

$$M_j^{new} = \frac{1}{N_j} \sum_{i=1}^{M} R(z_{ij}) f_i \quad (14)$$

$$C_j^{new} = \frac{1}{N_j} \sum_{i=1}^{M} R(z_{ij}) \left(f_i - M_j^{new}\right) \left(f_i - M_j^{new}\right)^T \quad (15)$$

$$w_j^{new} = \frac{\sum_{i=1}^{M} R(z_{ij})}{M} \quad (16)$$

(iv) Recalculate log likelihood given by Eq. 12 using new values of model parameters.

Repeat the steps from 2 to 4 until convergence criteria is met either for model parameters or log likelihood function.

4 Experimental Results and Discussions

In this section, we discuss datasets used for feature extraction and recognition purpose. Also we discuss experimental set up and implementation details.

4.1 Dataset of Digits and Operators

We conducted a preliminary experiment and evaluated arithmetic calculator experiments on a speech database of numbers and operators. We recorded audio samples for single digit numbers from 0–9 to 5 arithmetic operators in a sound proof room. Our speech database for arithmetic calculation contains around 450 audio samples. We stored 30 speech samples (15 male and 15 female) for every class of digits and operators as WAV files. Also in each class of audio samples utterances were recorded at three different paces-normal, fast and slow.

4.2 Feature Extraction Using MFCC

We sampled speech signals stored for digits and operators at the rate of 16,000 Hz. The signals are divided into frames of length 400 samples with overlapping of 160 samples. The size of the hamming window is 400 samples. It is shifted with a frame shift of 160 samples. Frequency components are extracted by applying 512 point FFT algorithm on zero padded frames. Due to the symmetric property of FFT, we retain only half part of the spectrum that is 257 components. FFT power spectrum is multiplied with a set of 26 Mel-spaced filter banks of vector length 257 and then we add up all these coefficients. The resulting vector will have 26 mel scaled energies. We take log of these energies and finally apply DCT to get 26 cepstral coefficients. Out of 26 cepstral coefficients, first 12 lower order coefficients are retained. These 12 coefficients will form MFCC of each frame.

4.3 *Training and Validation of GMM*

In this work, we train GMM using Expectation Maximization algorithm on a data set of 450 audio samples of digits and operators. The model obtained after training is used for predicting the class of input speech command. In order to verify the robustness of the GMM as classifier, we used 10 fold cross validation. The confusion matrix obtained after validating GMM with 10 fold cross validation is shown in Table 1.

4.4 *Text to Speech Conversion*

Our speech based arithmetic calculator receives arithmetic expressions in the form of speech commands. The voice command can be recognized by using GMM

Table 1 Confusion matrix for a set of 5 operators and 5 numbers

Speech	1. Wav	2. Wav	3. Wav	4. Wav	5. Wav	Plus. wav	Minus. wav	Mul. wav	Div. wav	Mod. wav
1. Wav	85	2	3	0	0	4	0	6	0	0
2. Wav	0	84	1	2	3	4	6	0	0	0
3. Wav	1	1	88	0	2	0	3	0	4	1
4. Wav	2	0	3	85	0	0	2	1	3	4
5. Wav	5	0	6	0	89	0	0	0	0	0
Plus. wav	6	5	0	0	0	85	0	0	0	4
Minus. wav	0	0	3	0	0	4	89	0	0	4
Mul. wav	5	2	1	0	0	4	0	83	0	5
Div. wav	5	1	2	0	4	4	0	0	84	0
Mod. wav	2	2	1	2	2	2	0	0	0	89

model and converted into machine understandable form. The arithmetic operation is performed on the input numbers and result will be stored in the form of text. Finally, result of arithmetic expression present in the form of text is converted into audio signal by using Text To Speech (TTS) API.

5 Conclusion

In our speech based arithmetic calculator, we have demonstrated extraction of MFCC feature vectors which are suitable for representing content of audio signal. We used MFCC features to train GMM by using EM algorithm. The model obtained after training is used for predicting the class of speech command. The speech commands are recognized as either digits or operators. Finally, the arithmetic expression is evaluated based on the input speech command and the result will be converted back to audio signal. As calculator is tested with natural speech commands, accuracy obtained is around 86 %. In future, recognition rate and robustness of the system can be improved by exploring HMMs and deep learning features. Further, the system can be extended to support calculation of complex expressions and recognition of local languages.

References

1. Rabiner, L., Juang, B.-H.: Fundamentals of speech recognition. In: Smith, T.F., Waterman, M. S. (eds.) Identification of Common Molecular. Prentice-Hall, Inc., Upper Saddle River (1993)
2. Gouvianakis, N., Xydeas, C.: Advances in analysis by synthesis lpc speech coders. J. Inst. Electron. Radio Eng. **57**(6), S272S286 (1987)
3. Hermansky, H.: Perceptual linear predictive (PLP) analysis of speech. J. Acoust. Soc. Am. **57** (4), 173852 (1990)
4. Davis, S., Mermelstein, P.: Comparison of parametric representations for monosyllabic word recognition in continuously spoken sentences. IEEE Trans. Acoust. Speech Sig. Process. **28** (4), 357366 (1980)
5. Rabiner, L.: A tutorial on hidden markov models and selected applications in speech recognition. IEEE Proc. **77**(2), 257286 (1989)
6. Juang, B., Levinson, S., Sondhi, M.: Maximum likelihood estimation for multivariate mixture observations of markov chains (corresp.). IEEE Trans. Inf. Theory. **32**(2), 307–309 (1986)
7. Dempster, A.P., Laird, N.M., Rubin, D.B.: Maximum likelihood from incomplete data via the em algorithm. J. Roy. Stat. Soc. B **39**(1), 138 (1977)
8. Duda, R.O., Hart, P.E., Stork, D.G.: Pattern Classification, 2nd edn. Wiley-Interscience (2000)
9. Bishop, C.M.: Pattern Recognition and Machine Learning (Information Science and Statistics). Springer-Verlag New York Inc, Secaucus (2006)
10. Hartigan, J.A., Wong, M.A.: A k-means clustering algorithm. JSTOR: Appl. Stat. **28**(1), 100108 (1979)

Comparative Analysis on Optic Cup and Optic Disc Segmentation for Glaucoma Diagnosis

Niharika Thakur and Mamta Juneja

Abstract Glaucoma is an eye disease which causes continuous increase in size of optic cup and finally the permanent vision loss due to damage to the optic nerve. It is the second most prevailing disease all over the world which causes irreversible vision loss or blindness. It is caused due to increased pressure in the eyes which enlarges size of optic cup and further blocks flow of fluid to the optic nerve and deteriorates the vision. Cup to disc ratio is the measure indicator used to detect glaucoma. It is the ratio of sizes of optic cup to disc. The aim of this analysis is to study the performance of various segmentation approaches used for optic cup and optic disc so far by different researchers for detection of glaucoma in time.

Keywords Segmentation · Cup to disc ratio (CDR) · Optic disc · Optic cup · Glaucoma

1 Introduction

Glaucoma is a primary cause of permanent blindness all over the world. It occurs due to compression and/or deteriorating blood flow through the nerves of the eyes. It is caused due to increase in optic cup size present on the optic disc. Optic disc is the location from where major blood vessels enter to supply the blood to retina. Optic cup is the central depth of variable size present on optic disc. A disc with disease condition varies in color from a pink or orange to white. In a survey conducted by World Health Organization (WHO) it has been ranked as the second major cause of blindness across the world. It influenced nearly 5.2 million population across the world i.e. 15 % of the total world population and is expected to

N. Thakur (✉) · M. Juneja
Computer Science and Engineering UIET, Panjab University, Chandigarh, India
e-mail: niharikathakur04@gmail.com

M. Juneja
e-mail: mamtajuneja@pu.ac.in

A. Nagar et al. (eds.), *Proceedings of 3rd International Conference on Advanced Computing, Networking and Informatics*, Smart Innovation, Systems and Technologies 43, DOI 10.1007/978-81-322-2538-6_23

increase by 11.2 million population by 2020 [1]. In country like Thailand, it affected 2.5–3.8 % of the total population of the country or approximately 1.7–2.4 million population over the country [2]. Currently, Cup to disc ratio (CDR) is a measure indicator of glaucoma. It is calculated as the ratio of the vertical diameters of optic cup to optic disc. CDR can be determined by analyzing the sizes of optic cup and optic disc [3]. As a researcher we can diagnose the glaucoma by segmenting the optic cup and optic disc and then calculating the ratio's of their vertical diameter for diagnosis of glaucoma. According to survey conducted this can be said that CDR for person with normal eye is less than 0.5 and that for eyes with disease Glaucoma is more than 0.5.

2 Methodologies Used with Performance Evaluation

In 2008, Liu et al. [4] gave thresholding followed by variational level-set approach for segmentation of optic cup and optic disc. This improved the performance of segmentation in comparison to color intensity based method but had a drawback that cup segmentation was not accurate.

In 2008, previous approach was further improved by Wong et al. [5] by adding ellipse fitting to the previous method in post processing which increased the accuracy of the segmentation.

In 2009, Wong et al. [6] further added support vector machine (SVM) for classification, neural network for training and testing. This further improved the accuracy and increased the acceptability of glaucoma diagnosis.

Evaluation criteria used for comparing these approaches were acceptability which is calculated as difference between some standard clinical CDR and calculated CDR. CDR values with a difference of less than 0.2 units are considered appropriate [4].

Table 1 below shows the comparison of above techniques on the basis of Acceptability.

In 2010, Joshi et al. [7] used bottom-hat transform, morphological closing operation, region-based active contour and thresholding to detect optic disc and optic cup boundary. This approach was good at handling gradient distortion due to the vessels. Evaluation criteria used for this approach was Mean CDR error calculated as mean of differences of CDR values .

In 2013, this approach was further improved by Cheng et al. [8] that used simple linear iterative clustering, contrast enhanced histogram, color maps, center surround statistics and support vector machines for classification. This approach improved the [7] by reducing the value of Mean CDR error.

Table 2 below shows the comparison of techniques [7] and [8] on the basis of Mean CDR error.

In 2011, Joshi et al. [9] used gradient vector flow (GVF), chan-vese (C-V) model, optic disc localization, Contour Initialization, and Segmentation in Multi-Dimensional Feature Space for optic disc detection. It used vessels r-bends information and 2D spline interpolation for optic cup detection. Proposed method

Table 1 Comparison on the basis of acceptability

Authors	Acceptability
Liu et al. [4]	62.5 %
Wong et al. [5]	75 %
Wong et al. [6]	87.5 %

Table 2 Comparison on the basis of Mean CDR error

Authors	Mean CDR error
Joshi et al. [7]	0.121
Cheng et al. [8]	0.107

Table 3 Comparison on the basis of Recall, Precision F-score, Accuracy

Authors		Recall	Precision	F-score	Accuracy
Joshi et al. [9]	Optic cup	–	–	0.84	–
	Optic disc	–	–	0.97	–
Rama Krishnan et al. [10]	Optic disc	0.91	0.93	0.92	93.4 %
Noor et al. [11]	Optic cup	0.35	1	0.51	67.25 %
	Optic disc	0.42	1	0.59	70.90 %
Vimala et al. [12]	Optic disc	–	–	–	90 %
Noor et al. [11]	Optic cup	0.806	0.999	0.82	90.26 %
	Optic disc	0.876	0.997	0.93	93.70 %

of optic disc segmentation performed better than GVF and C-V model and that of cup was better than threshold and ellipse fitting. It used F-score for evaluating its result and found that F-score for this approach was higher than other approaches and hence was considered better than other approaches.

In 2012, Rama Krishnan et al. [10] used intuitionistic fuzzy histon for optic disc segmentation, adaptive histogram equalization for pre-processing for disc and gabor response filter for retinal classification. This approach improved the accuracy of this method as compared to other by increasing value of F-score.

In 2013, Noor et al. [11] used region of interest analysis followed by color channel analysis and color multithresholding for segmentation. It also evaluated various parameters such as Precision, recall, F-score, accuracy and calculated the CDR value. With this approach 88 % of the CDR results agreed with those of ophthalmologists.

In 2013, Vimala et al. [12] used line operator and fuzzy c means clustering for optic disc segmentation. This approach detected the optic disc better than other by achieving the accuracy of 90 %.

In 2014, Noor et al. [11] used region of interest extraction, morphological operation and fuzzy c mean segmentation to detect optic cup and optic disc. With this approach accuracy achieved for optic cup and optic disc was increased. 88 % of CDR results agreed with that from standard CDR achieved by ophthalmologist.

Parameters such as Precision, recall, F-score and accuracy are calculated as given below [13]

$$\text{Precision} = \frac{\text{tp}}{\text{tp} + \text{fp}} \times 100 \tag{1}$$

$$\text{Recall} = \frac{\text{tp}}{\text{tp} + \text{fn}} \times 100 \tag{2}$$

$$\text{F score} = 2 \times \frac{\text{Precision} \cdot \text{Recall}}{\text{Precision} + \text{Recall}} \tag{3}$$

$$\text{Accuracy} = \frac{\text{tp} + \text{tn}}{\text{tp} + \text{fp} + \text{fn} + \text{tn}} \times 100 \tag{4}$$

where,

tp is the region segmented as Disk/Cup that proved to be Disk/Cup
tn is the region segmented as non Disk/Cup that proved to be non Disk/Cup
fp is the region segmented as Disk/Cup that proved to be non Disk/Cup
fn is the region segmented as non Disk/Cup that proved to be Disk/Cup

Table 3 below shows the comparison on the basis of Precision, Recall, F-Score and Accuracy.

3 Conclusion

In this study, we analysed various existing methods used for optic cup and optic disc segmentation for glaucoma detection used by different researcher from time to time based on their performance. Most of these techniques used methods such as region of interest extraction, histogram equalization and morphological operations for pre-processing to overcome the problems of segmentation due to presence of vessels, noise etc. The performance of all these methods varies depending upon the segmentation techniques used by different researchers. All these methods have their own importance depending upon the types of images taken. It has gained a great attention in recent years due to the growth of glaucoma rapidly and commonly. Glaucoma is detected by calculating the CDR values which is the ratio of optic cup to optic disc vertical diameter. This ratio is achieved by segmenting optic disc and optic cup. Segmentation is done only in retinal fundus images captured by fundus cameras. From this comparative study, we analysed that clustering techniques are more appropriate for segmenting optic cup and optic disc due to improved accuracy as compared to others techniques and hence can be modified and improved further to increase the accuracy. Region of interest extraction makes the segmentation fast. It can be seen that optic cup segmentation is more appropriate in green channel and that of optic disc in red channel due to clear visibility of cup and disc in these channels.

Acknowledgments I sincerely thank all those people who helped me in completing this comparative analysis.

References

1. World Health Organisation Media centre (April, 2011). Magnitude and Causes of Visual Impairment [Online] Available: http://www.who.int/mediacentre/factsheets/
2. Rojpongpan, P.: Glaucoma is second leading cause of blindness [Online]. March (2009) Available: http://www.manager.co.th/QOL/ViewNews/
3. Burana-Anusorn, C. et al.: Image processing techniques for glaucoma detection using the cup-to-disc ratio. Thammasat Int. J. Sci. Technol. **18**(1), 22–34 January-March (2013)
4. Liu et al.: Optic cup and disc extraction from retinal fundus images for determination of cup-to-disc ratio. IEEE, 1828–1832, (2008)
5. Wong et al.: Level-set based automatic cup-to-disc ratio determination using retinal fundus images in ARGALI. In: 30th Annual International IEEE EMBS Conference Vancouver, British Columbia, Canada, August 20–24, pp. 2266–2269 (2008)
6. Wong et al.: Intelligent Fusion of Cup-to-Disc Ratio Determination Methods for Glaucoma Detection in ARGALI. In: 31st Annual International Conference of the IEEE EMBS Minneapolis, Minnesota, USA, September 2–6, pp. 5777–5780 (2009)
7. Joshi et al.: Optic disc and cup boundary detection using regional information, IEEE, ISBI, 948–951 (2010)
8. Cheng et al.: Super pixel classification based optic disc and optic cup segmentation for glaucoma screening. IEEE Trans. Med. Imaging. **32**(6), 1019–1032 (2013)
9. Joshi, G.D. et al.: Optic disc and optic cup segmentation for monocular colour retinal images for glaucoma detection. IEEE Trans. Med. Imaging. **30**(6), 1192–1205 (2011)
10. Rama Krishnan et al., Application of Intuitionistic Fuzzy Histon Segmentation for the Automated Detection of Optic Disc in Digital Fundus Images, In: Proceedings of the IEEE-EMBS International Conference on Biomedical and Health Informatics (BHI), Hong Kong and Shenzhen, China, pp. 444–447, 2–7 Jan (2012)
11. Noor et al.: Optic Cup and Disc Colour Channel Multithresholding Segmentation. In: IEEE International Conference on Control System, Computing and Engineering, Penang, Malaysia, pp. 530–534, 29 November–1 Dececmber (2013)
12. Annie Grace Vimala, G. S., Kaja Mohideen, S.: Automatic Detection of Optic Disk and Exudate from Retinal Images Using Clustering Algorithm. In: Proceedings of the 7th International Conference on Intelligent Systems and Control (ISCO), pp. 280–284 (2013)
13. Khalid, N.E.A., et al.: Fuzzy C-Means (FCM) for optic cup and disc segmentation with morphological operation, Procedia Computer Science. In: International Conference on Robot PRIDE 2013–2014 Medical and Rehabilitation Robotics and Instrumentation, ConfPRIDE 2013–2014. Elsevier, vol. 42, pp. 255–262, Dec 2014

Part III
Advanced Image Processing Methodologies

A Secure Image Encryption Algorithm Using LFSR and RC4 Key Stream Generator

Bhaskar Mondal, Nishith Sinha and Tarni Mandal

Abstract The increasing importance of security of multimedia data has prompted greater attention towards secure image encryption algorithms. In this paper, the authors propose a highly secure encryption algorithm with permutation-substitution architecture. In the permutation step, image pixels of the plain image are shuffled using Linear Feedback Shift Register (LFSR). The output of this step is an intermediary cipher image which is of the same size as that of the plain image. In the substitution step, sequence of random numbers is generated using the RC4 key stream generator which is XORed with the pixel value of the intermediary cipher image to produce the final cipher image. Experimental results and security analysis of the proposed scheme show that the proposed scheme is efficient and secure.

Keyword Permutation-substitution · Linear feedback shift register (LFSR) · RC4 key stream

1 Introduction

Security of image data is one of the primary concerns with growing multimedia transmission. Confidentiality of image data in the medical, military and intelligence fields is indispensable. Encryption techniques are applied on sensitive image data to

B. Mondal (✉) · T. Mandal
Department of Mathematics, National Institute of Technology Jamshedpur, Jharkhand 831014, India
e-mail: bhaskar.cse@nitjsr.ac.in

T. Mandal
e-mail: tmandal.math@nitjsr.ac.in

N. Sinha
Department of Computer Science and Engineering, Manipal Institute of Technology, Manipal 576104, Karnataka, India
e-mail: nhs.loyola@gmail.com

A. Nagar et al. (eds.), *Proceedings of 3rd International Conference on Advanced Computing, Networking and Informatics*, Smart Innovation, Systems and Technologies 43, DOI 10.1007/978-81-322-2538-6_24

ensure its security. This paper presents a highly secure image encryption algorithm based on the permutation-substitution architecture [14]. The proposed algorithm uses a 32 bit Linear Feedback Shift Register (LFSR) [8, 10] for permutation and RC4 key stream generator for substitution. Statistical analysis of the cipher image produced after performing permutation and substitution show that cipher image is secure enough to be used for image encryption and transmission.

Over the years, researchers have proposed various algorithms for image encryption by modifying the statistical and visual characteristics of the plain image. In [2, 6, 7, 11, 12, 15, 17], researchers have proposed algorithms for image encryption by using chaotic systems. Image encryption schemes have also been developed centered around finite field transformation. Authors in [5] have used finite field cosine transformation in two stages. In the first stage, image blocks of the plain image are transformed recursively while in the second stage, positions of the image blocks are transformed. In [9], authors propose to use elliptical curve cryptography for image encryption by first encoding and then encrypting the plain image.

The proposed algorithm uses LFSR for permutation. The initial value of the LFSR is called the seed. In every iteration, each bit of the LFSR is shifted by one bit position towards the right. This operation results in the flushing out of the least significant bit (LSB) and no value present for the most significant bit (MSB). The value of the MSB is determined as a linear feedback function of the current state of the LFSR. This process results in the generation of a pseudorandom number which is then used for permutation. Since the pseudorandom number is generated using a deterministic algorithm, the feedback function must be selected carefully ensuring that the sequence of bits produced have a long period. This would make the output from the LFSR seem truly random.

The intermediary cipher image produced after permutation of pixels using LFSR is subjected to substitution using RC4 key stream generator. RC4 key stream generator uses a key, which is provided to it using a key generator to generate pseudorandom numbers for substitution. The key stream generation process can be divided into two phases: Key Scheduling Algorithm (KSA) and Pseudo Random Generation Algorithm (PRGA). Execution of the KSA and PRGA phase results in a pseudo-random sequence of numbers which is then used for substitution.

The rest of this paper is organized as follows. Section 2 describes the Preliminaries. Section 3 discusses the proposed algorithm and Sect. 4 presents the experimental results followed by Sect. 5 presenting performance of the proposed algorithm and security analyses. Finally, we summarize our conclusions in Sect. 6.

2 Preliminaries

In this section, we explain the preliminaries namely LFSR and RC4 key stream generator in detail which would serve as the base for the rest of the paper.

2.1 Linear Feedback Shift Register

Linear feedback shift register (LFSR) [1, 10, 16] is a n-bit shift register which scrolls between $2^n - 1$ values. The initial value of the LFSR is called the seed. In every iteration, each bit of the LFSR is shifted by one bit position towards the right. This operation results in the flushing out of the least significant bit (LSB) and no value present for the most significant bit (MSB) as shown in Fig. 1. The value of the MSB is determined as a linear feedback function of the current state of the LFSR. This process results in the generation of a pseudorandom number which is then used for permutation. Since the pseudorandom number is generated using a deterministic algorithm, the feedback function must be selected carefully ensuring that the sequence of bits produced by LFSR have a long period. This would make the output from the LFSR seem truly random.

2.2 RC4 Key Stream Generator

The RC4 algorithm was initially proposed by Ron Rivest in 1987. Today, it is one of the most important stream ciphers widely used in various security protocols such as Wi-Fi Protocol Access (WPA) and Wired Equivalence Privacy (WEP). Popularity of RC4 is mainly because it is fast, utilizes less resources and easy to implement [4]. RC4 [3, 13] is also used extensively for pseudo-random number generation. It takes a secret key as the input and produces a stream of random bits using a deterministic algorithm. This stream of random bits is known as key stream.

The key stream generation process can be divided into two phases: Key Scheduling Algorithm (KSA) and Pseudo Random Generation Algorithm (PRGA). Initialization of S-box is done by KSA using a variable length key as the input. Shuffling of S-box values takes place in the PRGA phase. Execution of the KSA and PRGA phase results in a pseudo-random sequence of numbers which is then used for encryption.

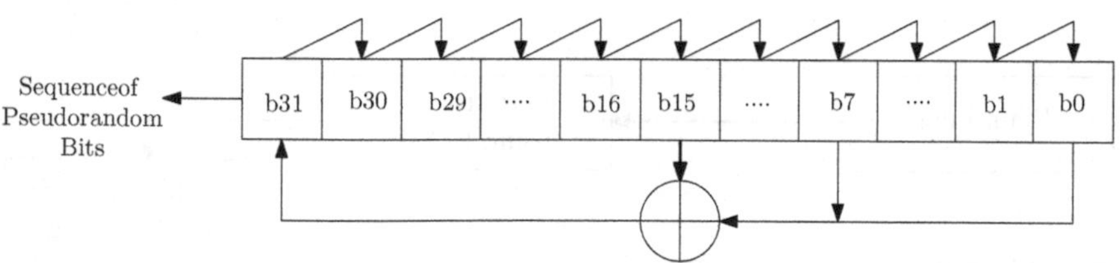

Fig. 1 32-bit LFSR as used in the proposed algorithm

3 The Proposed Scheme

The proposed algorithm can be divided into two parts-permutation using LFSR and substitution using RC4 key stream generator. Both these parts are described in detail in this section.

3.1 Permutation Using LFSR

Shuffling of pixel is done in the permutation process with the intent to modify the statistical and visual features of the plain image. A 32 bit seed is given as the input to the LFSR. Implementation of LFSR can be further divided into two phases. In the first phase, permutation of rows takes place while in the second phase permutation of columns are done.

Permutation process results in the intermediary cipher image which is of the same size as that of the plain image. Substitution is then performed on this intermediary cipher image to produce the final cipher image.

3.2 Substitution Using RC4 Key Stream Generator

Correlation among pixels of the plain image is destroyed by the shuffling of pixels in the permutation process, which is then subjected to substitution using RC4 key stream generator.

A sequence of N pseudorandom numbers is calculated using a key stream generator, where N is the total number of pixels. The pseudorandom number produced for each pixel is XORed with the present pixel value of the intermediary cipher image to produce the pixel value of that particular pixel in the final cipher image. Once, this process of substitution has been completed for all pixels of the intermediary cipher image, the final cipher image is obtained. Graphical representation of proposed algorithm is as given in Fig. 2.

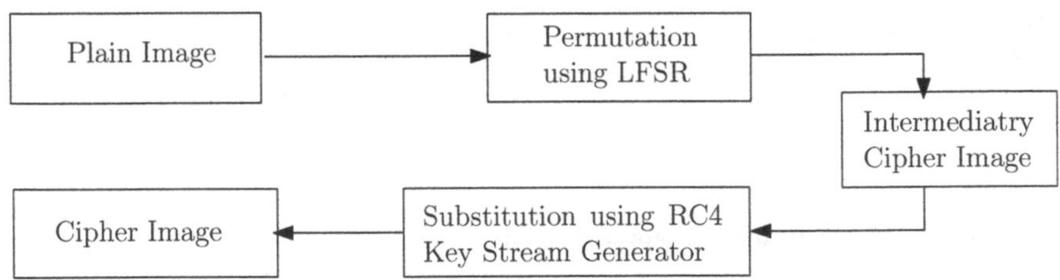

Fig. 2 Proposed encryption algorithm architecture

In the permutation phase, the pseudo random numbers generated from LFSR are XORed with the pixel locations of the plain image to determine the pixel to be swapped, first each row at a time then each column at a time. Similarly, the pseudo-random numbers generated for substitution from the RC4 key stream generator are XORed with the pixel values of the intermediary cipher image to produce the final cipher image.

4 Experimental Result

In this section, we establish the effectiveness of the proposed algorithm with the help of a couple of test cases. The two test cases are mentioned in detail below. Simulation results show that the visual characteristics of the cipher image are completely altered, thereby ensuring confidentiality of the image.

4.1 The Test Cases

Hre two images are use as test cases. The first test case Fig. 3, is an image of size $256 * 192$. The seed given to LFSR for permutation is 6571423141. For substitution, the key for the RC4 key stream generator is produced using the key generator. The value of $N = 49152$ and $Z = 2.0345 * 10^{-5}$ for the first test case.

In the second test case Fig. 6, we use an image of size $256 * 256$ pixels. The seed given to LFSR for permutation is 87253141. For substitution, the key for the RC4 key stream generator is produced using the key generator. The value of $N = 65536$ and $Z = 1.5258 * 10^{-5}$ for the second test case.

Fig. 3 Original image Test Case 1

5 Security Analysis of Test Results

One of the most important characteristics of any encryption algorithm is that it should be resistant to all kind of known attacks. In this section, we establish the robustness of the proposed algorithm against statistical, differential and brute force attacks.

5.1 Histogram Analysis

Image histogram exhibits the nature of distribution of pixels in an image. If the histogram of the encrypted image is uniform, the encryption algorithm is considered to be more resistant to statistical attacks. Figures 4 and 5 show the histogram of the

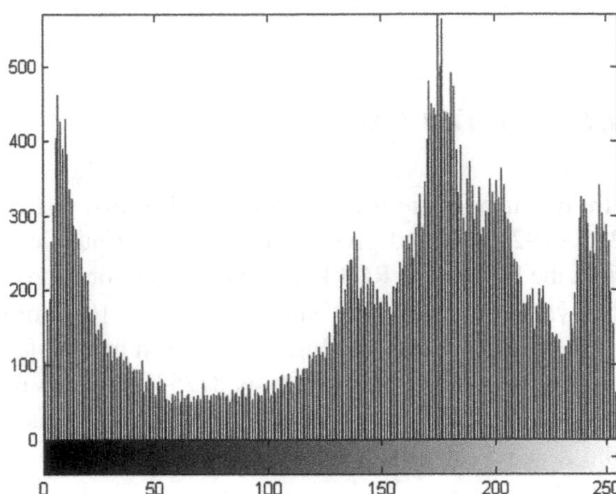

Fig. 4 Histogram of the original image-Test Case 1

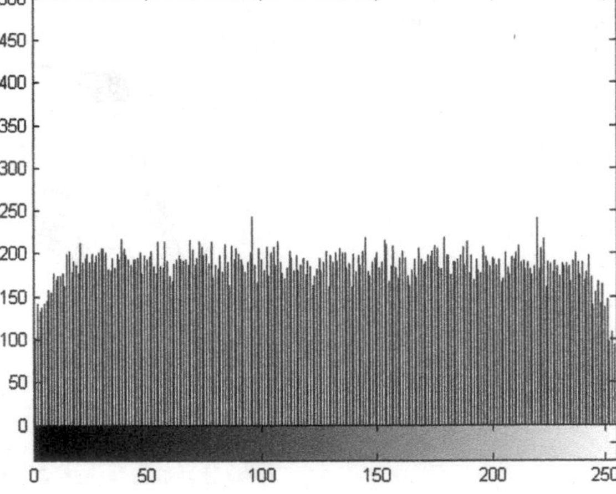

Fig. 5 Histogram, final cipher image-Test Case 1

Fig. 6 Original image Test Case 2

plain image and the final cipher image respectively used in Test Case 1 of Fig. 3 while Figs. 7 and 8 show the histogram of the plain image and the final cipher image respectively used in Test Case 2 of Fig. 6. It is evident that the distribution of pixels in the cipher image is uniform and is significantly different from that of the plain image establishing that the proposed algorithm is resilient against statistical and differential attacks.

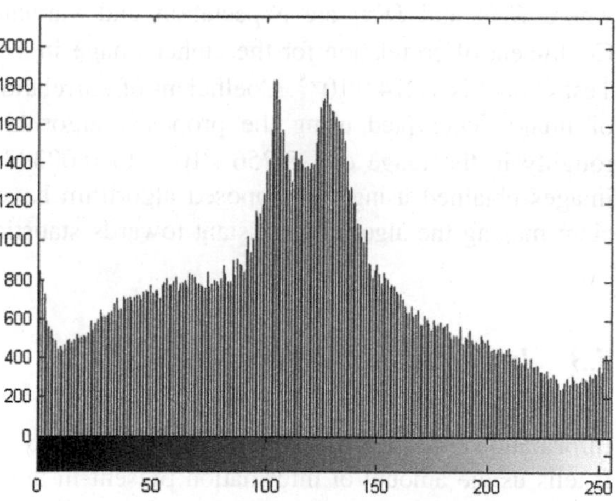

Fig. 7 Histogram of the original image-Test Case 2

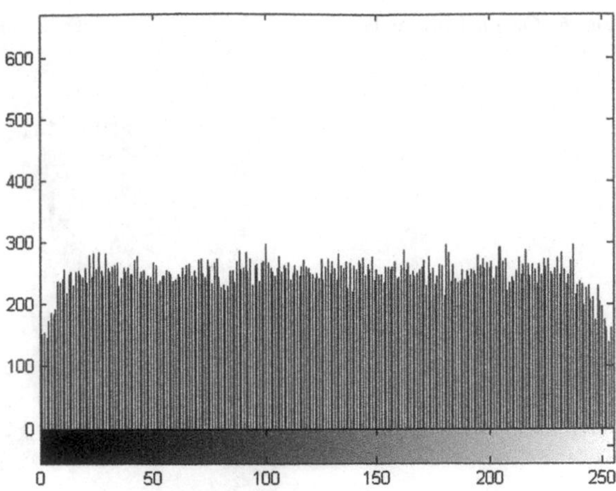

Fig. 8 Histogram of the final cipher image-Test Case 2

5.2 Correlation Analysis

Correlation between adjacent pixels of an image is an objective measure of the efficiency of encryption. Plain image has strong correlation between adjacent pixels. It is desired that the encryption algorithm removes the strong correlation in order to be resistant to statistical attacks. Correlation property can be computed by means of correlation coefficients as in Eq. 1:

$$r = \frac{E((x - E(x))(y - E(y))}{\sqrt{D(x)D(y)}} \tag{1}$$

where $E(x)$ and $D(x)$ are expectation and variance of variable x respectively. Coefficient of correlation for the cipher image in Test Case 1 is $8.138 * 10^{-4}$ and Test Case 2 is $1.274 * 10^{-5}$. Coefficient of correlation was calculated for a number of images encrypted using the proposed algorithm. The values obtained were roughly in the range of $-8.756 * 10^{-4}$ to 0.00397. This implies that the cipher images obtained using the proposed algorithm have fairly low correlation coefficient making the algorithm resistant towards statistical attacks.

5.3 Information Entropy Analysis

Information entropy is the degree of the uncertainty associated with a random event. It tells us the amount of information present in the event. It increases with uncertainty or randomness. It finds its application in various fields such as statistical inference, lossless data compression and cryptography. The entropy H(m) of m can be calculated as in Eq. 2:

$$H(m) = \Sigma_0^{L-1} p(m_i) \log_2 \frac{1}{p(m_i)} \tag{2}$$

where L is the total number of symbols, $m_i \in m$ and $p(m_i)$ is the probability of symbol m_i. In case of a random gray scale image, $H(m)$ should theoretically be equal to 8 as there are 256 gray levels. Information entropy value for the cipher image in Test Case 1 is 7.9536 and Test Case 2 is 7.9569. The values obtained were roughly in the range of 7.9536–7.9591 on random image test.

6 Sensitivity Analysis

Cipher image produced by the encrypting algorithm must be sensitive to both the plain image and the secret key to ensure resistance towards differential attacks. Number of pixel change rate (NPCR) in Eq. 3 and Unified Average Changing Intensity (UACI) in Eq. 4 are used to quantify sensitivity towards plaintext and secret key.

$$NPCR = (1/n) \sum_{i=1}^{n} D(x_i, y_i) \tag{3}$$

$$UACI = (1/n) \sum_{i=1}^{n} \frac{|x_i - y_i|}{255} \tag{4}$$

where $D(x_i, y_i) = 0$ if $x_i = y_i$ and $D(x_i, y_i) = 1$ if $x_i \neq y_i$. NPCR value for Test Case 1 is 0.9958 while the same for Test Case 2 is 0.9956. NPCR value was calculated for a number of images encrypted using the proposed algorithm. The values obtained were roughly in the range of 0.9956–0.9966. UACI for Test Case 1 is 0.2742 while the same for Test Case 2 is 0.1957. UACI value was calculated for a number of images encrypted using the proposed algorithm. The values obtained were roughly in the range of 0.19–0.27.

6.1 *Key Space Analysis*

The proposed algorithm uses a 32 bit seed for permutation and a 64 bit key for the substitution using a key generator. Thus the key space for the proposed algorithm is 2^{92}. Experimental results also validate that the suggested algorithm is highly sensitive to the secret key. Even a slight change to the secret key causes a substantial change to the cipher image formed. Hence, we can state that the proposed algorithm is resilient to brute force attacks.

7 Conclusion

In this paper, we present a novel image encryption algorithm based on the permutation-substitution architecture. In the proposed algorithm, LFSR is used for permutation. It takes a 32 bit seed as input and generates 32 bit pseudorandom number. Shuffling of pixels of the plain image takes place based on the generated pseudo-random number forming the intermediary cipher image. In the substitution phase, a 64 bit key is generated using a key generator which is fed to the RC4 key stream generator. Key stream is then used to alter the pixel values on the intermediary cipher image to form the final cipher image.

Simulation results using the proposed algorithm show that the visual characteristics of the plain image are completely altered ensuring confidentiality of the plain image. Various security analyses were performed on the cipher image obtained using the proposed algorithm. The results of these security tests proved that the proposed algorithm is resistant towards statistical, differential and brute force attacks demonstrating the security and validity of the proposed algorithm.

References

1. Ayinala, M., Parhi, K.K.: High-speed parallel architectures for linear feedback shift registers. IEEE Trans. Sig. Process. **59**(9), 4459–4469 (2011)
2. Fu, C., Chen, J.J., Zou, H., Meng, W.H., Zhan, Y.F., Yu, Y.W.: A chaos-based digital image encryption scheme with an improved diffusion strategy. Opt. Express **20**(3), 2363–2378 (2012)
3. Kwok, S.H.M., Lam, E.Y.: Effective uses of fpgas for brute-force attack on rc4 ciphers. IEEE Trans. Very Large Scale Integr. (VLSI) Syst. **16**(8), 1096–1100 (2008)
4. Lamba, C.S.: Design and analysis of stream cipher for network security. In: Communication Software and Networks, 2010. ICCSN'10. Second International Conference on, pp. 562–567 (2010)
5. Lima, J.B., Lima, E.A.O., Madeiro, F.: Image encryption based on the finite field cosine transform. Sig. Process. Image Commun. **28**(10), 1537–1547 (2013)
6. Mirzaei, O., Yaghoobi, M., Irani, H.: A new image encryption method: parallel sub-image encryption with hyper chaos. Nonlinear Dyn. **67**(1), 557–566 (2012)
7. Mondal, B., Priyadarshi, A., Hariharan, D.: An improved cryptography scheme for secure image communication. Int. J. Comput. Appl. **67**(18), 23–27 (2013). Published by Foundation of Computer Science, New York, USA
8. Mondal, B., Singh, S.K.: A highly secure steganography scheme for secure communication. Int. Conf. Comput. Commun. Adv. (IC3A), (2013)
9. Tawalbeh, L., Mowafi, M., Aljoby, W.: Use of elliptic curve cryptography for multimedia encryption. IET Inf. Secur. **7**(2), 67–74 (2013)
10. Wang, L.-T., McCluskey, E.J.: Linear feedback shift register design using cyclic codes. IEEE Trans. Comput. **37**(10), 1302–1306 (1988)
11. Wang, X.-Y., Wang, T.: A novel algorithm for image encryption based on couple chaotic systems. Int. J. Mod. Phys. B **26**(30), 1250175 (2012)
12. Wang, Y., Wong, K.-W., Liao, X., Chen, G.: A new chaos-based fast image encryption algorithm. Appl. Soft Comput. **11**(1), 514–522 (2011)

13. Weerasinghe, T.D.B.: An effective RC4 stream cipher. In: Industrial and Information Systems (ICIIS), 2013 8th IEEE International Conference on, pp. 69–74, Dec 2013
14. Wong, K.-W., Kwok, B.S.-H., Law, W.-S.: A fast image encryption scheme based on chaotic standard map. Phys. Lett. A **372**(15), 2645–2652 (2008)
15. Ye, G.: Image scrambling encryption algorithm of pixel bit based on chaos map. Pattern Recogn. Lett. **31**(5), 347–354 (2010)
16. Zadeh, A.A., Heys, H.M.: Simple power analysis applied to nonlinear feedback shift registers. IET Inf. Secur. **8**(3), 188–198 (2014)
17. Zhu, C.: A novel image encryption scheme based on improved hyperchaotic sequences. Opt. Commun. **285**(1), 29–37 (2012)

An Improved Reversible Data Hiding Technique Based on Histogram Bin Shifting

Smita Agrawal and Manoj Kumar

Abstract In this paper, we propose a novel reversible data hiding scheme, which can exactly recover the original cover image after the extraction of the watermark. The proposed scheme is based on the histogram bin shifting technique. This scheme utilizes peak point (maximum point) but unlike the reported algorithms based on histogram bin shifting, we utilize second peak point instead of zero point (minimum point) to minimize the distortion created by the shifting of pixels. Proposed scheme has low computational complexity. The scheme has been successfully applied on various standard test images and experimental results along with comparison with an existing scheme show the effectiveness of the proposed scheme. Higher Peak Signal to Noise Ratio (PSNR) values indicate that proposed scheme gives better results than existing reversible watermarking schemes.

Keywords Reversible watermarking · Histogram bin shifting · Peak point · PSNR

1 Introduction

In today's world, digital technology is growing by leaps and bounds. The revolution in digital technology has brought drastic changes in our lives. The recent advancement in digital technology has generated much chances for development of new and challenging things but has also raised the concern of protection of digital media such as audio, video, images etc. Digital watermarking is the art of covertly hiding secret information in digital media to protect and authenticate the media. In digital watermarking, the watermark is embedded into a multimedia data in such a

S. Agrawal (✉) · M. Kumar
Vidya Vihar, Babasaheb Bhimrao Ambedkar University, Raibareli Road, Lucknow, India
e-mail: smita.bbau@gmail.com

M. Kumar
e-mail: mkjnuiitr@gmail.com

A. Nagar et al. (eds.), *Proceedings of 3rd International Conference on Advanced Computing, Networking and Informatics*, Smart Innovation, Systems and Technologies 43, DOI 10.1007/978-81-322-2538-6_25

way that alteration of the multimedia cover data due to watermark embedding is perceptually negligible. In simple digital watermarking, there is always some loss of information due to distortion created by embedding of watermark bits which is unacceptable in some highly sensitive applications such as military, medical etc. Reversible watermarking, also called the 'lossless' or 'invertible' data hiding, is a special type of digital watermarking, which deals with such types of issues. In reversible watermarking, the watermark is embedded in such a way that at the time of extraction, along with watermark bits, original media is also recovered bit by bit. Reversible watermarking has many applications in various fields such as medical imaging, military etc., where even a slightest distortion in original media is not tolerable. It is a type of fragile watermarking, in which watermark is altered or destroyed even if any modification is made or any tampering is done to watermarked media and therefore original watermark and cover media cannot be recovered. Therefore, reversible data hiding is mainly used for content authentication. The main application of reversible data hiding is in the areas where distortion free recovery of original host media, after the extraction of watermark bits from the watermarked media, is of extreme importance.

In literature, various techniques have been proposed since the inception of reversible watermarking concept. Barton [1] was first to propose the reversible watermarking algorithm in 1997. Since then many researchers proposed various techniques and algorithms for reversible data hiding [2–10]. In reversible watermarking algorithms, main concern is to improve hiding capacity while maintaining or improving the visual quality of watermarked image. Reversible watermarking techniques can be categorized mainly into three categories [11], namely histogram modification based techniques [3], difference expansion based techniques [4, 7, 8] and compression based techniques [6]. Among all these proposed techniques, algorithms based on histogram modification belong to a simple but effective class of technique. Since the introduction of reversible data hiding algorithm based on histogram modification by Ni et al. [3], many variants have been proposed which utilize pixel value of most frequently occurring gray scale pixel point i.e. peak point and pixel value corresponding to which there is no pixel value in the image i.e. zero point. The main concept behind the histogram bin shifting based technique is to shift the pixels between peak point and zero point to create space next to the peak point for data embedding.

Histogram bin shifting based techniques are computationally very simple as compared to other techniques. Existing variants of histogram modification based algorithm utilize peak point and zero point to shift pixels between them. Although this technique is very simple and has various advantages such as no need of storing location map which contains information used to recover the original image, less distortion compared to many existing reversible data hiding techniques etc., yet there is a drawback that there are many pixel values between peak point and zero point and much distortion is caused due to the shifting of pixels between these two points. In our proposed work, we attempt to improve this shortcoming by minimizing the distortion by reducing the number of pixels to be shifted between peak point and zero point. Instead of utilizing peak point and zero point, we have used first peak point and second peak point to shift the pixels between them so that there are less number of pixels to be shifted for the embedding purpose. Due to this, there is less distortion in watermarked image and quality of watermarked image is improved in terms of perceptibility while maintaining the embedding capacity.

The rest of the paper is organized as follows: In Sect. 2, an overview of the histogram bin shifting based techniques is given through the scheme proposed by Ni et al. [3]. Section 3 describes proposed scheme. In Sect. 4, we discuss the experimental results and give comparison of the proposed scheme with the existing histogram modification based reversible watermarking technique [3]. In Sect. 5, Conclusions are drawn.

2 Existing Technique

In this section, we describe the histogram bin shifting based reversible data hiding technique proposed by Ni et al. [3] in which peak point and zero point or minimum point of histogram of given image are utilized and pixel values are altered slightly to embed the watermark.

2.1 Embedding Procedure

The basic histogram bin shifting technique [3] uses the histogram of original cover image. The main idea behind using the histogram is to utilize the peak point (the most frequently occurring pixel value) and zero point (the pixel value corresponding to which there is no gray scale value in the image) of the histogram of original image. The pixels between peak point and zero point are shifted by 1 unit to create space next to peak point and watermark bits are embedded in this space. For this process, histogram of given image is generated. Peak point and zero point of the histogram are stored. It is assumed that the value of peak point is always less than the value of zero point. Whole image is scanned in a sequence and all pixels between peak point and zero point are shifted to right by 1 to create space for data embedding next to the peak point. Again scan the image and where pixel value is found to be equal to peak point, check the to-be-embedded watermark bit sequence. If it is “1”, the grayscale pixel value is incremented by 1, otherwise pixel value remains as it is.

2.2 Extraction Procedure

For extraction of watermark and recovery of original cover image, watermarked image is scanned and if pixel value is found to be 1 greater than peak point value,

"1" is extracted as watermark bit. If pixel value is equal to the peak point value, "0" is extracted as watermark bit. In this way, watermark is extracted from the watermarked image. Whole image is scanned once again and all pixel values y, such that $y \in (peakpoint, zeropoint]$, are subtracted by 1. In this way, original image is recovered.

Example

For example, consider the Lena image shown in Fig. 1a. Figure 1b shows the histogram of Lena image before shifting. In Fig. 1b, peak point and zero point are shown. For creating the space for data embedding, pixels between peak point and zero point are right shifted by 1 and histogram after shifting is shown in Fig. 1c. In Fig. 1c, space created by shifting process can be seen next to the peak point. Now, image is scanned for pixel value equal to the peak point and watermark is embedded in the vacant space by incrementing the pixel value by 1 if watermark bit is "1" otherwise pixel value remains as it is. Figure 1d displays the histogram of watermarked Lena image.

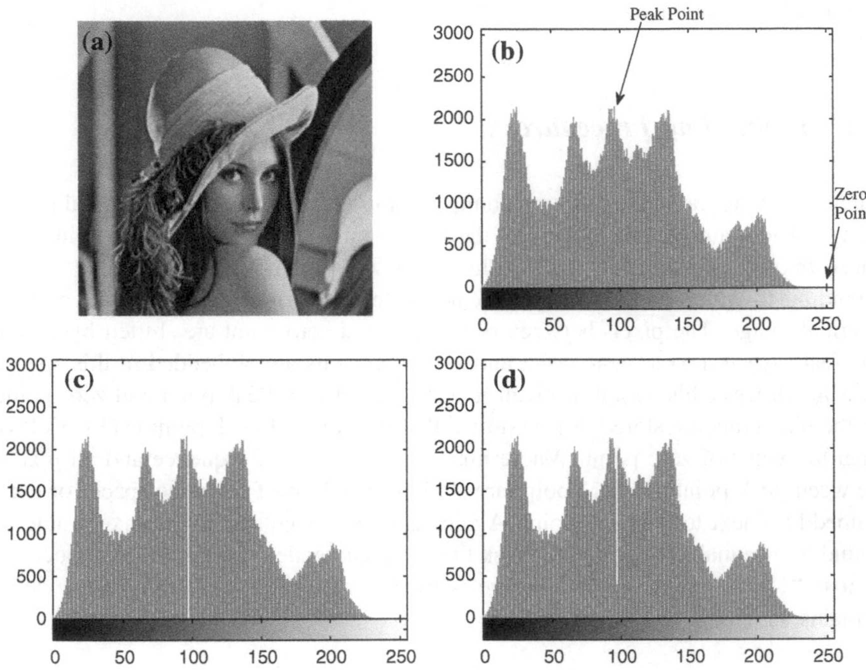

Fig. 1 Histogram bin shifting technique proposed by Ni et al. [3]. **a** Lena Image. **b** Histogram of Lena image before shifting. **c** Histogram of Lena image after shifting. **d** Histogram of Lena image after watermark embedding

3 Proposed Technique

In the proposed technique, we have considered first peak point and second peak point instead of peak point and zero point. First we have illustrated the proposed algorithm with the help of "Lena" image (512 × 512 × 8) shown in Fig. 1a and then proposed watermark embedding and extraction procedure is explained in detail.

We generate the histogram of Lena image and find the first peak point and second peak point (value just less than peak point). First peak point corresponds to the pixel value occurring most frequently in the given image, i.e. 96 and second peak point i.e. 25 in Fig. 2a. The main aim of finding peak point is to embed as much data as possible and aim of finding second peak point is the shifting of minimum number of pixels between first peak point and second peak point. Now, the whole image is scanned in a sequence, either by column wise or row wise. Here, value of second peak point (25) is less than first peak point (96), therefore all pixel values between 25 and 96 are decremented by "1", leaving the empty space at grayscale value 95 as shown in Fig. 2b. Once again, the whole image is scanned and if pixel value is found equal to 96, to be embedded watermark bits sequence is checked. If it is "1", decrement the corresponding pixel value, otherwise the pixel

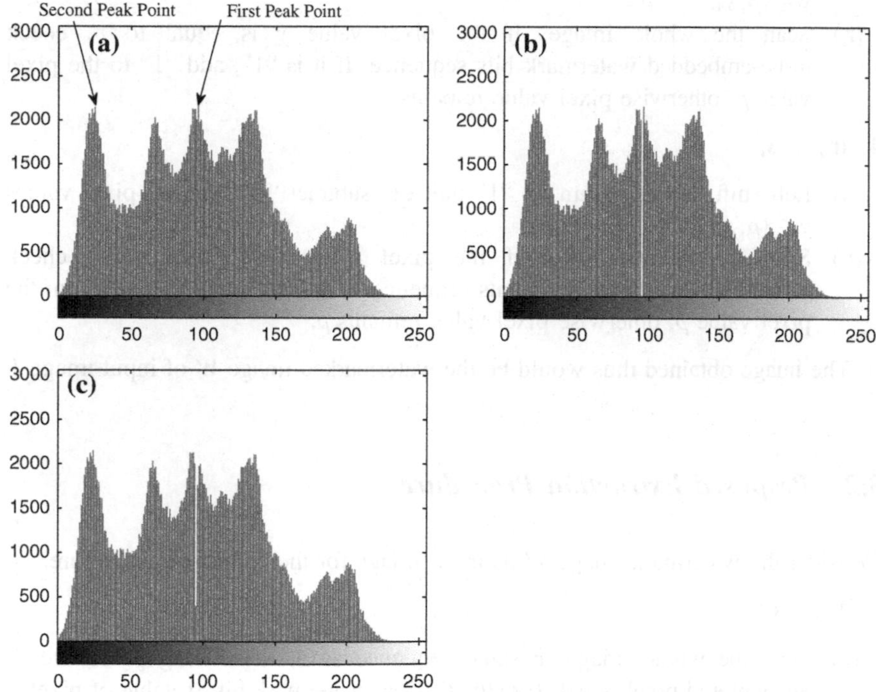

Fig. 2 Proposed scheme. **a** Histogram of Lena image before shifting. **b** Histogram of Lena image after shifting. **c** Histogram of Lena image after watermark embedding

value remains same. In this way, the whole embedding procedure is completed and the histogram of watermarked Lena image is shown in Fig. 2c.

For extracting the watermark and recovering original Lena image, scan the watermarked Lena image in same sequence as in embedding method. If a pixel is found having grayscale value 95 (i.e. 96-1), "1" is extracted as watermark bit and if pixel value is found equal to 96, "0" is extracted as watermark bit. In this way, watermark is extracted. For recovering the original image, whole watermark image is scanned once again, and if pixel value y is found such that $y \in [23, 96]$, the pixel value y is incremented by 1. In this way, original Lena image is recovered.

Following is the proposed embedding and extracting algorithms:

3.1 Proposed Embedding Procedure

1. Generate Histogram H of given image I.
2. Find first peak point p and second peak point s of histogram H, such that $p \in [0, 255]$ and $s \in [0, 255]$.
3. If $p < s$,
   (a) Right shift the histogram by "1" unit i.e. add "1" to all the pixel values $y \in (p, s)$.
   (b) Scan the whole image. If the pixel value y is equal to p, check to-be-embedded watermark bits sequence. If it is "1", add "1" to the pixel value p, otherwise pixel value remains p.
4. if $p > s$,
   (a) Left shift the histogram by "1" unit i.e. subtract "1" from all pixel values $y \in (p, s)$.
   (b) Scan the whole image. If the pixel value y is equal to p, check to-be-embedded watermark bits sequence. If it is "1", subtract "1" from the pixel value p, otherwise pixel value remains p.

The image obtained thus would be the watermarked image W of input image I.

3.2 Proposed Extraction Procedure

Consider the watermark image W as input image for the extraction procedure.

1. If $p < s$,
   (a) Scan the whole image in same sequence as in embedding procedure. If encountered pixel is $p + 1$, extract "1" as watermark bit. If value of pixel is p, extract "0" as watermark bit.

(b) Once again, scan the whole image and decrement the pixel value y by 1 if $y \in (p, s)$.

2. If $p > s$,

(a) Scan the whole image in same sequence as in embedding procedure. If encountered pixel value is $p - 1$, extract "1" as watermark bit. If value of pixel is p, extract "0" as watermark bit.
(b) Once again, scan the whole image and increment the pixel value y by 1 if $y \in (p, s)$.

4 Experimental Results

To demonstrate the effectiveness of the proposed scheme, we have implemented the proposed scheme on MATLAB for various standard grayscale test images shown in Fig. 3. Test images used for embedding the watermark are shown in Fig. 3a–d and the corresponding watermarked images are shown in Fig. 3e–h. We have also compared our proposed scheme with an existing scheme [3].

As it is evident from Figs. 1b and 2a that the number of pixel values between peak point and zero point are more than the number of pixel values between first peak point and second peak point for histogram of Lena image. In Fig. 1b, the peak point is 96 and zero point is 255. So the number of pixel values to be altered are 158 while in Fig. 2a, the pixel values to be altered are 70 because in this, first peak point is 96 and second peak point is 25. Therefore, there is less shifting by using proposed algorithm and hence less distortion.

Peak Signal to Noise Ratio (PSNR) is used as a measure for distortion. High PSNR values means less distortion and thus better visual quality. The formula for calculating PSNR is as follows:

$$PSNR = 10 * \log_{10}\left(\frac{255 \times 255}{MSE}\right)$$

where, Mean Square Error (MSE) is defined as-

$$MSE = \frac{1}{mn}\sum_{i=0}^{m-1}\sum_{j=0}^{n-1}[I(i,j) - W(i,j)]^2$$

where, I is the original image of size $m \times n$ and W is the watermarked image.

PSNR values for proposed algorithm and existing algorithms are shown in Table 1. Higher PSNR values of proposed scheme in comparison to existing scheme [3] show the effectiveness of proposed scheme. PSNR values for images having less pixel values between first peak point and second peak point are higher than others because there are less number of pixels that are shifted between these

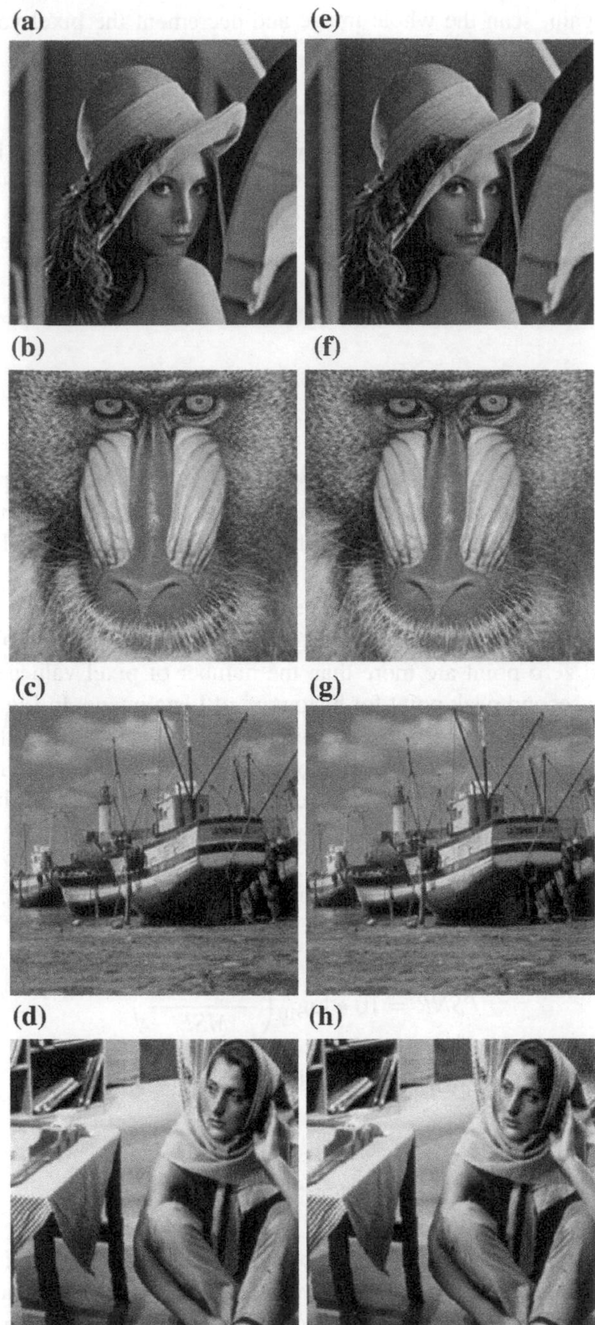

Fig. 3 Implementation of proposed scheme on various standard grayscale images of size 512 × 512. **a–d** Original images (Lena, Mandrill, Boat, Barbara). **e–h** Watermarked images (Lena, Mandrill, Boat, Barbara)

Table 1 PSNR of proposed scheme and existing scheme for various test images shown in Fig. 3a–d

S. no.	Test images	Embedding rate (bpp)	PSNR (proposed scheme)	PSNR (existing scheme)
1	Lena	0.0082	52.23	51.07
2	Mandrill	0.0103	52.00	50.48
3	Boat	0.0221	53.05	52.50
4	Barbara	0.0143	63.47	48.30

two points. Therefore less distortion is caused in the watermarked image and visual quality of watermarked image is improved while maintaining the embedding capacity. PSNR value for Barbara image using proposed scheme is much higher than the PSNR value using existing scheme because only 18 pixel values are changed using proposed scheme as compared to existing scheme where 236 pixel values between peak point and zero point that have been altered. So, it is evident from the above discussion that the proposed scheme is very useful for the images that are more textured.

5 Conclusions

Histogram modification is a technique used for embedding watermark in reversible manner. In this paper, a simple and improved version of existing histogram bin shifting technique has been proposed to minimize the distortion caused due to watermark embedding while maintaining embedding capacity. By choosing second peak point, we strive to minimize the number of pixels shifted during embedding process because there are always less number of pixel values between first peak point and second peak point in comparison to the number of pixel values between peak point and zero point or minimum point. Higher PSNR values demonstrate and verify the effectiveness of the proposed scheme. Experimental results show that the proposed scheme is better than existing reversible watermarking technique in terms of PSNR values.

References

1. Barton, J.M.: Method and apparatus for embedding authentication information within digital data. U.S. Patent No. 5646997. (1997)
2. Honsinger, C.W., Jones, P.W., Rabbani, M., Stoffel, J.C.: Lossless recovery of an original image containing embedded data. U.S. Patent No. 6278791. (2001)
3. Ni, Z., Shi, Y., Ansari, N., Su, W.: Reversible data hiding. IEEE Trans. Circuits Syst. Video Technol. **16**(3), 354–362 (2006)
4. Tian, J.: Reversible data embedding using a difference expansion. IEEE Trans. Circuits Syst. Video Technol. **13**, 890–896 (2003)

5. Yang, B., Schucker, M., Funk, W., Busch, C., Sun, S.: Integer-DCT based reversible watermarking technique for images using companding technique. Proc. SPIE **5306**, 405–415 (2007)
6. Sharma, C., Tekalp, S.: Lossless generalized-LSB data embedding. IEEE Trans. Image Process. **14**(2), (2005)
7. Thodi, D.M., Rodríguez, J.J.: Expansion embedding techniques for reversible watermarking. IEEE Trans. Image Process. **16**(3), 721–730 (2007)
8. Lee, C.C., Wu, H.C., Tsai, C.S., Chu, Y.P.: Adaptive lossless steganographic scheme with centralized difference expansion. Pattern Recogn. **41**, 2097–2106 (2008)
9. Luo, L., Chen, Z., Chen, M., Xeng, X., Xiong, Z.: Reversible image watermarking using interpolation technique. IEEE Trans. Inf. Forensics Secur. **5**(1), 187–193 (2010)
10. Leung, H., Cheng, L.M., Liu, F., Fu, Q.K.: Adaptive reversible data hididng based on block median preservation and modification of prediction errors. J. Syst. Softw. **86**, 2204–2219 (2013)
11. Feng, J.B., Lin, I.C., Tsai, C.S., Chu, Y.P.: Reversible watermarking: current status and key issues. Int. J. Netw. Secur. **2**(3), 161–171 (2006)

A Novel Approach Towards Detection and Identification of Stages of Breast Cancer

M. Varalatchoumy and M. Ravishankar

Abstract A robust and efficient CAD system to detect and classify breast cancer at its early stages is an essential requirement of radiologists. This paper proposes a system that detects, classifies and also recognizes the stage of the detected tumor which helps radiologists in reducing false positive predictions. A MRM image is preprocessed using histogram equalization and dynamic thresholding approach. Segmentation of the preprocessed image is carried out using a novel hybrid approach, which is a hybridization of PSO and K-Means clustering. Fourteen textural features are extracted from the segmented region in order to classify the tumor using Artificial Neural Network. If the tumor is classified as malignant then the stage of the tumor is identified using size as a key parameter.

Keywords Robust and efficient CAD system · Histogram equalization and dynamic thresholding · Novel hybrid approach of PSO and K-Means clustering · Textural features · Artificial neural network · Size of tumor

1 Introduction

The systems used for Medical Image Processing and Analysis [1] aids in visualization and analysis of medical images from various modalities. Researchers mainly use the Medical Image Processing systems to intensify their capability to diagnose and provide treatment for various medical disorders [2]. Computer Aided Diagnosis [3] systems are mainly developed to aid the radiologists in their analysis. As,

M. Varalatchoumy (✉)
Department of Information Science and Engineering, Dayananda Sagar College of Engineering, Kumaraswamy Layout, Bangalore, India
e-mail: Kvl186@gmail.com

M. Ravishankar
Vidya Vikas Institute of Engineering and Technology, Mysore, India
e-mail: ravishankarmcn@gmail.com

A. Nagar et al. (eds.), *Proceedings of 3rd International Conference on Advanced Computing, Networking and Informatics*, Smart Innovation, Systems and Technologies 43, DOI 10.1007/978-81-322-2538-6_26

identifying abnormalities proves to be a challenging task for well trained radiologists, they usually tend to result in more false positive predictions. A well developed, robust CAD system can be used by radiologists to help them in overcoming the above mentioned challenge.

This paper presents a robust and efficient CAD system, which is capable of detecting, classifying and numbering the stages of breast cancer. Magnetic Resonance Mammography is considered to be the best imaging modality [2] as it detects various abnormalities related to breast cancer [3]. The proposed CAD system analysis a MRM image in order to detect breast cancer at its early stages. The various modules of a CAD system, used for analyzing a MRM image are preprocessing [3], segmentation of ROI [3], Feature Extraction [3], classification [3] and detection of stages. The individual modules and the system on the whole has been tested on Mammography Image Analysis Society (MIAS) database [3] images.

The first module, namely, preprocessing is used to remove unwanted disturbances or noises [4] from a MRM image. This module also intensifies the features, which helps to improve the robustness of other modules.

A woman's age plays an important role in deciding the efficiency of a CAD system. MRM images of old women usually show the presence of bright objects inside the grey regions. On the other hand analysis of the MRM image of young women seems to be very challenging as the tumor region is found to be present with glandular-disc. This challenge is overcome by an efficient segmentation approach, which differentiates the abnormalities present in the breast region from the background. The efficiency of segmentation module is proved by the reduction of false positive [3] detections.

The segmentation module is followed by feature extraction module which is used to extract various textural features which plays a key role in classifying the detected abnormality as either benign or malignant tumor. Efficiency towards classification has been achieved by using renowned and simple classifier that is fast and accurate in performance. Finally the staging module is used to identify the severity and impact of the tumor by detecting the stage that the tumor belongs to. This serves to be a major support to radiologists as it can be used as a second tool, prior to directing the patients to undergo Biopsy, thereby avoiding the stress undergone by the patients.

The developed CAD system, when tested on multiple images, has proven to be highly robust and efficient when compared to all existing systems. The CAD system can be of major support to the medical field as it aids in reducing mortality rate.

2 Related Work

Egmont Peterson et al. [5] has reviewed the effect of neural networks on image processing. Applications of feed-forward neural networks, kohonen feature maps and Hopfield neural networks have been discussed in detail. Two-Dimensional taxonomy has been used to classify the applications. Advantages and disadvantages of various neural networks, along with the issues related to neural network has been discussed in detail.

Schaefer et al. [6] has analyzed how thermography can be used to detect breast cancer using statistical features and fuzzy classification. The novel method when tested on 150 cases has provided 80 % accurate results. The paper also provides a listing on several features that can be used for feature extraction and highlights the drawbacks of fuzzy classification.

Kocur et al. [7] has experimented how artificial neural networks can be used to select unique and best features to perform automatic detection. The paper also emphasis the criticality, of selecting unique features, which aids in improving the efficiency from 72 to 88 %. It has also been identified that, techniques used for feature selection, helps in analyzing risk factor data, to identify relations that are non linear between historical patient data and cancer.

Ahmed et al. [8] has experimented different types of wavelets to and identified the better type of wavelet with its optimal potential level of decomposition[] that helps in improving detection. An overall study of optimal level of decomposition has been performed to aid in automatic detection using multiresolution wavelet transform [9].

Tang et al. [1] briefs various CAD techniques involved in detection of breast cancer [9]. The principles of breast imaging and various types of mammograms has been discussed. Several algorithms used for classification and detection has been reviewed. The need for image enhancement, for a CAD system has also been explained in detail.

Nagi et al. [2] has analyzed the benefits and drawbacks of seeded region growing algorithm [9] and morphological preprocessing in segmentation of region of interest. The testing results of the algorithm over multiple mammograms reveals that seeded growing segmentation method fails to capture complete breast region.

Gopi Raju et al. [10] has experimented the effect of, combining few clustering algorithms with Particle Swarm Optimization for segmentation of mammograms. Various types of PSO along with their advantages and disadvantages have been discussed in detail. The performance of the proposed algorithms of PSO has been measured based on statistical parameters.

Ganesan et al. [3] has proposed a new algorithm for segmentation, by combining seeded region growing and PSO. It has been proved that PSO is highly capable of overcoming the similarity and seed selection problem of seeded region growing algorithm. The purpose of region merging has also been explained in detail.

Tandan et al. [11] has reviewed various PSO based methods, to automatically identify cluster centre in random data set. The survey has mainly considered PSO methods that do not require aprior knowledge of already existing regions in the image. Effect of Parallel PSO algorithm in robotics and neural networks has been highlighted.

3 Proposed Methodology

The proposed system consist of five major modules that performs various tasks such as preprocessing, segmentation, Feature extraction, classification of the tumor as benign or malignant and in case of malignancy further processing is done to identify the stage the tumor belongs to. The CAD system has been developed as a robust and efficient system which has achieved better results for individual modules when compared to all other existing algorithms and techniques.

The first module, preprocessing module, has be developed using histogram equalization and dynamic thresholding techniques. These techniques have played a major role in suppression of pectoral muscle and removal of various artifacts in a MRM image.

In the second module, namely, segmentation of Region of Interest, a novel approach has been used for segmentation which involves hybridization of two efficient approaches, namely, Particle Swarm Optimization and K-Means Clustering. The combination of these techniques has helped in achieving highly efficient segmentation, which has played a major role in attaining 95 % efficiency of the overall system.

Third module, Feature Extraction, is mainly used to extract the textural features to attain perfect classification. Texture can be defined as a sub pattern in an image, which is basically a group of pixels organized in a periodic manner in the spatial domain [9].

In the fourth module, classification, Artificial Neural Network has been used which classifies the detected region of abnormality as benign or malignant tumor based on the textural features.

Finally the fifth module, staging, is entered if the tumor is detected to be a malignant tumor. Based on input from experts, size has been used as a parameter in identifying the first two stages of the tumor. Figure 1 presents the overall design of the proposed Computer Aided Diagnosis system that is used to detect, classify and recognize the stage of breast cancer.

The following sub sections provide a detail description of the individual modules along with the samples of respective outcomes obtained.

3.1 Module I-Preprocessing

Preprocessing stage is considered to be the most crucial stage, as the detection of region of interest is highly related to the output of this stage. The more efficient preprocessing is, the more efficient segmentation would be. Denoising, pectoral muscle suppression and enhancing are considered to be the vital steps of preprocessing a MRM image. There are different types of noises present in a MRM image like label [4], tape artifacts [4]. Filters that exist for preprocessing includes adaptive median filter [4], Weiner filter [4], Median filter [4]. Although these filters are very

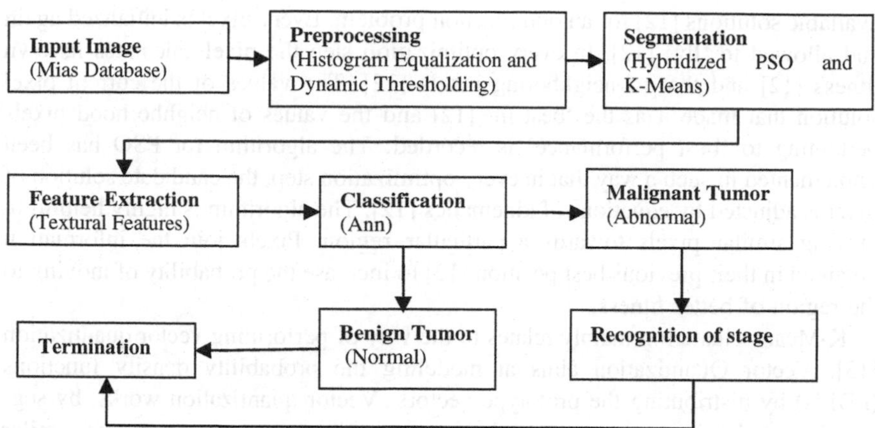

Fig. 1 Proposed computer aided diagnosis system for detection and recognition of stages of breast cancer

simple to implement, several drawbacks exists for all filters. The proposed system aims at achieving maximum accuracy in preprocessing by overcoming the drawbacks of existing filters. Hence, Histogram equalization and dynamic thresholding has been used for preprocessing. This approach overcomes all the drawbacks of the existing techniques. Horizontal and vertical histogram, representing the row and column histograms [4] are used for denoising the input MRM images. The results of these histograms are, sum of differences [4] of grey values existing among neighboring pixels [4] of an image, calculated row-wise [4] and column wise [4]. The results of row sum and column sum are stored in individual arrays. Usually, the difference between the neighboring pixels of a noisy mammogram image is high. The difference is estimated to be very high towards the edges. Survey of several preprocessing techniques suggests that, the values of horizontal and vertical histogram would always be high over a tumor region. Hence the regions having low histogram values are removed using a dynamic threshold. In the proposed system, dynamic threshold is calculated as the average value of a histogram.

3.2 Module II-Segmentation

Segmentation plays a key role in deciding the robustness and efficiency of a system. A hybrid combination of Particle Swarm Optimization and K-Means clustering has been used in segmenting the Region of Interest from the preprocessed image.

Particle Swarm Optimization (PSO) algorithm is a highly optimized, stochastic algorithm. The algorithm has been devised by simulating the social behavior of bird flocks [12]. Particle Swarm Optimization technique uses a group of pixels and each pixel would become a candidate solution [12] in turn to examine the range of available solutions [12] for an optimization problem. Every pixel is initialized again and allowed to 'fly' [12]. In every optimization step the pixel calculates its own fitness [12] and that of neighboring pixels [12]. The values of the current pixel solution that made it as the 'best fit' [12] and the values of neighborhood pixels pertaining to 'best performance' is recorded. The algorithm for PSO has been implemented in such a way that in every optimization step, the candidate solution of pixel is adjusted by equations of kinematics [12]. The algorithm is highly helpful in moving similar pixels towards a particular region. Pixels join the information obtained in their previous best position [12] to increase the probability of moving to the region of better fitness.

K-Means clustering mainly relates to the way of performing vector quantization [13]. Vector Quantization aims at modeling the probability density functions (PDF's) by distributing the prototype vectors. Vector quantization works, by segregating a voluminous set of vectors into clusters. Clusters are groups having similar number of values nearest to it. In K-means clustering each cluster is signified by its centroid point. It focuses on partitioning M observations into H clusters, wherein each observation maps to the cluster having nearest mean, which serves to be the prototype of the cluster. Hence k-means clustering ends up in partitioning the input values into cells. If the input consists of a collection of observations (Y1, Y2, Y3, ….., Ym), wherein every observation is represented as a D-dimensional vector, the aim of K-means clustering is to segregate the M observations into H sets, where H is always less than or equal to m, in order to reduce the within cluster sum of squares (WCSS) [13]. K-means algorithm uses continuous improvement process. The following algorithm is used in the proposed system.

Having an initial set of K means a1, a2, a3,….., ak, the algorithm works by changing between two steps, assignment step [12] and update step [11]. In assignment step each input data is mapped to the cluster whose mean yields the minimum within cluster sum of squares (WCSS) [11]. As the sum of squares [11] is usually the squared Euclidian distance, it is predicted to be the nearest mean. In this approach a point would be perfectly assigned to one set, although it matches to multiple sets.

In the update step the new means are calculated as the, "centroids" of observations in the newly formed cluster. As the arithmetic mean is the estimator of least squares it minimizes the within cluster sum of squares [11] objective. The algorithm always uses distance to map the objects to nearest clusters. This algorithm also focuses on reducing the WCSS objective, which exactly resembles to mapping by the least Euclidian distance.

The initialization approach used in the segmentation process is random partitioning. Initially, the method assigns a cluster randomly to each observation and performs the update step [11]. This method results in calculating the initial mean, as the centroid of the cluster.

In the proposed system K-Means clustering is mainly used to partition the input data, namely, the pixels of the MRM image, into K mutually exclusive cluster. The centroid of every cluster is calculated by identifying the minimal value for the summation of distances from all input data. Finally the index of each cluster to which the input data is assigned to, is returned. This approach creates clusters in

Table 1 Sample clustering output

Iteration 1	Iteration 2	Iteration 5	Iteration 7
Total data in cluster 1 = 96,126	Total data in cluster 1 = 109,409	Total data in cluster 1 = 718,670	Total data in cluster 1 = 37,202
Total data in cluster 2 = 61,659	Total data in cluster 2 = 775,041	Total data in cluster 2 = 162,493	Total data in cluster 2 = 85,563
Total data in cluster 3 = 80,4024	Total data in cluster 3 = 90,887	Total data in cluster 3 = 90,486	Total data in cluster 3 = 115,896
Total data in cluster 4 = 86,767	Total data in cluster 4 = 73,239	Total data in cluster 4 = 76,927	Total data in cluster 4 = 809,915

single level as it operates on the actual input data. The Euclidian distance measure used groups' similar input data into one cluster and distinguishes it separately from other clusters. Based on the partitioning criteria four clusters are being created for segmenting and identifying the region of interest. A Sample output of the clustering can be viewed in Table 1.

3.3 Module III-Feature Extraction

Feature extraction involves extracting features from segmented region of interest (ROI), in order to classify the tumor. Features are usually defined as various patterns, identified in the image. These patterns are usually used to gain knowledge about an image. Feature extraction serves to be an important stage for classification. The efficiency of classification is directly related to the feature extraction stage.

A set of 14 features has been used for classifying segmented data. The features have to be identified precisely to aid in categorizing information from a large data. Hence, textural features have been used for classification. Textural feature focuses on information pertaining to spatial distribution of tonal variation within a band [14]. The features are calculated in the spatial domain [14]. Feature extraction step mainly consists of computing spatially dependent, probability distribution matrices for the segmented region. Further each of these matrices is used to calculate the 14 features.

In an image, texture of a region is identified by the manner in which grey levels are spread out in the pixels of that region. Variance is usually defined as the measure of width of histogram, giving the deviation of gray levels from the mean [14]. Smoothness of an image is usually measured using Angular Second Moment. If the obtained ASM value is less, then the relative frequencies are uniformly distributed and the region is less smooth [14]. Local level variations [14] are measured using contrast. Contrast values are usually higher for higher contrast images. The extent to which pixels in two different directions are dependent on each other is measured using correlation feature. Key characteristic called randomness that characterizes texture in an image is measured using the feature called entropy. It also provides information about distribution variation [14] in a region.

Table 2 Sample feature values obtained for three images

Feature #	Image 1	Image 2	Image 3
1	27085.81395607725	29816.095561245937	57327.771659962906
2	1.1867717213769476E8	2.203826243329859E7	7.000760419686762E8
3	1.1867717213769476E8	2.203826243329859E7	7.000760419686762E8
4	1.1467426098785048E7	1.6214668999284925E7	3.469300633455667E7
5	1.3259688972773234	1.2904585237722201	132.42142677282305
6	2867695.3454549154	1259746.0444615707	8907643.358246382
7	7.430318364595469E12	2.416489350544253E11	3.368676843245445E14
8	−30516.876488997535	−8979.292436350104	−112183.6784169174
9	5842.28450665015	7729.806715227504	33527.460477181456
10	740.910447515173	188.20422051894133	1396.8859340675328
11	−30423.563146404304	−8941.453597244023	−118250.0129824238
12	237359.2533934947	294981.43497392413	1814942.2779983098
13	1.0	1.0	1.0
14	0.03051116594224797	0.017054274791225406	0.10819019836974018

Entropy values are always less for smooth images [14]. Correlation measures the degree of dependence of a pixel to its neighbor [14] over a region. The correlation value of a positively correlated image is 1 and negatively correlated image is −1. Correlation is usually calculated using mean, standard deviation and partial probability density function [14]. The dataset created using the values of the extracted features is fed into the Artificial Neural Network for classification. Table 2 below provides a list of sample features obtained for three random images.

3.4 Module IV-Classification

The features stored as dataset are fed as input to Artificial Neural Network (ANN). Neural networks basically resemble the way how human brain is organized and serves to be a mathematical model of the same. In the proposed system, ANN is used as a classifier, which is used to classify the extracted features of the ROI as either belonging to class I which corresponds to benign group also termed as normal, or class II which corresponds to malignant group also termed as abnormal. In order to achieve maximum efficiency in classification a Multilayer Perceptron with Back Propagation learning algorithm is used. The ANN is well trained and is highly efficient in terms of speed and accuracy. Using ANN, unknown samples [7] of MRM images are detected and classified in a faster manner when compared to other classifiers. Table 3 below, lists the results of classification of few random images chosen from the database, after having gone through the previous modules of the system. The table highlights the fact that 95 % efficiency has been achieved through classification. The output also proves that the developed CAD system mainly aids in reducing the false positive rates.

Table 3 Sample output of classification with accuracy

Sample output	Sample output
Input = Image 1/Output = Normal	Input = Image 2/Output = Abnormal
Input = Image 3/Output = Normal	Input = Image 10/Output = Abnormal
Input = Image 16/Output = Normal	Input = Image 15/Output = Abnormal
Input = Image 20/Output = Normal	Input = Image 19/Output = Abnormal
Input = Image 27/Output = Normal	Input = Image 32/Output = Abnormal
Input = Image 49/Output = Normal	Input = Image 63/Output = Abnormal
Input = Image 56/Output = Normal	Input = Image 97/Output = Abnormal
Input = Image 65/Output = Abnormal	Input Image 99/Output = Abnormal
Input = Image 89/Output = Normal	Input = Image 104/Output = Abnormal
Input = Image 94/Output = Normal	Input = Image 121/Output = Normal
Total errors for trained ANN = 2	Accuracy = 95 %

3.5 Module V-Stage Recognition

Stage of any particular cancer gives information about the size of the cancer and its severity. The proposed system is mainly developed to attain information regarding the stage of the detected tumor in order to aid for early and best treatment. TNM [7] is the basic criteria used by experts to decide upon the treatment. 'T' stands for size of the tumor, 'N' indicates whether it has spread to the lymph glands [7] and finally, 'M' stands for metasis which provides information about the extent to which the cancer has spread to other parts of the body. In the proposed system the first parameter is taken into account, that is, size of the tumor. According to the inputs collected from experts, the size parameter, T, can be further sub divided into three stages. Stage 1 or T1 indicates that the size of the tumor can range from 0.1 cm to less than 2 cm, Stage 2 or T2 indicates that the tumor size is greater than 2 cm but not more than 5 cm across and Stage 3 or T3 indicates the size of the tumor is more than 5 cm across.

4 Experimental Results and Discussion

The preprocessing stage that has been accomplished using histogram equalization and dynamic thresholding has proven to be a successful approach towards extraction of pectoral muscle, removal of noise and in enhancement of the image. Table 4 shows the snapshots of some images after preprocessing.

Sample Outputs of classification and staging module has been provided in Table 5 below. From the results it can be seen that the proposed method has achieved 95 % efficiency in classification of benign and malignant tumors, which

Table 4 Sample output of preprocessing showing denoizing and pectoral muscle suppression

Input image#	Input image	Preprocessed image
Mdb001	Input	
Mdb003	Input	
Mdb010	Input	

Table 5 Sample output of segmentation and classification module

Input image#	Segmentation result	Classification result
Mdb001	PSO - Kmeans clustering; Segemented region	Segemented region; Normal
Mdb002	PSO - Kmeans clustering; Segemented region	Segemented region; Abnormal stage 2
Mdb003	PSO - Kmeans clustering; Segemented region	Segemented region; Normal
Mdb010	PSO - Kmeans clustering; Segemented region	Segemented region; Abnormal stage 2

indirectly proves the efficiency of segmentation algorithm used, hybridized PSO and K-Means clustering. Performance of Artificial Neural Network and detection of stages is highly dependent on the perfection of detection of ROI. A detailed survey of all the segmentation and classification algorithms highlights the fact that the maximum achieved efficiency is around 92 %. Hence hybrid PSO and K-Means clustering aided with artificial Neural Network, has proven to be the best approach towards segmentation and classification as compared to all other algorithms. In order to identify the stage of cancer, size has been considered to be an important parameter. Based on experts input, tumors of size, less than or equal to 2 cm corresponds to stage 1 cancer and tumors of size less than or equal to 5 cm corresponds to stage 2 cancer. The efficiency of stage detection system has determined to be 87 % based on the evaluation of the output by experts.

Table 6 below provides the snapshots of few outputs highlighting the performance of the proposed system on the whole. The proposed system has been tested using several images, chosen from different mammograms of the MIAS database.

Table 6 Sample output of proposed system

Input image#	Output of proposed system	Input image#	Output of proposed system
Mdb001	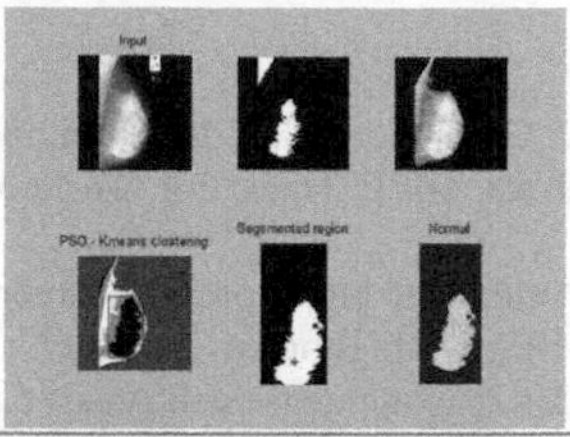	Mdb010	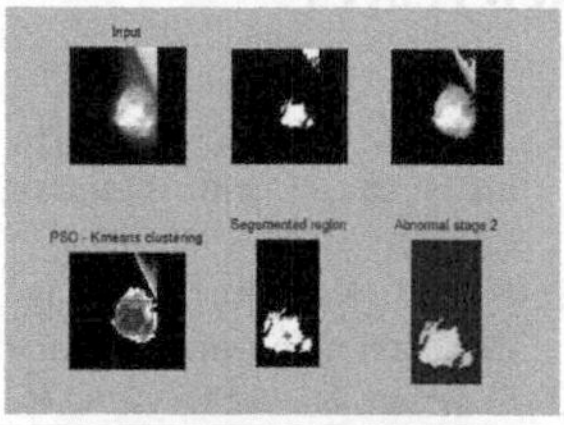
Mdb002	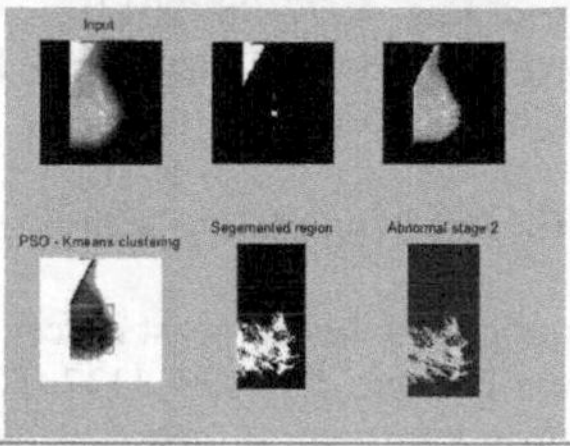	Mdb015	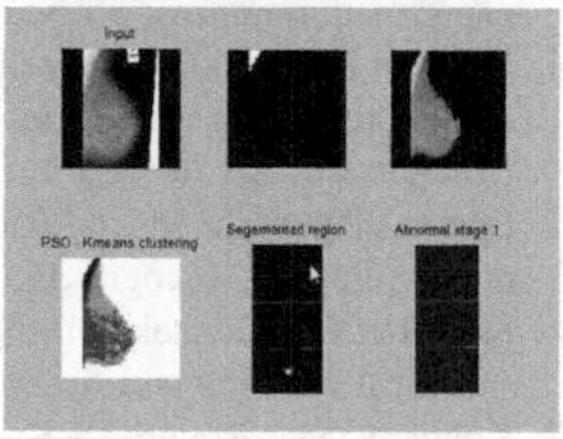

5 Conclusion and Future Work

The developed CAD system has proved to be highly efficient in detection, classification and recognition of stages of tumor. Particle Swarm optimization combined with K-Means clustering has achieved great efficiency when compared to other existing algorithms and PSO combined with Fuzzy C-means algorithm for segmentation. It has also been proved that fourteen textural features combined with ANN, has achieved better results with good accuracy and speed. Considering size as a parameter better efficiency has been attained in identifying the stages of the tumor which would aid the radiologists in further diagnosis and decisions.

The proposed system has made use of textural features in spatial domain. The future work can be extended to developing the system for frequency domain and also by checking the efficiency of staging through other parameters like shape, features, energy levels etc. Certain methods or techniques, uniquely used for detection and identification of stages of other types of cancer can be tested for applicability for breast cancer. Perfect Staging outputs can also be obtained by considering 3D images [15] from varying modalities.

References

1. Tang, J., Rangayyan, R.M., Xu, J., El Naqa, I.: Computer aided detection and diagnosis of breast cancer with mammography: recent advances. IEEE Trans. Inf. Technol. Biomed. **13** (2009)
2. Nagi, J., Kareem, S.A., Nagi, F., Ahmed, S.K.: Automated breast profile segmentation for ROI detection using digital mammograms. IEEE EMBS Conf. Biomed. Eng. Sci. (2010)
3. Ganesan, K., Acharya, U.R., Chua, C.K., Min, L.C.: Computer-aided breast cancer detection using mammograms: a review. IEEE Rev. Biomed. Eng. **6**, 77–98 (2013)
4. Narayan Ponraj, D., Evangeline Jenifer, M., Poongodi, P., Samuel Manoharan, J.: A survey on the preprocessing techniques of mammogram for the detection of breast cancer. J. Eng. Trends Comput. Inf. Sci. **2**, (2011)
5. Egmont-Petersen, M., de Ridder, D., Handels, H.: Image processing with neural networks-a review. Pattern Recogn. **35**, 2279–2301 (2002)
6. Schaefer, G., Zavisek, M., Nakashima, T.: Thermography based breast cancer analysis using statistical features and fuzzy classification. Pattern Recogn. **47**, 1133–1137 (2009)
7. Kocur, C.M., Rogers, S.K., Myers, L.R.: Thomas burns: using neural networks to select wavelet featuers for breast cancer diagnosis. IEEE Eng. Med. Biol. 0739–5175 (1996)
8. Hamed, N.B., Taouil, K., Bouhlel, M.S.: Exploring wavelets towards an automatic microclacification detection in breast cancer. In: IEEE (2006)
9. Malek, J., Sebri, A., Mabrouk, S.: Automated breast cancer diagnosis based on GVF-Snake segmentation, wavelet features extraction and fuzzy classification. J. Sign. Process Syst. (2008)
10. Gopi Raju, N., Nageswara Rao, P.: Particle swarm optimization method for image segmentation applied in mammography. Int. J. Eng. Res. Appl. **3**, (2013)
11. Tandan, A., Raja, R., Chouhan, Y.: Image segmentation based on particle swarm optimization. Int. J. Sci. Eng. Technol. Res. **3**(2), (2014)
12. Mohessen, F., Hadhoud, M., Mostafa, K., Amin, K.: A new image segmentation method based on particle swarm optimization. Int. Arab J. Inf. Technol. **9**, (2012)

13. Ghamisi, P., Couceiro, M.S., Benediktsson, J.A.: An efficient method for segmentation of images based on fractional calculus and natural selection. Expert Syst. Appl. (2012)
14. Haralick, R.M., Shanmugam, K.: Textural features for image classification. IEEE Trans.
15. Guardiola, M., Capdevila, S., Romeu, J., Jofre, L.: 3-D microwave magnitude combined tomography for breast cancer detection using realistic breast models. IEEE Antennas Wirel. Propogat. Lett. **11**, 1622–1625 (2012)

Analysis of Local Descriptors for Human Face Recognition

Radhey Shyam and Yogendra Narain Singh

Abstract Facial image analysis is an important and profound research in the field of computer vision. The prime issue of the face recognition is to develop the robust descriptors that discriminate facial features. In recent years, the local binary pattern (LBP) has attained a big attention of the biometric researchers, for facial image analysis due to its robustness shown for the challenging databases. This paper presents a novel method for facial image representation using local binary pattern, called augmented local binary pattern (A-LBP) which works on the consolidation of the principle of locality of uniform and non-uniform patterns. It replaces the non-uniform patterns with the mod value of the uniform patterns that are consolidated with the neighboring uniform patterns and extract pertinent information from the local descriptors. The experimental results prove the efficacy of the proposed method over LBP on the publicly available face databases, such as AT & T-ORL, extended Yale B, and Yale A.

Keywords Face recognition · Local binary pattern · Histogram · Descriptor

1 Introduction

Computer vision and Biometric systems have illustrated the significant improvement in recognizing and verifying faces in digital images. The face recognition methods that are performing well in constrained environments, includes principal component analysis [1], linear discriminant analysis [2], Fisherface [3], etc. In many applications, including facial image analysis, visual inspection, remote

R. Shyam (✉) · Y.N. Singh
Department of Computer Science & Engineering, Institute of Engineering and Technology, Lucknow 226021, India
e-mail: shyam0058@gmail.com

Y.N. Singh
e-mail: singhyn@gmail.com

A. Nagar et al. (eds.), *Proceedings of 3rd International Conference on Advanced Computing, Networking and Informatics*, Smart Innovation, Systems and Technologies 43, DOI 10.1007/978-81-322-2538-6_27

sensing, biometrics, motion analysis, etc. Mostly, these environments are not constrained. Therefore, we aim to develop an efficient method that accurately recognizes the individual from their unconstrained facial images.

In literature, the methods that work in unconstrained face images are mainly based on the texture descriptions. The local feature-based or multimodal approaches to face recognition have achieved attention in the scientific world [4, 5]. These local feature-based and multimodal methods are less sensitive to variations in pose and illumination than the traditional methods. In unconstrained environments, the local binary pattern (LBP) is one of the popular methods of face recognition. The intuition behind using the LBP operator for face description is that the face can be seen as a composition of various micro-patterns; and it is insensitive to variations, such as pose and illumination. Global description of the face image is obtained by consolidating these micro-patterns [6].

In literature, plenty of methods have been proposed to improve the robustness of the LBP operator in unconstrained face recognition. For example, Liao et al. [7] proposed dominant LBPs which make use of the frequently occurred patterns of LBP. Center-symmetric local binary pattern is used to replace the gradient operator used by the SIFT operator [8]. Multi-block LBP, replaces intensity values in the computation of LBP with the mean intensity value of image blocks [9]. Local ternary pattern (LTP) was initiated by Tan and Triggs [10], to add resistance to noise. However, LTP representation is still limited due to its hard and fixed quantization. Three-patch LBP and four-patch LBP utilize novel multi-patch sampling patterns to add sparse local structure into a composite LBP descriptor [11].

The prime issues of the LBP are that insensitive to the monotonic transformation of the gray-scale, they are still susceptible by illumination variations that generate non-monotonic gray-scale changes, such as self shadowing [4]. LBP may not work properly for noisy images or on flat image areas of constant graylevel. This is due to the thresholding scheme of the operator [12]. The remainder of the paper is organized as follows: Sect. 2, proposes the novel method for face recognition. Evaluation of the proposed method and their comparison with LBPs are presented in Sect. 3. Finally, the conclusions are outlined in the last section.

2 Proposed Method

In this section, we present a novel method that acts on the LBP, called *augmented local binary pattern*. Earlier work on the LBP have not given too much attention on the use of non-uniform patterns. They are either treated as noise and discarded during texture representation, or used in consolidation with the uniform patterns. The proposed method considers the non-uniform patterns and extract the discriminatory information available to them so as to prove their usefulness. They are used in consolidation to the neighboring uniform patterns and extract invaluable information regarding the local descriptors.

The proposed method uses a grid-based regions. However, instead of directly putting all non-uniform patterns into 59th bin, it replaces all non-uniform patterns with the mod of neighboring uniform patterns. For this, we have taken a kernel of size 3 × 3 that is moved on the entire LBP generated surface texture. In this filtering process, the central pixel's value (c_p) is replaced with the mode of a set in case of the non-uniformity of the central pixel. This set contains 8-closet neighbors of central pixel, in which non-uniform neighbors are substituted with 255. Here 255 is the highest uniform value.

The lookup table containing decimal values of 8-bit uniform patterns are $\boldsymbol{U}$ = {0, 1, 2, 3, 4, 6, 7, 8, 12, 14, 15, 16, 24, 28, 30, 31, 32, 48, 56, 60, 62, 63, 64, 96, 112, 120, 124, 126, 127, 128, 129, 131, 135, 143, 159, 191, 192, 193, 195, 199, 207, 223, 224, 225, 227, 231, 239, 240, 241, 243, 247, 248, 249, 251, 252, 253, 254, 255} [13]. The basics of filtering process is explained in Fig. 1.

The classification performance of the proposed method is evaluated with Chi square (χ^2) distance measure which are formally defined as follows:

$$\chi^2(p,q) = \sum_{i=1}^{N} \frac{(p_i - q_i)^2}{(p_i + q_i)} \tag{1}$$

where N is the dimensionality of the spatially enhanced histograms, p is the histogram of the probe image, q is the histogram of the gallery image, i represents the bin number and p_i, q_i are the values of the ith bin in the histograms p and q to be compared.

The schematic of the proposed A-LBP method is shown in Fig. 2.

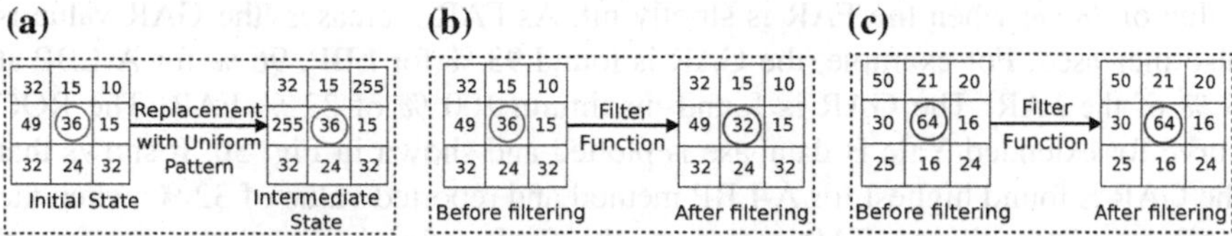

Fig. 1 **a** Neighboring non-uniform patterns are replaced with highest uniform pattern 255, **b** the central non-uniform pattern 36, replaced with mode value of 32, and **c** the next central pattern 64 is found to be uniform, therefore remains unchanged

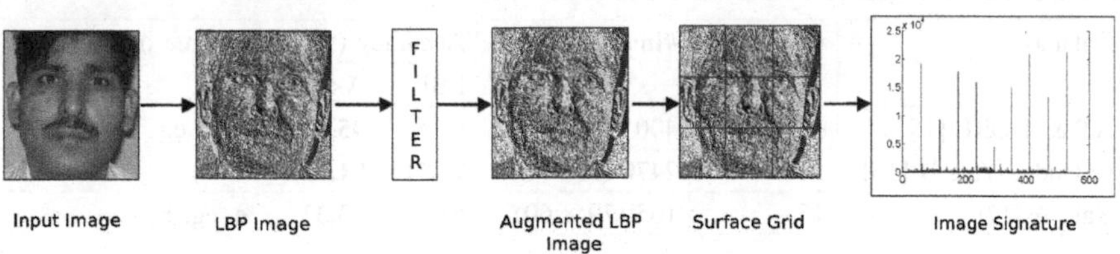

Fig. 2 Schematic of a A-LBP face recognition system [14]

3 Experimental Results

The efficiency of the proposed method is tested on the publicly available face databases, such as AT & T-ORL [15], extended Yale B [16], and Yale A [17]. These databases differ in the degree of variation in pose (p), illumination (i), expression (e) and eye glasses (eg) present in their facial images. The face recognition accuracy of the proposed A-LBP method is compared to the LBP method on the different face databases (see Table 1). The performance of the proposed A-LBP method is analyzed using equal error rate, which is an error, where the likelihood of acceptance assumed the same value to the likelihood of rejection of people who should be correctly verified. The performance of the proposed method is also confirmed by the receiver operating characteristic (ROC) curves. The ROC curve is a measure of classification performance that plots the genuine acceptance rate (GAR) against the false acceptance rate (FAR).

The face recognition accuracy of the proposed A-LBP method is compared to the LBP method on different face databases. The experimental results show that the A-LBP performs better than the LBP. For AT & T-ORL database, the A-LBP achieves a recognition accuracy of 95 %, whereas LBP reports an accuracy of 92.5 %. Similar trends are also observed for extended Yale B and Yale A databases. For extended Yale B database, proposed method performs better than LBP such as the accuracy values are reported to 81.22 % and 74.11 %, respectively. For Yale A database, the proposed method reported the better accuracy value of 73.33 % in comparison to LBP accuracy value of 61.19 % (see Table 1).

The ROC curve for AT & T-ORL database is plotted and shown in Fig. 3a. It shows that the GAR is found highest for the proposed A-LBP method and reported value of 78 %; when the FAR is strictly nil. As FAR increases, the GAR value is also increased. For example, the GAR is found 93 % for LBP, 96 % for A-LBP at 5 % of the FAR. The GAR is found maximum 100 % of 32 % FAR. The ROC curve for extended Yale B database is plotted and shown in Fig. 3b. It shows that the GAR is found highest for A-LBP method and reported value of 32 %; when the FAR is strictly nil. As FAR increases, the GAR value is also increased for all methods accordingly. For example, the GAR is found 62 % for LBP, 82 % for A-LBP of 20 % FAR. The GAR is found maximum 100 % at 83 % of FAR for LBP

Table 1 Face recognition accuracies of methods on different face databases

Databases	#Subjects	#Images (size)	Accuracy (%)		Degree of variation
			LBP	A-LBP	
AT & T-ORL [15]	40	400 (49 × 60)	92.50	**95.00**	p, e, eg
Extended Yale B [16]	38	2470 (53 × 60)	74.11	**81.22**	i
Yale A [17]	15	165 (79 × 60)	61.19	**73.33**	e, eg, i

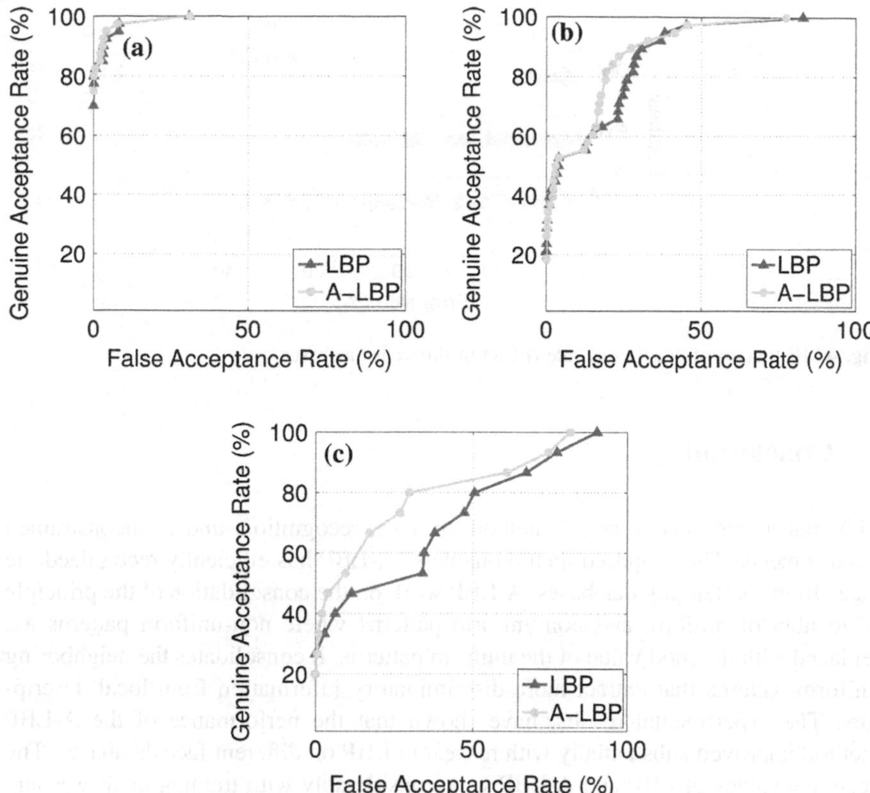

Fig. 3 ROC curves representing the performance of different face recognition methods on the databases: **a** AT & T-ORL, **b** extended Yale B, and **c** Yale A

and 78 % of FAR for A-LBP. The A-LBP method achieves better recognition accuracy, because it is insensitive to changes such as illumination.

The ROC curve for Yale A database is plotted and shown in Fig. 3c. It shows that the GAR is found highest for A-LBP method and reported value of 20 %; when the FAR is strictly nil. As FAR increases, the GAR value is also increased for all methods accordingly. For example, the GAR is found 50 % for LBP, 69 % for A-LBP at 20 % of the FAR. The GAR is found maximum 100 % at 90 % of FAR for LBP and 82 % of FAR for A-LBP. The A-LBP method achieves better recognition accuracy, because it is insensitive to changes such as illumination.

The histogram representation of recognition performance achieved by the LBP and proposed A-LBP methods using different face databases, such as Yale A, extended Yale B, and AT & T-ORL is shown in Fig. 4.

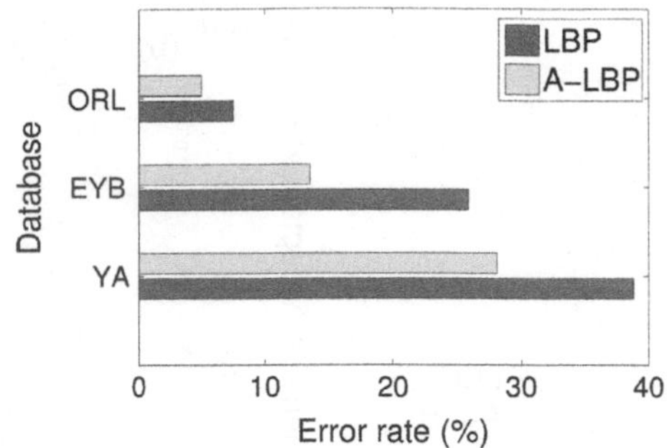

Fig. 4 Histogram of equal error rate (EER) of different databases

4 Conclusion

This paper presents a novel method of face recognition under unconstrained environments. The proposed method namely, A-LBP has efficiently recognized the faces from challenging databases. A-LBP work on the consolidation of the principle of locality of uniform and non-uniform patterns where non-uniform patterns are replaced with the mod value of the uniform patterns. It consolidates the neighboring uniform patterns that extract more discriminatory information from local descriptors. The experimental results, have shown that the performance of the A-LBP method improved substantially with respect to LBP on different face databases. The accuracy values of LBP and A-LBP vary considerably with training databases and the distance metrics preferred. When there are more variations in illumination of facial images in the databases, the A-LBP has shown promising recognition accuracy than the LBP. Similar trends are also observed for the databases that have variations in pose and expressions.

Acknowledgment The authors acknowledge the Institute of Engineering and Technology (IET), Lucknow, Uttar Pradesh Technical University (UPTU), Lucknow for their financial support to carry out this research under the Technical Education Quality Improvement Programme (TEQIP-II) grant.

References

1. Turk, M.A., Pentland, A.P.: Eigenfaces for recognition. J. Cogn. Neurosci. **3**(1), 71–86 (1991)
2. Lu, J., Kostantinos, N.P., Anastasios, N.V.: Face recognition using LDA-based algorithms. IEEE Trans. Neural Networks **14**(1), 195–200 (2003)
3. Belhumeur, P.N., Hespanha, J.P., Kiregman, D.J.: Eigenfaces vs. Fisherfaces: recognition using class specific linear projection. IEEE Trans. Pattern Anal. Mach. Intell. **19**(7), 711–720 (1997)

4. Shyam, R., Singh, Y.N.: Evaluation of Eigenfaces and Fisherfaces using Bray Curtis Dissimilarity Metric. In: Proceedings of 9th IEEE International Conference on Industrial and Information Systems (ICIIS 2014), pp. 1–6 (2014)
5. Shyam, R., Singh, Y.N.: Identifying individuals using multimodal face recognition techniques. In: Proceedings of International Conference on Intelligent Computing, Communication & Convergence (ICCC-2014). TBA, Elsevier (2014)
6. Shyam, R., Singh, Y.N.: A taxonomy of 2D and 3D face recognition methods. In: Proceedings of 1st International Conference on Signal Processing and Integrated Networks (SPIN 2014), pp. 749–754. IEEE (2014)
7. Liao, S., Law, M.W.K., Chung, A.C.S.: Dominant local binary patterns for texture classification. IEEE Trans. Image Process. **18**(5), 1107–1118 (2009)
8. Heikkila, M., Pietikainen, M., Schmid, C.: Description of interest regions with local binary patterns. Pattern Recogn. **42**(3), 425–436 (2009)
9. Zhang, L., Chu, R., Xiang, S., Liao, S., Li, S.: Face detection based on multiblock LBP representation. In: Proceedings of International Conference on Biometrics. (2007)
10. Tan, X., Triggs, B.: Enhanced local texture feature sets for face recognition under difficult lighting conditions. In: Proceedings of 3rd International Workshop on Analysis and Modelling of Faces and Gestures. Lecture Notes in Computer Science (LNCS), vol. 4778, pp. 168–182. Springer (2007)
11. Wolf, L., Hassner, T., Taigman, Y.: Descriptor based methods in the wild. In: Proceedings of Workshop Faces in Real-Life Images: Detection, Alignment, and Recognition, Marseille, France (2008). https://hal.inria.fr/inria-00326729
12. Pietikainen, M., Hadid, A., Zaho, G., Ahonen, T.: Computer vision using local binary patterns. In: Proceedings of Computational Imaging and Vision vol. 40, pp. 13–43. Springer (2011). http://dx.doi.org/10.1007/978-0-85729-748-8_2
13. Shyam, R., Singh, Y.N.: Face recognition using augmented local binary patterns in unconstrained environments. In: Proceedings of 8th IAPR/IEEE International Conference on Biometrics (ICB 2015). TBA, Phuket, Thailand (2015)
14. Shyam, R., Singh, Y.N.: Face recognition using augmented local binary patterns and bray curtis dissimilarity metric. In: Proceedings of 2nd International Conference on Signal Processing and Integrated Networks (SPIN 2015). TBA, IEEE (2015)
15. Samaria, F., Harter, A.: Parameterisation of a stochastic model for human face identification. In: Proceedings of 2nd IEEE Workshop on Applications of Computer Vision, Sarasota, FL (1994)
16. Lee, K.C., Ho, J., Kriegman, D.: Acquiring linear subspaces for face recognition under variable lighting. IEEE Trans. Pattern Anal. Mach. Intell. **27**(5), 684–698 (2005)
17. UCSD: Yale. http://vision.ucsd.edu/content/yale-face-database

Implementation and Comparative Study of Image Fusion Methods in Frequency Domain

Keyur N. Brahmbhatt and Ramji M. Makwana

Abstract Complementary multi-focus and/or multi-model data from two or more different images are combined into one new image is called Image fusion. The main objective is to decrease vagueness and minimizes redundancy in the output while enhancing correlate information specific to a task. Medical images coming from different resources may often give different data. So, it is challenging task to merge two or more medical images. The fused images are very useful in medical diagnosis. In this paper, image fusion has been performed in discrete wavelet transform (DWT) and Contourlet transform (CNT). As a fusion rule, spatial techniques like Averaging, Maximum Selection and PCA is used. Experiments are performed on CT and MRI medical images. For evaluation and comparative analysis of methods, a set of standard performance measures are used. This paper's results show that, the Contourlet method gives a good performance in medical image fusion, because it provides parabolic scaling and vanishing moments.

Keywords Image fusion · Spatial domain · DWT (discrete wavelet transform) · Contourlet transform

1 Introduction

The process of combining multiple images into a one image that contains a more useful data than provided by any of the resource images known as image fusion. Image fusion is more appropriate for the purpose of human visual perception [1]. The research work of image fusion can be classified into the following three stages:

K.N. Brahmbhatt (✉)
Department of Computer Engineering, CHARUSET, Changa, Gujarat, India
e-mail: keyur.brahmbhatt@bvmengineering.ac.in

R.M. Makwana
Department of Computer Engineering, ADIT, V.V. Nagar, Gujarat, India
e-mail: ramjimmakwana@gmail.com

A. Nagar et al. (eds.), *Proceedings of 3rd International Conference on Advanced Computing, Networking and Informatics*, Smart Innovation, Systems and Technologies 43, DOI 10.1007/978-81-322-2538-6_28

- Primitive methods
- Discrete Wavelet Transform
- Contourlet Transform

Primitive methods include all spatial domain methods like averaging, PCA and Max. Selection, but the disadvantage of it is, it will not provide the directional singularity [2]. So frequency domain can be used with image fusion, which provides directional singularity. Discrete Wavelet Transform provides directionality but it is limited. It will provide only horizontal, vertical and diagonal directionality. To enhance the directionality Contourlet transform is helpful. Contourlet transform provides C^2 directional singularity which gives good result along with curves [3]. Though work has been carried out in DWT and CNT, in these paper basic methods are implemented in combination with Averaging, Maximum Selection and PCA methods to do comparative study and analysis of methods.

2 Image Fusion in Spatial Domain

The primitive fusion schemes do the fusion on the resource images. Operations like averaging, addition etc. are to be fused in primitive fusion, which gives some disadvantages also like reducing contras etc. but, it also having advantage that it works better for input images which have an overall high brightness and high contrast [4]. The primitive fusion methods are:

- Averaging [5, 6]
- Select Maximum [4, 6]
- PCA [4–6]

3 Image Fusion in Frequency Transform

Transform domain techniques are based on updating the Fourier transform of image. Every pyramid transform has three parts—Decomposition, Formation and Recomposition [5].

The Discrete Wavelet Transform (DWT) is spatial frequency decomposition. DWT gives a flexible multi-resolution analysis of an image [7]. DWT having disadvantages which degrades some image processing applications. The disadvantages are Shift sensitive, Absence of phase information, Produces sub bands in three directions 0° (Horizontal), 90° (vertical) and 45° (diagonal) only [2, 8].

A new multiscale, rich directional selectivity transform is introduced called Contourlet Transform (CT). It bases on an efficient 2D non-separable filter banks and gives an elastic multi-resolution, local and directional approach for image processing [3, 9]. The properties of it is Multiresolution, Localization, Critical sampling, Directionality and Anisotropy [10]. CT is better than DWT in dealing with the singularity in higher dimensions, it provides a rich directional selectivity and can stand for different directional smooth contours in natural images [3, 11].

CT is also called Pyramidal Direction Filter Bank (PDFB), which combines Laplacian Pyramid (LP) and Directional Filter Bank (DFB) into a double filter bank construction. Here images are decomposed by LP into one low frequency sub-band and different high frequency sub-bands, and then high frequency sub-bands are nourished into DFB and segmented into multiple directional sub-bands [12].

Majority of Contourlet coefficients are close to zero that's why it is a sparse. It also gives detailed information in any arbitrary direction, as the number of directional sub-bands in each scale is usually 2^n, which is fairly elastic when different n is selected [12].

4 Frequency Domain Methods

Following combination of methods has been implemented in frequency domain:

1. Averaging, Maximum selection and PCA in Discrete Wavelet Transform
2. Various combination of methods like
   - Averaging and Maximum selection fusion rule
   - Averaging and PCA fusion rule
   - Maximum selection and Averaging fusion rule
   - Maximum selection and PCA fusion rule
   - PCA and Averaging fusion rule
   - PCA and Maximum selection fusion rule

have been applied in Contourlet transform. Here first method has been applied on Low coefficients and second method has been applied on high coefficients.

5 Experimental Results and Discussion

Image fusion methods have been applied on multimodality medical image data to derive useful information. Here Computer Tomography (CT) and Magnetic Resonance Imaging (MRI) images are used as shown in Fig. 1a, b.

Figure 2 shows the resultant fused images. Here we have implemented Averaging, Maximum selection and PCA fusion rule in DWT.

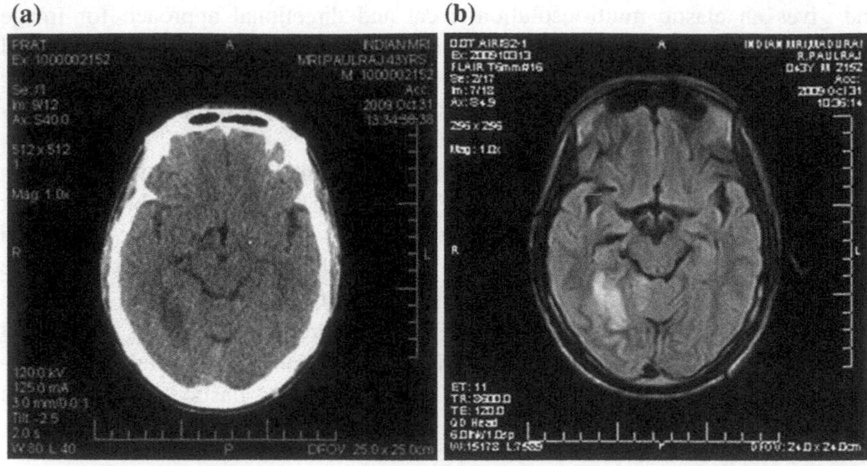

Fig. 1 **a** CT image (*input image 1*). **b** MRI image (*input image 2*) [15]

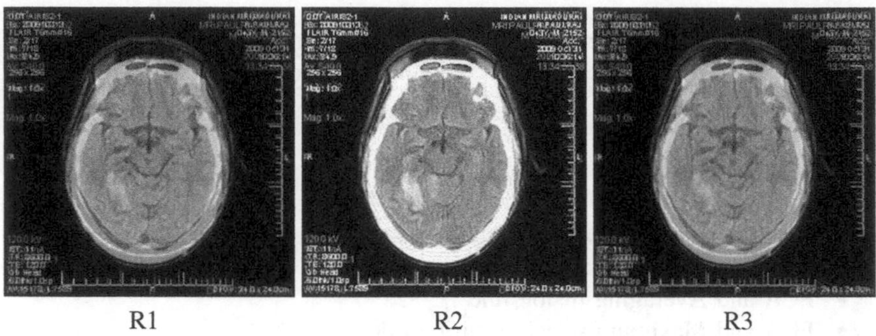

Fig. 2 Medical image fusion in DWT. **R1** Averaging in discrete wavelet transform. **R2** Maximum selection in DWT. **R3** PCA in DWT

Figure 3 shows the resultant fused images. We have implemented Averaging and Max selection, Averaging and PCA, Max selection and PCA, Max selection and Averaging, PCA and Averaging and PCA and Max selection fusion rule in Contourlet transform. Here Qualitative analysis has been done. It shows that Averaging and Max. selection gives good fusion information.

Performance of image fusion methods have been measured using five standard metrics RMSE [13], PSNR [4, 6], NCC [4, 6], Standard deviation (SD) [3, 13] and Degree of Distortion (DD) [14]. All fusion methods are implemented in MATLAB 9.

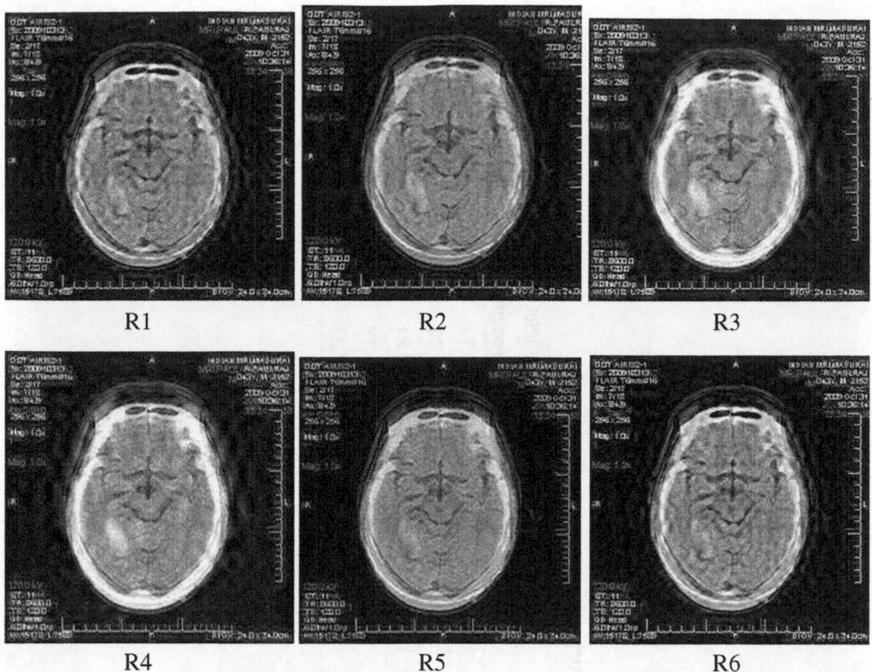

Fig. 3 Medical image fusion in Contourlet transform. **R1** Averaging and maximum selection in Contourlet transform. **R2** Averaging and PCA in Contourlet transform. **R3** Maximum selection and PCA in Contourlet transform. **R4** Maximum selection and averaging in Contourlet transform. **R5** PCA and averaging in Contourlet transform. **R6** PCA and maximum selection in Contourlet transform

Table 1 shows that in Maximum selection good contras is achieved in all the dataset. While less degree of distortion (DD), High NCC and high PSNR are obtained in averaging method.

Table 2 shows that in Maximum selection with combination of Averaging and PCA good contras is obtained in all the dataset. While less degree of distortion (DD), High NCC and high PSNR are obtained in PCA method.

Table 1 Comparative result of different fusion rule on CT and MRI images in DWT

Evaluation parameters	DataSet 1			DataSet 2			DataSet 3		
	Averaging	Max. selection	PCA	Averaging	Max. selection	PCA	Averaging	Max. selection	PCA
RMSE	34.69	54.22	38.09	38.79	59.54	44.85	43.78	70.93	56.1
PSNR	32.72	30.78	32.32	32.24	30.38	31.61	31.71	29.62	30.6
NCC	0.90	0.86	0.88	0.86	0.84	0.82	0.84	0.83	0.77
SD	59.95	80.03	60.81	58.24	83.16	60.15	64.79	93.71	70.8
DD	19.97	23.27	21.93	22.90	26.01	26.49	27.25	34.64	34.9

Table 2 Comparative result of different fusion rule on CT and MRI images in CNT

Fusion rule	RMSE	PSNR	NCC	SD	DD
Dataset 1					
Averaging and max. selection	9.4916	38.3574	0.9992	45.8996	3.8724
Averaging and PCA	11.5710	37.4971	0.9988	45.2420	4.3222
Max. selection and PCA	15.3848	36.2599	0.9978	41.3981	4.9753
Max. selection and averaging	15.0246	36.3628	0.9979	40.7901	4.7984
PCA and averaging	10.9969	37.7181	0.9989	44.7545	4.0664
PCA and max. selection	9.3771	38.4101	0.9992	45.9643	3.8259
Dataset 2					
Averaging and max. selection	43.9206	31.7041	0.8549	70.0413	27.9190
Averaging and PCA	39.5431	32.1601	0.8763	66.0551	24.5120
Max. selection and PCA	57.1955	30.5572	0.8444	83.3786	29.9443
Max. selection and averaging	58.5684	30.4542	0.8219	79.8799	31.2198
PCA and averaging	48.9377	31.2344	0.8141	67.6949	30.1354
PCA and max. selection	51.0299	31.0526	0.8185	75.4910	32.4004
Dataset 3					
Averaging and max. selection	50.5641	31.0924	0.8362	77.6684	33.1135
Averaging and PCA	49.5516	31.1802	0.8321	73.8123	31.7824
Max. selection and PCA	71.6050	29.5814	0.8098	94.3906	41.9669
Max. selection and averaging	69.3595	29.7197	0.8068	89.9417	40.3021
PCA and averaging	57.7873	30.5125	0.7894	78.3785	37.3587
PCA and max. selection	61.3323	30.2539	0.7948	86.9000	40.2977

6 Conclusion

Various combinations of methods like Averaging, Max selection and PCA have been developed in multi model image fusion using Contourlet transform. In DWT Averaging, Max selection and PCA have been implemented in multi model image fusion. For evaluation and comparative analysis of the methods RMSE, PSNR, NCC, SD and DD measuring parameters have been used. In DWT highest PSNR 32.7278 is obtained with Averaging method while in Contourlet transform, the highest PSNR 38.3574 is obtained in Averaging and Maximum Selection. Good contras are also achieved with Maximum Selection because it chooses one highest intensity value of pixel from both the image. Here results shows that, Contourlet transform is, better than discrete wavelet transform when it deals with the higher dimensions singularity, such as line, curve, edge and etc. Contourlet transform also provides rich directional selectivity.

References

1. Wu, T., Wu, X.-J., Luo, X.-Q.: A study on fusion of different resolution images. Int. Workshop Inf. Electron. Eng. (IWIEE) **29**, 3980–3985 (2012) (Procedia Engineering, Elsevier)
2. Sapkal, R.J., Kulkarni, S.M.: Image fusion based on wavelet transform for medical application. Int. J. Eng. Res. Appl. (IJERA) **2**(5), 624–627 (2012). ISSN: 2248–9622
3. Liu, S., Wang, M., Fang, Y.: A contourlet transform based fusion algorithm for nighttime driving image. In: Wang, L., et al. (eds.) FSKD, LNAI 4223, pp. 491–500. ©Springer, Berlin (2006)
4. Shivsubramani Krishnamoorthy, K., Soman, P.: Implementation and comparative study of image fusion algorithms. Int. J. Comput. Appl. **9**(2), 0975–8887 (2010)
5. Singh, N., Tanwar, P.: Image fusion using improved Contourlet transform technique. Int. J. Recent Technol. Eng. (IJRTE) **1**(2), (2012). ISSN: 2277–3878
6. Brahmbhatt, K.N., Makwana, R.M.: Comparative study on image fusion methods in spatial domain. Int. J. Adv. Res. Eng. Technol. (IJARET) **4**(2), (2013) ISSN 0976–6480 (Print), ISSN 0976–6499 (Online)
7. Na, T., Manfred, E., Yang, W.: Remote sensing image fusion with multi-wavelet transform. In: SPIE Proceedings Series, vol. 5574, pp. 14–22 (2004). ISBN 0-8194-5521-0
8. Lee, C.-H., Zhou, Z.-W.: Comparison of image fusion based on DCT-STD and DWT-STD. In: Proceeding of International multi Conference of Engineers and Computer Scientist, vol. I. Hong Kong, 14–16 March 2012
9. He, Z.H., Bystrom, M.: Reduced feature texture retrieval using Contourlet decomposition of luminance image component. In: International Conference on Communications, Circuits and Systems, pp. 878–882 (2005)
10. Do, M.N., Vetterli, M.: The Contourlet transform: an efficient directional multiresolution image representation. In: IEEE Transactions on Image Processing
11. Chen, Y., Rick, S.B.: Experimental tests of image fusion for night vision. In: 7th International Conference on Information fusion, pp. 491–498 (2005)
12. Chen, S., Wu, Y.: Image fusion based on Contourlet transform and fuzzy logic. In: The National Natural Science Foundation of China (No. 60872065), 978-1-4244-4131-0/09
13. Deshmukh, M., Bhosale, U.: Image fusion and image quality assessment of fused images. Int. J. Image Process. (IJIP) **4**(5)
14. Shulei, W.U., Zhizhong, Z.H.A.O., Huandong, C.H.E.N., Chunhui, S.O.N.G., Minjia, L.I.: An improved algorithm of remote sensing image fusion based on wavelet transform. J. Comput. Inf. Syst. **8**(20), 8621–8628 (2012)
15. Prakash, C., Rajkumar, S., Chandra Mouli, P.V.S.S.R.: Medical image fusion based on redundancy DWT and Mamdani type min-sum mean-of-max techniques with quantitative analysis. In: International Conference on Recent Advances in Computing and Software Systems (2012)
16. Cao, W., Li, B., Zhang: A remote sensing image fusion method based on PCA and wavelet packet transform. In: IEEE International Conference Neural Networks and Signal Processing, pp. 976–981
17. Guest editorial. Image fusion: Advances in the state of the art. Inf. Fus. **8**, 114–118 (2007) (Elsevier)
18. Sabari Banu, R.: Medical Image Fusion by the analysis of Pixel Level Multi-sensor Using Discrete Wavelet Transform. In: Proceedings of the National Conference on Emerging Trends in Computing Science, pp. 291–297 (2011)
19. Rahman, S.M.M., Ahmad, M.O., Swamy, M.N.S.: Contrast-based fusion of noisy images using discrete wavelet transform. J. Image Process. **4**(5), 374–384 (2010)
20. Zhong, Z.: Investigations on image fusion. PhD Thesis, University of Lehigh, May 1999

Iris Recognition Using Textural Edgeness Features

Saiyed Umer, Bibhas Chandra Dhara and Bhabatosh Chanda

Abstract A method for feature extraction from an iris image based on the concept of textural edgeness is presented in this paper. Here for authentication purpose we have used two textural edgeness features namely: (1) a modified version of Gray Level Auto Correlation (GLAC) and (2) Scale Invariant Feature transform (SIFT) descriptors over dense grids in the image domain. Extensive experimental results using MMU1 and IITD iris databases demonstrate the effectiveness of the proposed system.

Keywords Biometric · Iris classification · HOG · SIFT · GLAC

1 Introduction

In the field of personal authentication biometric characteristics play a vital role. Among the various biometric traits, iris is most useful and stable trait for person identification [1]. It is full of texture characteristic for the analysis of different types of image regions with in an image. To obtain information from these iris patterns various statistical and structural approaches [2] are used. The structural approaches work well for regular patterns whereas statistical approaches are easy to use in practice. Among many statistical approaches, textural edgeness of a texture image analyzes the texture by its edges per unit area. These textural edgeness helps to

S. Umer (✉) · B. Chanda
Electronics and Communication Sciences Unit, Indian Statistical Institute, Baranagar, India
e-mail: saiyedumer@gmail.com

B. Chanda
e-mail: chanda@isical.ac.in

B.C. Dhara
Department of Information Technology, Jadavpur University, Kolkata, India
e-mail: bibhas@it.jusl.ac.in

A. Nagar et al. (eds.), *Proceedings of 3rd International Conference on Advanced Computing, Networking and Informatics*, Smart Innovation, Systems and Technologies 43, DOI 10.1007/978-81-322-2538-6_29

define the fineness and coarseness properties of the image texture. Using textural edgeness properties of image texture, the useful features for iris patterns are also obtained.

Daugman [3, 4] developed iris recognition system with iris phase structure encoded by multi-scale quadrature wavelets. Bowyer et al. [5] explained many existing iris recognition works. Here we discuss only those iris recognition works which are related to textural edgeness feature models like HOG [6], SIFT [7], GLAC [8] with their different variants. Fernandez et al. [9] proposed recognition system using SIFT features without transforming iris pattern to polar coordinates where as Yang et al. [10] extracted SIFT features from enhanced normalized iris and used region based approach [11] for comparison. Mehrotra et al. [12] obtained SURF features from sectored normalized iris. The identification by exploiting iris and periocular features based on HOG, SIFT, LBP and LMF were demonstrated by Tan and Kumar [13]. Unconstrained iris recognition using F-SIFT on annular region of iris is proposed by Mehrotra and Majhi [14].

This paper presents the recognition system in which a modified GLAC and SIFT based features are employed to extract features from iris pattern to authenticate the person. The organization of the paper is as follows. The basic concepts of textural edgeness are discussed in Sect. 2. Proposed iris classification method is described in Sect. 3. Section 4 explains the results and discussions of the experimentation and conclusions are reported in Sect. 5.

2 Textural Edgeness

To analyze the texture property of a region, first of all we have to compute the edgeness of the region (R). Edgeness of R is considered as the ratio of the edge pixels and number of pixels in R. Again, the edgeness is further extended to include the orientation of edges as $f_{m,\theta} = (H_m(R), H_\theta(R))$ where $H_m(R)$ and $H_\theta(R)$ are normalized histogram for gradient magnitude and gradient direction, respectively [15]. The subsequent topics discuss below are some traditional methods for textural edgeness features like HOG, SIFT and GLAC.

Histogram of oriented gradient (HOG): HOG [6] features are extracted by taking orientation histograms of edge intensity of each block of size $n \times n$ of an image. The filters like Prewitt are used to obtain magnitude of gradients m_p and orientation θ_p for each pixel $p(x, y)$ of that block and then these are normalized for all blocks and finally obtain HOG descriptors for that normalized block.

Scale invariant feature transform (***SIFT***): SIFT [7] is used to extract local features which are invariant to the geometric operations (like scaling, rotation) and changes in viewpoint. To compute SIFT features: (i) the input image is convolved with Gaussian filters at different scales, (ii) compute difference-of-Gaussian (DOG) images, (iii) each pixel of DOG is compared with 9 neighbors of both adjacent scales and 8 neighbors of the same scale and the pixel with local optimum,

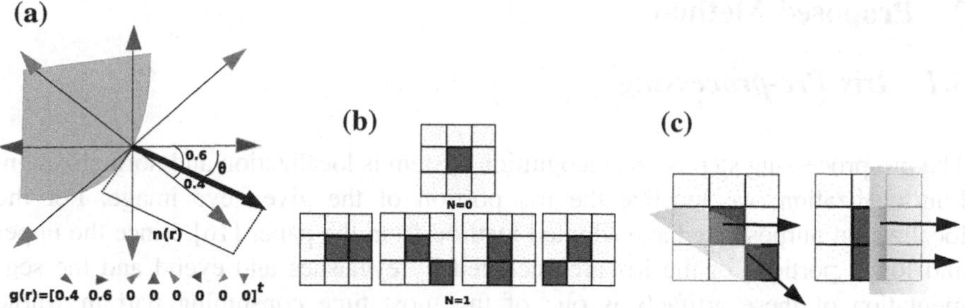

Fig. 1 **a** Gradient orientation vector f, **b** mask patterns, **c** calculation of auto-correlation of f, weighted by gradient magnitude m (images are taken from [8])

is considered as key-point. (iv) the key-point with low contrast is removed, and from the rest key-points the gradient orientation histograms are used as the SIFT descriptor (Fig. 1).

Gradient Local Auto-Correlation (***GLAC***): GLAC [8] is a feature which computes orientational as well as spatial auto-correlation of the local gradient. For an image I at position p, the magnitude of the gradient is m_p and orientation of the gradient is θ_p, say. A sparse vector $f(\in R^k)$ (shown in Fig. 2a) is computed with considering k bins of the orientations. The 0th order and 1st order GLAC features are defined as:

$$\begin{aligned} \zeta_{oth} &= \sum\nolimits_{p \in I} m(r) f_{d_0}(p) \\ \zeta_{1th} &= \sum\nolimits_{p \in I} \min[m(p), m(p+a_1)] f_{d_0}(p) f_{d_1}(p+a_1) \end{aligned} \tag{1}$$

where a_1 is displacement vector from p and f_d is dth element of f. The ζ_{oth} corresponds the histogram of the gradient orientation like HOG and SIFT and ζ_{1th} is the joint histogram of the orientation pairs and computed by quantizing the distribution into $k \times k$ bins. The 1st order GLAC is obtained by convolving the mask pattern over the image (see Fig. 2b–c).

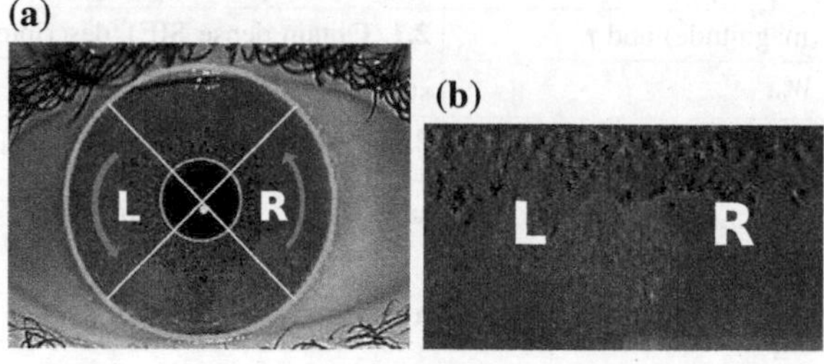

Fig. 2 Illustration of iris preprocessing: **a** division of iris portion into four parts where regions L and R are used for normalization. **b** Normalized iris

3 Proposed Method

3.1 Iris Pre-processing

The pre-processing step of iris recognition system is localization and normalization. For localization, we localize the iris portion of the given eye image. For the localization purpose we have adopted method from the paper [16]. Since the upper and lower portions of the iris are occluded by eyelashes and eyelid and the segmentation of these artifacts is one of the most time consuming part in whole process. So we normalize on the parts of the iris that are not usually occluded by eyelid and eyelashes (as shown in Fig. 1a). To normalize the circular iris portion, Daugman's Rubber sheet model [17] is applied. The normalized image of Fig. 1a is shown in Fig. 1b.

3.2 Feature Extraction

It is observed that for each individual, iris pattern is unique [18]. To extract features we analyze textures from these iris pattern. In this paper we modify the scheme of GLAC, and use that modified version to extract iris features says $f_{IRIS-GLAC}$. For this modification we reformulate Eq. (1) as Eq. (2) and Eq. (3) respectively.

$$\begin{aligned} f^{(0)}(p) &= \sum m(p) \times \wp(p) \\ f^{(0)}(p_{next}) &= \sum m(p) \times \wp(p_{next}) \end{aligned} \tag{2}$$

Table 1 Algorithms for proposed iris features

Algorithm 1: $f_{IRIS-GLAC}$	Algorithm 2: $f_{IRIS-SIFT}$
Input: Normalized iris I	**Input**: Normalized iris I
Output: $f_{IRIS-GLAC}$	**Output**: $f_{IRIS-SIFT}$
1. Partition I into # of patches W.	**1.** Partition I into # of patches W.
2. For each patch W_p at position p.	**2.** For each patch W_p at position p
2.1. Compute m (magnitude) and f	**2.1.** Obtain dense SIFT descriptor d_{W_p}
(G-O vector) for W_p.	such that $d_{W_p} = [d_1, \ldots, d_\beta]^T$.
2.2. Obtain $f_{W_p} = \{f_{W_p}^{(0)}, f_{W_p}^{(1)}\}$ where	**2.2.** Reduce the dimension of d_{W_p} to e_{W_p}
$f_{W_p}^{(0)}$ and $f_{W_p}^{(1)}$ are computed by using	such that $e_{W_p} = \sum_{i=1}^{\beta} d_i$.
Equations (2) and (3) respectively.	
3. $f_{IRIS-GLAC} = \{f_{W1}, \ldots, f_{W\phi}\}$.	**3.** $f_{IRIS-SIFT} = \{e_{W1}^T, \ldots, e_{W\psi}^T\}$.

$$
\begin{aligned}
&f^{(1,1)}_{(\Delta_x,\Delta_y)}(p+(p+a_1)\times k)=\sum m'\times\wp(p)\times\wp(p+a_1)\\
&f^{(1,2)}_{(\Delta_x,\Delta_y)}(p_{\text{next}}+(p+a_1)\times k)=\sum m'\times\wp(p_{\text{next}})\times\wp(p+a_1)\\
&f^{(1,3)}_{(\Delta_x,\Delta_y)}(p+(p+a_1)_{\text{next}}\times k)=\sum m'\times\wp(p)\times\wp((p+a_1)_{\text{next}})\\
&f^{(1,4)}_{(\Delta_x,\Delta_y)}(p_{\text{next}}+(p+a_1)_{\text{next}}\times k)=\sum m'\times\wp(p_{\text{next}})\times\wp((p+a_1)_{\text{next}})
\end{aligned}
\tag{3}
$$

where $m' = \min(m(p), m(p+a_1))$. Here $m(p)$ is the magnitude of current pixel p, $\wp(p)$ is the element of G–O vector of p obtained by voting weights to its nearest bins and (Δ_x, Δ_y) are the mask patterns as shown in Fig. 2b. $f^{(0)}$ contains 0th order GLAC features where as $f^{(1)}$ contains 1st order GLAC features.

At first, a small non overlapping patch $W_{\alpha\times\alpha}$ is taken from I whose gray level pattern is consider as the local descriptor of the texture. Then for each position p of W, the gradient m and the orientation θ which is further used to obtain G–O vector f. Using these magnitude m and vector f, we obtain $f^{(0)}_{W_p}$ (0th order GLAC features) by using Eq. (2) and $f^{(1)}_{W_p}$ (1st order GLAC features) by using Eq. (3) we find $f^{(1)}_{W_p} = (f^{(1,1)}_{W_p} + f^{(1,2)}_{W_p} + f^{(1,3)}_{W_p} + f^{(1,4)}_{W_p})$ and the feature vector of window W at position p is considered as $f_{W_p} = \{f^{(0)}_{W_p}, f^{(1)}_{W_p}\}$.

The entire image is used to obtain SIFT descriptors. These descriptors are global in nature and retain less information where the texture has less discriminating features. So, to enhance the power of discriminating features of the normalized iris image we have partitioned the normalized iris into non overlapping patches (dense grids) and apply SIFT descriptors from each patch to obtain local features. Using these local features are used to represent a global iris feature says $f_{IRIS-SIFT}$. To evaluate this feature we consider all non overlapping patches ($W_{\alpha\times\alpha}$) from I and then from each W at position p, $d_{W_p} = [d_1, \ldots, d_\beta]$ dense SIFT descriptors are obtained where the dimension of d_{W_p} is $\lambda \times \beta$. Now reduce the dimension of d_{W_p} to $e_{W_p} = \sum_{i=1}^{\beta} d_i$ that results $\lambda \times 1$ dimensional vector e_{W_p} which is the feature representation of W_p. Finally feature $f_{IRIS-SIFT} = \{e^T_{W1}, \ldots, e^T_{W\psi}\}$ is obtained. Table 1 shows the algorithmic sketch for the above proposed iris features.

3.3 Conversion to Two Class Problem

In this experiment we use dichotomy model [19] to convert multi-class problem to a two-class problem. To understand the two-class approach, we consider N subjects where each subject has q samples. Then we first generate the feature vectors F for all samples of all subjects, and compute mean vectors V_i over $q-1$ training samples of the subject S_i. This $V_i s$ is called as the template of the i-th class and is included in that i-th class. Then from two vectors F_i and F_j we compute a difference vector $u_{(i,j)}$ such that its k-th element is the absolute difference between k-th elements of F_i and F_j, i.e., $u_{(i,j)}(k) = |F_i(k) - F_j(k)|$.

If F_i and F_j are taken from the samples of same subject, we consider $u_{(i,j)}$ as intra-class vector, and its collection $\{u_{(i,j)}\}$ is known as V_{intra}. Now, if F_i and F_j are prototypes V_i and V_j of the i-th and j-th subjects respectively, $u_{(i,j)}$ is called the inter-class vector and its collection forms a set V_{inter}. Thus we get $q_2 \times N$ number of intra-class vectors and N_2 number of inter-class vectors.

4 Experimental Results and Discussions

In this paper, the performance of the proposed method is evaluated by using two benchmark iris databases, namely, MMU1 [20] and IITD [21]. The MMU1 database has 45 subjects and each subject has 10 (5 left and 5 right) iris images. The IITD iris database contains 224 subjects and each subject has 10 images with 5 left iris and 5 right iris. In this experimental setup, left iris images and right iris images are treated separately. So, in our experimentation, MMU1 has 90 subjects and IITD has 448 subjects.

We have implemented the iris recognition system in MATLAB on fedora O/S of version 14.0 with a Intel Core i3 processor. During normalization process, we use two distinct part of the iris which gives a normalized image of size 100×180 pixels. The size of V_{Intra} and V_{Inter} class vectors for each iris database are shown in Table 2. To implement the recognition system, LIBSVM package [22] based on RBF-kernel using leave one out method with five-fold cross validation is used to train the SVM classifier. Here for SIFT feature we select $\alpha = 4$ and $h = 8$ and thus $\lambda = 128$. The test for both verification and identification performances have been adopted for this proposed iris classification.

During verification, we authenticate the test sample by finding the similarity score. Identification of the test sample is based on N different scores obtained by comparing sample with each of the N different representative subjects. The subject with maximum score (rank-one) is declared as the identity of the test sample.

Table 2 Intra-inter class vectors using feature vectors for iris databases

Iris database	MMU1	IITD
Subjects	90	448
Samples	5	5
V_{intra}	$\binom{5}{2} \times 90$	$\binom{5}{2} \times 448$
V_{inter}	$\binom{90}{2}$	$\binom{448}{2}$
Image size	320×280	320×240

Table 3 The performance for proposed iris classification

	MMU1			
	$f_{IRIS-GLAC}$		$f_{IRIS-SIFT}$	
Patch	Verify	Identify	Verify	Identify
30	97.41	97.77	98.80	98.88
40	96.62	88.88	97.82	95.55
50	95.31	88.88	97.76	95.55
	IITD			
	$f_{IRIS-GLAC}$		$f_{IRIS-SIFT}$	
Patch	Verify	Identify	Verify	Identify
30	96.99	91.74	98.21	93.30
40	95.66	88.39	97.43	92.86
50	95.93	87.50	96.02	89.95

Table 3 shows both verification and identification performances for MMU1 and IITD iris databases for Algorithms 1 and 2 using patches of different sizes i.e. $\alpha \in \{30, 40, 50\}$. The patch size i.e. $\alpha < 30$ results large feature dimension which is time consuming for classification purposes. Here we see that for patch size $\alpha = 30$, we achieve high recognition scores both for MMU1 and IITD databases using $f_{IRIS-GLAC}$ and $f_{IRIS-SIFT}$ feature models and the models are referred as $Prop_{IRIS-GLAC}$ and $Prop_{IRIS-GLAC}$ respectively.

Table 4 shows recognition performances of HOG and GLAC feature models with the proposed models ($Prop_{IRIS-GLAC}$ and $Prop_{IRIS-GLAC}$). Here to extract HOG features we have adopted the code from [6] and for GLAC features from [8]. Here we see that proposed features achieve better performances using less feature dimension with respect to HOG and GLAC features. The performance of the proposed methods are compared with the existing ones and comparative performances are given in Table 5 with Correct Recognition Rate (CRR) as rank-one identification rate where as for verification performance EER is used. For comparison purpose we have quoted the results from the respective papers. From Table 5 it is evaluated that for MMU1

Table 4 The performance with different textural models

Database	Method	Dimension	Verify	Identify (*rank* 1)
MMU1	HOG	3780	94.23	93.33
	GLAC	3990	95.55	84.44
	$Prop_{IRIS-GLAC}$	**2376**	**97.41**	**97.77**
	$Prop_{IRIS-SIFT}$	**3072**	**98.80**	**98.88**
IITD	HOG	3780	87.48	81.03
	GLAC	3990	95.97	85.26
	$Prop_{IRIS-GLAC}$	**2376**	**96.99**	**91.74**
	$Prop_{IRIS-SIFT}$	**3072**	**98.21**	**93.30**

Table 5 Performance comparison for MMU1 and IITD databases

Database	Methods	Subject	CRR(%)	EER(%)
MMU1	Masood [23]	45	95.90	0.040
	Rahulkar [24]	45	–	1.880
	Harjoko [25]	45	82.90	0.280
	$Prop_{IRIS-GLAC}$	90	**97.77**	**0.022**
	$Prop_{IRIS-SIFT}$	90	**98.88**	**0.009**
IITD	Zhou [26]	224	–	0.530
	Rahulkar [24]	90	–	0.150
	Elgamal [27]	80	99.50	0.040
	$Prop_{IRIS-GLAC}$	411	**100**	**0.000**
	$Prop_{IRIS-SIFT}$	418	**100**	**0.000**

database, Masood [23], Rahulkar [24] and Harjoko [25] have obtained their results only for 45 subjects where as the proposed method obtain recognition scores for 90 subjects with higher CRR and better EER. For IITD database, Zhou [26] and Rahulkar [24] obtained results only for verification purposes where as Elgamal [27] and proposed methods develop system for both verification and identification purposes but the proposed method obtained recognition scores for 448 subjects which is higher then other methods in the table.

The verification accuracy is evaluated through *true positive rate* and *false positive rate* and for this purpose ROC (Receiver Operating Characteristic) curves [28, 29] are drawn (see. Figure 2a and b) and the performance evaluation for identification system is expressed through cumulative match curve (CMC) [28] (see. Figure 3a and b) with x-axis denotes rank values and the y-axis the probability of correct identification at each rank (Fig. 4).

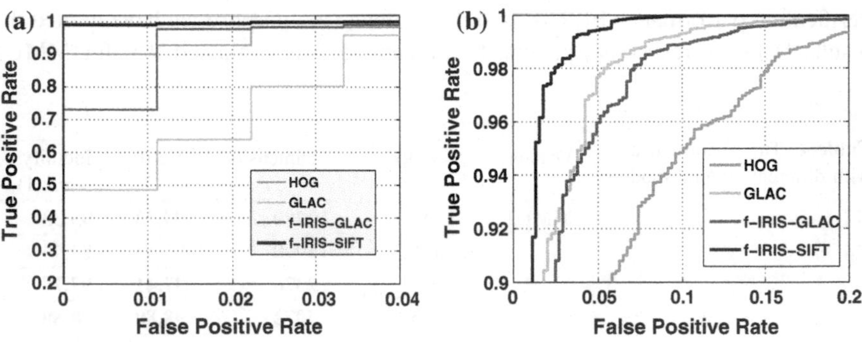

Fig. 3 ROC curves for **a** MMU1 and **b** IITD database

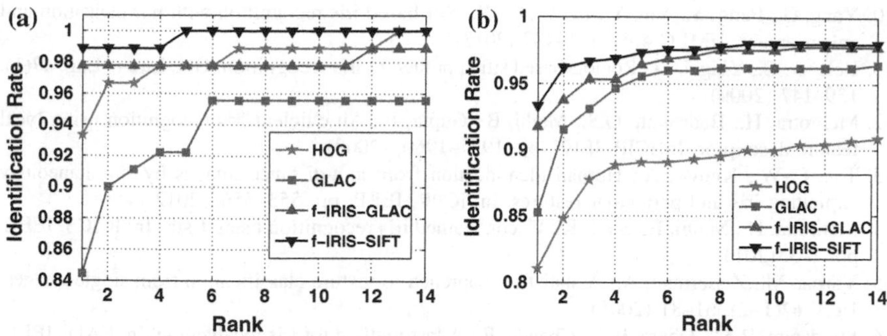

Fig. 4 CMC curves for **a** MMU1 and **b** IITD database

5 Conclusion

In this paper, we propose a biometric system that verify as well as identify the subjects with improved performance. The experimental results show that the proposed system based on textural edgeness features out performs better in case of MMU1 and IITD databases. By using only a part of iris, we obtain efficient feature representation for an iris and develop recognition system for nearly large iris database as compared to the existing method as reported in the literature. In future, we plan to incorporate more biometric characteristics to develop multimodal biometric system design.

References

1. Miyazawa, K., Ito, K., Aoki, T., Kobayashi, K., Nakajima, H.: An efficient iris recognition algorithm using phase-based image matching. In: ICIP, vol. 2, p. II–49, IEEE (2005)
2. Haralick, R.M.: Statistical and structural approaches to texture. Proc. IEEE **67**(5), 786–804 (1979)
3. Daugman, J.: The importance of being random: statistical principles of iris recognition. PR **36** (2), 279–291 (2003)
4. Daugman, J.: How iris recognition works. CSVT, IEEE Trans. **14**(1), 21–30 (2004)
5. Bowyer, K.W., Hollingsworth, K., Flynn, P.J.: Image understanding for iris biometrics: a survey. CVIU, **110**(2), 281–307 (2008)
6. Dalal, N., Triggs, B.: Histograms of oriented gradients for human detection. In: CVPR, vol. 1, pp. 886–893. IEEE (2005)
7. Lowe, D.G.: Distinctive image features from scale-invariant keypoints. IJCV. **60**(2):91–110 (2004)
8. Kobayashi, T., Otsu, N.: Image feature extraction using gradient local auto-correlations. In: ECCV, Springer, pp. 346–358 (2008)
9. Alonso-Fernandez, F., Tome-Gonzalez, P., Ruiz-Albacete, V., Ortega-Garcia, J.: Iris recognition based on sift features. In BIdS, pp. 1–8 IEEE (2009)

10. Yang, G., Pang, S., Yin, Y., Li, Y., Li, X.: Sift based iris recognition with normalization and enhancement. IJMLC **4**(4), 401–407 (2013)
11. Belcher, C., Yingzi, D.: Region-based sift approach to iris recognition. Opt. Lasers Eng. **47**(1), 139–147 (2009)
12. Mehrotra, H., Badrinath, G.S., Majhi, B., Gupta, P.: An efficient iris recognition using local feature descriptor. In ICIP, IEEE, pp. 1957–1960, (2009)
13. Tan, C.-W., Kumar, A.: Human identification from at-a-distance images by simultaneously exploiting iris and periocular features. In: ICPR, IEEE, pp. 553–556. (2012)
14. Mehrotra, H., Majhi, B., Sa, P.K.: Unconstrained iris recognition using f-sift. In: ICICS, IEEE, pp. 1–5, (2011)
15. Varma, M., Zisserman, A.: A statistical approach to texture classification from single images. IJCV **62**(1–2), 61–81 (2005)
16. Sundaram, R.M., Dhara, B.C., Chanda, B.: A fast method for iris localization. In: EAIT, IEEE, pp. 89–92 (2011)
17. Masek et al. Recognition of human iris patterns for biometric identification. PhD thesis, Masterâ€™s thesis, University of Western Australia (2003)
18. Ma, L., Wang, Y., Tan, T.: Iris recognition using circular symmetric filters. In: ICPR, IEEE vol. 2, pp. 414–417 (2002)
19. Yoon, S., Choi, S.-S., Cha, S.-H., Lee, Y., Tappert, C.C.: On the individuality of the iris biometric. In Image analysis and recognition, pp. 1118–1124 Springer (2005)
20. Multimedia university iris database [online]. Available:http://pesona.mmu.edu.my/ ccteo/
21. Kumar, A., Passi, A.: Comparison and combination of iris matchers for reliable personal authentication. PR **43**(3), 1016–1026 (2010)
22. Chang, C.-C., Lin, C.-: Libsvm: a library for support vector machines. TIST **2**(3), 27 (2011)
23. Masood, K., Javed, M.Y., Basit, A.: Iris recognition using wavelet. In ICET, pp. 253–256. IEEE (2007)
24. Rahulkar, A.D., Holambe, R.S.: lf-iris feature extraction and recognition using a new class of biorthogonal triplet half-band filter bank and flexible k-out-of-n: A postclassifier. IFS. IEEE Trans. **7**(1), 230–240 (2012)
25. Harjoko, A., Hartati, S., Dwiyasa, H.: A method for iris recognition based on 1d coiflet wavelet. World Acad. Sci. Eng. Technol. **56**(24), 126–129 (2009)
26. Zhou, Y., Kumar, A.: Personal identification from iris images using localized radon transform. In ICPR, pp. 2840–2843, IEEE (2010)
27. Elgamal, M., Nasser, A.: An efficient feature extraction method for iris recognition based on wavelet transformation. Int. J. Comput. Inf. Technol. *2*(03), 521–527 (2013)
28. Bolle, R.M., Connell, J.H., Pankanti, S., Ratha, N.K., Senior, A.W.: The relation between the roc curve and the cmc. In AIAT, pp. 15–20 IEEE (2005)
29. Fawcett, T.: An introduction to roc analysis. PRL **27**(8), 861–874 (2006)

Z-Transform Based Digital Image Watermarking Scheme with DWT and Chaos

N. Jayashree and R.S. Bhuvaneswaran

Abstract Digital Image Watermarking has recently been used widely to address issues concerning authentication and copyright protection. Chaos is one of the promising techniques implemented in image watermarking schemes. In this paper, a new watermarking algorithm based on chaos along with discrete wavelet transform and z-transformation is proposed. The host image is decomposed into 3-level Discrete Wavelet Transform (DWT). The HH3 and HL3 sub-bands are then converted to z-domain using z-transformation (ZT). Arnold Cat Map (ACM), a chaotic map is applied to the watermark image and it is divided into two equal parts. The sub-bands HH3 and HL3 are chosen for embedding the watermark image. One part of the watermark is embedded in the z-transformed HH3 sub-band, and the other part is embedded in the z-transformed HL3 sub-band. The watermarked image is obtained by taking the inverse of the ZT and the inverse of DWT. The experimental results and the performance analysis show that the proposed method is efficient and can provide practical invisibility with additive robustness.

Keywords Digital watermarking · Chaotic mapping · Z-transform · Arnold cat map · Discrete wavelet transform

1 Introduction

The security of digital media content is of primary concern owing to the advancement in the digital information and communication technology. Digital watermarking method is one of the solutions to overcome this problem. Information

N. Jayashree (✉) · R.S. Bhuvaneswaran
Ramanujan Computing Centre, Anna University, Chennai, India
e-mail: jaisri8@gmail.com

R.S. Bhuvaneswaran
e-mail: bhuvan@annauniv.edu

A. Nagar et al. (eds.), *Proceedings of 3rd International Conference on Advanced Computing, Networking and Informatics*, Smart Innovation, Systems and Technologies 43, DOI 10.1007/978-81-322-2538-6_30

hiding is the basis of digital watermarking which plays a crucial role in the copyright protection of digital data.

A watermark is a sequence, carrying information, embedded into the digital image [1]. According to the working domain, watermarking techniques can broadly be categorized into the spatial and the frequency domain. The spatial domain methods directly change the pixel values in the host image to incorporate the watermark. Although methods in this class are easier to implement, sophisticated operations on the digital information require more computing resources and time making schemes in this category unsuitable for real-time situations. Transform domain schemes were proposed to overcome the limitations of the spatial domain methods. The transform domain systems convert the original data, which is in the spatial domain, into transform domain and vice versa. These conversions can be done by utilizing various mathematical transformations such as Discrete Fourier Transform (DFT), Discrete Cosine transform (DCT), Z-transform and DWT [2–8].

A robust watermarking algorithm based on Arnold cat map (ACM) in combination with DWT and Z-transform is proposed in this paper. ACM, DWT, and the Z-transform are briefly discussed in Sect. 2. The proposed algorithm is discussed in detail in Sect. 3, and experimental results and performance analysis are included in Sect. 4 before concluding the paper in Sect. 5.

2 Background

DWT is extensively used in image processing, especially in watermarking, owing to its excellent time-frequency features and became the underlying technology behind the development of the JPEG 2000 standard [4]. Many schemes have been proposed using DWT and the commonly used technique is the Cox's watermark embedding scheme [1]. Compared to other transforms such as Fourier, Wavelet and so on, Z-transform is seldom used in image watermarking. A small variation of the pixel values may ultimately result in drastic changes in the zeros of the Z-transform. This characteristic of Z-transform provides a better way for data hiding and watermarking. These features are mostly employed for designing fragile watermarking, for authentication and tamper detection.

Chaotic based systems are quite extensively used in digital watermarking due to their complex structure, and its inherent properties of sensitivity to initial conditions and difficult to analyze [6]. Chaotic based systems are simple, fast and also these properties are highly favorable of providing high security. A chaotic map is a map that contains chaotic behavior [7, 9, 10]. Though chaotic maps are usually used in analyzing dynamical systems, they are widely employed to enhance the efficiency of watermarking algorithms [7]. Sensitivity to initial conditions and parameters, topological transitivity, the density of the periodic points are the basic attributes of a chaotic map. According to Shanon's theory, on the constraints of confusion and diffusion, the typical characteristics of the chaotic maps provide an excellent basis for its application in the areas of watermarking and encryption [11].

2.1 Arnold Cat Map (ACM)

In recent times, many schemes have been proposed using various chaotic maps such as logistic map, Arnold's cat map, tent map, etc. [7, 9–12]. The proposed algorithm uses Arnold's cat map.

ACM is widely used in applications which employ chaos; particularly in chaotic watermarking algorithms [10]. This map includes ergodic and mixing properties. This map is an invertible one, and it is also area preserving.

$$\begin{pmatrix} f(x+1) \\ g(x+1) \end{pmatrix} = \begin{pmatrix} 1 & 1 \\ 1 & 2 \end{pmatrix} \begin{pmatrix} f(x) \\ g(x) \end{pmatrix} \mod 1 \tag{1}$$

where $f(x), g(x) \in [0, 1]$. When the ACM is generalized and converted to its discrete form, it becomes

$$\begin{pmatrix} f(x+1) \\ g(x+1) \end{pmatrix} = \begin{pmatrix} 1 & a \\ b & ab+1 \end{pmatrix} \begin{pmatrix} f(x) \\ g(x) \end{pmatrix} \mod N \tag{2}$$

where $f(x), g(x) \in \{0, \ldots, \mathrm{N}-1\}$. The pixels of the image are iterated for n number of times. Let the initial position of the pixels of the image be $[x(0), y(0)]^T$ and after the nth iteration of the map, the pixel position becomes $[x(n), y(n)]^T$. This n is termed as the period T of the ACM.

2.2 Discrete Wavelet Transform

Wavelets are the useful mathematical tool used in a wide variety of applications like image denoising, data compression, watermarking, fingerprint verification, etc. [3]. The wavelets also provide a better representation of data—using Multi-Resolution Analysis (MRA). The advantages of using wavelet transform are

- The simultaneous localization of data, in both time and frequency domain
- A few coefficients are used to get the fine approximation of the data.

2.3 Z-Transform

The z-transform plays a significant role in processing discrete data [13]. It is a suitable tool to represent, design and analyze discrete-time signals and systems [13]. The z-transform can be defined as

$$Z[x] = \sum_{x=-\infty}^{\infty} f[x]t^{-x} \tag{3}$$

where t is a complex variable. For image pixels, the values of $f(x)$ are always greater than zero, and these values lie between $0 \leq n \leq \infty$. Hence, Eq. (3) becomes:

$$Z[x] = \sum_{x=0}^{\infty} f[x]t^{-x} \tag{4}$$

Similarly, the inverse z-transform is defined as

$$Z'[x] = Z^{-1}[Z(x)] \tag{5}$$

where $Z[x]$ is the z-transform and Z^{-1} is the inverse function. A method for calculating z-transform of a sequence containing N samples is proposed in [14]. The above method is termed as chirp z-transform (CZT), and it is used for efficiently computing *z-transform at points in z-plane*. The values of z-transform on a circular or a spiral plane can be shown as discrete convolution, forming the basis for the chirp z-transform. Computing z-transform along the path that corresponds to the unit circle corresponds to Fourier Transform. Hence, z-transform is suitable for many applications including correlation, filtering, and interpolation. There are many potential applications where the CZT algorithm has been implemented. For example, this algorithm has been applied to the spectral analysis of speech and voice data efficiently [14]. Z-transform evaluates the points both inside and outside the unit circle. This property can be used to analyze both high resolution and narrow frequency spectrum bands. The application of this algorithm in digital watermarking improves the robustness and the imperceptibility.

3 The Proposed Scheme

In this paper, a chaos-based robust watermarking method is proposed using z-transform in combination with Discrete Wavelet Transform. Z-transform has been widely used in several image processing applications such as authentication, copyright protection, and tamper detection [8, 13–21]. However, fragile watermarking was implemented in all these schemes. The embedding and extracting steps of the algorithm are discussed below.

3.1 Watermark Embedding Scheme

The steps for embedding the watermark are given below:

1. The watermark image W is transformed using Arnold transformation and it is divided into two parts: W1′ and W2′.

$$W' = ACM2\ (W)$$

2. Apply 3-level DWT on the original image I, and it is decomposed into four sub-bands LL3, LH3, HL3, HH3.

$$[LL3,\ LH3,\ HL3,\ HH3] = DWT\ (I)$$

3. Perform the Z-Transformation on the sub-bands HL3 and HH3.

$$I1 = ZT(HL3); I2 = ZT(HH3)$$

4. The watermark is embedded by altering the Z-transformed HL3 and HH3 sub-bands using a scaling factor α to manage the strength of the watermark image

$$I1' = I1 + \alpha\ W1'; I2' = I2 + \alpha\ W2'$$

5. Perform the inverse Z-transform on the modified frequency sub-bands and execute inverse DWT (3 level) to generate the watermarked image I′.

$$HL31 = iZT\ (I1'); HH31 = iZT\ (I2')$$
$$I' = 3\text{-level iDWT}\ (LL3, LH3, HL31, HH31)$$

3.2 Watermark Extraction Scheme

The procedure for extracting the watermark is given below:

1. The watermarked image I′ is decomposed into sub-bands by applying 3-level DWT

$$[LL3, LH3, HL3, HH3] = DWT\ (I')$$

2. The sub-bands HL3 and HH3 are taken, and z-transform is computed for these two sub-bands

$$I'1 = ZT(HL3);\ I'2 = ZT(HH3)$$

3. The keys viz., key1 and key2, generated while computing the z-transformation during the embedding process is employed to extract the watermark from the z-transformed sub-bands I′1 and I′2. Subtract the values of I′1 and I′2 from key1 and key2 as follows:

$$g\_1 = \mathrm{key1} - I'1; g\_2 = \mathrm{key1} - I'2$$

4. The values g_1 and g_2 are combined and the watermark image is generated by applying the Arnold transformation.

$$G = [g\_1\ g\_2]; W = \mathrm{ACM2}\ (G)$$

where W is the extracted watermark.

4 Performance Evaluation

Many experiments were conducted to evaluate the invisibility and the robustness of the proposed algorithm. Matlab version 7 is used for this evaluation on standard test images such as Lena, Baboon, Bridge, etc. Watermarking algorithms are evaluated with reference to its invisibility and robustness. Imperceptibility or the invisibility is the measure that gives the quality of the watermarked image. In other terms, by embedding the watermark, the original image should not be distorted.

Robustness is the measure of a watermark to resist against intentional and unintentional attacks. The various attacks include common signal processing attacks such as Gaussian noise, salt and pepper noise, median filtering, Blurring, Histogram Equalization, Gamma Correction and geometric attacks like scaling and rotation. Peak Signal to Noise Ratio (PSNR), Correlation Coefficient (CC), Bit Error Rate (BER) and Structural Similarity Index (SSIM) were used to evaluate the imperceptibility and robustness of the proposed scheme.

The PSNR is widely used to measure the quality of reconstruction of images. The correlation coefficient is a quantity to determine the similarity between two images—the original and the extracted watermark images and the value usually lie between −1 and 1. The Bit Error Rate is a measure that gives the ratio of erroneously detected bits to the total number of embedded bits. The structural similarity index proposed by [22] as a solution for image quality assessment. It combines three factors, namely the image luminance, contrast and the structural changes in images. The values of SSIM lie between 0 and 1. It is calculated by using the formula:

$$SSIM = \frac{(2\mu_x\mu_y + c_1)(2\sigma_{xy} + c_2)}{\left(\mu_x^2 + \mu_y^2 + c_1\right)\left(\sigma_x^2 + \sigma_y^2 + c_2\right)} \quad (6)$$

where c_1 and c_2 are zero corresponding to the universal image quality index, μ_x denotes mean of x, μ_y mean of y, σ_x and σ_y denotes the standard deviation of x and y respectively, and σ_{xy} denotes the cross-correlation. The experimental results showing the different values of PSNR, BER, CC and SSIM for the images Lena, Peppers, Baboon, and Barbara are given in the Table 1.

Various attacks were applied on the watermarked image to check the robustness of the proposed algorithm. Both host and watermark images under various attacks are depicted in Table 2. The results# compared with the method proposed by [14]* are given in Tables 3, 4 and 5.

Table 1 Values of PSNR, BER, CC and SSIM

Image	PSNR	BER	CC	SSIM
Lena	53.6491	0.1343	1	1.0000
Peppers	54.6146	0.1091	1	1.0000
Baboon	53.4593	0.1375	1	1.0000
Barbara	54.6998	0.0958	1	1.0000

Table 2 Watermarked images under various attacks and the extracted watermark

(a) No Attack

(b) Additive White Gaussian Attack

(c) Blurring Attack

(d) Contrast Enhancement Attack

(e) Flipping Attack

Table 3 Comparison of SSIM values of attacked watermark images

Image	Lena		Peppers		Baboon	
Kind of attack	1*	2#	1*	2#	1*	2#
No attack	0.867	1.0000	0.9067	1.0000	0.8815	1.0000
AWGN	0.1955	0.6308	0.2091	0.6309	0.2755	0.7919
Blurring	0.6373	0.8930	0.7666	0.9218	0.3535	0.6584
Contrast enh.	0.8181	0.9115	0.8625	0.9154	0.7968	0.7967
Flipping	0.2258	0.2518	0.396	0.2102	0.1322	0.1429
Gamma corr.	0.6051	0.9174	0.4174	0.9941	0.614	0.9627

Table 4 Comparison of SSIM values of extracted watermark images

Image	Lena		Peppers		Baboon	
Kind of attack	1*	2#	1*	2#	1*	2#
AWGN	0.1414	0.3911	0.222	0.3928	0.1509	0.3867
Blurring	−0.0787	0.3612	−0.0594	0.4886	−0.0843	0.2286
Contrast enh.	0.6704	0.4312	0.5326	0.5688	0.5572	0.1956
Flipping	1	1.0000	1	1.0000	1	1.0000
Gamma corr.	0.7392	0.7581	0.828	0.9836	0.916	0.6047

Table 5 Comparison of correlation coefficient of extracted watermark images

Image	Lena		Peppers		Baboon	
Kind of attack	1*	2#	1*	2#	1*	2#
No attack	1	1	1	1	1	1
AWGN	0.6017	0.9228	0.633	0.9276	0.5803	0.9222
Blurring	−0.6073	0.4468	−0.5979	0.6020	−0.579	0.2931
Contrast enh.	0.9892	0.5376	0.97	0.7022	0.9724	0.2532
Flipping	1	1	1	1	1	1
Gamma corr.	0.9743	0.8492	0.9894	0.9963	0.9963	0.7295

5 Conclusion

In this paper, an image watermarking technique based on Arnold cat map was presented. This algorithm combines the strengths of the DWT, Z-transform, and the chaotic ACM making the scheme more robust. Several experiments were conducted using standard test images, and the results were analyzed. The PSNR values, CC, BER and the SSIM values of the experimental results exhibit that the algorithm is secure as well as robust.

References

1. Cox, I.J., Kilian, J., Leighton, F.T., Shamoon, T.: Secure spread spectrum watermarking for multimedia. IEEE Trans. Image Process. **6**, 1673–1687 (1997)
2. Barni, M., Bartolini, F., Cappellini, V., Piva, A.: A dct-domain system for robust image watermarking. Sig. Process. **66**, 357–372 (1998)
3. Sifuzzaman, M., Islam, M.R., Ali, M.Z.: Application of wavelet transform and its advantages compared to Fourier transform. J. Phys. Sci. **13**, 121–134 (2009)
4. Dawei, Z., Guanrong, C., Wenbo, L.: A chaos based robust wavelet domain watermarking algorithm. Chaos, Solitons Fractals **22**(1), 47–54 (2004)
5. Yang, Q., Cai, Y.: A digital image watermarking algorithm based on discrete wavelet transform and discrete cosine transform. In: 2012 International Symposium on Information Technology in Medicine and Education
6. Liu, B., Liu, N., Li, J.X., Liand, W.: Research of image encryption algorithm base on chaos theory. In: IEEE 6th International Forum on Strategic Technology (2011)
7. Wu, X., Guan, Z.H.: A novel digital watermark algorithm based on chaotic maps. Phys. Lett. A **365**(5–6), 403–406 (2007)
8. Rabiner, L.R., Schafer, R.W., Rader, C.M.: The chirp z-transform algorithm and its applications. Bell Syst. Tech. J. 1249–1292 (1969)
9. Vinod, P., Pareek, N.K., Sud, K.K.: A new substitution-diffusion based image cipher using chaotic standard and logistic maps. Elsevier (2008)
10. Fei, C., Kwok-Wo, W., Xiaofeng, L., Tao, X.: Period distribution of the generalized discrete arnold cat map for N = 2e. IEEE Trans. Inform. Theory **59**(5) (2013)
11. Pisarchik, A.N., Zanin, M.: Chaotic map cryptography and security. In: Encryption: Methods, Software and Security
12. Zhang, L., Liao, X., Wang, X.: An image encryption approach based on chaotic maps. Chaos, Solitons Fractals **24**, 759–765 (2005)
13. Ho, A.T.S.,, Xunzhan, Z., Jun, S., Pina, M.: Fragile watermarking based on Encoding of the Zeros of the z-Transform. IEEE Trans. Inform. Forensics Secur. **3**(3) (2008)
14. Ho, A.T.S., Xunzhan, Z., Jun, S.: Authentication of biomedical images based on zero location watermarking. In: 8th International Conference on Control, Automation, Robotics and Vision, Kunming, China (2004)
15. Vich, R.: Z-Transform theory and applications. D. Reidel Publishing Company, Dordrechi (1987)
16. Stillings, S.A.: A Study of the Chirp Z-Transform Algorithm and its Applications, Technical Report. Kansas State University (1972)
17. Kolchi, S., Masataka, K., Tomohisa, M.: Applications of image reconstruction by means of Chirp z-Transform. MVA'90, IAPR Workshop on Machine Vision Applications, Tokyo Nov. 28–30 1990
18. Khan, M.S., Boda, R., Vamshi Bhukaya, V.: A copyright protection scheme and tamper detection using z transform. Int. J. Comput. Electric. Adv. Commun. Eng. **1**(1) (2012)
19. Mary, A., Erbug, Ç., Gholamreza, A.: A watermarking algorithm based on chirp z-transform, discrete wavelet transform, and singular value decomposition. SIViP. Springer-Verlag, London (2014). doi:10.1007/s11760-014-0624-9
20. Mandal, J.K.: A frequency domain steganography using Z transform (FDSZT). In: International Workshop on Embedded Computing and Communication System (IWECC 2011), Rajagiri School of Engineering & Technology, Kochin 22–23 Dec. 2011
21. Cox, I.J., Matthew, L.M., Jeffrey, A.B., et al.: Digital watermarking and steganography, 2nd edn. Morgan Kaufmann Publishers (Elsevier), Burlington (2007)
22. Wang, Z., Bovik, A.C., Sheikh, H.R., Simoncelli, E.P.: Image quality assessment: from error visibility to structural similarity. IEEE Trans. Image Process. **13**(4), 600–612 (2004)

Panoramic Image Mosaicing: An Optimized Graph-Cut Approach

Achala Pandey and Umesh C. Pati

Abstract Panoramic images have numerous important applications in the area of computer vision, video coding/enhancement, virtual reality and surveillance. The process of building panorama using mosaicing technique is a challenging task as it requires consideration of various factors such as camera motion, sensor noise and illumination difference, that deteriorate the quality of the mosaic. This paper proposes a feature based mosaicing technique for creation of high quality panoramic image. The algorithm mitigates the problem of seam by aligning the images employing Scale Invariant Feature Transform (SIFT) features and blending the overlapping region using optimized graph-cut. The results show the efficacy of the proposed algorithm.

Keywords Panorama · Mosaicing · Blending · Features · Geometric transformations · SIFT · RANSAC

1 Introduction

A panorama contains a wide field-of-view (FOV) of the scene surrounding an observer and therefore, has more feature and information as compared to the normal input images of a scene. Earlier in 1900, panoramic cameras were used for general purpose panoramic photography for aerial and architectural panoramas. However, the introduction of computer technology and digital images simplified the process and digital panoramas achieved wide acceptance. Panoramic image mosaicing aims at reconstructing the whole scene from the limited images or video frames in order

A. Pandey (✉) · U.C. Pati
Department of Electronics and Communication Engineering, National Institute of Technology, Rourkela 769 008, Orissa, India
e-mail: achala.pandey13@gmail.com

U.C. Pati
e-mail: ucpati@nitrkl.ac.in

A. Nagar et al. (eds.), *Proceedings of 3rd International Conference on Advanced Computing, Networking and Informatics*, Smart Innovation, Systems and Technologies 43, DOI 10.1007/978-81-322-2538-6_31

to give better visualization and complete view of the environment. There are many applications such as computer vision/graphics [1], image based rendering [2], virtual travel, virtual environment [3], scene change detection, video indexing/compression [4] and satellite/aerial imaging [5]. Anyone can generate a panorama these days having a computer and a camera with the help of available software.

A panorama can be captured in many possible ways, either single camera rotated around its optical centre or multiple cameras covering different part of the scene. Sometimes, large FOV lens e.g. fish-eye lens or parabolic mirror [6] are used to capture panoramic images but these are very expensive as well as hardware intensive means of panorama generation. Mosaicing algorithms [7–9] provide simplified solution to overcome the above mentioned problems.

The main emphasis of this work is on panoramic image mosaic generation. It has been implemented using feature based method of alignment and optimized graph-cut blending to generate high quality panoramic mosaic from a sequence of input images.

The paper organization is as follows: Sect. 2 gives an overview of the fundamental steps required for panoramic image mosaicing. The proposed method is described in Sect. 3 with the help of different steps and flow chart. Section 4 demonstrates the results obtained on implementation of the proposed algorithm on scenery images. The paper concludes in Sect. 5.

2 Panoramic Image Mosaicing

Panoramic mosaic generation process can be described in three main steps: image capturing system, pre-processing of the images and image mosaicing.

Panoramic image from multiple images can either be generated using a camera mounted on a tripod and rotated about its axis or using multiple cameras mounted together covering different directional views of the scene. In some cases, omnidirectional image sensors with fisheye lenses or mirrors are used, but their short focal length, high cost and lens distortion limits their access.

The images captured using panoramic imaging system are pre-processed before mosaicing for removing noise or any other intensity variations in the image. The processed images are then mosaiced by using the fundamental steps of mosaicing to get a seamless panoramic image.

3 Proposed Method of Panoramic Image Mosaicing

The images acquired using panoramic image capturing systems are pre-processed before stitching them together and then features are extracted from the images. The images are then re-projected over global frame which may result in creation of seam in the overlapping region of the images. A new optimized graph-cut method is employed for blending the individual images to create a single seamless panorama.

3.1 Feature Detection and Matching

In the proposed method of mosaicing, SIFT (Scale invariant feature transform) features [10] have been used for feature detection. SIFT features are stable, accurate, invariant to scaling and rotation which provides robust matching.

After feature point detection, next step is selection of corresponding features between the images. RANdom SAmple Consensus (RANSAC) [11] is used for robust estimation of homography between images for matching inliers.

3.2 Transformation of Images and Re-Projection

Based on the presence of different type of geometric distortion, transformations such as rigid, affine or projective are used. In case of panoramic image mosaics, the transformation used is projective transform between the images [12]. The projective model using a small number of parameters is most efficient and accurate parametric model for panoramic mosaic generation [13]. The images are transformed so as to align them on a common compositing surface or re-projection manifold. There are mainly three types of compositing manifold planar, cylindrical and spherical. In the proposed algorithm, planar manifold has been considered.

3.3 Blending of Images

After projecting the images over the compositing manifold, the next operation performed over it is blending of images. Various factors such as change in illumination or exposure artifacts cause intensity differences over the region of overlap resulting in the generation of visible edges in the final mosaic. These regions are handled differently from the other regions of the mosaic by applying an algorithm that helps in removing the seam while preventing loss of information. Image blending algorithms are categorized into two classes which are *transition smoothing* and *optimal seam finding*. In the proposed method, an optimized graph-cut algorithm is used for blending of images. It is based on optimal seam finding approach. Optimized graph-cut performs better for seam smoothing as compared to simple graph-cut and other existing blending techniques.

3.4 Optimized Graph-Cut

In graph cut technique, a seam is searched in such a way that the region between the images has a minimum variation of intensities. The search is in terms of minimization

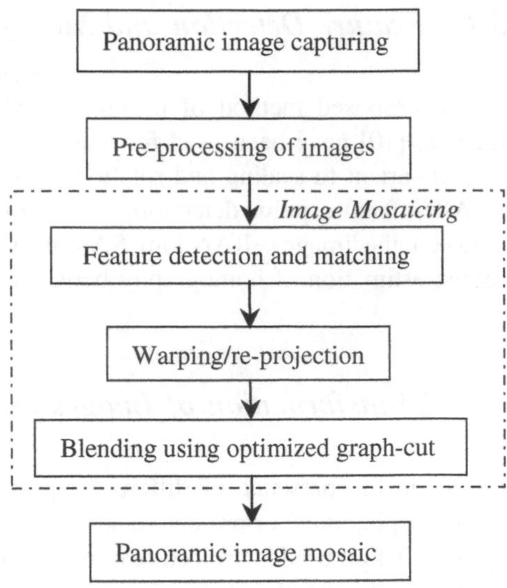

Fig. 1 Proposed algorithm for panoramic image mosaic generation

of energy of certain energy function. Earlier, algorithm such as dynamic programming [14] could not explicitly improve the existing seams as it was memoryless. However, graph-cut overcomes this limitation by treating each pixel uniformly and placing patches on the existing images. A graph is constructed for the overlapping region with edge weight corresponding to error across the seam cutting that edge and thus, creating a seam with least error.

In the algorithm, matching quality measure calculates the intensity difference between the pairs of pixels. Let 's' and 't' be the position of two adjacent pixels in the overlapping region, $A(s)$ and $B(s)$ be the pixel intensities at 's' for old and new patch. Similarly $A(t)$ and $B(t)$ be the pixel intensities at position 't'. The matching quality [15] cost can now be defined as,

$$M(s,t,A,B) = \|A(s) - B(s)\| + \|A(t) - B(t)\| \tag{1}$$

where $\|\bullet\|$ denotes the appropriate norm. All the steps of the proposed panoramic mosaicing algorithm have been summarized in Fig. 1.

4 Results and Discussions

Different set of panoramic images have been considered and the algorithm is employed over these sets for generation of seamless panoramic mosaics. The first set consists of three input images as shown in Fig. 2a, b, c. In the first step, SIFT features are extracted and matched using RANSAC. Figure 2d shows the

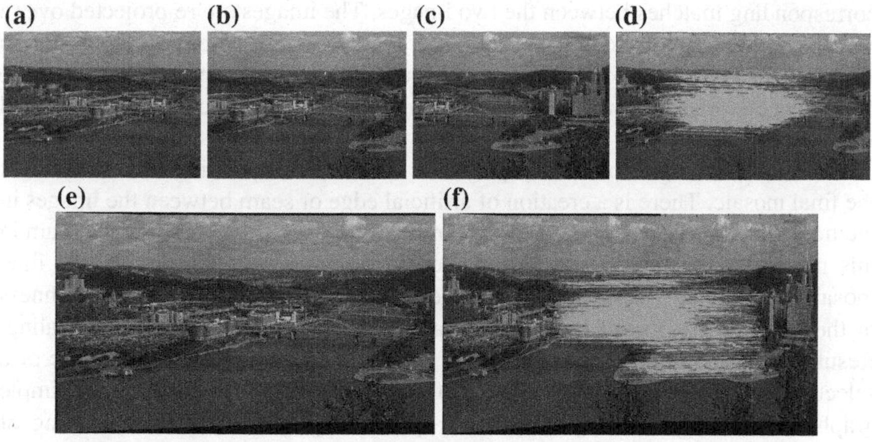

Fig. 2 Panoramic mosaicing. **a, b, c** Input images. **d** Matching between image (a) and (b). **e** Mosaic of (a) and (b). **f** Matching between generated mosaic (e) and input image (c)

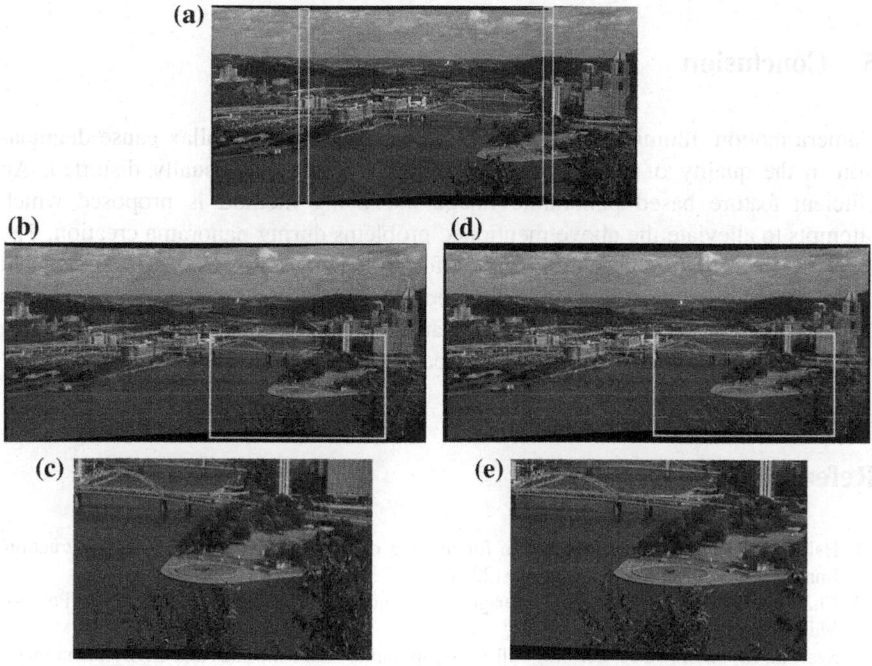

Fig. 3 Seam visibility. **a** *Red boxes* show the seam generated between the image in the transition region, **b** seam smoothing using graph-cut showing horizontal discrepancies, **c** zoomed view to show the *horizontal marks* (artifacts) of the mosaic in (**b**). **d** Panoramic mosaic using proposed method, **e** zoomed view to show the *removed horizontal marks* (artifacts) of the mosaic

corresponding matches between the two images. The images are re-projected over a planar compositing surface for mosaic generation. The mosaic of two input images is shown in Fig. 2e which becomes an input image for the upcoming steps. The third image is matched with the generated mosaic for feature correspondence and alignment (Fig. 2f).

The aligned images are finally projected over the compositing manifold to get the final mosaic. There is a creation of artificial edge or seam between the images in the transition area as can be clearly seen in Fig. 3a. The red boxes show the seam in this mosaic. If the images are not blended for removing the seam, then the final mosaic will contain visible artifacts. Blending ensures a high degree of smoothness in the transition region of the mosaic making it seamless and visually appealing. Results using graph-cut is shown in Fig. 3b and the zoomed view in Fig. 3c of a selected part of the scene shows the artifacts remaining in the image after simple graph-cut blending. The algorithm uses optimized graph-cut for the blending of images. The results of optimized graph-cut are shown in Fig. 3d. The horizontal line artifacts Fig. 3c have been removed using proposed technique.

The zoomed view of the selected portion of the image is shown in Fig. 3e. It can be observed that there is no horizontal artifact in the image mosaic as in the case of graph-cut.

5 Conclusion

Camera motion, illumination difference, sensor noise and parallax cause degradation in the quality of panoramic image mosaic making it visually distorted. An efficient feature based panoramic image mosaicing method is proposed which attempts to alleviate the above mentioned problems during panorama creation. The algorithm uses feature based registration for aligning the images. The seam created in the area of overlap is removed using optimized graph-cut method which provides a feasible solution to the problem of seamless panoramic image mosaicing. The results show the effectiveness of the proposed algorithm.

References

1. Hsieh, J.W.: Fast stitching algorithm for moving object detection and mosaic construction. Image Vis. Comput. **22**(4), 291–306 (2004)
2. Chan, S.C., Shum, H.Y., Ng, K.T.: Image based rendering and systhesis. IEEE Signal Process. Mag. **24**(6), 22–33 (2007)
3. Szeliski, R., Shum, H.Y.: Creating full view panoramic image mosaics and environment maps. In: Proceedings of SIGGRAPH, pp. 251–258 (1997)
4. Irani, M., Anandan, P.: Video indexing based on mosaic representation. Proc. IEEE **86**(5), 905–921 (1998)

5. Kekec, T., Vildirim, A., Unel, M.: A new approach to real-time mosaicing of aerial images. Robot. Auton. Syst. **62**, 1755–1767 (2014)
6. Gledhill, D., Tian, G.Y., Taylor, D., Clarke, D.: Panoramic imaging—a review. Comput. Graph. **27**, 435–445 (2003)
7. Brown, M., Lowe, D.: Recognising panoramas. In: Proceedings of Ninth IEEE International Conference on Computer Vision, Nice, France, vol. 2, pp. 1218–1225 (2003)
8. Marzotto, R., Fusiello, A., Murino, V.: High resolution video mosaicing with global alignment. In: Proceedings of IEEE Computer Society Conference on Computer Vision and Pattern Recognition, Washington, DC. vol. 1, pp. 692–698 (2004)
9. Robinson, J.A.: A simplex-based projective transform estimator. In: Proceedings of International Conference on Visual Information Engineering, U.K., pp. 290–293 (2003)
10. Lowe, D.: Distinctive image features from scale-invariant keypoints. Int. J. Comput. Vision **60** (2), 91–110 (2004)
11. Fischler, M., Bolles, R.: Random sample consensus: A paradigm for model fitting with application to image analysis and automated cartography. Commun. ACM **24**, 381–395 (1981)
12. Kim, D.H., Yoon, Y.I., Choi, J.S.: An efficient method to build panoramic image mosaics. Pattern Recogn. Lett. **24**, 2421–2429 (2003)
13. Mann, S., Picard, R.W.: Video orbits of the projective group: a simple approach to featureless estimation of parameters. IEEE Trans. Image Process. **6**(9), 1281–1295 (1997)
14. Efros, A.A., Freeman, W.T.: Image quilting for texture synthesis and transfer. In: Proceedings 28th Annual Conference SIGGRAPH, Los Angeles, CA, USA, pp. 341–346 (2001)
15. Kwatra, V., Schodl, A., Essa, I., Turk, G., Bobick, A.: Graphcut textures: image and video synthesis using graph cuts. ACM Trans. Graphics 277–286

Spatial Resolution Assessment in Low Dose Imaging

Akshata Navalli and Shrinivas Desai

Abstract Computed tomography has been reported as most beneficial modality to mankind for effective diagnosis, planning, treatment and follows up of clinical cases. However, there is a potential risk of cancer among the recipients, who undergoes repeated computed tomography screening. This is mainly because the immunity of any living tissue can repair naturally the damage caused due to radiation only up-to a certain level. Beyond which the effort made by immunity in the natural repair can lead to cancerous cells. So, most computed tomography developers have enabled computed tomography modality with the feature of radiation dose management, working on the principle of as low as reasonably achievable. This article addresses the issue of low dose imaging and focuses on the enhancement of spatial resolution of images acquired from low dose, to improve the quality of image for acceptability; and proposes a system model and mathematical formulation of Highly Constrained-Back Projection.

Keywords Computed tomography · Low dose · Cancer · Radiation · Image quality

A. Navalli (✉)
Department of Computer Science, B.V.B College of Engineering & Technology, Hubli 580031, India
e-mail: akshatanavalli06@gmail.com

S. Desai
Department of Information Science and Engineering, B.V.B College of Engineering & Technology, Hubli 580031, India
e-mail: shree.desai07@gmail.com

A. Nagar et al. (eds.), *Proceedings of 3rd International Conference on Advanced Computing, Networking and Informatics*, Smart Innovation, Systems and Technologies 43, DOI 10.1007/978-81-322-2538-6_32

1 Introduction

The medical imaging has been benefited by the development of Computed Tomography (CT) and has contributed in effective diagnosis, treatment and managing complications in clinical domain. However, the concern towards the probable occurrence of cancer due to repeated CT scanning cannot be overlooked. Recently, a survey for estimating the risk of cancer due to usage of CT is carried out and reported to be 0.4 % in the United States alone. It is also estimated to be 1.5–2 % increment in this risk by the end of 2010 considering the trend in usage of CT [1]. The radiation doses during CT screening are approximately 50–100 times more, compared to the radiation doses in x-ray examinations [2]. As the dosage used during CT is much higher than conventional adding to risk of cancer, the CT developers are designing and coming out with a low dose protocol enabled CT units which work on the principle of As Low As Reasonably Achievable (ALARA) [3]. The challenge with low dose imaging is the degradation of image quality with poor spatial resolution. Even with accurate acquisition settings, set during screening process the presence of noise is observed, which is mainly due to low dose imaging protocol. Some of the low dose imaging protocol suggests to acquire the projection data at a sparse angle leading to under-sampling due to lesser radiation dose, and then apply suitable compensating reconstruction algorithm to get better quality images (on par with standard dose) that compensates the under-sampling.

From the literature and official websites of leading CT unit developers, we observe various compensating algorithms which are intended to address poor spatial resolution in CT images which are acquired by following low dose imaging protocol. Most predominantly used methods are Adaptive Statistical Iterative Reconstruction (ASIR), Iterative Reconstruction in Image Space (IRIS) and Model Based Iterative Reconstruction (MBIR) [4]. In case of IRIS, the de-noising property of iterative function helps to achieve a 50 % reduction in radiation dose [5]. In case of MBIR, number of samples, allowing multiple projections of the same object allows construction of CT image with better image quality and improved spatial resolution [1, 6]. Iterative algorithm is reported to construct image with better quality even with incomplete data, preserving higher spatial resolution of the CT images [2]. Comparison of various techniques is shown in the Table 1. Highly Constrained Back Projection (HYPR) was introduced recently for the reconstruction of under sampled (sparse view angle) CT/MRI (Magnetic Resonance Imaging) images [7]. In any given clinical case, both temporal and spatial resolution is of clinical interest as they provide definite and clear anatomical structures. To have better spatial resolution the acquisition need to be complete in all respect while, to achieve higher temporal resolution acquisition system should be tuned to acquire projections in lesser time, but, there is a trade-off between acquisition time and quality of constructed image. To address this trade-off one possible approach is to scan the object with a sparse view (achieving lesser acquisition time). And later apply compensating algorithms as aforementioned to achieve improved image quality. The HYPR method works on this principle.

Table 1 Comparison of various reconstruction techniques

Reconstruction techniques	Disadvantages	Advantages
ASIR	Computation time is higher (approximately 30 % higher for a standard CT) and artificial over-smoothing of the image	Allow significant radiation dose reduction
IRIS	Very high computational cost, which can be 100–1000 times higher than filtered back projection (FBP)	Permit the detection process to be accurately modelled
MBIR	Complicated algorithm, uses multiple iterations and multiple models. The reconstruction time is very high	Significant dose reduction
HYPR	Poor spatial resolution for dynamic low dose CT images	Better computation time

2 Existing Image Reconstruction Techniques

2.1 Adaptive Statistical Iterative Reconstruction (ASIR)

GE Healthcare has introduced new CT unit called BrightSpeed Elite© with the feature of low dose imaging. Internally Adaptive Statistical Iterative Reconstruction (ASIR) algorithm works for this feature [8]. Due to the low dose imaging feature, images are captured with reduced tube current. The ASIR technique uses Filtered Back Projection (FBP) constructed standard image as the primary building block [9, 10].

2.2 Iterative Reconstruction in Image Space (IRIS)

Siemens has introduced new CT modality called SAFIRE© to enable low dose imaging, which works on the principle of Iterative Reconstruction in Image Space (IRIS) [11]. Basically, IRIS uses both data pertaining to sinogram space and image space. The number of iterations is dependent on the requirement of a specific scan. This technique is recorded to provide higher Contrast Noise Ratio (CNR) and Signal-to-Noise Ratio (SNR) in low dose imaging as well as some exceptional clinical cases such as paediatric and obese patients [12].

2.3 Model Based Iterative Reconstruction (MBIR)

MBIR improves the image quality, which are generated by low dose imaging protocol. Comparatively, MBIR significantly removes the image noise and artifacts over ASIR technique [13].

3 Proposed Image Reconstruction Technique (HYPR)

In HYPR method all the time slices are subjected to constrained forward and backward projections. The composite image is generated by integrating all the images of previous results. The resulting composite image exhibits good spatial characteristics. However, the composite image exhibits poor temporal characteristics. Hence a weight image is generated by calculating the ratio of unfiltered back projections of original projections to the unfiltered back projection of the composite image. The good temporal resolution is expected in weight image as it considers both original images and composite images. When weight image is multiplied with composite image, the result is a HYPR frame with good signal to noise ratio (SNR) and good temporal characteristics. Figure 1 presents the system model of HYPR. Original HYPR calculates composite image and weight factor and try to address spatial resolution improvement. The projection S_t is obtained by applying radon transform R on the image I_t at some angle ϕ_t

$$S_t = R_{\phi_t}[I_t] \tag{1}$$

Next, the composite image C is found from the filtered back projection applied to all the S_t is as follows:

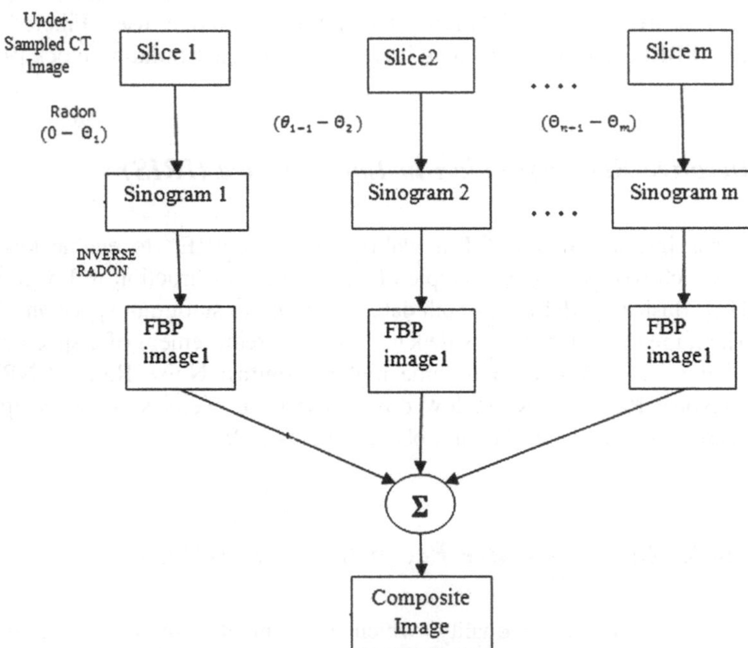

Fig. 1 System model of highly constrained back projection (HYPR)

$$C = \sum_{i=1}^{N} R^{f}_{\phi_{t_i}}[S_{t_i}] \quad (2)$$

3.1 Quality Evaluation Parameter: Peak Signal-to-Noise Ratio (PSNR)

In medical imaging, the quality of reconstructed images is subjective in nature. They are accepted to provide clear and distinct visualization of anatomical structures. However, to access the proposed HYPR method of quantifying the image quality we preferred to choose Peak Signal-to-Noise Ratio (PSNR). It provides the presence of the signal component against the unwanted noise component. In usual practice, we calculate PSNR by Mean Square Error (MSE) and represent using decibel unit. Normally, CT images with PSNR value greater than 40 dB are considered to be clinically useful, while images with lesser than 20 dB are of no much use. If K is the noisy approximation of noise-free $m * n$ monochrome image I, then MSE can be defined as [14, 15]:

$$MSE = \frac{1}{xy}\sum_{i=0}^{m-1}\sum_{j=0}^{n-1}[I(i,j) - K(i,j)]^2 \quad (3)$$

The PSNR is defined as:

$$PSNR = 10 \cdot \log\left(\frac{255}{\sqrt{MSE}}\right) \quad (4)$$

4 Results and Discussions

We have considered a MATLAB platform for simulation purpose. The experimental setup consists of different CTA (Computed tomography Angiography) datasets collected from online database and then classified based number of time frames, intensity variation, dynamic nature and noise level. For the study purpose, we have considered the original image as shown in the Fig. 2 and have obtained the under-sampled images by varying the incremental angle of 1.5°, 2°, 4°, 8° as shown in the Fig. 2b–e. Composite image of the original dataset is obtained by initially applying radon and radon filter back projections and summation of all original images at a sparse angle (Shown in Fig. 3a).

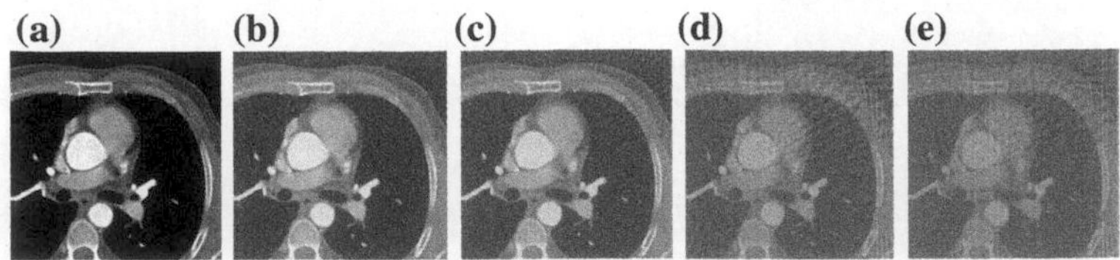

Fig. 2 **a** Original standard dose image, **b** undersampled image with an incremental angle 1.5°, **c** undersampled image with an incremental angle 2°, **d** undersampled image with an incremental angle 4°, **e** Undersampled image with an incremental angle 8°

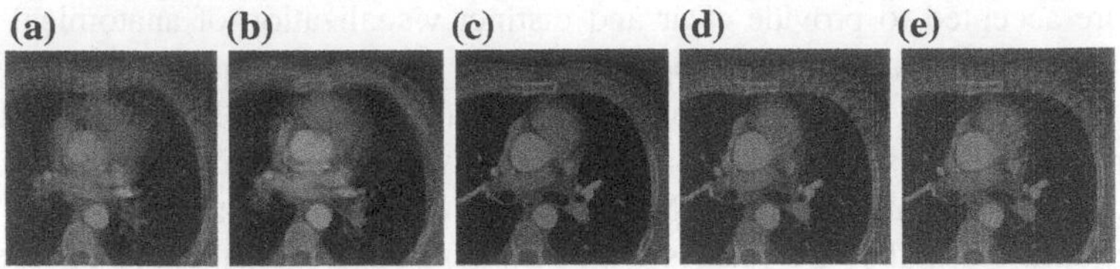

Fig. 3 **a** Composite image of the original image, **b** composite image with an incremental angle 1.5°, **c** composite image with an incremental angle 2°, **d** composite image with an incremental angle 4°, **e** composite image with an incremental angle 8°

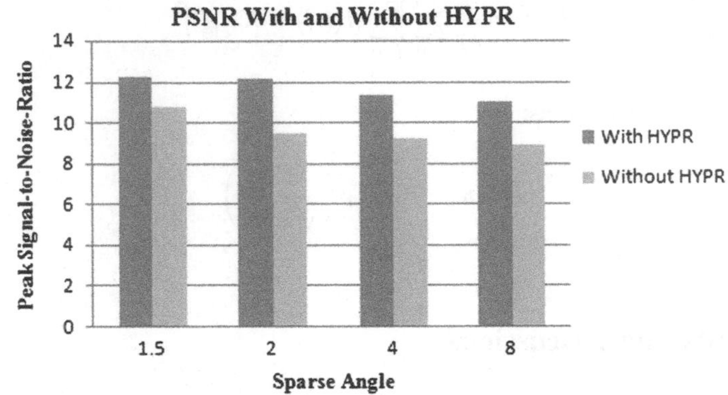

Fig. 4 PSNR (with and without HYPR) at various sparse angles

Incremental Angle datasets are then subjected to radon and iradon filter-back projection and summation of all the images are performed to obtain a composite image. Composite images obtained by varying the incremental angle of 1.5°, 2°, 4°, 8° are as shown in the Fig. 3b–e.

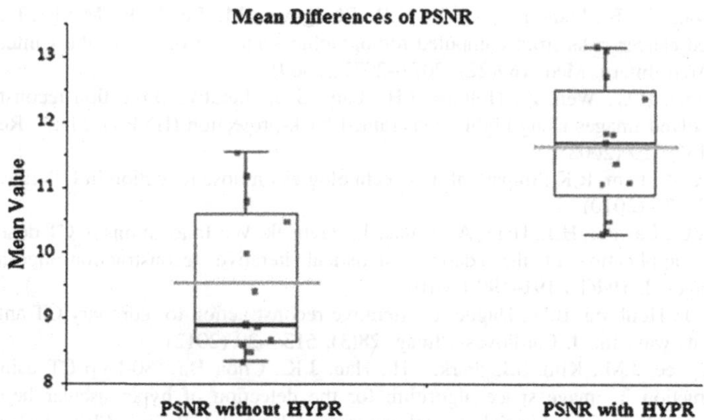

Fig. 5 Mean differences of PSNR (with and without HYPR)

5 Conclusion

Comparative study of the dataset using standard dose (without HYPR) and low dose (with HYPR) by considering the sparse, angular projections are made and the quality evaluation parameter, PSNR is calculated. From the Fig. 4 it is clear that the quality of the images that were not acceptable with the PSNR below 10 dB, was enhanced significantly when HYPR was applied and it also shows the extent to which low dose is achievable without compromising the image quality. The mean difference of PSNR with HYPR and without HYPR is depicted in the Fig. 5 conducted for thirteen experiments shows a significant improvement in the PSNR value. An increase in the PSNR value will eventually lead to the increase in the spatial resolution of the images. Hence, we were able to assess the spatial resolution of the images obtained using the low dose.

References

1. Brenner, D.J., Hall, E.J.: Computed tomography -an increasing source of radiation exposure. N. Engl. J. Med. **357**(22), 2277–2284 (2007)
2. Tubiana, M., Nagataki, S., Feinendegen, L.E.: Computed tomography and radiation exposure. N. Engl. J. Med. **358**(8), 850 (2008)
3. Entrikin, D.W., Leipsic, J.A., Carr, J.J.: Optimization of radiation dose reduction in cardiac computed tomographic angiography. Cardiol. Rev. **19**(4), 163–176 (2011)
4. European Society of Radiology (ESR): ECR 2012 Book of Abstracts-B-Scientific Sessions. Insights into imaging **3**(Suppl 1), 135 (2012)
5. Friedland, G.W., Thurber, B.D.: The birth of CT. AJR Am. J. Roentgenol. **167**(6), 1365–1370 (1996)

6. de Gonzalez, A.B., Mahesh, M., Kim, K.P., Bhargavan, M., Lewis, R., Mettler, F., Land, C.: Projected cancer risks from computed tomographic scans performed in the United States in 2007. Arch. Intern. Med. **169**(22), 2071–2077 (2009)
7. O'Halloran, R.L., Wen, Z., Holmes, J.H., Fain, S.B.: Iterative projection reconstruction of time-resolved images using highly-constrained back-projection (HYPR). Magn. Reson. Med. **59**(1), 132–139 (2008)
8. Lee, T.Y., Chhem, R.K.: Impact of new technologies on dose reduction in CT. Eur. J. Radiol. **76**(1), 28–35 (2010)
9. Silva, A.C., Lawder, H.J., Hara, A., Kujak, J., Pavlicek, W.: Innovations in CT dose reduction strategy: application of the adaptive statistical iterative reconstruction algorithm. Am. J. Roentgenol. **194**(1), 191–199 (2010)
10. Leipsic, J., Heilbron, B.G., Hague, C.: Iterative reconstruction for coronary CT angiography: finding its way. Int. J. Cardiovasc. Imag. **28**(3), 613–620 (2012)
11. Hur, S., Lee, J.M., Kim, S.J., Park, J.H., Han, J.K., Choi, B.I.: 80-kVp CT using iterative reconstruction in image space algorithm for the detection of hypervascular hepatocellular carcinoma: phantom and initial clinical experience. Korean J. Radiol. **13**(2), 152–164 (2012)
12. Grant, K., Raupach, R.: SAFIRE: Sinogram affirmed iterative reconstruction. siemens medical solutions whitepaper. http://www.medical.siemens.com/siemens/en_US/gg_ct_FBAs/files/Definition_AS/Safire.pdf. Accessed Nov 2012
13. Dutta, J., Ahn, S., Li, Q.: Quantitative statistical methods for image quality assessment. Theranostics, **3**(10), 741 (2013)
14. Peak Signal to Noise Ratio: http://en.wikipedia.org/wiki/Peak_signal-to-noise_ratio
15. Gupta, S., Dua, M.: A hybrid approach for dehazing images. Int. J. Hybrid Inform. Technol. **7**(3) (2014)

A Probabilistic Patch Based Hybrid Framework for CT/PET Image Reconstruction

Shailendra Tiwari, Rajeev Srivastava and Arvind Kumar Tiwari

Abstract Statistical image reconstruction for computed tomography and positron emission tomography (CT/PET) play a significant role in the image quality by using spatial regularization that penalizes image intensity difference between neighboring pixels. The most commonly used quadratic membrane (QM) prior, which smooth's both high frequency noise and edge details, tends to produce an unfavourable result while edge-preserving non-quadratic priors tend to produce blocky piecewise regions. However, these edge-preserving priors mostly depend on local smoothness or edges. It does not consider the basic fine structure information of the desired image, such as the gray levels, edge indicator, dominant direction and frequency. To address the aforementioned issues of the conventional regularizations/priors, this paper introduces and evaluates a hybrid approach to regularized ordered subset expectation maximization (OSEM) iterative reconstruction technique, which is an accelerated version of EM, with Poisson variability. Regularization is achieved by penalizing OSEM with probabilistic patch-based regularization (PPB) filter to form hybrid method (OSEM+PPB) for CT/PET image reconstruction that uses neighborhood patches instead of individual pixels in computing the non-quadratic penalty. The aim of this paper is to impose an effective edge preserving and noise removing framework to optimize the quality of CT/PET reconstructed images. A comparative analysis of the proposed model with some other existing standard methods in literature is presented both qualitatively and quantitatively using simulated test phantom and standard digital image. An experimental result indicates that the proposed method yields significantly improvements in quality of reconstructed images from the projection data. The obtained results justify the applicability of the proposed method.

Keywords Reconstruction algorithms · Computed tomography (CT) · Positron emission tomography (PET) · Ordered subset expectation-maximization algorithms (OSEM) · Probabilistic patch based prior (PPB) · Acceleration techniques

---

S. Tiwari (✉) · R. Srivastava · A.K. Tiwari
Department of Computer Science & Engineering, Indian Institute of Technology (Banaras Hindu University), Varanasi 221005, India
e-mail: stiwari.rs.cse@iitbhu.ac.in

A. Nagar et al. (eds.), *Proceedings of 3rd International Conference on Advanced Computing, Networking and Informatics*, Smart Innovation, Systems and Technologies 43, DOI 10.1007/978-81-322-2538-6_33

1 Introduction

Statistical image reconstructions (SIR) have been increasingly used in computed tomography and positron emission tomography (CT/PET) to substantially improve the image quality as compared to the conventional filtered back-projection (FBP) method [1] for various clinical tasks. SIR based maximum likelihood expectation maximization (MLEM) algorithm [2] produces images with better quality than analytical techniques. It can better use of noise statistics, accurate system modeling, and image prior knowledge. MLEM estimates the objective function that is being maximized (log-likelihood) when the difference between the measured and estimated projection is minimized. There have been further refinements of the SIR with introduction of ordered subset expectation maximization (OSEM) [3] that uses a subset of the data at each iteration, there by producing a faster rate of conversion.

Nowadays OSEM has become the most important iterative reconstruction techniques for emission computed technology. Although, likelihood increases, the images reconstructed by classical OSEM are still very noisy because of ill-posed nature of iterative reconstruction algorithms. During reconstruction process, poisson noise effectively degrades the quality of reconstructed image. Regularization is therefore required to stabilize image estimation within a reconstruction framework to control the noise propagation and to produce a reasonable reconstruction. Generally, the penalty term is chosen as a shift-invariant function that penalizes the difference among local neighbouring pixels [4]. The regularization term incorporates prior knowledge or expectations of smoothness or other characteristics in the image, which can help to stabilize the solution and suppress the noise and streak artifacts. Various regularizations have been presented in the past decades based on different assumptions, models and knowledge. Although some of them were initially proposed for SIR of CT and PET, they can be readily employed for CT. This regularization term is used to stabilize the image estimation. To incorporate prior knowledge or expectations of smoothness in the image, which encourage preservation of the piecewise contrast region while eliminating impulsive noise, but the reconstructed images still suffer from streaking artifacts and poisson noise.

Numerous edge preserving priors have been proposed in the literature [5–12] to produce sharp edges while suppressing noise within boundaries. A wide variety of methods such as the quadratic membrane (QM) [5] prior, Gibbs prior [6], entropy prior [7], Huber prior function [8] and total variation (TV) prior [9] which smoothes both high frequency noise and edge details tends to produce an unfavourable results while edge-preserving non-quadratic [10] priors tend to produce blocky piecewise regions. In order to suppress the noise and preserve edge information simultaneously, image reconstruction based on AD has also become the interesting area of research [11].

The main reason for the instability of traditional regularizations is that the image roughness is calculated based on the intensity difference between neighbouring pixels, but the pixel intensity differences may not be reliable in differentiating sharp edges from random fluctuation due to noise. When the intensity values contain noise, the measure of roughness is not robust. To address this issue, [12] proposed patch-based regularizations which utilize neighborhood patches instead of individual pixels to measure the image roughness. Since they compare the similarity between patches, the patch-based regularizations are believed to be more robust in distinguishing real edges from noisy fluctuation.

Here in this paper, we introduces and evaluates a hybrid approach to regularize which dominate in CT/PET images. Our model is looking equivalent to that proposed in [12], but it's different in the sense that we focus on edge-preserving regularizer (PPB) with accelerated version of MLEM i.e. OSEM, which produces fast reconstructed results in an efficient manner. However, unlike [9, 11] which treat post-processing reconstruction steps our approach is based on an elegant formulation that use priors (filters) within the reconstruction process rather than using at the end after the reconstructed image is ready.

This paper is divided into the following sections. Section 2 formulates the backgrounds of reconstruction problem and introduces some notations of the OSEM method. Section 3 describes the proposed hybrid method using fusion of regularization term PPB with OSEM Sect. 4 presents simulation and results of the qualitative and quantitative experiments. It also verifies that the proposed method yields best results by comparing with other standard method using simulated data. A conclusion is in Sect. 5.

2 Backgrounds

Ordered Subset Expectation Maximization (OSEM) is one of the most widely used iterative methods for CT/PET reconstruction. Here a standard model of photon emission tomography as described in [13] is used and the measurements follow independent Poisson random distribution as follows:

$$y_i \sim Poisson(\bar{y}_i(f)), \quad i = 1, \ldots, I \tag{1}$$

where y_i is the measured projectional data which are counted by the ith detector during the data collection, f represents the estimated image vector and the element of f denotes the activity of image. In iterative methods, the calculation of the system matrix during the reconstruction process is essential and given as follows:

$$\bar{y}_i(f) = \sum_{j}^{J} a_{ij} f_j \tag{2}$$

where $A = \{a_{ij}\} \in R^{n_i \times n_j}$ is the system matrix which describes the relationship between the measured projection data and the estimated image vector, with a_{ij}

denoting the probability of detecting an event originated at pixel j by detector pair I. The probability distribution function (*pdf*) of the Poisson noise reads:

$$P(y|f) = \prod_{i}^{I} \frac{\bar{y}_i(f)^{y_i}}{y_i!} \exp(-\bar{y}_i(f)), \tag{3}$$

and the corresponding log-likelihood can be described as follow:

$$L(f) = \log P(y|f) = \sum_{i=1}^{I} \left(y_i \log\left(\sum_{j=1}^{J} a_{ij} f_j \right) - \sum_{j=1}^{J} a_{ij} f_j \right) \tag{4}$$

where I is the number of detector pairs, J is the number of the objective image pixels, and $P(y|f)$ is the probability of the detected measurement vector y with image intensity f. The penalized likelihood reconstruction estimates image by maximizing the following objective function:

$$f^* = \arg\max_{f \geq 0} (L(y|f) - \beta U(f)) \tag{5}$$

where $U(f)$ is the image roughness penalty.

$$f^* = \arg\max_{f \geq 0} [L(y|f) - \beta U(f)] = \arg\min_{f \geq 0} \left[\frac{1}{2} (y - Af)^T \Lambda (y - Af) + \beta U(f) \right] \tag{6}$$

Conventionally the image roughness is measured based on the intensity difference between neighboring pixels:

$$U(f) = \frac{1}{4} \sum_{j=1}^{n_j} \sum_{k \in \mathbb{N}_j} w_{jk} \varphi(f_j - f_k) \tag{7}$$

where $\varphi(t)$ is the penalty function. The regularization parameter β controls the trade-off between data fidelity and spatial smoothness. When β goes to zero, the reconstructed image approaches the ML estimate.

A common choice of $\varphi(t)$ in PET image reconstruction is the quadratic function:

$$\varphi(t) = \frac{1}{2} t^2 \tag{8}$$

A disadvantage of the quadratic prior is that it may over-smooth edges and small objects when a large β is used in order to smooth out noise in large regions. Huber penalty [12] is an example of non-quadratic penalty that can preserve edges and small objects in reconstructions. It is defined as:

$$\varphi(t) = \begin{cases} t^2/2, & |t| \leq \delta \\ \delta|t| - \delta^2/2 & |t| \geq \delta \end{cases} \quad (9)$$

where δ is the hyper-parameter to control the shape of the non-quadratic penalty. The parameter clearly delineates between the "non-edge" and "edge" regions, and is often referred as the "edge threshold" or "transition point". Other family of convex potential functions is described in [12].

3 Methods and Model

In this paper, a new hybrid framework (here referred to as: OSEM+PPB) to reduce number of iterations as well as improve the quality of reconstructed images is proposed. Finally, hybrid method is applied to CT/PET tomography for obtaining optimal solutions. Generally, the SIR methods can be derived from the maximum a posteriori (MAP) estimation, which can be typically formulated by an objective function consisting of two terms named as "data-fidelity" term, models the statistics of projection measurements, and "regularization" term, penalizes the solution. It is an essential criterion of the statistical iterative algorithms that the data-fidelity term provides an accurate system modelling of the projection data. The regularization or penalty term play an important role in the successful image reconstruction. The proposed reconstruction is a hybrid combination of iterative reconstruction and a prior part as shown in Fig. 1.

The proposed model works in conjunction to provide one iterative cycle of objective function and prior part. This is repeated a number of times till we get the required result. The use of prior knowledge within the secondary reconstruction enables us to tackle noise at every step of reconstruction and hence noise is tackled in an efficient manner. Using Probabilistic patch based prior (PPB) prior [12] inside reconstruction part gives better results than working after the reconstruction is over. It has been widely used for image denoising, image enhancement, image segmentation [13] and often obtains better quality than other methods.

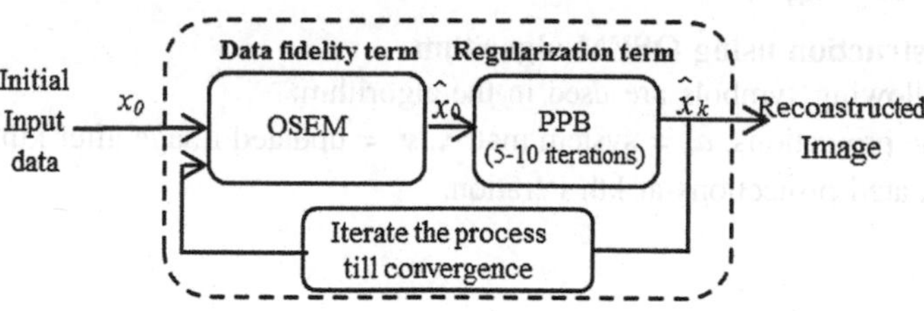

Fig. 1 The proposed hybrid model

The patch-based roughness regularizations are defined as:

$$f_{j,Smooth}^{n+1} = U(f) = \sum_{j=1}^{n_j} \sum_{k \in \mathbb{N}_j} \varphi\left(\left\|g_j(f) - g_k(f)\right\|_w\right) \tag{10}$$

where $g_j(f)$ is the feature vector consisting of intensity values of all pixels in the patch centered at pixel j. The patch based similarity between the pixel j and k is measured by

$$\left\|g_j(f) - g_k(f)\right\|_w = \sqrt{\sum_{l=1}^{n_l} w\left(f_{jl} - f_{kl}\right)^2} \tag{11}$$

where jl denotes the lth pixel in the patch of pixel j and wl is the corresponding weight coefficient with $w_{jk} = 1, or\ w_{jk} = 1/d_{jk}$. The weighting coefficient is smaller if the distance between the patch of a neighboring pixel and the patch of the concerned pixel is larger. By this way, the regularization can better preserve edges and boundaries. The basic idea of PPB is to choose a convex function that is unique and stable, so that regions are smoothed out and edges are preserved as compared to non-convex functions. The basic OSEM model as:

$$f_{j,OSEM}^{n+1} = f_j^n \left(\frac{1}{\sum_{j \in S_n} a_{ij}} \sum_{j \in S_n} \frac{y_j a_{ij}}{\sum_{i'=1}^{I} f_{i'}^{(n)} a_{ij}} \right), \text{ for pixels } i = 1, 2, \ldots, I. \tag{12}$$

where $f_j^{(n+1)}$ is the value of pixel j after the nth iteration of OSEM correction step.

Finally the proposed model is given as follows:

$$f_j^{(n+1)} = \arg\max_f \left(\left(f_{j,OSEM}^{n+1} \right) - \beta \left(f_{j,Smooth}^{n+1} \right) \right) \tag{13}$$

Towards the end, we refer to the proposed algorithm as an efficient hybrid approach for CT/PET image reconstruction and outline it as follows.

The Proposed Algorithm:

A. Reconstruction using OSEM algorithm

Let the following symbols are used in the algorithm:

X = true projections, a_{ij} = system matrix, y^k = updated image after kth iteration, x_j^k = calculated projections at kth iteration.

1. Set k = 0 and put:

$$y^0 = g_{final} \tag{14}$$

2. Repeat until convergence of $\hat{x}^m$

   (a) $x^1 = \hat{x}^m,\ m = m + 1$ (15)

   (b) For subsets t = 1, 2,…, n

   Calculate Projections: find projections after kth iterations using updated image

$$x(j)^k = \sum_{i=1}^{I} a_{ij}^t \times y^k,\ \text{for detectors}\ j \in S_n \tag{16}$$

   Error Calculation: Find error in calculated projection (element-wise division)

$$x_{error}^k = \frac{X}{x_j^k} \tag{17}$$

   Back projection: back project the error onto image

$$x(i)^{k+1} = x(i)^k \frac{\sum_{j \in S_n} \frac{y(j)a_{ij}}{\mu(j)^t}}{\sum_{j \in S_n} a_{ij}},\ \text{for pixels}\ i = 1, 2, \ldots, I. \tag{18}$$

   (c) $X_{error}^k = a_{ij} * x_{error}^k$ (19)

3. Normalization: normalize the error image(element-wise division)

$$X_{norm}^k = \frac{X_{error}^k}{\sum_j a_{ij}} \tag{20}$$

4. Update: update the image:

$$y^{k+1} = y^k. * X_{norm}^k \tag{21}$$

B. Prior: Use *PPB* as prior

5. Set m = 0 and apply Probabilistic patch based filter:

$$y_{m+1}^{k+1} = PPB\left(y_m^{k+1}\right) \tag{22}$$

   Put m = m + 1 and repeat till m = 3;

6. Put k = k + 1, repeat with OSEM reconstruction.

In our algorithm, we monitor the SNR during each loop of secondary reconstruction. The processing is stopped when SNR begins to saturate or degrade from any existing value.

4 Results and Discussions

In this simulation study, only two- dimensional (2-D) simulated phantoms were considered because our main aim here is to compare proposed method with other algorithms and to demonstrate that the proposed method was applicable to different ECT imaging modalities such as CT/PET, where 2-D phantoms were sufficient for this purpose. The comparative analysis of the proposed method is also presented with other standard methods available in literature such as OSEM [3], OSEM+QM, OSEM+Huber, OSEM+TV and OSEM+AD. For simulation study MATLAB 2013b software was used on a PC with Intel(R) Core (TM) 2 Duo CPU U9600 @ 1.6 GHz, 4.00 GB RAM, and 64 bit Operating system. For quantitative analysis the various performance measures used include signal-to-noise ratio (SNR), peak signal-to-noise ratio (PSNR), the root mean square error (RMSE), and the correlation parameter (CP) [14]. The SNR, PSNR and RMSE give the error measures in reconstruction process. The correlation parameter (CP) is a measure of edge preservation after the reconstruction process, which is necessary for medical images.

The brief description of the various parameters used for generation and reconstruction of the two test cases are as follows: The first test case is a Modified Shepp-Logan Phantom of size 64 × 64 and 120 projection angles was used. The simulated data was all Poisson distributed and all assumed to be 128 radial bins and 128 angular views evenly spaced over 180°. The second test case used for simulation was a gray-scale standard medical thorax image of size 128 × 128. For this test case, the projections are calculated mathematically with coverage angle ranging from 0 to 360° with rotational increment of 2° to 10°.

For both the test cases, we simulated the sinograms with total counts amount 6×10^5. A Poisson noise of magnitude 15 % is added to projections. The proposed algorithm was run for 500 to 1000 iterations for simulation purposes and the convergence trend of the proposed method and other methods were recorded. However, the proposed and other algorithms converged in less than 500 iterations. Also, this was done to ensure that the algorithm has only single maxima and by stopping at the first instance of stagnation or degradation, we are not missing any further maxima which might give better results. The corresponding graphs are plotted for SNR, PSNR, RMSE, and CP. The graphs support the fact as shown in Figs. 3 and 6. From these plots, it is clear that proposed method (OSEM+PPB) gives the better result in comparison to other methods by a clear margin. Using OSEM with PPB prior brings the convergence much earlier than the usual algorithm. With proposed method, result hardly changes after 300 iterations whereas other methods converge in more than 300 iterations. Thus we can say that using PPB prior with accelerated version of EM brings the convergence earlier and fetches better results. The visual results of the resultant reconstructed images for both the test cases obtained from different algorithms are shown in Figs. 2 and 5. The experiment reveals the fact that proposed hybrid framework effectively eliminated Poisson noise and it performs better even at limited number of projections in comparison to other standard methods and has better quality of reconstruction in term of SNRs, PSNRs, RMSEs, and CPs. At the same time, it is also observed that the hybrid cascaded method overcomes the short coming of streak artifacts existing in other iterative algorithms and the reconstructed image is more similar to the original phantom (Figs. 2 and 5).

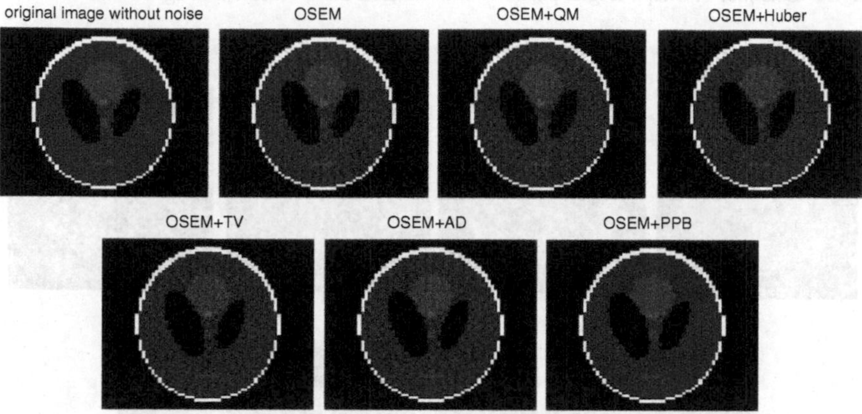

Fig. 2 The modified Shepp-Logan phantom with different reconstruction methods projection including 15 % uniform Poisson distributed background events

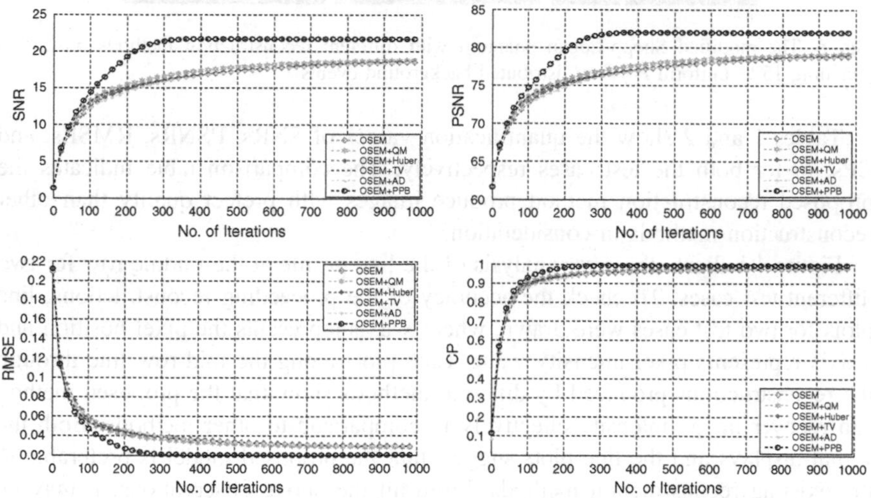

Fig. 3 The plots of SNR, PSNR, RMSE, and CP along with no. of iterations for different algorithms for test case 1

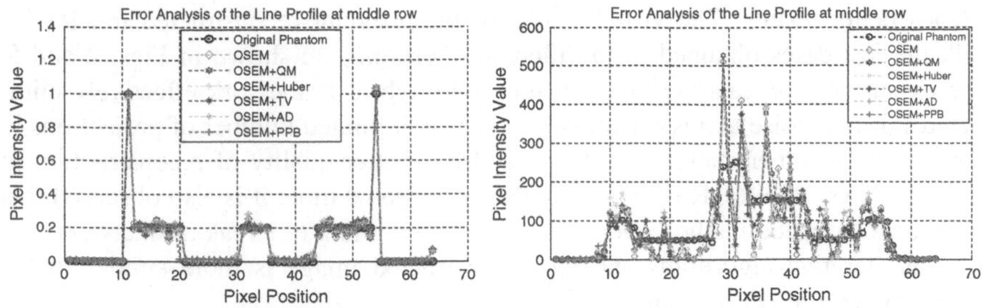

Fig. 4 Line plot of Shepp-Logan phantom and standard thorax medical image

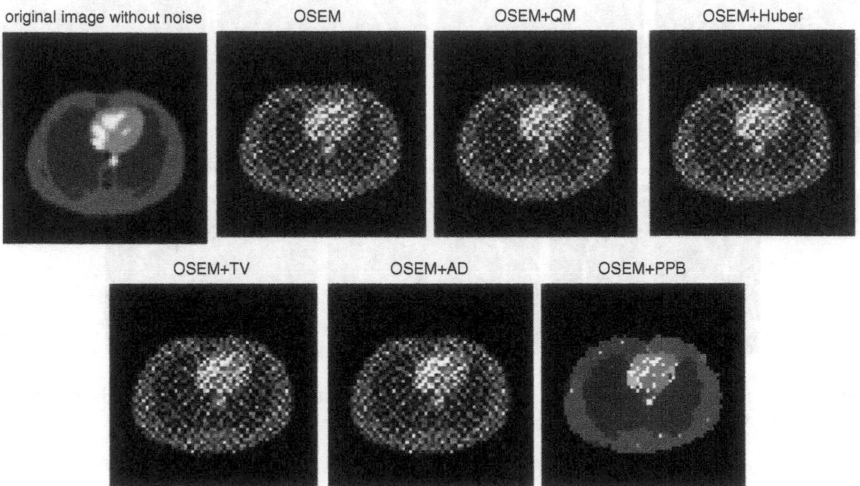

Fig. 5 The modified Shepp-Logan phantom with different reconstruction methods projection including 15 % uniform Poisson distributed background events

Tables 1 and 2 show the quantification values of SNRs, PSNRs, RMSEs, and CPs. in for both the test cases respectively. The comparison table indicates the proposed reconstruction method produce images with prefect quality than other reconstruction methods in consideration.

Figure 4 indicate the error analysis of the line profile at the middle row for two different test cases. To check the accuracy of the proceeding reconstructions, line plots for two test cases were drawn, where x-axis represents the pixel position and y-axis represents pixel intensity value. Line plots along the mid-row line through the reconstructions produced by different methods show that the proposed method can recover image intensity effectively in comparison to other methods. Both the visual-displays and the line plots suggest that the proposed model is preferable to the existing reconstruction methods. From all the above observations, it may be concluded that the proposed model is performing better in comparison to its other counterparts and provide a better reconstructed image.

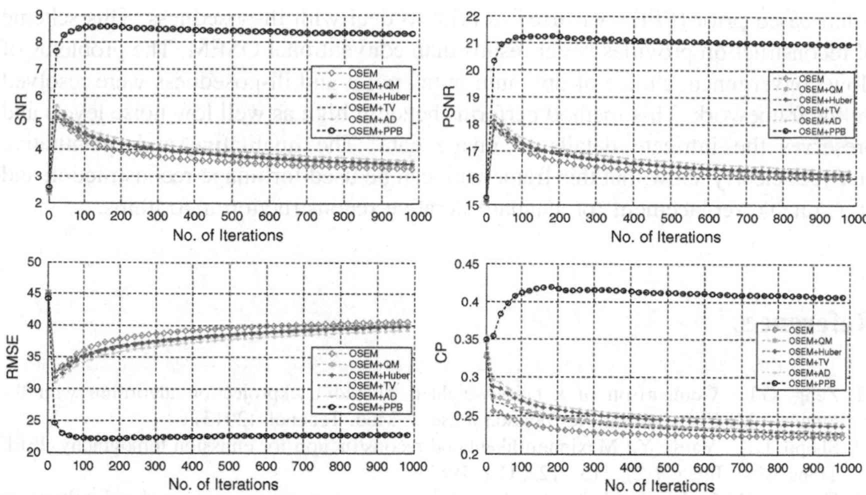

Fig. 6 The Plots of SNR, PSNR, RMSE, and CP along with no. of iterations for different algorithms for Test case 2

Table 1 Different performance measures for the reconstructed images in Fig. 2

Performance measures	OSEM	OSEM +QM	OSEM +Huber	OSEM +TV	OSEM +AD	OSEM+PPB (The proposed method)
SNR	18.5809	18.8318	18.5728	18.6459	18.5569	21.6899
PSNR	78.8701	79.1211	78.8620	78.9351	78.8462	81.9791
RMSE	0.0292	0.0283	0.0292	0.0289	0.0292	0.0204
CP	0.9773	0.9783	0.9776	0.9771	0.9770	0.9924

Table 2 Different performance measures for the reconstructed images in Fig. 5

Performance measures	OSEM	OSEM +QM	OSEM +Huber	OSEM +TV	OSEM +AD	OSEM+PPB (The proposed method)
SNR	5.1612	5.1928	5.4384	5.2234	5.6546	8.5621
PSNR	17.8484	17.8800	18.1255	17.9106	18.3417	21.2492
RMSE	32.7959	32.6770	31.7660	32.5621	30.9852	21.2492
CP	0.3265	0.3306	0.3469	0.3327	0.3543	0.4190

5 Conclusion

In this paper, we have demonstrated a hybrid framework for image reconstruction which consists of two stages during reconstruction process. The reconstruction was done using Ordered Subset Expectation Maximization (OSEM) while probabilistic patch based prior (PPB) was used as prior to deal with ill-posedness. This scheme of reconstruction provides better results than conventional OSEM. The problems of slow convergence, choice of optimum initial point and ill-posedness were resolved in this framework. This method performs better at high as well low noise levels and preserves the intricate details of image data. The qualitative and quantitative analyses clearly show that this framework can be used for image reconstruction and is a suitable replacement for standard iterative reconstruction algorithms.

References

1. Zeng, G.L.: Comparison of a noise-weighted filtered backprojection algorithm with the Standard MLEM algorithm for poisson noise. J. Med. Technol. (2013)
2. Shepp, L.A., Vardi, Y.: Maximum likelihood reconstruction for emission tomography. IEEE Trans. Med. Imag. **1**(2), 113 –122, Oct. 1982
3. Hudson, H.M., Larkin, R.S.: Accelerated image reconstruction using ordered subsets of projection data. IEEE Trans. Med. Imag. **13**(4) (1994)
4. Ling, J., Bovik, A.C.: Smoothing low-SNR molecular image via anisotropic median-diffusion. IEEE Trans. Med. Imag. **21**(4), 377–384 (2002)
5. Chen, Y., Chen, W.F., Feng, Y.Q., Feng, Q.J.: Convergent Bayesian reconstruction for PET using new MRF quadratic membrane-plate hybrid multi-order prior, lecture notes in computer science. Med. Imag. Aug. Real. (2006)
6. Lange, K.: Convergence of EM image reconstruction algorithms with Gibbs smoothness. IEEE Trans. Med. Imag. **9**, 439 (1990)
7. Denisova, N.V.: Bayesian reconstruction in SPECT with entropy prior and iterative statistical regularization. IEEE Trans. Nucl. Sci. (2004)
8. Chlewicki, W., Hermansin, F., Hansen, S.B.: Noise reduction and convergence of Bayesian algorithms with blobs based on the Huber function and median root prior. Phys. Med. Biol. (2004)
9. Panin, V.Y., Zeng, G.L., Gullberg, G.T.: Total variation regulated EM algorithm. IEEE Trans. Nucl. Sci. **46**, 2202–2210 (1999)
10. Sukovic, P., Clinthorne, N.H.: Penalized weighted least-squares image reconstruction in single and dual energy X-ray computed tomography. IEEE Trans. Med. Imaging **19**(11), 1075–1081 (2000)
11. Kazantsev, D., Arridge, S.R., Pedemonte, S., et al.: An anatomically driven anisotropic diffusion filtering method for 3D SPECT reconstruction. Phys. Med. Biol. (2012)
12. Wang, G., Qi, J.: Penalized likelihood PET image reconstruction using patch-based edgepreserving regularization. IEEE Trans. Med. Imag. **31**(12), 2194–2204 (2012)
13. Wernick, M.N., Aarsvold, J.N.: Emission Tomography: The Fundamentals of PET and SPECT. Academic Press (2004)
14. Srivastava, Rajeev, Srivastava, Subodh: Restoration of Poisson noise corrupted digital images with nonlinear PDE based filters along with the choice of regularization parameter estimation. Pattern Recogn. Lett. **34**, 1175–1185 (2013)

Low Cost Eyelid Occlusion Removal to Enhance Iris Recognition

Beeren Sahu, Soubhagya S. Barpanda and Sambit Bakshi

Abstract The Iris recognition system is claimed to perform with very high accuracy in given constrained acquisition scenarios. It is also observed that partial occlusion due to the presence of the eyelid can hinder the functioning of the system. State-of-the-art algorithms consider that the inner and outer iris boundaries are circular and thus these algorithms do not take into account the occlusion posed by the eyelids. In this paper, a novel low-cost approach for detecting and removing eyelids from annular iris is proposed. The proposed scheme employs edge detector to identify strong edges, and subsequently chooses only horizontal edges. 2-means clustering technique clusters the upper and lower eyelid edges through maximizing class separation. Once two classes of edges are formed, one indicating edges contributing to upper eyelid, another indicating lower eyelid, two quadratic curves are fitted on each set of edge points. The area above the quadratic curve indicating upper eyelid, and below as lower eyelid can be suppressed. Only non-occluded iris data can be fetched to the further biometric system. This proposed localization method is tested on publicly available BATH and CASIAv3 iris databases, and has been found to yield very low mislocalization rate.

Keywords Iris recognition · Personal identification · Eyelid detection

B. Sahu (✉) · S.S. Barpanda · S. Bakshi
Department of Computer Science and Engineering, National Institute of Technology Rourkela, Rourkela, India
e-mail: beeren.sahu@gmail.com

S.S. Barpanda
e-mail: soubhagya1984@gmail.com

S. Bakshi
e-mail: sambitbaksi@gmail.com

A. Nagar et al. (eds.), *Proceedings of 3rd International Conference on Advanced Computing, Networking and Informatics*, Smart Innovation, Systems and Technologies 43, DOI 10.1007/978-81-322-2538-6_34

1 Introduction

The iris is located between the white sclera and the black pupil and it is an annular portion of the eye [5]. The motivation behind 'Iris Recognition', a biometrical based innovation for individual recognizable proof and check, is to perceive an individual from his/her iris prints. An anatomy of iris reveals that iris patterns are highly random and has high level of stability. Being a biometric trait, it is highly distinctive. Iris patterns are different for every individual; the distinction even exists between indistinguishable twins. Even the iris patterns of left and right eye of the same individual differ from each other.

A common iris recognition framework incorporates four steps: acquisition, preprocessing, feature extraction and matching. Preprocessing comprises of three stages: segmentation, normalization and enhancement. During acquisition, a iris images are acquired successively from the subject with the help of a specifically designed sensor. Preprocessing involves in segmentation of the region of interest (RoI) i.e. detecting the pupillary and limbic boundary followed by converting the RoI to polar form from Cartesian form. Feature extraction is a typical procedure used to bring down the span of iris models and enhance classifier exactness. Then, feature matching is done by comparing the extracted feature of probe iris with a set of (gallery) features of candidate iris to determine a match/non-match. Segmentation of iris is a very crucial stage in an iris feature extraction. The aim is to segment the actual annular iris region from the raw image. The segmentation process highly affect the overall procedure of iris recognition i.e. normalization, enhancement, feature extraction and matching. Since segmentation process determines the overall performance of the recognition system, it is given utmost importance. Most segmentation routines in the literature describe that the pupillary and limbic boundaries of the iris are circular. Hence, model parameters are focused on determining the model that best fit these hypotheses. This leads to error, since the iris boundaries are not exactly circles.

2 Related Work

Daugman [4] has proposed a widely used iris segmentation scheme. In this scheme, his assumption is based on that both the inner boundary (pupillary boundary) and outer boundary (limbic boundary) of the iris can be represented by a circle in Cartesian plane. The circles in 2D plane can be described using the parameters (x_0, y_0) (coordinate of the center) and the radius r. Hence, integro-differential operator can be used effectively to determining these parameters and is described by the equation below:

$$\max_{(r,x_0,y_0)} \left| G_\sigma(r) * \frac{\partial}{\partial r} \oint_{r,x_0,y_0} \frac{I(x,y)}{2\pi r} ds \right|$$

where $I(x, y)$ is the eye image and $G_\sigma(r)$ is a smoothing function determined by a Gaussian of scale σ. Basically, the IDO looks for the center co-ordinates i.e. the parameters (x_0, y_0, r) over the entire image domain (x, y) for the maximum Gaussian blurred version of the partial derivatives (with respect to a different radius r). Hence, it works like a circular edge detector, which searches through the parameter space (x, y, r) to segment the most salient circular edge (limbic boundary).

Wildes [12] proposed a marginally diverse calculation which is generally in view of Daugman's strategy. Wildes has performed a two stage method to the contour fitting. First, a gradient based edge detector is utilized to produce an edge-map from the crude eye picture. Second, the edge-map, which contains positive points can vote to instantiate specific roundabout form parameters. This voting plan is actualized by Hough transform.

Ma et al. [7] generally focus the iris area in the original image, and then uses edge detection technique and Hough transform to exactly detect the inner and outer boundaries in the determined region. The image is then projected in the vertical and horizontal direction to compute the center coordinates (X_c, Y_c) of the pupil approximately. Since the pupil is by and large darker than its surroundings, the center coordinates of the pupil is calculated as the coordinates corresponding to the minima of the two projection profiles.

$$X_c = \arg\min_x \left(\sum_y I(x,y) \right)$$

$$Y_c = \arg\min_y \left(\sum_x I(x,y) \right)$$

where X_c and Y_c represent the coordinates of the pupil' center in the original image $I(X, Y)$. Then a more accurate estimate of the center coordinates of the pupil is computed. Their proposed technique considers a region of size 120×120, centered at the point by selecting a reasonable threshold by applying adaptive method and using the gray level histogram of this region, followed by binarization. The resulting binary region, from where the centroid is found, is considered as the new estimate of the pupil coordinates. Then, the exact parameters of these two circles using edge detection are calculated. Canny operator is also considered by the authors in their experiments along with Hough transform.

Similar works, Masek [10] has proposed a method for the eyelid segmentation in which the iris and the eyelids have been separated through Hough transformation. In contrast to [8], Mahlouji and Noruzi [9] have used Hough transform to localize all the boundaries between the upper and lower eyelid regions. Cui et al. [3] have proposed two algorithms for the detection of upper and lower eyelids and tested on CASIA dataset. This technique has been successful in detecting the edge of upper and lower eyelids after the eyelash and eyelids were segmented. Ling et al. proposed an algorithm capable of segmenting the iris from images of very low quality [6] through eyelid detection. The algorithm described in [6] being an unsupervised one, does not require a training phase. Radman et al. have used the live-wire technique to localize the eyelid boundaries based on the intersection points between the eyelids and the limbic boundary [11].

3 Proposed Work

Most segmentation methods assume that iris is circular, but in practical iris is not completely circular as often occluded by eyelids [4, 7, 12]. For accurate iris recognition it is needed that these occluded regions must be removed. The procedure of iris segmentation starts from detecting papillary and limbic boundaries of the iris. Next the upper and lower eyelids are detected for accurate segmentation of the iris using 2-means clustering with the canny edge detector. The proposed scheme can be shown in Fig. 1 and the overall algorithm given in Algorithm 1.

Algorithm 1 Eyelid_Detection

Require: *im*: input image
Ensure: [*eyelidUp*, *eyelidLow*]: detected eyelid boundaries

```
A ← smooth(im)
B ← edge(A)
cc1 ← getConnectedComponents(B)
no_of_edges ← count(cc1)
for i = 1 to no_of_edges do
    hor_spr ← horizontal spread of the edge
    nop ← number of pixel in i^th edge
    horizontality(i) ← nop / hor_spr
end for
cc2 ← cc1
for i = 1 to max(cc1) do
    if horizontality(i) < 90% of horizontality of other edges then
        cc2 ← cc1 − i^th edge
    end if
end for
num_of_clusters ← 2
cl ← kmeans(cc2, num_of_clusters)
eyelidUp = quadraticCurveFit(cl[1])
eyelidLow = quadraticCurveFit(cl[2])
return [eyelidUp, eyelidLow]
```

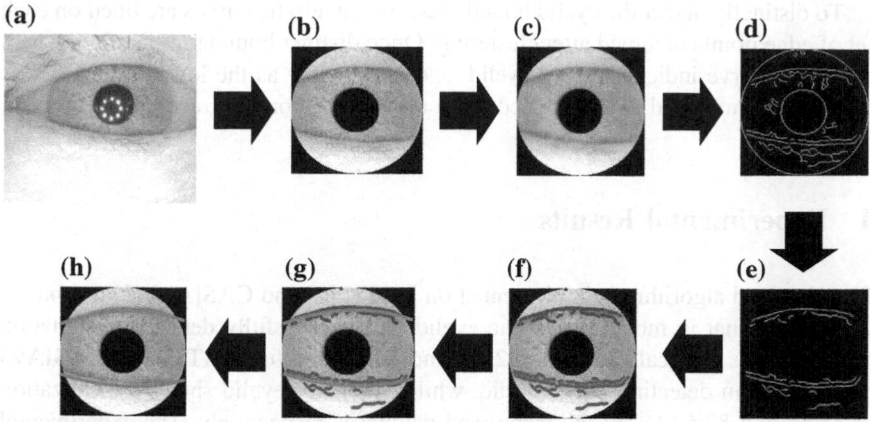

Fig. 1 Proposed eyelid detection scheme: **a** input image; **b** annular iris segmented image with occlusion; **c** iris smooth image; **d** Edge detection using canny filter; **e** horizontal edges retained; **f** 2-means clustering; **g** quadratic curve fitting; **h** upper and lower eyelid detected

After pupil and iris boundaries are detected, we get an annular iris image as shown in Fig. 1b. Wiener filter is used for smoothening the annular iris image. It is smoothed in order to reduce noise and to detect only prominent edges like eyelid boundaries. Then canny edge detector is applied on smoothed image. Next we apply a horizontal filter to the connected components formed by each edge. This filter computes the horizontality of each connected components and keeps only those edges which have high horizontality factor and neglecting others. Edges formed by eyelid boundaries have high horizontality factor.

After the horizontal filter applied we have roughly detected the eyelid boundaries. To categorize them as upper and lower eyelid boundaries we apply 2-means clustering technique, a specific case of k-means clustering [8] where $k = 2$. This is a fast and simple clustering algorithm, which is found to be adopted for many applications including image segmentation. The algorithm can be defined for the observations $\{x_i : i = 1, \ldots, L\}$, where the goal is to partition the observations into K groups with means $\bar{x}_1, \bar{x}_2, \ldots, \bar{x}_K$ such that

$$D(K) = \sum_{i=1}^{L} \min_{1 \leq j \leq K} (\bar{x}_1 - \bar{x}_j)^2$$

is minimized and K is then increased gradually and the algorithm terminates when a criterion is met. In our problem number of groups to be formed is two and thus we take $K = 2$.

To distinctly detect the eyelid boundaries, two quadratic curves are fitted on each set of edge points obtained after clustering. Once distinct boundaries are found, area above the curve indicates upper eyelid, and area below as the lower eyelid. These regions are unwanted and thus need to be removed before the recognition process.

4 Experimental Results

The proposed algorithm is implemented on BATH [1] and CASIAv3 [2] databases, we observe that in most images, the eyelid was successfully detected. Simulation results shows a localization of 92.74 and 94.62 % for BATH and CASIAv3 respectively in detecting upper eyelid, while the lower eyelid shows a localization of 83.15 and 87.64 % for the mentioned databases respectively. The experimental data have been tabulated in the Table 1 and Fig. 2 demonstrates some eyelid detection results for the stated databases.

Table 1 Localization results on BATH and CASIAv3 databases

Accuracy →	Upper eyelid	Lower eyelid
Databases ↓	Localization (%)	Localization (%)
BATH	92.74	83.15
CASIAv3	94.62	87.64

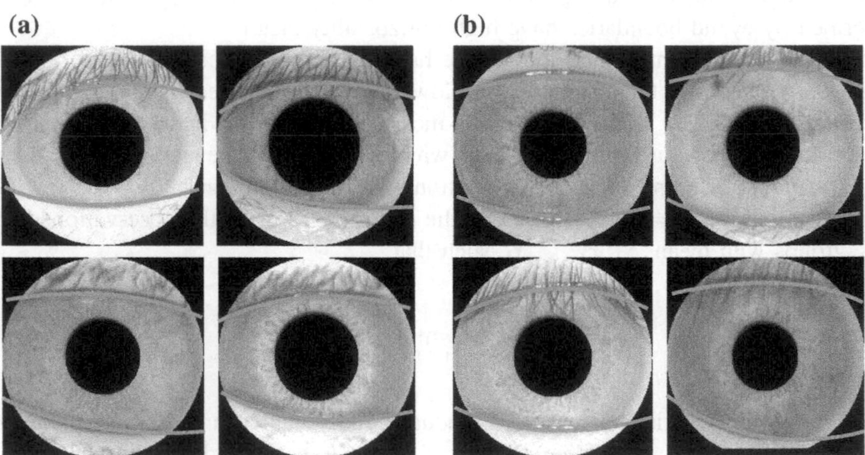

Fig. 2 Iris localization for databases: **a** CASIAv3, **b** BATH

5 Conclusion

Occlusion due to eyelids affects the performance of an iris biometrics. Hence to improve the accuracy of the system, with the proposed method, eyelid boundaries are detected and subsequently can be removed before the recognition process. In this paper, 2-means clustering algorithm is adopted in iris localization. An algorithm is presented that can efficiently detect the eyelid boundaries. Upon implementing the proposed eyelid detection algorithm on publicly available iris databases, CASIAv3 and BATH, it is observed that both the upper and lower eyelids are properly detected for most of the images.

References

1. Bath University Database: http://www.bath.ac.uk/elec-eng/research/sipg/irisweb
2. Chinese academy of sciences' institute of automation (casia) iris image database v3.0: http://www.cbsr.ia.ac.cn/english/IrisDatabase.asp
3. Cui, J., Wang, Y., Tan, T., Ma, L., Sun, Z.: A fast and robust iris localization method based on texture segmentation. In: Defense and Security, pp. 401–408. International Society for Optics and Photonics (2004)
4. Daugman, J.G.: High confidence visual recognition of persons by a test of statistical independence. IEEE Trans. Pattern Anal. Mach. Intell. **15**(11), 1148–1161 (1993)
5. Jain, A.K., Ross, P.F., Arun, A.: Handbook of Biometrics. Springer (2007)
6. Ling, L.L., de Brito, D.F.: Fast and efficient iris image segmentation. J. Med. Biol. Eng. **30**(6), 381–391 (2010)
7. Ma, L., Tan, T., Wang, Y., Zhang, D.: Personal identification based on iris texture analysis. IEEE Trans. Pattern Anal. Mach. Intell. **25**(12), 1519–1533 (2003)
8. MacQueen, J.: Some methods for classification and analysis of multivariate observations. In: Proceedings of the Fifth Berkeley Symposium on Mathematical Statistics and Probability, vol. 1, pp. 281–297. University of California Press (1967)
9. Mahlouji, M., Noruzi, A.: Human iris segmentation for iris recognition in unconstrained environments. IJCSI Int. J. Comput. Sci. Issues **9**(3), 149–155 (2012)
10. Masek, L.: Recognition of human iris patterns for biometric identification. Ph.D. thesis, Masters thesis, University of Western Australia (2003)
11. Radman, A., Zainal, N., Ismail, M.: Efficient iris segmentation based on eyelid detection. J. Eng. Sci. Technol. **8**(4), 399–405 (2013)
12. Wildes, R.P.: Iris recognition: an emerging biometric technology. Proc. IEEE **85**(9), 1348–1363 (1997)

Part IV
Next Generation Optical Systems

On Improving Static Routing and Wavelength Assignment in WDM All-Optical Mesh Networks

Abhishek Bandyopadhyay, Debdutto Chakraborty, Uma Bhattacharya and Monish Chatterjee

Abstract The method in which Routing and Wavelength Assignment (RWA) of connection requests is performed in optical WDM networks, can appreciably affect resource consumption. The blocking probability for lightpath requests increases significantly with increase in resource consumption. Thus the method of performing RWA should be such that it minimizes consumption of network resources. RWA in all-optical networks is an NP-Complete problem. This paper proposes six new heuristic algorithms for static RWA in all-optical mesh networks that are not only efficient in terms of resource conservation but can also solve the RWA problem effectively in polynomial time. Comparisons show that the proposed algorithms perform better than some earlier well-known strategies.

Keywords Static · Routing · Wavelength assignment · Lightpath requests · Resource consumption · All-optical · WDM · Mesh networks

A. Bandyopadhyay (✉) · D. Chakraborty · M. Chatterjee
Department of Computer Science and Engineering,
Asansol Engineering College, Asansol, West Bengal, India
e-mail: ab.bandyopadhyay@gmail.com

D. Chakraborty
e-mail: debdutto.chakraborty@gmail.com

M. Chatterjee
e-mail: monish_chatterjee@yahoo.com

U. Bhattacharya
Department of Computer Science and Technology,
Indian Institute of Engineering Science and Technology Shibpur,
Howrah, West Bengal, India
e-mail: ub@cs.becs.ac.in

A. Nagar et al. (eds.), *Proceedings of 3rd International Conference on Advanced Computing, Networking and Informatics*, Smart Innovation, Systems and Technologies 43, DOI 10.1007/978-81-322-2538-6_35

1 Introduction

Due to their huge transmission bandwidth, Optical Wavelength Division Multiplexed (WDM) mesh networks are ideal candidates to provide the transport backbone of present day Internet. In such networks, bandwidth of an optical fiber is split into many wavelength channels to support simultaneous transmission from different users through the same fiber at separate wavelengths [1, 2]. In WDM networks *lightpaths*, which are established between two end-nodes serve as optical communication links or channels for carrying data at a rate of 40 Gbps in modern networks [1]. A lightpath must be assigned a wavelength in each fiber that it traverses from its source to the destination. In *all-optical* networks, the data stays in the optical domain from its source to the destination and hence lightpaths are assigned the same wavelength in all fiber-links that it traverses. This is called as *wavelength continuity constraint* [1]. As there is no need for wavelength conversion, wavelength converters find no application all-optical WDM networks.

The Routing and Wavelength Assignment (RWA) problem is considered to be the problem of assigning lightpaths to each request in a given set of connection requests [1]. Researchers have addressed two variants of the problem. In *static* RWA problem, the set of lightpaths to be established is known in advance and is not expected to change whereas in *dynamic* RWA, lightpaths need to be established and terminated on request. In this work we address the issue of *static* RWA. The mode in which RWA is carried out can have a substantial effect on network performance. For example, the method of performing RWA should be such that it minimizes resource consumption, since it means reducing blocking probability for future lightpath requests [1].

Researchers have studied both static and dynamic versions of the RWA problem in optical WDM networks extensively. The work [3] proposes a star network having a hub node with fixed wavelength conversion capability and which can sustain all sets of lightpath requests having a maximum load of W, W being the link capacity in terms of number of wavelengths. Authors [4] propose RWA strategies in a two-fiber ring without wavelength conversion. Researchers [5] study the performance of two methods for multicast RWA in tree of rings namely shortest path tree and minimum spanning tree methods. Literature [6] studies the average-case performance of four dynamic RWA algorithms by considering separate copies of the physical topology, one for each wavelength. The work [7], which extends [6] shows that static RWA and throughput maximization algorithms perform better than dynamic RWA algorithms studied in [6]. Work [8] proposes an efficient graph decomposition technique for wavelength assignment, given a routing for the lightpaths in topologies based on de Bruijn graph. The authors [9] propose static RWA techniques based on decomposing a de Bruijn graph into rings. Two efficient static RWA algorithms are proposed for de Bruijn graphs [10] based on request selection strategy and categorizing requests that traverse a link. Heuristic routing strategies are presented [11] for reducing congestion in presence of arbitrary single link failure for WDM networks with de Bruijn graph topology.

The organization of the rest of this paper is described next. In Sect. 2 we present the contributions of this paper. In Sect. 3 we discuss the RWA algorithms, in Sect. 4 we provide the time complexity analysis, in Sect. 5 we compare the performance of the RWA strategies and analyze our results and finally we conclude in Sect. 6.

2 Our Contributions

The RWA problem is NP-Complete in absence of wavelength conversion [2]. As *all-optical* networks do not employ wavelength conversion, RWA in such networks is an NP-Complete problem. So one has to use solutions that are sub-optimal but efficient and solves the problem in polynomial time. The static RWA problem for *all-optical networks* can be defined in the following manner [7]. If a set of connection requests $R = \{r_1, r_2, r_3, \ldots, r_n\}$ and wavelengths $\lambda_1, \lambda_2, \lambda_3\ldots$ is known, the problem is that for each request r_i find a lightpath route p_i and select wavelength λ_j for assigning in every link that p_i traverses satisfying the condition that lightpaths traversing the same link are assigned different wavelengths and that the number of wavelengths used for assignment is minimized. In this paper we propose *six* new *static* RWA algorithms that can provide solutions to the RWA problem in polynomial time for all-optical WDM mesh networks. Our algorithms perform better than the *static* RWA strategies studied in [7].

3 Proposed Static RWA Algorithms

In this section we first discuss the four RWA strategies [6] namely First-Fit (FF), Best-Fit (BF), Random-Fit (RF) and Densest-Fit (DF) which can be used for *both* static and dynamic RWA. In [7], the author studies the effect of ordering requests by the length of their lightpath routes on the physical topology and then performing RWA in that order using the four algorithms proposed in [6] which resulted in eight *static* RWA algorithms, namely First-Fit Decreasing (FFD), First-Fit Increasing (FFI), Best-Fit Decreasing (BFD), Best-Fit Increasing (BFI), Random-Fit Decreasing (RFD), Random-Fit Increasing (RFI), Densest-Fit Decreasing (DFD) and Densest-Fit Increasing (DFI).

We also discuss the *six* proposed *static* RWA algorithms namely, BestMaxLoad-Fit (BMLF), BestMaxMinLoad-Fit (BMMLF) and the four strategies obtained by ordering requests by the length of their lightpath routes on the physical topology and then performing RWA using BMLF and BMMLF namely BestMaxLoad-Fit Decreasing (BMLFD), BestMaxLoad-Fit Increasing (BMLFI), Best Max MinLoad-Fit Decreasing (BMMLFD) and BestMaxMinLoad-Fit Increasing (BMMLFI). In a way similar to [7], we replicate physical topology $G = (V, E)$ W times $G_1, G_2, \ldots, G_w$, generating a copy for each wavelength. Lightpath requests that are routed using G_i are assigned wavelength λ_i. Lightpath routes on the same copy G_i are link-disjoint. Initially there is only one copy G_1 but additional copies may be generated as required. The two sub-problems of the RWA problem, namely routing sub-problem and wavelength assignment sub-problem are solved concurrently. Let the copies of $G = (V, E)$ that are currently generated and available be $G_1, G_2, \ldots, G_n$. A lightpath request r_i is served by selecting an existing copy G_i, finding a lightpath route p_i for r_i in G_i and assigning the wavelength λ_i to route p_i. The wavelength-links assigned to p_i are then deleted from G_i so that these wavelengths are not assigned to any other request.

Each physical link is assumed to be bidirectional and having capacity W i.e. W wavelengths are available in both directions (forward and reverse) for RWA of lightpaths. We define traffic load *TrLoad*, of a physical link in a specific direction (forward or reverse) as the count of the number of wavelength-links used in the link in that direction. We define maximum traffic load *maxTrLoad* of a route p as maximum *TrLoad* in a link of p over all links in p. Similarly we define minimum traffic load *minTrLoad* of a route p as the minimum *TrLoad* in a link of p over all links in p. We associate each physical link in $G = (V, E)$ with a two variables *Wcount*0 and *Wcount*1 which stores the value of *TrLoad* of a link in the forward and reverse directions respectively. *Wcount*0 or *Wcount*1 is updated whenever a wavelength-link is consumed in a physical link in the forward or reverse direction respectively. We next discuss the RWA algorithms of the work [6] and the proposed algorithms.

3.1 First-Fit (FF)

Find the shortest lightpath route that connects source and destination of request r_i in the first copy G_1 using links that are still intact. If found, then perform RWA. If not found, find the shortest route in G_2, then $G_3,\ldots$, and so on, until a lightpath route is found.

3.2 Best-Fit (BF)

Find shortest lightpath routes that connects source and destination of request r_i in all existing copies $G_1, G_2, \ldots, G_n$ using wavelength-links that are still intact. Let the routes so found be $p_1, p_2, \ldots, p_m$. The shortest lightpath route p_i among $p_1, p_2, \ldots, p_m$ is chosen as the required lightpath route.

3.3 Random-Fit (RF)

Find the shortest lightpath route that connects source and destination of request r_i in a randomly selected copy G_i among all copies $G_1, G_2, \ldots, G_m$ which can provide shortest lightpath routes for r_i.

3.4 Densest-Fit (DF)

Find the shortest lightpath route that connects source and destination of request r_i in a copy G_i having maximum number of wavelength-links intact among all existing copies $G_1, G_2, \ldots, G_n$. If not found, then find the shortest route in the copy with second most links, then in the copy with the third most links,…., and so on, until a lightpath route is found.

3.5 BestMaxLoad-Fit (BMLF)

In BMLF a lightpath request r_i is processed in the following manner. Shortest lightpath routes connecting source and destination of request r_i is searched in all existing copies G_1, G_2,…., G_n using wavelength-links that are still intact using Breadth First Search (BFS) [12]. A lightpath route may be successfully found in many existing copies or it may exist in a single copy or none of the existing copies. If the route, say p_i is found in a single copy G_i, RWA is performed by assigning wavelength λ_i to p_i and then deleting the wavelength-links used by p_i from G_i. The values of *Wcount*0 or *Wcount*1 (whichever applicable depending on the direction of p_i) are incremented in the corresponding physical links for each wavelength-link deleted of p_i. If more than one copy say $G_1, G_2, \ldots, G_m$ have lightpath routes say $p_1, p_2, \ldots, p_m$ for r_i then the shortest route p_i among $p_1, p_2, \ldots, p_m$ is selected as the required lightpath route. If there is a tie i.e. if two or more routes say $p_1, p_2, \ldots, p_k$ have the same minimum number of hops then for each route p_j having minimum number of hops, *maxTrLoad* is calculated using the values of *Wcount*0 or *Wcount*1 (whichever applicable depending on route direction). The route p_j selected is the one having *minimum* value of *maxTrLoad*. If more than one route is found with the same minimum value of *maxTrLoad*, then ties are broken by choosing the route that comes *first in the order*. RWA is performed on the chosen route p_j in the already mentioned way. If a lightpath route does not exist in any of the existing copies, a new copy is created and a lightpath route is searched in it by using BFS and RWA is performed in the said way.

In case of BestMaxLoad-Fit Decreasing (BMLFD) and BestMaxLoad-Fit Increasing (BMLFI), routes of each lightpath request is first computed on the physical topology $G = (V, E)$ and then the requests are sorted in decreasing and increasing order respectively in the two algorithms based on the length of their lightpath routes. The lightpath requests are then processed in that order using BMLF as already discussed.

3.6 *BestMaxMinLoad-Fit (BMMLF)*

Algorithm BMMLF differs from BMLF in that, when more than one route is found with the same *minimum* value of *maxTrLoad* then ties are broken by selecting the route having *minimum* value of *minTrLoad*. If a tie still persists, then choosing the route that comes *first in the order* breaks it. The rest of the algorithm in BMMLF is same as that of BMLF.

In case of BestMaxMinLoad-Fit Decreasing (BMMLFD) and BestMaxMinLoad-Fit Increasing (BMMLFI), routes of each lightpath request is first computed on the physical topology $G = (V, E)$ and then the requests are sorted in decreasing and increasing order respectively in the two algorithms based on the length of their lightpath routes. The lightpath requests are then processed in that order using BMMLF as already discussed.

4 Time Complexity Analysis

We analyze the worst-case time complexity of the proposed algorithms. In both BMLF and BMMLF, for every lightpath request, lightpath routes are searched in all existing copies of the physical topology. Since Breadth First Search (BFS) [12] is used for finding shortest lightpath route in an existing copy, the time taken to find the route is $O(|V| + |E|)$. There can be W copies (W is the link capacity) at the most and so the time consumed to find routes is $W^*(|V| + |E|)$. We assume the worst-case i.e. lightpath route existed in each of the W copies. Then the total number of routes found is W, say $p_1, p_2, \ldots, p_W$. To find the Best-Fit path a total number of W comparisons are required. Let us again assume the worst-case i.e. all the W routes have the same number of hops. In such a case a total number of W operations are further required for finding *maxTrLoad* and *minTrLoad* for each of the W routes as both *maxTrLoad* and *minTrLoad* can be calculated by a single scan of a route. Similarly W number of comparisons are further required for finding the routes having the *minimum* value of *maxTrLoad* and the *minimum* value of *minTrLoad* among the W copies as a route is scanned at most once. Thus the total number of operations required in both BMLF and BMMLF for each lightpath request is $W^*(|V| + |E|) + 3W$. If n lightpath requests are to be served, then a total of $n^*(W^*(|V| + |E|) + 3W) \approx n^*W^*(|V| + |E|)$ operations, neglecting smaller terms are needed. So the worst case time complexity of both algorithms is $O(n^*W^*(|V| + |E|))$.

In case of BMLFD, BMLFI, BMMLFD and BMMLFI, first lightpath routes are computed on the physical topology using BFS which requires $n * (|V| + |E|)$ operations for n lightpath requests. Then the n-computed routes are sorted in ascending or descending order. Since we use Heap Sort [12], the number of operations required is $n \log_2^n$. So the total number of required operations in the worst case is $n^*(|V| + |E|) + n \log_2^n + n^*(W^*|V| + |E|) + 3W) = n^*(|V| + |E| + \log_2^n + W^*(|V| + |E|) + 3W) \approx n^*(W+1)^*(|V| + |E|)$, neglecting the smaller terms. Thus the worst-case time complexity of the four algorithms is $O(n^*(W+1)^*(|V| + |E|))$. Thus all the six proposed algorithms are able to provide polynomial-time solutions to the RWA problem.

5 Performance Comparisons

We analyze the performance of the eighteen (six new and twelve existing) static RWA algorithms using results obtained from extensive simulations performed using programs written in JAVA (JDK 1.6), run on a Pentium 4 processor in Windows environment. We provide the results for the following networks: 11-node NJLATA with 50 requests, 14-node NSFNET with 90 requests, 20-node EON with 150 requests, 24-node ARPANET like network with 250 requests and 46-node USANET with 1000 requests. Due to space limitations the topologies mentioned are not included. For each topology, 20 experiments were conducted and we report the average. Simulations were carried out in a *non-blocking* scenario i.e. with sufficient number of wavelengths in links to estimate the link capacity required by each algorithm to support all the lightpath requests. The link capacity required by an algorithm can be estimated by the *number of copies* of the physical topology generated for RWA of all the lightpath requests. Table 1 shows the comparison of the required link capacity for all the algorithms. In Table 1 we use the notation X-Topology-Y to denote that the Topology used has X number of nodes and it was tested with Y number of requests. The entries in the tables are fractions as we have reported the average of 20 experiments. Table 2 shows the average and maximum percentage decrease in required link capacity obtained by using the proposed algorithms over the existing ones considering all the conducted experiments. Result analysis shows the following two key observations.

Observation 1: All the proposed algorithms are efficient than the existing algorithms in terms resource conservation. This is evident from Table 1, which shows that the proposed algorithms have managed to decrease the link capacity requirement for serving the entire lightpath requests. Table 2 also shows that an appreciable average (maximum) percentage decrease in resource consumption is obtained by using the proposed algorithms over the existing algorithms.

Observation 2: The Increasing–Algorithms perform the worst whereas the Decreasing-Algorithms perform the best. This is because wavelengths in each copy

Table 1 Comparisons of required link capacity for the eighteen static RWA algorithms

Required link capacity					
Algorithm	Number of nodes—topology—size of request set				
	11-NJLATA-50	14-NSFNET-90	20-EON-150	24-ARPA-250	46-USANET-1000
FF	4.6	13.0	19.6	40.8	62.5
BF	4.9	11.9	16.9	37.6	58.4
RF	4.8	13.3	19.8	38.6	64.1
DF	4.9	13.1	20.8	40.7	67.1
BMLF	3.5	9.9	13.8	33.6	54.0
BMMLF	3.3	9.6	13.5	33.1	51.7
FFI	4.8	15.1	20.6	42.6	65.5
BFI	5.4	13.9	17.5	38.9	62.6
RFI	5.3	15.3	20.6	39.8	66.4
DFI	5.2	15.6	21.8	41.5	67.4
BMLFI	3.6	10.6	14.6	35.2	58.7
BMMLFI	3.3	10.2	14.3	34.1	54.4
FFD	4.6	12.9	18.5	39.7	60.9
BFD	4.6	11.5	16.5	36.5	57.3
RFD	4.7	13.8	19.5	40.6	61.6
DFD	4.6	13.1	19.9	40.8	63.2
BMLFD	3.4	9.5	13.5	32.6	52.5
BMMLFD	3.2	9.1	13.1	31.5	50.4

Table 2 Comparisons of average (maximum) percentage decrease in required link capacity

Average (maximum) percentage decrease in required link capacity				
Using algorithm	*Over algorithms*			
	FF	BF	RF	DF
BMLF	21.71 (29.59)	16.37 (28.57)	22.44 (30.30)	24.72 (33.65)
BMMLF	24.40 (31.12)	19.10 (32.65)	24.89 (31.81)	27.21 (35.09)
	FFI	BFI	RFI	DFI
BMLFI	22.33 (29.80)	17.87 (33.33)	23.00 (32.07)	24.78 (33.02)
BMMLFI	26.23 (32.45)	21.83 (38.88)	26.80 (37.73)	28.53 (36.53)
	FFD	BFD	RFD	DFD
BMLFD	22.72 (26.35)	16.14 (26.08)	24.79 (31.15)	24.54 (32.16)
BMMLFD	25.35 (30.43)	18.08 (30.43)	27.84 (34.05)	27.63 (34.17)

G_i are better utilized when longer lightpath routes are assigned first. The proposed BMMLFD however outperforms all the other seventeen algorithms in most of the cases.

6 Conclusion

We study the problem of static RWA in all-optical mesh WDM networks in this paper. As the problem is NP-Complete we have suggested heuristic algorithms to provide solutions to the problem. We have proposed six algorithms and compared their performance with existing ones. Performance analysis shows that the proposed algorithms outperform the existing ones in terms of network resource conservation. Time complexity analysis shows that the algorithms are indeed able to solve the problem effectively in polynomial time.

References

1. Ramaswami, R., Sivarajan, K.N.: Optical networks—a practical perspective. Morgan Kaufmann Publishers Inc, San Francisco (1998)
2. Chlamtac, I., Ganzand A., Karmi, G.: Lightpath communications: an approach to high-bandwidth optical WANs. IEEE Trans. Commun. **40**, 1171–1182 (1992)
3. Ramaswami, R., Sasaki, G.H.: Multiwavelength optical networks with limited wavelength conversion. Proc. IEEE INFOCOM **1997**, 489–498 (1997)
4. Konke, M., Hartmann, H.L.: Fast optimum routing and wavelength assignment for WDM ring transport network. In: Proceedings IEEE International Conference on Communications (ICC 2002), vol. 5, pp. 2740–2744 (2002)
5. Jia, X., Hu, X., Ruan, L., Sun, J.: Multicast routing, load balancing, and wavelength assignment on tree of rings. IEEE Commun. Lett. **6**(2), 79–81 (2002)

6. Li, K.: Experimental average-case performance evaluation of online algorithms for routing and wavelength assignment and throughout maximization in WDM optical networks. ACM J. Experimental Algorithmics (JEA) **12** (2008)
7. Li, K.: Heuristic algorithms for routing and wavelength assignment in WDM optical networks. In: Proceedings of IEEE International Parallel and Distributed Processing Symposium (IPDPS 2008), pp. 1–8 (2008)
8. Chatterjee, M., Sanyal, S., Nasipuri, M., Bhattacharya, U.: A wavelength assignment algorithm for de Bruijn WDM networks. Int. J. Parallel, Emergent Distributed Syst. (IJPEDS), Taylor and Francis, U.K. **26**(6), 477–491 (2011)
9. Chatterjee, M., Goswami, A., Mukherjee, S., Bhattacharya, U.: Heuristic for routing and wavelength assignment in de Bruijn WDM networks based on graph decomposition. In: Proceedings of 5th IEEE International Conference on Advanced Networks and Telecommunication Systems (IEEE ANTS 2011), pp. 1–6 (2011)
10. Chatterjee, M., Barat, S., Majumder, D., Bhattacharya, U.: New strategies for static routing and wavelength assignment in De Bruijn WDM networks. In: Proceedings of 3rd International Conference on Communication Systems and Networks (COMSNETS 2011), pp. 1–4 (2011)
11. Chatterjee, M., Bhattacharya, U.: Heuristic routing for reducing congestion in presence of link fault in de Bruijn WDM networks. In: Proceeding of 2011 IEEE Region 10 International Conference (IEEE TENCON 2011), pp. 544–548 (2011)
12. Cormen, T.H., Leiserson, C.E., Rivest, R.L., Stein, C.: Introduction to algorithms. MIT Press (2001)

A Randomized *N*-Policy Queueing Method to Prolong Lifetime of Wireless Sensor Networks

Maneesha Nidhi and Veena Goswami

Abstract The increasing interest in Wireless Sensor Networks (WSN) can be understood to be a result of their wide range of applications. However, due to uneven depletion of their batteries, the nodes often face premature failure. In this paper, we explore a queue based method for the improvement of the WSNs. We propose a *N* threshold lifetime improvement method which considers the probability variation for the various states of a WSN—sleep, idle, start-up, and busy. The experimental results validate our theoretical analysis and prove that our approach is indeed an efficient method for improving the lifetime of Wireless Sensor Networks.

Keywords Wireless sensor networks · *N* policy · Queue · Start-up · Lifetime · Energy consumption · Battery allocation

1 Introduction

Wireless Sensor Networks have made themselves indispensable in various fields, including those of environmental monitoring, energy monitoring, transportation, industrial monitoring, machine condition monitoring and distributed temperature monitoring. A WSN consists of several hundreds (and at times thousands) of small, self-powered nodes, with each node being associated with one or more sensors. Each sensor node consists of four basic components: a sensing unit, a processing unit, a transceiver unit and a power unit [1–3]. Though these sensor networks have

M. Nidhi (✉)
Department of Computer Science and Engineering, National Institute of Technology, Rourkela 769008, India
e-mail: maneeshanidhi@yahoo.com

V. Goswami
School of Computer Application, KIIT University, Bhubaneswar 751024, India
e-mail: veena_goswami@yahoo.com

A. Nagar et al. (eds.), *Proceedings of 3rd International Conference on Advanced Computing, Networking and Informatics*, Smart Innovation, Systems and Technologies 43, DOI 10.1007/978-81-322-2538-6_36

opened up endless opportunities, they do pose some formidable challenges, the foremost being that of energy scarcity. In order to sustain these nodes, we require the non-renewable resource-energy. However, these nodes are prone to unequal dissipation of energy due to workload variations, and heterogenous hardware. Sensor networks usually have many-to-one traffic networks, leading to an increase in heterogeneity of node power consumption.

The energy that is consumed by the nodes serves as a limiting factor for lifetime. In various analysis, it is assumed that most of the power is consumed during the busy state, in which packet transmission takes place. However, this assumption is flawed [4, 5]. A significant amount of power is consistently consumed during signal detection and processing in the idle mode. Even the sleep mode has an associated amount of power loss. The difference in the energy levels of the different states has led to various literatures that concentrate on sleep-wake-up scheduling [6]. A dynamic sleep scheduling algorithm has been presented by Yuan et al. [7]. Their algorithm seeks to balance the energy consumed by a sensor node. Lin et al. [8] have elaborated on dynamic power management in WSN. Thus, studies on energy efficiency are gaining momentum [9, 10].

The recent times have seen a rise in literature that considers N threshold mechanism as an efficient method to conserve energy. The variations of N policy queue have been discussed in [11–13]. The advantage of this mechanism is that there need to be N data-packets for a server to switch states to busy. This decreases the probability of an inadvertent transition from sleep to busy state. Also, threshold policy brings flexibility and simplicity. The utilization of servers can be changed by changing the threshold manually, and, this can be used to root a number of jobs in a specific way [14–17]. An optimal resource replication, with the usage of queueing theory, for wireless sensor networks has been presented in [18].

In this paper, we propose a power-saving technique using a N-threshold $M/M/1$ queueing model with start-up time. Initially, when the system does not have any data packets, the server is assumed to be in sleep state. As the data packets start arriving, the server enters the idle mode. Associated with this change is an increase in energy loss. As the technique developed is N-threshold, the server never starts operating when number of packets is less than N. Also, the server requires an exponential start-up time, denoted by η in order to start providing services. This is represented by introducing a start-up state between the idle and busy states. When the number of data packets in the system reaches the value $N-1$, and another data packet is introduced into the system, the server enters the start-up state with a probability q and the server is left off with probability $(1-q)$. If the server is left off, it acquires N data packets and then has to enter the start-up state according to the definition of proposed model. After the start up state, the server enters the busy state which depletes more energy per unit time than all the other states. In the busy state, the server provides the requisite service. After all the data packets have been transmitted, the server again enters the sleep state. Thus, this forms a cycle which begins when data packets enter the system.

The rest of the paper is organised as follows. Section 2 provides a description of the system model and its analysis. In Sect. 3, we present the performance indices of the proposed model. The numerical results and its discussion has been carried out in Sect. 4. Section 5 concludes the paper.

2 System Description

In our paper, we consider the working of a single server, with the assumption that each server follows half-duplex communication. The half-duplex mode of channel is adopted as it consumes less power and allows for both down and up data streams. The model defines four operational states: sleep, idle, start-up and busy. The proposed model has been represented by a flowchart in Fig. 1.

Initially, when the system is empty, the server is in sleep state, in order to minimize energy loss. When the data packets start arriving, the state of server

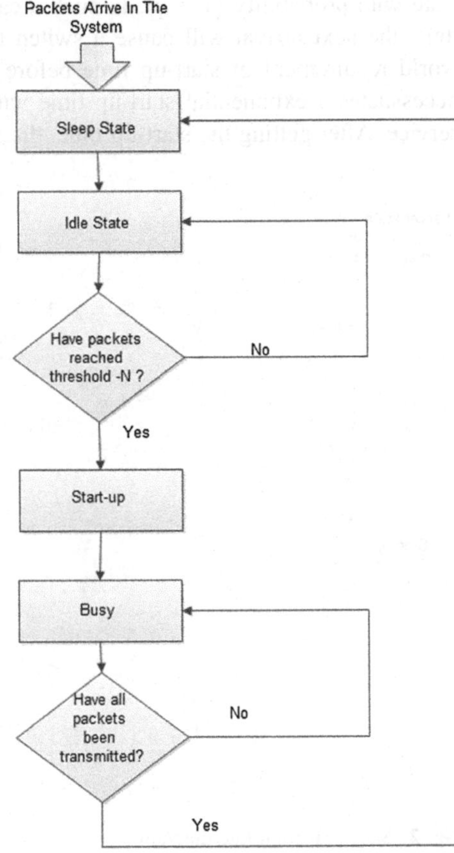

Fig. 1 Proposed system model

switches to idle state, in order to listen for and receive the incoming packets. It is assumed that the data packets arrive according to a Poisson distribution with parameter λ. The service times of data packets are considered to be independent random variables having an exponential distribution with mean $1/\mu$. The data packets are considered to form a queue in order of their arrival. We assume that the server can transmit only one data packet at a time, and the service is provided in a first-come, first-served basis. We simulate the real world scenario by modelling the nodes to have finite buffer capacity, that is a finite queue size, denoted as K. We have represented the state transition for each state, along with its associated probability in the birth-death graph of Fig. 2.

The probability that the node reaches sleep mode is denoted as p, while the probability of reaching idle mode is $1-p$. Thus, the product of $(1-p)$ and λ is the probability that a particular data packet could fulfill the transmission between states. The N-policy model presented in this paper deals with the control of service in a queueing system. For the number of data packets ranging from 1 to $N-1$, the state of server remains as idle. When the number of data packets in the system attains N, there are two possibilities. There may be a state change to start-up, which has a probability q associated with it. Another option is that the server may remain in idle state with probability $(1-q)$. In such a case, the threshold value is reached, and thus, the next arrival will cause a switch to the start-up state. It models the real world requirement of start-up time before a server can start serving. The server necessitates a exponential start-up time with parameter η before allowing for the service. After getting the start-up time, the server starts its service normally till the

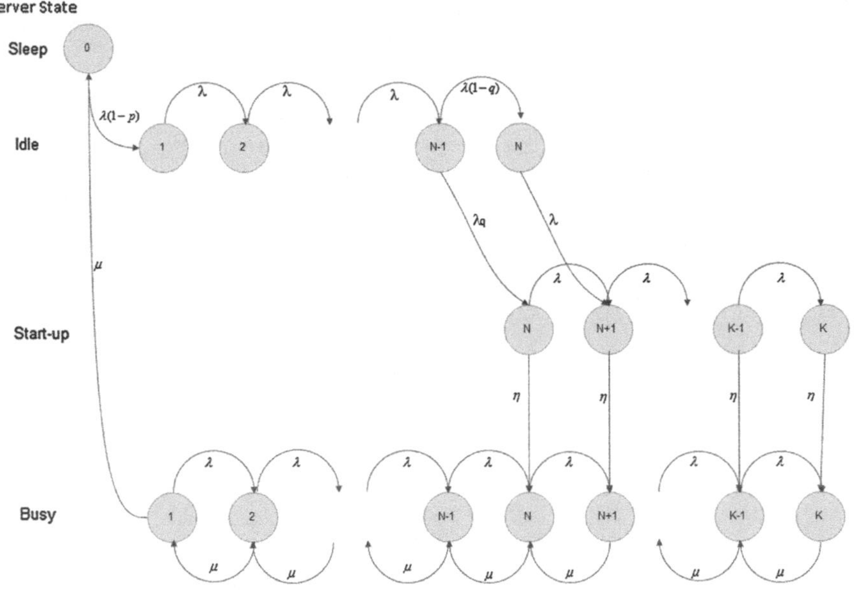

Fig. 2 State transition rate diagram

system becomes empty again at which time the above process is repeated all over again. Once all the data packets are transmitted, and number of data packets reduces to zero, the server switches back to the sleep mode. Let us denote the states of the system by (i,j), where $j = 0, 1, 2$ denotes that the server is in idle, startup, busy state, respectively; while i represents the number of data packets in the sensor node. Let us assume the state probability for sleep, idle, start-up and busy as $P_{0,0}$, $P_{i,0}(1 \le i \le N)$, $P_{i,1}(N \le i \le K)$ and $P_{i,2}(1 \le i \le K)$, respectively. Using the one-step transition analysis, the steady-state equations can be written as

$$\lambda(1-p)P_{0,0} = \mu P_{1,2}, \tag{1}$$

$$\lambda P_{1,0} = \lambda(1-p)P_{0,0}, \tag{2}$$

$$\lambda P_{n,0} = \lambda P_{n-1,0}, 2 \le n \le N-1, \tag{3}$$

$$\lambda P_{N,0} = \lambda(1-q)P_{N-1,0}, \tag{4}$$

$$(\lambda + \eta)P_{N,1} = \lambda q P_{N-1,2}, \tag{5}$$

$$(\lambda + \eta)P_{N+1,1} = \lambda P_{N,0} + \lambda P_{N,1}, \tag{6}$$

$$(\lambda + \eta)P_{n,1} = \lambda P_{n-1,1}, N+2 \le n \le K-1, \tag{7}$$

$$\eta P_{K,1} = \lambda P_{K-1,1}, \tag{8}$$

$$(\lambda + \mu)P_{1,2} = \mu P_{2,2}, \tag{9}$$

$$(\lambda + \mu)P_{n,2} = \lambda P_{n-1,2} + \mu P_{n+1,2}, 2 \le n \le N-1, \tag{10}$$

$$(\lambda + \mu)P_{n,2} = \lambda P_{n-1,2} + \mu P_{n+1,2} + \eta P_{n,1}, N \le n \le K-1, \tag{11}$$

$$\mu P_{K,2} = \lambda P_{K-1,2} + \eta P_{K,1}. \tag{12}$$

Solving Eqs. (1), (2) and (3), we obtain

$$P_{n,0} = \frac{\lambda(1-p)}{\lambda} P_{0,0}, 1 \le n \le N-1,$$
$$P_{1,2} = \frac{\lambda(1-p)}{\mu} P_{0,0}.$$

Equation (4), yields

$$P_{N,0} = \frac{\bar{q}\lambda(1-p)}{\lambda} P_{0,0}.$$

where $\bar{q} = 1 - q$ and $\omega = \frac{\bar{\lambda} q}{\lambda + \eta}$. Solving Eqs. (5)–(8) recursively, we get

$$\begin{aligned} P_{N,1} &= \frac{\lambda(1-p)q}{\lambda+\eta} P_{0,0}, \\ P_{n,1} &= \psi \phi^{n-N-1} P_{0,0}, N+1 \leq n \leq K-1, \\ P_{K,1} &= \frac{\psi \lambda \phi^{K-N-2}}{\eta} P_{0,0}. \end{aligned}$$

where $\psi = \frac{\lambda(1-p)(\lambda+\bar{q}\eta)}{(\lambda+\eta)^2}$ and $\phi = \frac{\lambda}{\lambda+\eta}$. Solving Eqs. (9)–(12), yields

$$\begin{aligned} P_{n,2} &= \frac{\lambda(1-p)(1-\rho^n)}{\mu(1-\rho)} P_{0,0}, 1 \leq n \leq N, \\ P_{N+1,2} &= \frac{\lambda(1-p)(1-\rho^n)}{\mu(1-\rho)} P_{0,0} - \frac{\eta\omega}{\mu} P_{0,0}, \\ P_{n,2} &= \frac{\lambda(1-p)(1-\rho^n)}{\mu(1-\rho)} P_{0,0} - \frac{\eta P_{0,0}}{\mu} \left[\frac{\omega(1-\rho^{n-N})}{1-\rho} + \psi \left\{ \frac{1-\rho^{n-N-1}}{1-\rho} \right. \right. \\ &\quad \left. \left. + \frac{\phi(1-\rho^{n-N-2})}{1-\rho} + (1+\rho+\phi)\phi^{n-N-3} \right\} \right], N+2 \leq n \leq K. \end{aligned}$$

Using normalization condition, $\sum_{n=0}^{N} P_{n,0} + \sum_{n=N}^{K} P_{n,1} + \sum_{n=1}^{K} P_{n,2} = 1$, we determine $P_{0,0}$ as

$$\begin{aligned} P_{0,0} = &\left[\frac{\lambda + (N-q)\lambda(1-p)}{\lambda} + \frac{\lambda(1-p)q}{\lambda+\eta} + \psi \sum_{n=N+1}^{K-1} \phi^{n-N-1} \right. \\ &\left. + \frac{\lambda\psi\phi^{K-N-2}}{\eta} + \sum_{n=1}^{K} P_{n,2} \right]^{-1}. \end{aligned}$$

3 Performance Indices

The expected values of data packets in the system for idle state (L_{idle}), setup state (L_{setup}) and busy state (L_{busy}) as

$$L_{idle} = \sum_{n=1}^{N} nP_{n,0}, L_{setup} = \sum_{n=N}^{K} nP_{n,1}, L_{busy} = \sum_{n=1}^{K} nP_{n,2}.$$

The total expected value of data packets in the system (L_s) is equal to the $L_s = L_{idle} + L_{setup} + L_{busy}$. Using Little's rule, the expected waiting time of a data packet spent in the system is given by $W_s = \frac{L_s}{\lambda^*}$, where $\lambda^* = \lambda(1 - P_{K,1} - P_{K,2})$ is the effective data packet arrival rate and $P_{loss} = P_{K,1} + P_{K,2}$ represents the probability of loss or blocking. The probability that the state of the server is in sleep state ($P_{0,0}$), in idle state (P_I), in startup state ($P_{startup}$) and in busy state (P_B) is given by

$$P_I = \sum_{n=0}^{N} P_{n,0}, P_{startup} = \sum_{n=N}^{K} P_{n,1}, P_B = \sum_{n=1}^{K} P_{n,2}.$$

Total energy consumption rate

To compute the power consumption for each cycle, the equation can be written as

$$F(N) = C_H L_S + C_S P_S + C_I P_I + C_B P_B + C_\mu \mu + C_\eta \eta.$$

where C_H is the energy consumed in order to hold a data packet, and C_S, C_I, C_B denote the energy consumed in sleep, idle and busy states, respectively. C_μ represents the energy consumed during the process of data transmission and C_η is energy consumed during the process of start-up.

4 Numerical Results

In this section, some numerical results have been illustrated in the form of tables and graphs. It gives managerial insights on optimal decisions to bring out the qualitative aspects of the queueing system under consideration. In Tables 1 and 2,

Table 1 Sensitivity analysis for various N when $K = 12, \mu = 4.0, \lambda = 3.0, \eta = 1.0,\ q = 0.25,\ p = 0.5$

N	L_s	W_s	P_{loss}	λ^*	P_B	P_1	P_s	F
2	4.93674	1.71006	0.96229	2.88688	0.68693	0.08117	0.13914	272.593
3	5.20615	1.81154	0.95796	2.87388	0.68831	0.11060	0.12066	273.081
4	5.48394	1.91923	0.95245	2.85736	0.68756	0.13390	0.10712	273.546
5	5.75880	2.02990	0.94566	2.83699	0.68502	0.15345	0.09692	273.969
6	6.02278	2.14161	0.93742	2.81227	0.68080	0.17074	0.08908	274.331
7	6.26968	2.25317	0.92753	2.7826	0.67490	0.18676	0.08300	274.617
8	6.4944	2.36389	0.91578	2.74733	0.66726	0.20225	0.07829	274.814
9	6.69287	2.47343	0.90197	2.70591	0.65781	0.21776	0.07466	274.909
10	6.86248	2.58177	0.88602	2.65805	0.64654	0.23364	0.07189	274.897

Table 2 Sensitivity analysis for various q when $K = 12, \mu = 4.0, \lambda = 3.0, \eta = 1.0,\ N = 5,$ $p = 0.5$

q	L_s	W_s	P_{loss}	λ^*	P_B	P_1	P_s	F
0.1	5.79744	2.04569	0.94466	2.83398	0.68460	0.15611	0.09558	274.027
0.2	5.7718	2.03521	0.94533	2.83598	0.68488	0.15435	0.09647	273.988
0.3	5.74568	2.02455	0.94600	2.83801	0.68516	0.15255	0.09738	273.949
0.4	5.71907	2.01370	0.94669	2.84008	0.68545	0.15072	0.09830	273.909
0.5	5.69194	2.00266	0.94740	2.84219	0.68574	0.14886	0.09924	273.868
0.6	5.66429	1.99143	0.94811	2.84434	0.68603	0.14696	0.10020	273.827
0.7	5.6361	1.97999	0.94884	2.84653	0.68634	0.14503	0.10118	273.784
0.8	5.60735	1.96834	0.94959	2.84876	0.68665	0.14305	0.10218	273.741
0.9	5.57802	1.95648	0.95035	2.85104	0.68696	0.14104	0.10320	273.697

the parameters are taken as $K = 12, \lambda = 3.0, \mu = 4.0, \eta = 1.0, p = 0.5, C_H = 2, C_I = 10, C_S = 5, C_B = 30, C_\mu = 50$ and $C_\eta = 10$.

Table 1 shows the impact of the threshold value N on the state of the system. When N increases, the average time for which the system stays in idle mode increases. This results in an increase in the parameters P_I, L_s, W_s and F. Correspondingly, any increment in the threshold value N means a decrease in P_{loss}, λ^*, P_B and P_S. Table 2 describes how the system is affected by the value of q. When q increases, so does the probability of state change into start-up phase, resulting in a decrease in the average time that the system stays in the idle mode. This results in a decrease in the parameters P_I, L, W and F. Therefore, there is a corresponding increase in the quantities P_{loss}, λ^*, P_B and P_S.

Figure 3 considers dependence of the total energy consumption rate (F) on threshold value N and probability q. It is seen that for fixed threshold value N the total energy consumption rate decreases when the probability q increases. Further, with fixed probability q the total energy consumption rate increases when the threshold value N increases. To ensure the minimum total energy consumption rate, we can carefully establish the threshold value N and the probability q in the system.

Figure 4 shows dependence of the expected waiting time of a data packet in the system (W_s) on probability q and threshold value N. It is observed that for fixed N, the expected waiting time of a data packet spent in the system decreases when the probability q increases. Further, with fixed probability q, the expected waiting time of a data packet spent in the system increases when the threshold N increases. To achieve this, we can carefully setup the threshold value N and the probability q in order to ensure the minimum expected waiting time of a data packet in the system.

Figure 5 plots the effect of η on the expected number of customers in the system (L_s) for various values of N. It can be seen that L_s monotonically decrease as η increase. When N increases it leads to the increase of L_s for fixed η. Figure 6 provides the impact of buffer size (K) on blocking probability (P_{loss}) for various

Fig. 3 *F* versus *q* and *N*

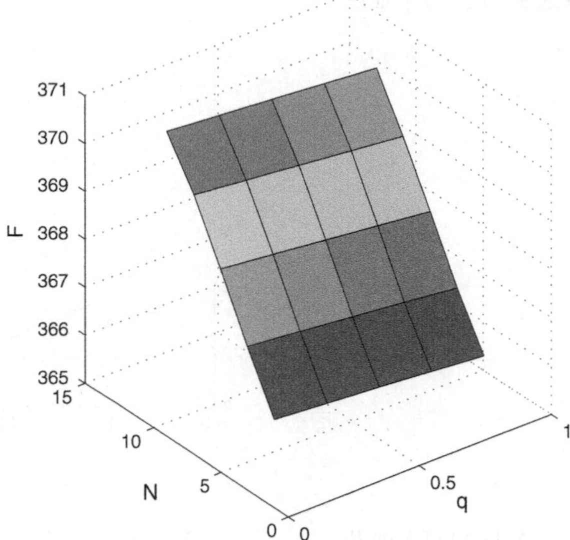

Fig. 4 W_s versus *q* and *N*

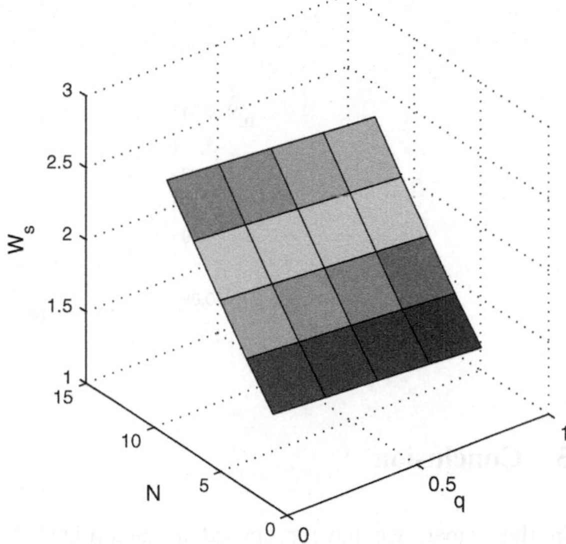

values of threshold *N*. For a fixed threshold *N*, blocking probability increases with increase of buffer size *K*. But, for fixed buffer size *K*, blocking probability decreases with increase of threshold *N*. Therefore, one can adjust the threshold *N* and the buffer size (K) in order to minimize the blocking probability.

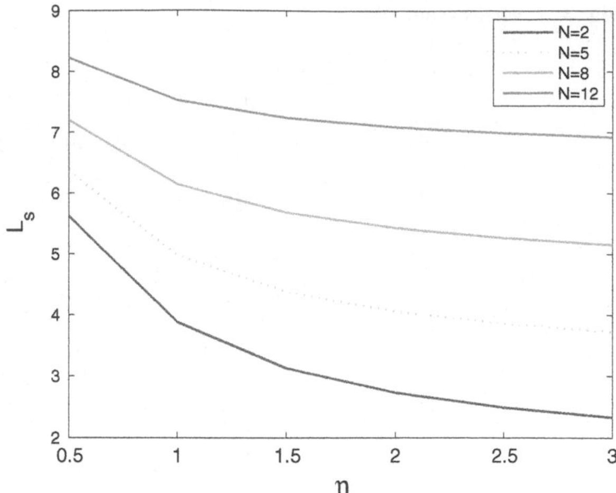

Fig. 5 Effect of η on L_s

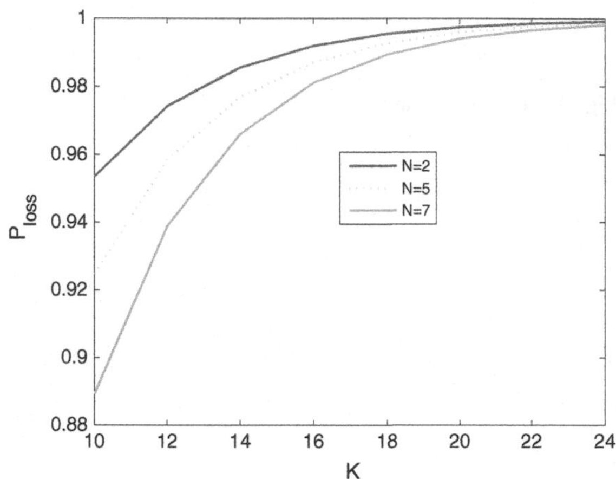

Fig. 6 Effect of K on P_{loss}

5 Conclusion

In this paper, we have proposed a randomized N-policy queuing model with start-up time, to prolong the lifetime of Wireless Sensor Networks. The working of a single server has been described with the assumption that each server follows half-duplex communication. The probabilities of sleep, idle, start-up and busy states have been derived using the proposed model. The performance indices of the same have been presented. We have also calculated the value of total energy consumed. The results that we obtained validate our theoretical analysis and prove that our approach is indeed an efficient method for improving the lifetime of Wireless Sensor Networks.

References

1. Akyilidiz, I.F., Su, W., Sankarasubramaniam, Y., Cayirci, E.: Wireless sensor networks: a survey. Comput. Networks **38**(4), 393–422 (2002)
2. Culler, D., Hong, W.: Wireless sensor networks. Commun. ACM **47**(6), 30–33 (2004)
3. Jones, C.E., Sivalingam, K.M., Argawal, P., Chen, J.C.: A survey energy efficient network protocols for wireless networks. Wirel. Netw. **7**, 343–358 (2001)
4. Liu, M., Cao, J., Zheng, Y.: An energy-efficient protocol for data gathering and aggregation in wireless sensor networks. J. Supercomput. **43**, 107–125 (2008)
5. Rajendran, V., Obraczka, K., Garcia-Luna-Aceves, J.J.: Energy-efficient, collision-free medium access control for wireless sensor networks. Wirel. Netw. **12**, 63–78 (2006)
6. Yang, X., Vaidya, N.H.: A wakeup scheme for sensor networks: achieving balance between energy saving and end-to-end delay. In: Proceedings of the 10th IEEE Real-time and Embedded Technology and Applications Symposium (RTAS04) (2004)
7. Yuan, Z., Wang, L., Shu, L., Qin, T.H.Z.: A balanced energy consumption sleep scheduling algorithm in wireless sensor networks. In: The 7th International Wireless Communications & Mobile Computing Conference (IWCMC 2011), Istanbul, Turkey, July 58 2011
8. Lin, C., Xiong, N., Park, J.H., Kim, T.H.: Dynamic power management in new architecture of wireless sensor networks. Int. J. Commun. Syst. **22**, 671–693 (2009)
9. Miller, M.J., Vaidya, N.H.: A MAC protocol to reduce sensor network energy consumption using a wakeup radio. IEEE Trans. Mob. Comput. **4**(3), 228–242 (2005)
10. Wang, L., Xiao, Y.: A survey of energy-efficient scheduling mechanisms in sensor networks. Mobile Networks Appl. **11**, 723–740 (2006)
11. Wang, K.H., Ke, J.C.: A recursive method to the optimal control of an M/G/1 queueing system with finite capacity and infinite capacity, Appl. Math. Modell. **24**, 899–914 (2000)
12. Kuo, C.C., Wang, K.H., Pearn, W.L.: The interrelationship between *N*-policy *M/G/1/K* and *F*-policy *G/M/1/K* queues with startup time. Qual. Technol. Quantitat. Manage. **8**, 237–251 (2011)
13. Yang, D.Y., Wang, K.H.: Interrelationship between randomized F-policy and randomized N-policy queues. J. Indus. Prod. Eng. **30**, 30–43 (2013)
14. Huang, D.C., Tseng, H.C., Deng, D.J., Chao, H.C.: A queue-based prolong lifetime methods for wireless sensor node. Comput. Commun. **35**(9), 1098–1106 (2012)
15. Huang, D.C., Lee, J.H.: A dynamic *N* threshold prolong lifetime method for wireless sensor nodes. Math. Comput. Modell. **57**(11), 2731–2741 (2013)
16. Jiang, F.C., Huang, D.C., Yang, C.T., Leu, F.Y.: Lifetime elongation for wireless sensor network using queue-based approaches. J Supercomput **59**, 1312–1335 (2012)
17. Jiang, F.C., Wu, H.W., Huang, D.C., Lin, C.H.: Lifetime security improvement in wireless sensor network using queue-based techniques. In: 2010 International Conference on Broadband, Wireless Computing, Communication and Applications (BWCCA), pp. 469–474. IEEE (2010)
18. Mann, C.R., Baldwin, R.O., Kharoufeh, J.P., Mullins, B.E.: A queueing approach to optimal resource replication in wireless sensor networks. Perform Eval. **65**, 689–700 (2008)

Overview on Location Management in PCS Network: A Survey

Abantika Choudhury, Abhijit Sharma and Uma Bhattacharya

Abstract Location Management is an important and key issue in any wireless communication network to deliver the services of mobile user and the network needs to keep track of that mobile user constantly. Two methods for location management exist-Location update (LU) and Paging. Both location update and paging takes huge cost in terms of bandwidth, battery power, memory space, computing time etc. More location updates reduce cost of paging and the vice versa is also true. Since the LU cost is much higher than the cost of paging, this paper focuses on location management using LU only. A thorough study of different existing LU procedures of location management process in Personal Communication Service (PCS) network and its related research work are presented in this paper.

Keywords Location update · Paging · PCS · HLR · VLR

A. Choudhury (✉)
Department of Information Technology, RCC Institute of Information Technology, Kolkata 700015, India
e-mail: abantika_choudhury@rediffmail.com

A. Sharma
Department of Computer Application, National Institute of Technology, Durgapur 713209, India
e-mail: abhijit.cst@gmail.com

U. Bhattacharya
Department of Computer Science and Technology, Indian Institute of Engineering Science and Technology, Shibpur, Howrah 711103, India
e-mail: ub@cs.iiests.ac.in

A. Nagar et al. (eds.), *Proceedings of 3rd International Conference on Advanced Computing, Networking and Informatics*, Smart Innovation, Systems and Technologies 43, DOI 10.1007/978-81-322-2538-6_37

1 Introduction

Nowadays it becomes necessary to communicate one another residing far from each other, where wireless communication may play a great role. In personal communication service (PCS) network, we use different wireless nodes like mobile phones, laptops etc., which can communicate via wireless communication. Here, we mainly draw the attention on the location management in PCS network, which is a process to locate the mobile stations (MS) and deliver the various services required. Location update (LU) means, the mobile user periodically update their location in the network, so that the network can update their database profile of that particular user. Paging is the process in which network searches for a mobile terminal by sending polling signal depending upon the last updated location. Again higher is the LU cost, lower is the paging cost and vice versa. LU cost is very much higher than that of paging cost (near about 10 times). So minimizing the LU cost is much more profitable than minimizing a paging cost. In this paper, we study about the location management (LM) process occurs in PCS network. Several types of dynamic LM schemes (in this scheme, MS always updates their current location to VLR) and their performances have been observed. Dynamic scheme's are distance based location management (DBLMS) [1], movement based location management (MBLMS) [1], *pointer based* [1] and *time based location management* [1]. Comparative studies of above mentioned approaches of dynamic LM schemes are given in Tables 1 and 2.

Related research works on different approaches in location management are discussed in Sect. 2. In Sect. 3, we draw the conclusion followed by references.

Table 1 Comparison between major static LU schemes

LU schemes	Update cost	Paging cost	Location accuracy	Major drawback
Always update	High	Low	One cell	No. of updates is too high
Never update	Low	High	One LA	Whole LA needs to be paged
Paging cell	Low	High	Several cells	Long time delay in large network
Static-interval based	Constant	High	Several cells	Unnecessary updates by stationary users
Reporting centres	Low	High	Several cells	High computational overhead

Table 2 Comparison between major dynamic LU schemes

LU schemes	Update cost	Paging cost	Location accuracy	Major drawback
Time based	Low	High	Several cells	Unnecessary updates by stationary users
Movement based	Low	High	Several cells	High computational overhead on MS side
Distance based	Low	High	Several cells	High computational overhead on MS side
HLR-level replication	Low	High	One cell to several cells	• Extra burden on current HLR • Increased call establishment delay
VLR-level	Low	High	Several cells	Overhead of regularly updating distributed mobility information
Pointer based	Low	High	Several cells	Increased call establishment delay

2 Related Research on Different Approaches of Location Management in PCS Network

Sections 2.1, 2.2 and 2.3 contain related research work on various approaches of dynamic LM. Research work in 3G network is described in Sect. 2.4.

2.1 Location Management Using Pointer Based Approach

A MHA-PB [2] method preserves the location information of different MSs concurrently by using DHA [2, 3] based scheme. Parallel registration of CoAs [2] is occurred in this method [2]. FA maintains the cache of CoAs of outgoing MS. UMTS-First [2] and WLAN-First [2] are two approaches proposed by the method [2]. An algorithm is provided to analyze and design the LDA [4]. When the MS enters in a LA under a new LDA, MS is registered with the current LA instead of Home Location Register (HLR). When the MS moves from its own LDA to next LDA a pointer is kept from home LDA to the current LDA. LDR also keeps track of its inter LDA movement frequency of the user. If it is found that in inter LDA movement LA of a user is not in FVLA [4] list, LDR sends update request to HLR. Here, the proposed methods minimize the registration cost.

2.2 Location Management Using Movement Based Approach

A method [5] is proposed to predict the current location of MS with the help of common mobility patterns model [5] from a collection of MSs. The model [5] finds

ULP [5] and learns cell residence time for each ULP. ULP provides important information about mobility behavior of MSs. UATs [5], associated with each UAP [5] are stored in ULP. SMS [6] model characterizes the smooth movement of mobile users in accordance with the physical law of motion. The model [6] is formulated by semi markov process [6] to analyze the steady state properties. To reduce the overhead in MBLMS, a simple scheme [7] of cell-ID assignment is proposed. In this scheme [7], network sends only the ID of centre cell of a LA-ring [7] and a threshold value [7] to MS. The MS can compute IDs of all other cells in its location area. This saves a significant amount of bandwidth by minimizing the signaling traffic at VLR level. Two different call handling models i.e. CPLU [8] model and CWLU [8] model are proposed for cost analysis and minimizing of MBLMS scheme. The exact LU costs of a MBLMS under CPLU and CWLU models are analyzed using a renewal process approach. This is found that a LU cost under CWLU is lower than that in CPLU. Author [8] also proposed some theorem for the simulation technique. A model [9] is able to predict the current location of MS in the network by using a hybrid model [9] of MLP [9] and SOFM [9]. This approach reduces the cost of repeated registration and predicts the location of MS. A non optimal search method [10] yields a low search delay and low search cost, where 2D Markov-walk [10] mobility model is used to describe a broad mobility pattern. Each cell is assigned a co-ordinate value by which the mobile station's location is denoted and by Markov method the probability to go to some other cell is found out. In a Markov model [11] for the PCS network, the cells are allocated by co-ordinates and six directions are determined for the possible movements of any MS. Probability of moving towards any direction is found out and the system can predict the location of a MS.

2.3 Location Management in Distance Based Approach

A ring level Markov chain mobility model [12] describes the movement of MS depending upon the mobility pattern of a mobile user. The analysis done by the model [12] makes possible to find the optimal distance threshold that minimizes the total cost of location management in DBLMS. A framework [13] is proposed and the location update cost is analyzed by considering Impact of call arrivals and the initial position of the MS.

2.4 Other Approaches in Location Management

A MS can calculate its distance from the base station, by using the RSS [14] and finds out its location by means of an AGA [14] method. Some metrics have been proposed in [15] to find out optimized LU cost. The method [15] introduces cell structure and the shortest route to reach the HLR from the last registered MSC/VLR.

An analytical model is developed [16] to investigate the busy line effect on inter service time distribution. In this approach, closed form analytical formulae for inter service time is calculated. Based on the analytical results the influence of busy line effect on modeling the portable movements is observed. Cost analysis for location management without considering the busy line effect may lead to unreliable result.

2.5 3G PCS Network

A new scheme [17] of 3G location management is proposed, where GLR [17] works as intermediate register between HLR and VLR. GLR contains roamer subscriber profile and location Information. The service area of GLR is called GLA [17] which consists of no. of LAs. When MS moves within a GLA, the registration takes place in GLR. Otherwise, registration is done in HLR. A GR [18] location tracking process is proposed for 2-level distributed database architecture in 3G network. The network consists of a number of radio areas (RA) [18]. Each RA updates their database according to its necessity. The database will update to its higher authority after getting a group updation of all the RAs under a single database. Authors [19] applied the some mathematical lemmas in three level architecture [17, 19] and studied DYNAMIC-3G [19] and STATIC-3G [19] schemes. A binary search algorithm [19] is proposed to find the optimal threshold which minimizes the total cost function of DYNAMIC-3G. A new GR location tracking strategy [20] with 2-level distributed database architecture is introduced in 3G wireless networks. The strategy [20] reduces the total LU cost in RA [20]. The RWL [20] list holds IDs of all newly moving MSs for group registration. The signaling cost of LU is expected to reduce significantly.

3 Conclusion

Motivation of this survey is to study on location management in PCS network. Location management consists of two different processes-location update and paging. The fact that location update is at least 10 times costlier than paging leads us to focus our survey work on location update process. Location updating can be of two types- static and dynamic. Three types of updating such as distance based, movement based and time based occur repeatedly in static one. Updating in dynamic one uses approaches based on threshold based, replication based and pointer based. Various user mobility models are random walk model, fluid flow model, shortest path model and selective prediction model. Research work on 2G and 2.5G PCS network deal location management using pointer forwarding approach, movement based approach, distance based approach and also some other approaches. Another scheme of 3G location management scheme also exists. Two tables in this paper show comparison between major dynamic LU schemes.

References

1. Bhadauria, V.S., Sharma, S., Patel, R.: A Comparative study of location management schemes: challenges and guidelines. Int. J. Comput. Sci. Eng. (IJCSE), **3**, 7 (2011)
2. Biswash, S.Kr., Kumar, C.: Multi home agent and pointer based (MHA-PB) location management scheme in integrated cellular wireless LAN networks for frequent moving users. Comput. Commun. **33**(Elsevier 7), 2260–2270 (2010)
3. Sangheon, P., Wojun, L.: Dual home agent (DHA)-based location management scheme in integrated cellular-WLAN networks. Comput. Netw. **52**, 3273–3283 (2008)
4. Chakraborty, S., De, D., Dutta, J.: An efficient management technique using frequently visited location areas and limited pointer forwarding Technique. In: International Conference on Advanced Communication Control and Computing Technologies (ICACCCT). IEEE (3) (2012)
5. An, N.T., Phuong, T.M.: A Gaussian Mixture Model for Mobile Location Prediction. ICACT (6) (2007)
6. Zhao, M., Wang, W.: A Unified Mobility Model for Analysis and Simulation of Mobile Wireless Network. Springer (10) (2007)
7. Bhadauria, V.S., Sharma, S., Patel, R.: Reducing Overhead in Movement Based Dynamic Location Management Scheme for Cellular Networks. Springer, New York (9) (2013)
8. Li, K.: Cost analysis and movement based location management schemes in wireless communication networks: a renewal process approach. Springer Science, Business Media (11) (2011)
9. Majumder, K., Das, N.: Mobile user tracking using a hybrid neural network. Wireless Netw. **11**(13), 275–284 (2005) (Springer Science)
10. Yuan, Y., Zhang, Y., Hu, L., Huang, Y., Qian, M., Zhou, J., Shi, J.: An efficient multicast search scheme under 2D markov walk model. In: Proceedings in GLOBECOM IEEE (17) (2009)
11. Zheng, J., Zhang Y., Wang, L., Chen, J.: Adaptive location update area design for PCS network under 2D Markov walk model. 1-4244-0419-3/06, IEEE (18) (2006)
12. Li, K.: Analysis of distance based location management in wireless communication network. IEEE Comput. Soc. (15) (2013)
13. Zhao, Q., Liew, S.C.: Location update cost of distance based scheme for PCS networks with CTRW model. IEEE Commun. Lett. **13**(6) (2009)
14. Jie, Y., Kadhim, D.J.: Performance evaluation of location management in GSM networks. IEEE (14) (2012)
15. Biswas, S., Kumar, C.: An effective metric based location management scheme for wireless cellular networks. J. Netw. Comput. Appl. 34, Elsevier.(2), ISM Dhanbad, India (2011)
16. Wang, X.: Impact of busy-line effect on the interservice time distribution and modeling of portable movements in PCS networks. IEEE Trans. Vehcular Technol. **59**(2) (2010)
17. Ali, S., Ismail, M., Mat, K.: Development of mobility management simulator for 3G cellular network. IEEE International Conferenc Telecommunication Malayasia International Conference on Communications, Malayasia (1) (2007)
18. Vergados, D.D., Panoutsakopoulos, A., Douligeris, C.: Location management in 3G network using 2 level distributed database architecture. In: 18th Annual IEEE International Symposium on Personal Communication, Indoor and Mobile Radio communications (PIMRC) 1-4244-1144-0 (16) (2007)
19. Xiao, Y., Pan, Y., Li, J.: Design and analysis of location management for 3G cellular network. IEEE Trans. Parallel Distrib. Syst. **15**(4), 12 (2004)
20. Vergados, D.D., Panoutsakopoulos, A., Douligeris, C.: Group registration with distributed databases for location tracking in 3G wireless network. Comput. Netw. **52**(8), 1521–1544 (2008) (Sience Direct, Elsevier)

Self-Organized Node Placement for Area Coverage in Pervasive Computing Networks

Dibakar Saha and Nabanita Das

Abstract In pervasive computing environments, it is often required to cover a certain service area by a given deployment of nodes or access points. In case of large inaccessible areas, often the node deployment is random. In this paper, given a random uniform node distribution over a 2-D region, we propose a simple distributed solution for self-organized node placement to satisfy coverage of the given region of interest using least number of active nodes. We assume that the nodes are identical and each of them covers a circular area. To ensure coverage we tessellate the area with regular hexagons, and attempt to place a node at each vertex and the center of each hexagon termed as *target points*. By the proposed distributed algorithm, unique nodes are selected to fill up the target points mutually exclusively with limited displacement. Analysis and simulation studies show that proposed algorithm with less neighborhood information and simpler computation solves the coverage problem using minimum number of active nodes, and with minimum displacement in 95 % cases. Also, the process terminates in constant number of rounds only.

Keywords Area coverage · Node deployment · Pervasive computing · Wireless sensor networks · Hexagonal tessellation

1 Introduction

In many applications of pervasive computing from home and health care to environment monitoring and intelligent transport systems, it is often required to place the sensors or computing nodes or access points to offer services over a predefined

D. Saha (✉) · N. Das
Advanced Computing and Microelectronics Unit, Indian Statistical Institute, Kolkata, India
e-mail: dibakar.saha10@gmail.com

N. Das
e-mail: ndas@isical.ac.in

A. Nagar et al. (eds.), *Proceedings of 3rd International Conference on Advanced Computing, Networking and Informatics*, Smart Innovation, Systems and Technologies 43, DOI 10.1007/978-81-322-2538-6_38

area. In some cases, like mobile surveillance, vehicular networks, mobile ad hoc networks, wireless sensor networks etc., the nodes are mobile and have limited energy, limited storage and limited computation and communication capabilities. These networks are often self-organized, and can take decision based on their local information only. In wireless sensor networks (WSN), large number of sensor nodes are spatially distributed over an area to collect ground data for various purposes such as habitat and ecosystem monitoring, weather forecasting, smart health-care technologies, precision agriculture, homeland security and surveillance. For all these applications, the active nodes are required to cover the area to be monitored. Hence, for these networks, the coverage problem has emerged as an important issue to be investigated. So far, many authors have modeled the coverage problem in various ways but most of them considered static networks. For deterministic node deployment, centralized algorithms can be applied to maximize the area coverage assuming the area covered by each node to be circular or square [9, 11, 12]. Many authors solved the coverage problem by the deterministic node placement techniques to maximize the network lifetime and to minimize the application-specific total cost. In paper [3], authors investigated the node placement problem and formulated a constrained multi-variable nonlinear programming problem to determine both the locations of the nodes and data transmission pattern in order to optimize the network lifetime and the total power consumption. Authors in [10, 15] proposed random and coordinated coverage algorithms for large-scale WSNs. But unfortunately, in many potential working areas, such as remote harsh environments, disaster affected regions, toxic regions etc., sensor deployments cannot be done deterministically. For random node deployment, virtual partitioning is often used to decompose the query region into square grid blocks and the coverage problem of each block by sensor nodes is investigated [6, 13, 14].

Whether the node deployment be deterministic or random, there is little scope of improving the coverage once the nodes are spatially distributed and they are static. Hence, mobility-assisted node deployment for efficient coverage has emerged as a more challenging problem. Many approaches have been proposed so far, based on virtual force [5, 18, 20], swarm intelligence [7, 8], and computational geometry [16], or some combination of the above approaches [4, 17]. In [19], a movement-assisted node placement method is proposed based on Van Der Waal's force where the relationship of adjacency of nodes was established by Delaunay Triangulation and force is calculated to produce acceleration for nodes to move. However, the computation involved is complex and it takes large number of iterations to converge. Authors in [1] proposed a distributed algorithm for the autonomous deployment of mobile sensors called *Push and Pull*, where sensors autonomously coordinate their movements in order to achieve a complete and uniform coverage. In [16], based on Voronoi diagram, authors designed and evaluated three distributed self-deployment algorithms for controlling the movement of sensors to achieve coverage. In these protocols, sensors move iteratively, eventually reaching the final destination. This procedure is also computation intensive and may take longer time to converge. Moreover each node requires the location information of its every neighbor to execute the algorithm.

In this paper, given a random node deployment over a 2-D region, a simple self-organized distributed algorithm is proposed to satisfy coverage of a given region of interest using minimum number of nodes with limited mobility. To avoid an iterative procedure, here some target points are specified deterministically by tessellating the area with regular hexagons. After deployment, each node computes the nearest target point. Next, a unique node closest to a target point is selected in a distributed fashion based on local position information only, to move towards the target point. In this way, nodes attempt to fill up the target points mutually exclusively with minimum displacement. The set of nodes selected is made active to cover the area. It is evident that compared to the works in [16, 19], the computation involved in this algorithm is significantly simple, it requires no location information of its neighbors and it converges faster in two rounds only. Analysis and simulation results show that proposed algorithm with less neighborhood information and simpler computation solves the coverage problem using minimum number of active nodes, and with minimum displacement in 95 % cases. Also, the process terminates in constant number of rounds only.

The rest of the paper is organized as follows. Section 2 defines the problem and introduces the basic scheme of area coverage. Section 3 presents the distributed algorithm for self-deployment. Section 4 evaluates the performance of the proposed protocol by simulation. Finally, Sect. 5 concludes the paper.

2 Movement Assisted Area Coverage

2.1 Problem Formulation

Let a set of n nodes $\mathcal{S} = \{s_1, s_2, \ldots, s_n\}$ be deployed randomly over a 2-D region A. It is assumed that each node is homogeneous, and covers a circular area with fixed sensing radius r. The goal of this paper is that given the random uniform distribution of a set of n nodes over a 2-D plane, to select a subset $P \subseteq S$ and to place them at nearest target points such that the cardinality of P is minimum and it covers the area. Our objective is to develop a light weight self-organized distributed algorithm for node rearrangement to reduce the amount of computation and rounds of communication, and the average distance traversed by a node, to maximize coverage utilizing minimum number of nodes. This in turn helps us to conserve energy better and hence to enhance the network lifetime.

2.2 Area Tessellation

Given a rectilinear area to be covered by nodes, Fig. 1 shows a typical regular placement of nodes such that the overlapped region is minimum and the area is fully covered using minimum number of nodes. The positions of all the nodes basically defines a set of regular hexagons that tessellates the area as shown in Fig. 2. The sensor nodes are to be placed exactly on the vertices and the centers of the hexagons, termed here as the *target points*. In [2], it is proved that such node placement technique maximizes the area coverage using minimum number of nodes. In this case, the minimum number of nodes corresponds to the total number of *target points*, and can be computed easily as a function of the sensing radius r as shown below.

Let A be a 2-D axis-parallel rectangle $L \times W$ with $(0, 0)$ as the bottom-left corner point, termed here as the origin. For any arbitrary bottom-left corner point (x_0, y_0), the co-ordinate system is to be translated appropriately. The tessellation of A with

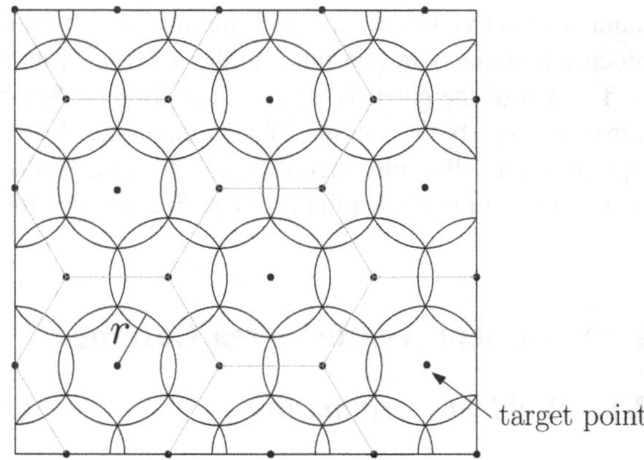

Fig. 1 Target points for node placement

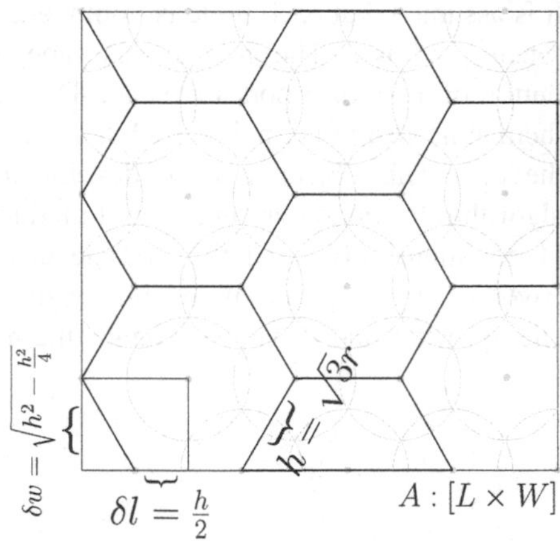

Fig. 2 Hexagons to tessellate the given area

regular hexagons of side h is shown in Fig. 2, where $h = \sqrt{3}r$ [1]. It is to be noted that the target points lie along some rows and columns parallel to the x-axis and y-axis respectively. Rows are separated by a distance:

$$\delta w = \sqrt{h^2 - \frac{h^2}{4}} = \frac{3}{2}r.$$

Similarly, columns are separated by a distance:

$$\delta l = \frac{h}{2} = \frac{\sqrt{3}}{2}r.$$

From Fig. 2, it is clear that, each even row-i starts with a target point $(0, i.\delta w)$, whereas each odd row-j starts with a target point $(\delta l, j.\delta w)$. Hence given the area A, the total number of rows is given by: $N_{row} = \lceil \frac{W}{\delta w} \rceil + 1$,. The total number of target points is,

$$N = N_{row} \cdot \left\lceil \frac{L}{2\delta l} \right\rceil + \left\lceil \frac{N_{row}}{2} \right\rceil.$$

Therefore, to cover the area A, the number of nodes to be deployed is $n \geq N$, to fill up the target points exclusively. However, in practice, with random distribution of nodes, the area to be monitored is over-deployed, and $n >> N$, providing sufficient redundant nodes to ensure coverage.

2.3 Nearest Target Point Computation

Let a set of n nodes $\mathcal{S} = \{s_1, s_2, \ldots, s_n\}$ be deployed randomly over a 2-D region A. Each node-i only have the information of its physical location (x_i, y_i) and the sensing range r. Now to estimate the location of its nearest target point, it should have the knowledge of the origin, i.e. the bottom-left point of the area. The sink may directly broadcast it to all nodes for a static area. In case, the area of interest is dynamic, or depends on the deployment, the nodes may determine the origin as the point with minimum abscissa and ordinate of all the nodes deployed as described below. Here, during initialization, each node-i broadcasts its own location (x_i, y_i), and maintains two variables initiated as $x_{min} = x_i$, and $y_{min} = y_i$ to keep the minimum abscissa and ordinate of all of the deployed nodes. It receives the messages with locations (x_j, y_j) from other nodes-j, and if $x_j \leq x_{min}$ and/or $y_j \leq y_{min}$ the values of x_{min} (y_{min}) are changed appropriately, and if there is any update, the new message is broadcasted again, otherwise it is ignored. In this way, after sufficient time, say T, all the nodes will acquire the same value of x_{min} and y_{min} and consider it as the origin of the area under consideration. In case of an event, the affected nodes may define the event area in terms of this origin dynamically. Now, the initialization phase is completed and the next phase starts. In the worst case, each node may have to transmit n messages to complete the procedure.

After initialization phase, the nearest target point is to be computed by each node. We assume that each node knows the origin (x_{min}, y_{min}) of A. Next, each node i at location (x_i, y_i), attempts to find out its nearest target point. It computes

$$t_x(i) = NI\left(\frac{|x_i - x_{min}|}{\delta h}\right)$$

and

$$t_y(i) = NI\left(\frac{|y_i - y_{min}|}{\delta w}\right).$$

Here $NI(x)$ denotes the nearest integer value of x. Next, it finds the location of its nearest target point $T_i(x_{Ti}, y_{Ti})$ as:

$$\begin{aligned} y_{Ti} &= t_y(i) \cdot \delta w \\ x_{Ti} &= t_x(i) \cdot h, \quad \mathit{when}\, t_x(i) \; \mathit{is\ even}, \\ &= t_x(i) \cdot h + \frac{h}{2} \;\; \mathit{otherwise}. \end{aligned}$$

Thus, each node finds its nearest target point and broadcasts it to its neighbors which lie within its communication range to select a unique node in a distributed fashion to be moved to a given target point exclusively. So, it is important to define the communication range of a node to define its set of neighbors.

2.4 Role of Communication Range

So far, we have mentioned the sensing range of the sensor node-i that defines the circular area with radius r, centered at node-i to be the area covered by node-i. When a node executes a distributed algorithm, it is very important to identify its neighborhood with which it can communicate directly. For that we should specify the communication range r_c of a node-i which indicates that when a node-i transmits, a node-j can receive the packet if and only if, the distance between the nodes $d(i,j) \leq r_c$. It is important to note that for all practical purposes, r_c is independent of r, since the transceiver hardware of the sensor node determines r_c and r is the property of the sensing hardware. For the proposed algorithm, it is required that each node should cooperate with its neighboring nodes which are within a distance of $2(h+r) = 2(1+\sqrt{3})r$. So, here we assume that $r_c \geq 2(1+\sqrt{3})r$, where r is the sensing range.

3 Algorithms for Area Coverage

In wireless sensor networks, since nodes have limited computing and communication capabilities, it is always better to adopt distributed algorithms where nodes may take decisions with simple computation based on their local information only to produce a global solution.

Algorithm 1: Target Point Computation

Input: node: $i(x, y)$
Output: Target point: $T_i(x_{Ti}, y_{Ti})$
$t_y \leftarrow NI(\frac{y}{l})$;
if t_y *is even number* **then**
  $t_x \leftarrow NI(\frac{x}{h})$;
  $x_{Ti} \leftarrow t_x \times h$;
else
  $t_x \leftarrow |x - \frac{h}{2}|$;
  $t_x \leftarrow NI(\frac{t_x}{h})$;
  $x_{Ti} \leftarrow (t_x \times h) + \frac{h}{2}$;
$y_{Ti} \leftarrow t_y \times l$;

Here initially, each node is in *Active* mode and knows its location and the origin (x_{min}, y_{min}) of the area to be covered. In the first round, each node-i assumes a virtual tessellation of the area with hexagon tiles and compute its nearest target point by Algorithm 1. It broadcasts a *Target* message with data $(t_i(x,y), d_i)$, where $t_i(x,y)$ is its nearest target point and d_i is its distance from $t_i(x,y)$. Each node-i waits till it receives *Target* messages from all of its neighbors. Then it checks if it is at minimum distance from the target $t_i(x,y)$. Then node-i is selected to fill up the target $t_i(x,y)$. The case of tie may be resolved by node-id. It broadcasts *Selected(i,* $t_i(x,y)$*)* message and moves to $t_i(x,y)$ point with displacement d_i and goes to *Active* state. Otherwise, it goes into the *Idle* state. It is clear that within the circle C_i of radius r around a target point $t_i(x,y)$, if there exists at least one node, it will be filled up by it. The problem arises if any circle C_i is originally empty due to initial random node deployment. In that case, the *Idle* nodes will execute a second round of computation. Each idle node-i finds if there is any unfilled target node around it, i.e. within the six adjacent circles overlapping with C_i. Next it finds the unfilled target points and sort them according to the distance from it. Next it follows the same procedure described above for each target unless it becomes *selected* or all the targets are filled up.

The distributed **Algorithm 3**, describes the sequence of steps of the procedure.

Algorithm 2: Node Selection for Target Point

```
Input: node: i, STATUS (active =1 or idle=0)
Output: movement: true/false
movement=true;
for each neighbor node do
    if receives target(t^j(x, y), d^j) message then
        if t^i(x, y) == t^j(x, y) // same target point then
            if d^i > d^j then
                movement=false;
                STATUS(i) ← 0;
            if d^i == d^j then
                if i > j then
                    movement=false;
                    STATUS(i) ← 0;
if movement==true then
    Move towards target point t^i(x, y);
    STATUS(i) ← 1;
    broadcasts Selected(t^i(x, y)) message;
```

Algorithm 3: Distributed Algorithm for Adjacent Target Points

```
Input: free node n_i
Output: Active or idle
for each node i do
    Compute nearest target point t^i(x, y) (call Algorithm 1);
    Compute all six neighbor target points of t^i(x, y) and distances from its position.
    Include all the target points in L_i sorted by distance D_i;
    for L_i ≠ {φ} do
        Take first point t^i(x, y) and d_i from L_i and D_i respectively;
        broadcasts target(t^i(x, y), d_i) message;
        Wait and listen until receives all Target message from the neighbors;
        Call Algorithm 2;
        if STATUS(i) == 1 then
            Goto Active Mode;
            Free L_i and D_i;
            Terminate;
        else
            if receives Selected(t^j(x, y)) message then
                Remove target point t^j(x, y) from L_i;
    Terminate and goto idle Mode;
```

3.1 Complexity Analysis

It is evident that to find the target points **Algorithm 1** is computed in constant time. To take decision for selecting a unique node for each target point, each node waits until it receives all the *target* messages from its d neighbors. Each node takes $O(d)$ time to get the minimum distance from its d neighbors. Therefore, **Algorithm 2** computes in $O(d)$ time, where d is the maximum number of neighbors of a node. Finally, each node attempts to fill up at most seven target points. Therefore, the total time complexity of the distributed algorithm (**Algorithm 3**) is $O(d)$. In the distributed algorithm nodes broadcast at most seven *Target* messages and only one *Selected* message. Therefore, per node at most eight messages are needed to complete the procedure. Hence, the message complexity per node is $O(1)$ only.

4 Simulation Results

In our simulation study, we assume that n nodes, $250 \leq n \leq 300$, are distributed randomly over a 500×500 area with radius $r = 28.86$ and side of the hexagon $h = 50$. All the target points associated with the circles are computed by nodes. After executing the distributed algorithm, each node moves to its target point. The performance of the proposed algorithm is evaluated in terms of coverage, rounds of computation needed and displacement of nodes. The graphs show the average value of 20 runs for 20 independent random deployments of nodes.

Figure 3 shows the variation of coverage percentage with n, the total number of nodes deployed. If the total number of target points is 105, for $n = 50, 100, 150$, the coverage percentage is found to be 47, 84.57 and 99.36 % respectively. It gives an idea that how much an area should be over deployed to achieve 100 % coverage with random node deployment.

In Fig. 4, the variation in the number of computation rounds with n is presented. It is to remember that with random deployment, if there is no empty circle C_i with center at a target t_i and radius r, the procedure completes in a single round only. This fact is exactly revealed in Fig. 4. For n = 100, 150, 200, target points are filled

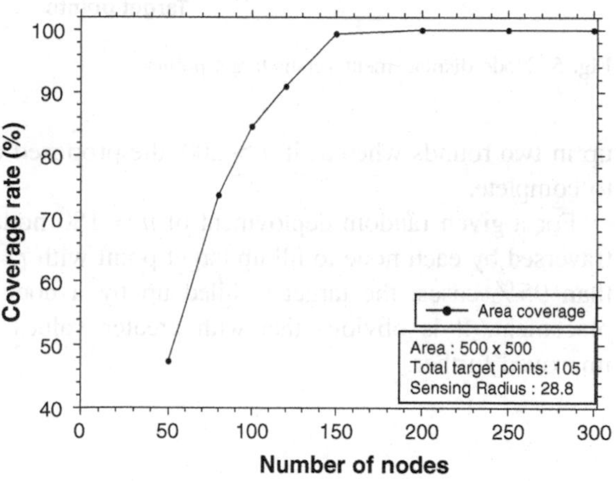

Fig. 3 Coverage rate with number of nodes

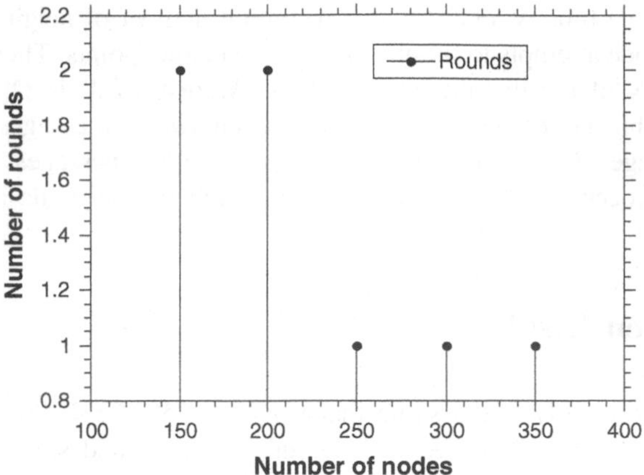

Fig. 4 Number of rounds for filling the target points

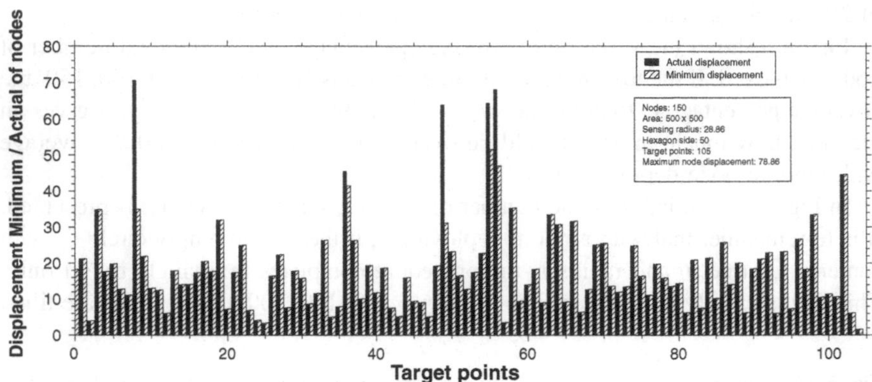

Fig. 5 Node displacement versus target points

up in two rounds whereas if $n > 200$, the proposed technique takes a single round to complete.

For a given random deployment of $n = 150$ nodes, Fig. 5 shows the distances traversed by each node to fill up target point with $r = 28.86$. It shows that in more than 95 % cases, the target is filled up by a node with minimum possible displacement. It is obvious that with greater values of n, this percentage can be improved further.

5 Conclusion

In this paper, we propose a self-organized node placement algorithm to satisfy the coverage of a given region of interest in Wireless Sensor Networks. The area is logically tessellated by regular hexagonal tiles starting from an origin. To get full coverage with random deployment of n nodes over a $2D$ region, we need to place unique nodes on every target point, which are essentially the vertices and the centers of the hexagons. With just the knowledge of its own location and the origin, each node executes a simple self-organized distributed algorithm $O(d)$ time complexity (d is the maximum number of neighbors of a node) and constant message complexity, to fill up all the target points mutually exclusively with minimum possible displacement. In case of failure of nodes, existing free nodes may take necessary action to fill up the empty target points to make the system fault-tolerant. We evaluate the performance of our proposed model by simulation. It shows that with sufficient node density the algorithm attains full coverage using minimum number of nodes and terminates in one round only. Also, in more than 95 % cases the displacement of an individual node is minimum.

References

1. Bartolini, N., Calamoneri, T., Fusco, E., Massini, A., Silvestri, S.: Push and pull: autonomous deployment of mobile sensors for a complete coverage. Wireless Netw. **16**(3), 607–625 (2010)
2. Brass, P.: Bounds on coverage and target detection capabilities for models of networks of mobile sensors. ACM Trans. Sens. Netw. (TOSN) **3**(2), 9 (2007)
3. Cheng, P., Chuah, C.N., Liu, X.: Energy-aware node placement in wireless sensor networks. In: Global Telecommunications Conference, GLOBECOM, vol. 5, pp. 3210–3214. IEEE (2004)
4. Han, Y.H., Kim, Y.H., Kim, W., Jeong, Y.S.: An energy-efficient self-deployment with the centroid-directed virtual force in mobile sensor networks. Simulation **88**(10), 1152–1165 (2012)
5. Heo, N., Varshney, P.K.: An intelligent deployment and clustering algorithm for a distributed mobile sensor network. IEEE Int. Conf. Syst. Man Cybernet. **5**, 4576–4581 (2003)
6. Ke, W.C., Liu, B.H., Tsai, M.J.: The critical-square-grid coverage problem in wireless sensor networks is np-complete. Comput. Netw. **55**(9), 2209–2220 (2011)
7. Kukunuru, N., Thella, B.R., Davuluri, R.L.: Sensor deployment using particle swarm optimization. Int. J. Eng. Sci. Technol. **2**(10), 5395–5401 (2010)
8. Liao, W.H., Kao, Y., Li, Y.S.: A sensor deployment approach using glowworm swarm optimization algorithm in wireless sensor networks. Expert Syst. Appl. **38**(10), 12180–12188 (2011)
9. Luo, C.J., Tang, B., Zhou, M.T., Cao, Z.: Analysis of the wireless sensor networks efficient coverage. In: International Conference on Apperceiving Computing and Intelligence Analysis (ICACIA), pp. 194–197 (2010)
10. Poe, W.Y., Schmitt, J.B.: Node deployment in large wireless sensor networks: coverage, energy consumption, and worst-case delay. In: Conference on Asian Internet Engineering. pp. 77–84. AINTEC, ACM, USA (2009)

11. Saha, D., Das, N., Bhattacharya, B.B.: Fast estimation of coverage area in a pervasive computing environment. In: Advanced Computing, Networking and Informatics, vols. 2, 28, pp. 19–27. Springer (2014)
12. Saha, D., Das, N., Pal, S.: A digital-geometric approach for computing area coverage in wireless sensor networks. In: 10th International Conference on Distributed Computing and Internet Technologies (ICDCIT), pp. 134–145. Springer (2014)
13. Saha, D., Das, N.: Distributed area coverage by connected set cover partitioning in wireless sensor networks. In: First International Workshop on Sustainable Monitoring through Cyber-Physical Systems (SuMo-CPS), ICDCN. India (2013)
14. Saha, D., Das, N.: A fast fault tolerant partitioning algorithm for wireless sensor networks. In: Third International Conference on Advances in Computing and Information Technology (ACITY). pp. 227–237. CSIT, India (2013)
15. Sheu, J.P., Yu, C.H., Tu, S.C.: A distributed protocol for query execution in sensor networks. In: IEEE Wireless Communications and Networking Conference, vol. 3, pp. 1824–1829 (2005)
16. Wang, G., Cao, G., Porta, T.L.: Movement-assisted sensor deployment. IEEE Trans. Mob. Comput. **5**(6), 640–652 (2006)
17. Wang, X., Wang, S., Ma, J.J.: An improved co-evolutionary particle swarm optimization for wireless sensor networks with dynamic deployment. Sensors **7**(3), 354–370 (2007)
18. Yu, X., Huang, W., Lan, J., Qian, X.: A van der Waals force-like node deployment algorithm for wireless sensor network. In: 8th International Conference on Mobile Ad-hoc and Sensor Networks (MSN), pp. 191–194. IEEE (2012)
19. Yu, X., Liu, N., Huang, W., Qian, X., Zhang, T.: A node deployment algorithm based on van der Waals force in wireless sensor networks. Distrib. Sens. Netw. **3**, 1–8 (2013)
20. Zou, Y., Chakrabarty, K.: Sensor deployment and target localization based on virtual forces. In: Twenty-Second Annual Joint Conference of the IEEE Computer and Communications (INFOCOM). vol. 2, pp. 1293–1303 (2003)

SNR Enhancement of Brillouin Distributed Strain Sensor Using Optimized Receiver

Himansu Shekhar Pradhan and P.K. Sahu

Abstract This paper presents an improvement on signal to noise ratio (SNR) of long range Brillouin distributed strain sensor (BDSS). Differential evolution (DE) algorithm is used for receiver (avalanche photo diode (APD)) optimization. We have extracted the strain information of the proposed sensor using Fourier deconvolution algorithm and Landau Placzek ratio (LPR). SNR of the proposed system is realized using Indium Gallium Arsenide (InGaAs) APD detector over 50 km sensing range. We have achieved about 30 dB improvement of SNR using optimized receiver compared to non-optimized receiver at 25 km of sensing distance for a launched power of 10 mW. The strain resolution is observed as $1670\mu\varepsilon$ at a sensing distance of 50 km. Simulation results show that the proposed strain sensor is a potential candidate for accurate measurement of strain in sharp strain variation environment.

Keywords SNR · DE · APD · BDSS

1 Introduction

Nowadays, Brillouin distributed sensors become increasingly popular because of its ability to sense both temperature as well as strain. The distributed fibre sensors are more attractive toward the sensing applications due to their many advantages such as: distributed sensing replaces complex integration of thousands of electric sensor with one optical fibre system. These sensors offer the ability of being able to

H.S. Pradhan · P.K. Sahu (✉)
School of Electrical Sciences, IIT Bhubaneswar, Bhubaneswar, Odisha, India
e-mail: pks@iitbbs.ac.in

H.S. Pradhan
e-mail: hsp10@iitbbs.ac.in

A. Nagar et al. (eds.), *Proceedings of 3rd International Conference on Advanced Computing, Networking and Informatics*, Smart Innovation, Systems and Technologies 43, DOI 10.1007/978-81-322-2538-6_39

measure physical and chemical parameters along their whole length of the fibre on a continuously manner. The optical fibre is used as sensing element because it is cheap, light weight, flexible and immune to electromagnetic interference (EMI) [1, 2]. The distributed sensing system can be a cost-effective as well as flexible solution because of the use of normal telecommunication fibre as sensing element. A temperature measurement system using the signature analysis of Raman anti-Stokes and Stokes backscattered signal is demonstrated by Dakin et al. [3]. Raman scattering based temperature sensor are very popular because Raman signal separation is easier and conventional silica-based optical fibres can be used as the sensing element. Another advantage is that the temperature sensitivity of Raman sensor is about 0.8 %/K [4]. However, the downside is that the anti-Stokes Raman backscattered signal is extremely weak and about 30 dB weaker than the Rayleigh backscattered signal as a result a high sensitive receiver is required to receive weak backscattered signal. In order to mitigate the above mentioned difficulties, Brillouin scattering based distributed sensor is developed. The frequency shift of the Stokes Brillouin backscattered signal with strain was reported by Horiguchi et al. [5] in 1989. After their reported work, lots of research works on Brillouin distributed sensor are carried out by the researchers and industries. The mostly focused areas of research on Brillouin distributed sensor (BDS) is that the performance improvement such as: sensing range, sensing resolution, spatial resolution and SNR etc. SNR enhancement of BDS using different optical pulses coding such as simplex code [6], bipolar Golay codes [7], colour simplex coding [8] are reported in the literature. In this paper, we have proposed a BDSS system and extracted the strain profile using optical time domain reflectometer (OTDR) technique and LPR. In the proposed BDSS system, we have injected a short pulse to the sensing fibre and a spontaneous Brillouin backscattered signal is detected at the input fibre end. The received Brillouin backscattered signal intensity can be expressed as the convolution of the input pulse profile and the strain distribution along the fibre. We have calculated the SNR of the detecting signal using non-optimized and optimized parameter values of APD. We have included both thermal as well as shot noise of APD for calculation of SNR. In particular, we have shown the improvement of SNR of the proposed system using receiver optimization.

2 Theory and Simulation Model

The schematic of proposed SBSS system for 50 km sensing range is shown in Fig. 1. We have measured Brillouin backscattered signal using OTDR technique of the proposed sensor. The principle of OTDR technique is that an optical pulse is launched into fibre and the backscattered signal is detected at the input fibre end. The location information can be found out by the delay time and the light velocity in the fibre.

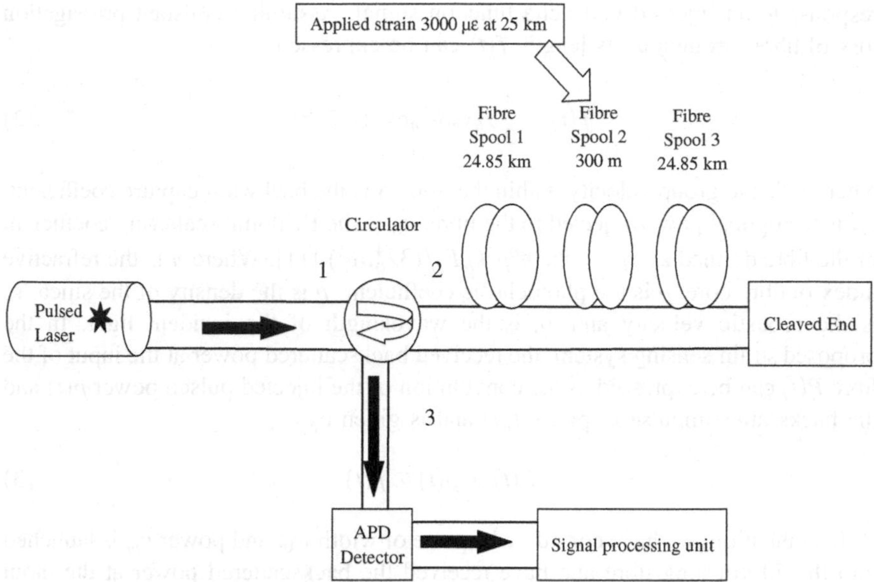

Fig. 1 The schematic of the proposed BDSS system

We have optimized the parameters of APD such as Responsivity R, dark current I_d and ionisation ratio k_A using DE algorithm. The optimum gain of the APD used as cost function for the optimization process and is given by [9]

$$G_{\text{opt}} = \left[\frac{4k_B T F_n}{k_A q R_L (R P_{in} + I_d)}\right]^{1/3} \tag{1}$$

In the above expression, k_B is the Boltzmann constant, T is the absolute temperature equal to 300 K, F_n is the excess noise factor, k_A is the ionization ratio or coefficient, q is the charge of an electron, R_L is the load impedance, R is the Responsivity, P_{in} is the input power to the receiver. The typical value of the other parameters of Eq. 1 are taken as $R_L = 1\,\text{k}\Omega$ and $F_n = 2$ [9]. We have used DE algorithm to maximize the cost function. In DE algorithm, we have used the mutation factor 0.5, number of populations 10 and number of generations 100 in our proposed algorithm. The maximization of APD optimum gain in Eq. 1 is done by taking three variables as R (from 1 to 9 A/W) I_d (from 1 to 10 nA) and k_A (from 0.01 to 1). We have considered the processes involved in DE algorithm [10] to maximize the optimum gain of APD. In the proposed system, we have considered the backscatter impulse response $f(t)$ defined as the backscattered signal power in response to an injected unit delta function signal. Assuming constant propagation loss of fibre throughout its length, $f(t)$ can be expressed as

$$f(t) = \frac{1}{2}\alpha_B v_g S p_{in} \exp(-2\alpha z) \tag{2}$$

where v_g is the group velocity within the fibre, S is the backward capture coefficient, p_{in} is the optical power injected to the fibre, α_B is the Brillouin scattering coefficient of the fibre defined as $\alpha_B = (8\pi^3 n^8 p^2 k_B T)/(3\lambda_0^4 \rho v_a^2)$ [11]. Where n is the refractive index of fibre core, p is the photoelastic coefficient, ρ is the density of the silica, v_a is the acoustic velocity and λ_0 is the wavelength of the incident light. In the proposed strain sensing system, the received backscattered power at the input of the fibre $P(t)$ can be expressed as the convolution of the injected pulsed power $p(t)$ and the backscatter impulse response $f(t)$ and is given by

$$P(t) = p(t) \otimes f(t) \tag{3}$$

In simulation, we have considered a pulse of width w_0, and power p_{in} is launched into the 50 km long fibre and have received the backscattered power at the input fibre end with the addition of white Gaussian noise. Similarly, for calculation of LPR, which is the ratio of Rayleigh signal intensity to Brillouin signal intensity, we have calculated the Rayleigh backscattered power. Rayleigh backscattered power P_R, with the function of fibre length z is given by [12]

$$P_R(z) = \frac{1}{2} p_{in} \gamma_R w_0 S v_g \exp(-2\gamma_R z) \tag{4}$$

where γ_R is the Rayleigh scattering coefficient. We have considered the strain dependence of the Brillouin backscattered signal intensity is given by $I_B = (I_R T)/(T_f(\rho v_a^2 \beta_T - 1))$ [13]. Where I_R and I_B are the Rayleigh and Brillouin backscattered signal intensities respectively, T_f is the fictive temperature, β_T is the isothermal compressibility and v_a can be expressed as $v_a = \surd((E(1-\sigma))/(\rho(1+\sigma)(1-2\sigma)))$ [13]. Where E is the Young's modulus and σ is the Poisson's ratio. The variation of Young's modulus of silica with the strain is given by $E = E_0(1 + 5.75\varepsilon)$ [14]. Where E_0 is the Young's modulus in unstrained fibre and ε is the tensile strain applied to the fibre. Assuming the Poisson's ratio is independent of strain, I_B can be rewritten as

$$I_B = \frac{k_1 I_R}{(k_2(1 + 5.75\varepsilon)) - 1} \tag{5}$$

where $k_1 = T/T_f$ and $k_2 = (E_0\beta_T(1-\sigma))/(((1+\sigma)(1-2\sigma)))$. To obtain the strain profile along the sensing fibre, we have considered the LPR at the unknown strain ε is compared with the known reference strain ε_R, and given by

$$\varepsilon = \frac{1}{K_s}\left(1 - \frac{\mathrm{LPR}(\varepsilon)}{\mathrm{LPR}(\varepsilon_R)}\right) + \varepsilon_R \tag{6}$$

The strain sensitivity of the proposed sensor is K_s. For calculation of SNR, we have used (InGaAs) APD receiver. The photo current of the receiver can be expressed as [9]

$$I_P = MRP_{in} \tag{7}$$

where R is the responsivity of the photo receiver and P_{in} is the input power to photo receiver. We have considered the noise powers such as: Gaussian noise, (σ_G^2) as well as thermal noise, (σ_T^2) and shot noise, (σ_S^2) of APD for calculation of total power. The shot noise power of APD can be calculated by the given expression $\sigma_S^2 = 2qM^2F_A(RP_{in} + I_d)\Delta f$ [15]. In the above expression, Δf is the effective noise bandwidth and F_A is the excess noise factor can be expressed as $F_A = k_AM + (1 - k_A)(2 - (1/M))$. Similarly, the thermal noise power of APD can be calculated by the given expression $\sigma_T^2 = 4(kT/R_l)F_n\Delta f$ [16]. Where F_n is the amplifier noise figure. The SNR of the proposed sensing system over a 50 km sensing range can be expressed as:

$$SNR = \frac{I_P^2}{\sigma_G^2 + \sigma_S^2 + \sigma_T^2} = \frac{M^2R^2P_{in}^2}{\sigma_G^2 + \sigma_S^2 + \sigma_T^2} \tag{8}$$

where I_P^2 is the Brillouin backscattered signal power at the output of APD and $\sigma_G^2 + \sigma_S^2 + \sigma_T^2$ is the total noise power of the proposed system. We have realized SNR of the proposed sensor using non-optimized as well as optimized receiver.

3 Simulations and Results

The optimum gain of APD is maximized using DE algorithm and the best solution shown in Fig. 2. We have calculated the Raleigh backscattered power and Brillouin backscattered power with additive white Gaussian noise of variance of $\sigma_G^2 = 10^{-7}$W for 50 km sensing range. A laser source operating at 1550 nm with 10 MHz linewidth is used for simulation. A rectangular pulse of width 100 ns and power 10 mW is launched to the sensing fibre and the backscattered signal is received at the input fibre end. The other parameters based on silica fiber such as $\alpha = 0.2\,\mathrm{dB/km}$, $k = 1.38 \times 10^{-23}\,\mathrm{J/K}, S = 1.7 \times 10^{-3}, n = 1.45, \gamma_R = 4.6 \times 10^{-5}\,1/\mathrm{m}, p = 0.286$, $\rho = 2330\,\mathrm{kg/m^3}$, $T_f = 1950\,\mathrm{K}$, $E_0 = 71.7\,\mathrm{GPa}$, $\beta_T = 7 \times 10^{-11}\,\mathrm{m^2/N}$ and $\sigma = 0.16$ are used for simulation. We have simulated the proposed 50 km strain sensor using Eq. 3. In the simulation process, we modeled a strained source with an artificial rectangular pulse variation of strain distribution around the point z = 25 km from

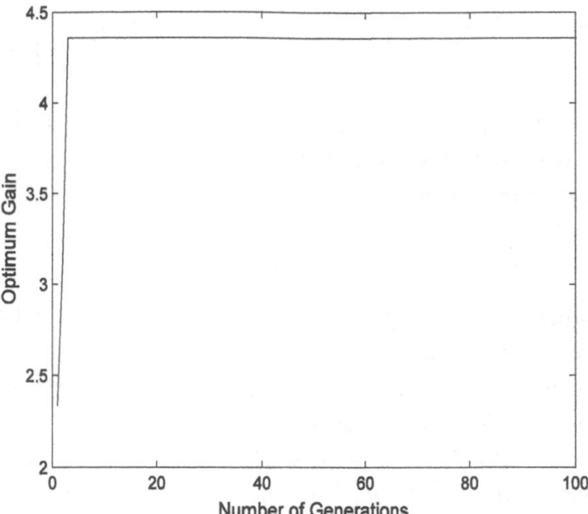

Fig. 2 The best solution of DE algorithm

the end point of the optical fibre whereas the rest of the fibre maintained at zero strain. We have calculated $LPR(\varepsilon_R)$ for unstrained fibre by taking $\varepsilon_R = 0\mu\varepsilon$. Similarly, $LPR(\varepsilon)$ for strained fibre is calculated by taking $\varepsilon = 3000\mu\varepsilon$. We have extracted the strain of proposed sensing system using FourD algorithm. The strain sensitivity K_s is 9.1×10^{-4} $\%\mu\varepsilon^{-1}$ [17] with respect to Brillouin intensity is considered for simulation process. The strain profile of the proposed system extracted using Eq. 6 and shown in Fig. 3. We have estimated the strain resolution by the exponential fit of the standard deviation of measured strain distribution versus distance. The strain resolution using FourD algorithm is shown in Fig. 4. We have observed the strain resolution of $1670\mu\varepsilon$ at 50 km distance using FourD algorithm. We have calculated SNR of the proposed

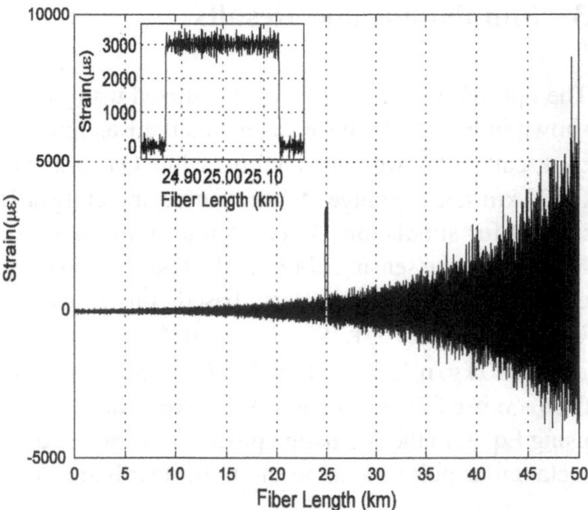

Fig. 3 Strain profile of the proposed system using FourD algorithm

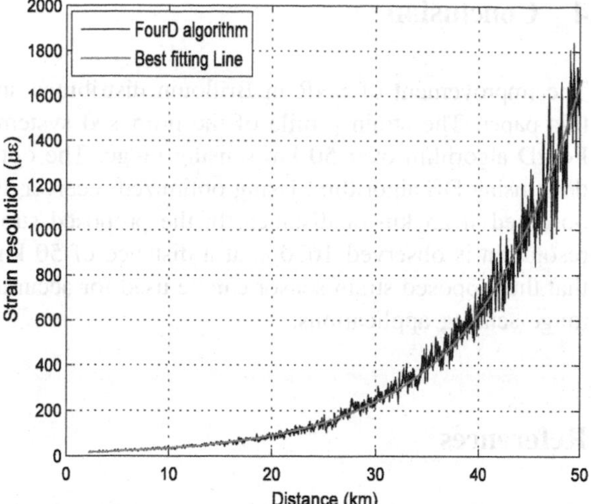

Fig. 4 Strain resolution of proposed system using FourD algorithm

system using InGaAs APD using non-optimized and optimized optimum gain for 50 km sensing range. The non-optimized parameter values 1, 0.9, 10 nA are used for R, k_A and I_d respectively in simulation process. The SNR of the proposed system is calculated using Eq. 8. Figure 5 shows the SNR of the proposed sensor using both non-optimized receiver and optimized receiver. The 30 dB improvement of the SNR is observed in Fig. 5 at 25 km of distance for optimized receiver compared to non-optimized receiver.

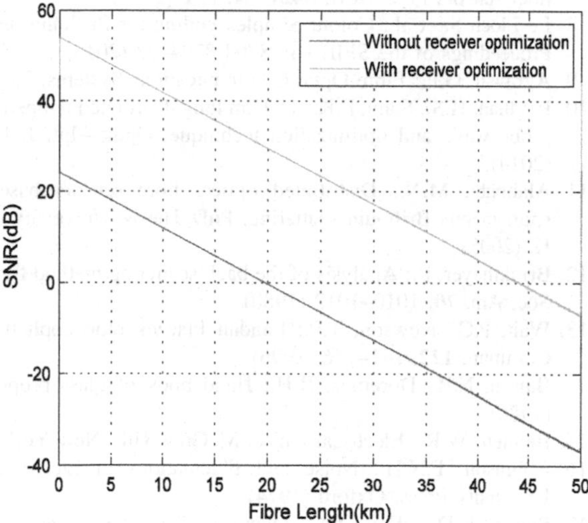

Fig. 5 SNR with and without receiver optimization of the proposed sensor

4 Conclusion

The improvement of SNR in Brillouin distributed strain sensor is investigated in this paper. The strain profile of the proposed system is extracted using LPR and FourD algorithm over 50 km sensing range. The optimization of receiver APD is done using DE algorithm. Using optimized receiver 30 dB improvement of SNR is observed at 25 km of distance. In the proposed strain sensing system, the strain resolution is observed $1670\mu\varepsilon$ at a distance of 50 km. Simulation results indicate that the proposed strain sensor can be used for accurate strain measurement in long range sensing applications.

References

1. Lopez-Higuera, J.M.: Handbook of Optical Fibre Sensing Technology. Wiley, New York (2002)
2. Lee, B.: Review of the present status of optical fibre sensors. Opt. Fibre Technol. **9**, 57–79 (2003)
3. Dakin, J.P., Pratt, D.J., Bibby, G.W., Ross, J.N.: Distributed optical fibre Raman temperature sensor using a semiconductor light source and detector. Electron. Lett. **21**, 569–570 (1985)
4. Kee, H.H., Lees, G.P., Newson, T.P.: 1.65 μm Raman-based distributed temperature sensor. Electron. Lett. **35**, 1869–1871 (1999)
5. Horiguchi, T., kurashima, T., Tateda, M.: Tensile strain dependence of Brillouin shift in silica optical fibres. IEEE Photon. Technol. Lett. **1** 107–108 (1989)
6. Soto, M.A., Sahu, P.K., Bolognini, G., Pasquale, F.: Di.: Brillouin based distributed temperature sensor employing pulse coding. IEEE Sens. J. **8**, 225–226 (2008)
7. Soto, M.A.: Bipolar pulse coding for enhanced performance in Brillouin distributed optical fiber sensors Proc. SPIE **8421**, 84219Y (2012)
8. Le Floch S. et al.: Colour Simplex coding for Brillouin distributed sensors. In: 5th EWOFS, Proceedings of the SPIE vol. 8794 8794–33 (2013)
9. Agrawal, G.P.: Fibre-Optic Communications Systems, 3rd edn. Wiley, New York (2002)
10. Pradhan, H.S., Sahu, P.K.: 150 km long distributed temperature sensor using phase modulated probe wave and optimization technique. Optik—Int. J. Light Electron Opt. **125**, 441–445 (2014)
11. Alahbabi, M.N.: Distributed optical fibre sensors based on the coherent detection of spontaneous Brillouin scattering. PhD Thesis, University of Southampton, Southampton, U. K. (2005)
12. Brinkmeyer, E.: Analysis of the back-scattering method for single-mode optical fibres. J. Opt. Soc. Am. **70**, 1010–1012 (1980)
13. Wait, P.C., Newson, T.P.: Landau Placzek ratio applied to distributed fibre sensing. Opt. Commun. **122**, 141–146 (1996)
14. Bansal, N.P., Doremus, R.H.: Hand book of glass properties. Academic Press, Chapter 2 (1986)
15. Bennett, W.R.: Electrical Noise. McGraw-Hill, New York (1960)
16. Robinson, F.N.H.: Noise and Fluctuations in Electronic Devices and Circuits. Oxford University Press, Oxford (1974)
17. Souza, K.D., Wait, P.C., Newson, T.P.: Characterisation of strain dependence of the Landau-Placzek ratio for distributed sensing. Electron. Lett. **33**, 615–616 (1997)

Comparative Performance Analysis of Different Segmentation Dropping Schemes Under JET Based Optical Burst Switching (OBS) Paradigm

Manoj Kumar Dutta

Abstract Optical burst switching (OBS) is the most promising technique to explore the huge bandwidth offered by the optical fiber. In OBS network sender does not wait for an acknowledgment of path reservation and uses only one-way reservation for data transmission. In spite of being very promising this one-way reservation policy of OBS technology may create contention during data transmission. Contention in the network leads to loss of packets as a result the efficiency of the optical network deteriorates. To achieve optimum performance of an OBS network it is essential to employ proper contention resolution techniques. This article investigates the contention resolution capability of segmentation based dropping scheme for JET based OBS network. Performances of three different segmentation burst dropping policies such as head dropping; tail dropping and modified tail dropping are discussed and compared for above mentioned networks. Both finite and infinite input cases are considered here.

Keywords OBS network · Contention resolution · Segmentation based dropping · Throughput · Carried traffic · Offered traffic

1 Introduction

OBS is assumed to be the most efficient switching paradigm for wavelength division multiplexing (WDM) system to explore the huge raw bandwidth available by an optical fiber [1–3]. Different signaling protocols are associated with optical burst switching scheme viz. just-enough-time (JET), just-in-time (JIT), tell-and-go (TAG), and tell-and-wait (TAW) [1, 4]. JET provides more efficient bandwidth utilization compared to JIT and TAW schemes, and a better QoS compared to TAG

M.K. Dutta (✉)
BIT Mesra, Deoghar Campus, Jharkhand 814142, India
e-mail: mkdutta13@gmail.com

A. Nagar et al. (eds.), *Proceedings of 3rd International Conference on Advanced Computing, Networking and Informatics*, Smart Innovation, Systems and Technologies 43, DOI 10.1007/978-81-322-2538-6_40

protocol, so JET-based OBS scheme is the best approach for deployment of optical burst switched networks [5].

OBS networks uses one-way reservation protocol and does not wait for the reservation acknowledgement of the entire optical path so there is always a chance of occurring contention in the intermediate nodes. Loss of burst means loss of data which in turn essentially degrades the performance of the network. So reduction of burst loss probability is essential in OBS network. Many dropping policy algorithms are reported in literature by scientific community. Conventional reactive contention resolution technique uses wavelength conversion, optical buffers and space domain approach [4]. However for these policies additional resources are required which may increase network cost and complexity. To avoid this problem another technique namely segmentation based dropping policy was adopted [6, 7]. In this scheme only the contending portion of any one bursts is dropped instead of dropping the entire burst. The advantage of this scheme is that the burst loss probability reduced significantly without using any addition hardware but by utilizing fragmented burst. Segmentation based dropping scheme involves mainly three different types of dropping schemes namely head dropping, tail dropping and modified tail dropping. In the present analysis all the three types of dropping policies are estimated to find the best possible one to achieve the optimum efficiency of a JET based OBS network using segmentation based dropping scheme [3].

2 Burst Segmentation

In OBS networks the data packets are first assembled to form a burst and then transmitted. Control packets are sent out-of band. The control packet includes different information such as destination address, length of the burst, quality of service requirements and offset time etc. In segmentation dropping scheme the data burst is re-divided into small segments which may include one or more than one packet. Every segment of a burst contains information such as portioning point, checksum, length and offset time.

If any contention occurs in the intermediate nodes only overlapped portion of any one burst is dropped instead of complete burst. The commonly used burst dropping policies includes head dropping and tail dropping. In case of head dropping the header part of the contending burst is dropped (Fig. 1a), and in case of tail dropping the tail part of the original burst is dropped (Fig. 1b). Head dropping is advantageous in sense of preemption less by next one [8, 9].

Advantage of tail dropping scheme over head dropping is that in case of tail dropping scheme the corresponding header packet is never being dropped so there is a very less chance of out of sequence delivery of the packets. The dropped portion of the segment is retransmitted later on. In case of modified tail-dropping

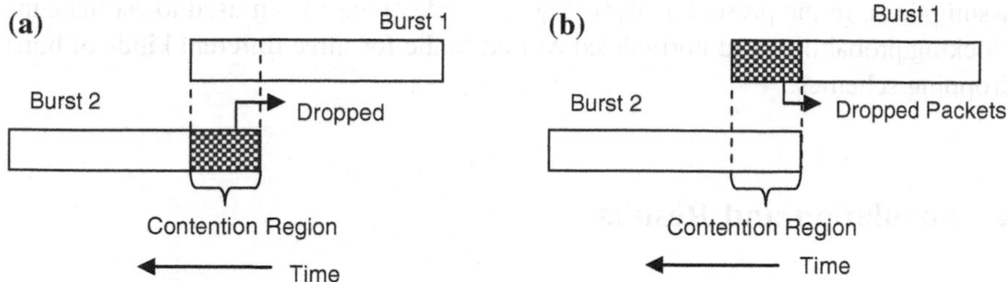

Fig. 1 Contention and corresponding segment dropping policy. **a** Head-dropping, **b** Tail-dropping

scheme, if the contending burst is having less number of segments than the original burst then the contending burst will be entirely dropped and vice versa. In this approach packet loss probability during the contention reduces significantly [10].

3 Mathematical Calculations for Blocking Probabilities

Considering Poisson arrival process the packet loss probability of a JET segmentation dropping with infinite channel is given by [7],

$$P_{JETseg(Infinite\ Input\ Channels)} = \frac{\sum_{i=1}^{\alpha} i.\gamma^{n+i} \frac{e^{-\gamma}}{(n+i)!}}{\gamma} \tag{1}$$

where γ is n.ρ ($\rho = \lambda/\mu$ is the offered load, $n = N_O W_O$ is total number of output channels), while remaining being the virtual channels. The virtual channels are used by the burst when all the real channels are busy. For finite input channels ($N_E W_E$) the packet loss probability of JET segmentation dropping scheme is given by,

$$P_{JETseg(Finite\ Input\ Channels)} = \sum_{k=n+1}^{N_E W_E} \left(\exp\left(-\frac{R}{b\beta}\right).\frac{k-n}{n}\frac{N_E W_E!}{k!.(N_E W_E - k)!}\frac{\left(\frac{\lambda}{\mu}\right)^k}{(1+\frac{\lambda}{\mu})^{N_E W_E}} \right) \tag{2}$$

where exp ($-R/b\beta$) is the node ideality factor and can be modeled as a function of bandwidth utilization (b), incoming data rate (R) and available bandwidth (β) [11].

These equations are used to find out the performance of an OBS network employing burst segmentation techniques. Efficiency of OBS network is usually measured by calculating blocking probability under the appropriate node and traffic assumptions. In the present analysis Eqs. (1) and (2) have been used to estimate the blocking probability and normalized carried traffic for three different kinds of burst dropping schemes.

4 Simulation and Results

Equations (1) and (2) have been used to carry out the simulations using MATLAB tools to measure the blocking probability reduction capacity of the segmentation based dropping scheme for different network parameters. Probability of blocking vs the offered load for three different dropping policies under consideration has been presented in Fig. 2, the result reveals that modified tail dropping scheme works most efficiently because modified tail dropping policy is having the advantage of very less out of sequence packet delivery probability and the dynamic dropping capacity of the smaller burst between the original and the contending burst.

For example, it can be seen that up to normalized offered traffic value of 0.4 the modified tail dropping scheme provides negligible blocking probability. The qualitative nature of blocking probability curves for all three dropping policies are almost similar but with a quantitative difference. Hence JET segmentation based modified tail dropping scheme is the best dropping technique among the three discussed.

Performance analysis of head dropping, tail dropping and modified tail dropping schemes of segmentation dropping based contention resolution technique for finite input channels and infinite input channels have been displayed in Figs. 3 and 4 respectively. The analysis was based on carried traffic vs normalized offered traffic. Result shows that all the dropping schemes are performing better for a system with finite input channels.

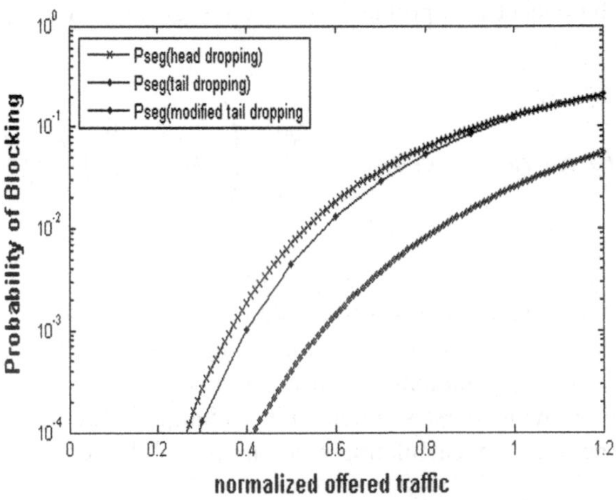

Fig. 2 Blocking probability versus normalized offered traffic for different dropping schemes

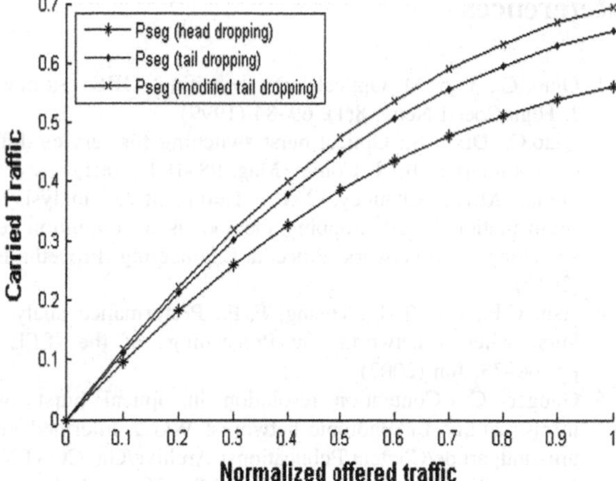

Fig. 3 Carried traffic versus offered traffic for Pseg (finite input channels)

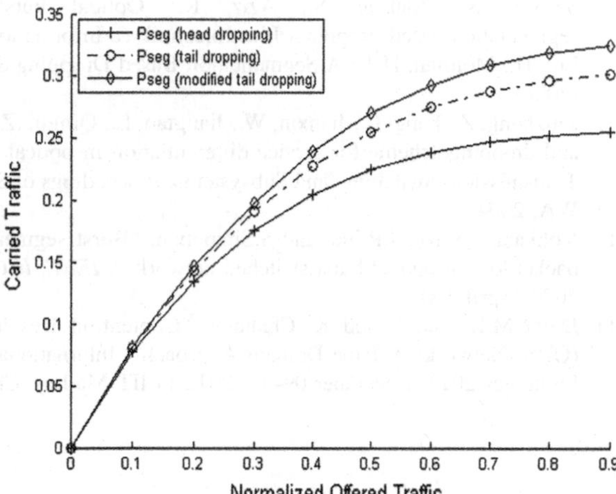

Fig. 4 Carried traffic versus offered traffic for Pseg (infinite input channels)

5 Conclusions

In this article the contention resolution by using segmentation based dropping scheme for JET based OBS network has been discussed. There are generally different types of burst dropping policies are possible. First by dropping the header of the contending burst, second by dropping the tail of the original burst and finally by dropping the smaller segment out of the contending and original burst. Effort has been given to compare the performance of three above mentioned dropping policies. Simulation result shows that the modified tail dropping provides the best contention resolution.

References

1. Qiao, C., Yoo, M: Optical burst switching (OBS)—a new paradigm for an optical Internet. J. High Speed Netw. **8**(1), 69–84 (1999)
2. Qiao,C., Dixit, S.: Optical burst switching for service differentiation in the next–generation optical internet. IEEE Comm. Mag. 98–104 (2001)
3. Dutta, M.K., Chaubey, V.K.: Comparative analysis of wavelength conversion and segmentation based dropping method as a contention resolution scheme in optical burst switching (obs) network. Procedia Engineering. Procedia Eng. Elsevier Publ. **30**, 1089–1096 (2012)
4. Hsu, C.F., Liu, T.-L., Huang, F.-F.: Performance analysis of deflection routing in optical burst-switched networks. In: Proceedings of the IEEE INFOCOM, vol. 1, New York, pp. 66–73, Jun (2002)
5. Gauger, C.: Contention resolution in optical burst switching networks. In: advanced infrastructures for photonic networks: WG 2 Intermediate Rep., Available: http://www.ikr.uni-stuttgart.de/Content/Publications/ Archive/Ga_COST266WG2_ 34 734.ps.gz, 2002
6. Haridos, K., Vokkarance, V., Jue, J.P.: Threshold-based burst assembly policies for QoS support in optical burst-switched networks. Opticomm, pp. 125–136, (2002)
7. Sarwar, S., Aleksic, S., Aziz, K.: Optical burst switched(OBS) system with segmentation-based dropping. Electrotechnik & Informationtechnik. **125**(7–8), (2008)
8. Lui, H., Mouftah, H.T.: A Segmentation-Based Dropping Scheme in OBS Networks. ICTON (2006)
9. Zhizhong, Z., Fang, C., Jianxin, W., Jiangtao, L., Qingji, .Z, Xuelei.X, "A new burst assembly and dropping scheme for service differentiation in optical burst switched networks", Optical Transmission, Switching, and Subsystems, Proceedings of SPIE Vol. 5281, SPIE, Bellingham, WA, 2004
10. Vokkarane V.M., J.P.Jue, and S.Sitaraman, "Burst segmentation: An approach for reducing packet loss in optical burst switched networks", *IEEE, ICC 2002*, New York, NY, pp.2673–2677, April 2002
11. Dutta M.K. and Vinod K. Chaubey. "Contention Resolution in Optical Burst Switching (OBS) Network: A Time Domain Approach", International Conference on Fiber Optics and Photonics 2012, December 09–12, 2012 in IIT Madras, Chennai, India

Free-Space Optical Communication Channel Modeling

G. Eswara Rao, Hara Prasana Jena, Aditya Shaswat Mishra and Bijayananda Patnaik

Abstract Free-space optical (FSO) systems have proved as the best technologies for communication and surveillance systems. These systems are merged with other technologies for providing robust applications. In FSO systems, when laser light is transmitted through the atmospheric channel, severe absorption is observed especially because of fog, snow, heavy rain etc. Therefore, we have discussed here different free space optical channel modelling techniques for mitigating these effects.

Keywords Free-space optics · Atmospheric turbulence · Rayleigh distribution, Gama–Gama modeling · Probability distribution function

1 Introduction

Out of many recent technologies free space optics (FSO) has proved as a vital technology in wireless communication systems. FSO is useful for outdoor links, short range wireless communications, and backbone of existing fiber optic communication systems [1, 2]. FSO technology is limited to the channel impairments. Unfortunately, the atmosphere has large number of noises, hence not an ideal communication channel [2]. Because of pressure and temperature variations of the atmosphere, it fluctuates the received signal amplitude and phase in the FSO systems and hence the bit error in the digital communication links [3]. Detailed

G. Eswara Rao (✉)
VITAM, Berhampur, India
e-mail: eswararaoskota@gmail.com

H.P. Jena · A.S. Mishra · B. Patnaik
IIIT, Bhubaneswar, India

A. Nagar et al. (eds.), *Proceedings of 3rd International Conference on Advanced Computing, Networking and Informatics*, Smart Innovation, Systems and Technologies 43, DOI 10.1007/978-81-322-2538-6_41

challenges of FSO communication systems in atmospheric turbulence effects given by Henniger and Wilfert [4]. Zhu and Kahn [5] chronicled assorted different techniques to minimise the intensity fluctuations caused by atmospheric turbulence and shown spatial diversity technique to nullify the fading effect, at different receivers and they have chronicled the use of maximum feasibility detection technique. Differential phase-shift keying (DPSK) technique is proposed by Gao at el. [6] to improve the receiver sensitivity of FSO systems. They have shown that under same channel constrains DPSK technique is better than on–off-keying (OOK) based techniques. Using the same bandwidth, theoretical analysis and numerical results are illustrated, OOK has a lower sensitivity than DPSK system, and thus DPSK format is very satisfactory for atmospheric channels and in wireless optical communication systems it has the broad anticipation [6].

2 Free Space Optical Communication Systems

In the proposed optical wireless communication system the major subsystems are shown in Fig. 1. LASER (light amplification by stimulated emission of radiation) beam is used as a wireless connectivity for the FSO system; free-space is used as the communication channel for carrying the data. In the received signal indiscriminate fluctuations, which is called as scintillation, is observed due to the atmospheric turbulence [7]. The performance of the FSO system is mainly affected by the free space channel. Thus the analysis of the free space channel is an important role for measuring the performance of the system. The three predominant parameters that affects the optical wave propagation in FSO systems are scattering, absorption, and changes in refractive index [8].

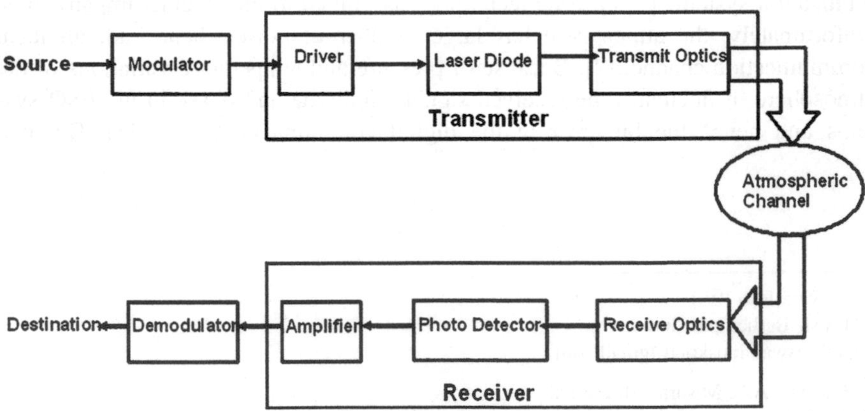

Fig. 1 Block diagram of FSO communication system

3 FSO Channel Modelling

The fading strength of the FSO channel depends on the (i) link distance, (ii) refractive index structure parameter C^2 of the medium and (iii) wavelength of the optical radiation. This model generally depends on Rytov variance σ^2. It is given by [7],

$$\sigma_R^2 = 1.23 C_n^2 p^{7/6} Z^{11/6} \quad (1)$$

where, p = 2π/λ is the optical wave number, C_n^2 is the refractive index structure parameter, Z is propagation link distance, which is a constant for horizontal paths.

3.1 Rayleigh Distribution

The Rayleigh model is used in FSO systems to describe the communication channel gain. For deeply faded channels the density function of the Rayleigh is highly concentrated. The scintillation index is 1 for the Rayleigh channel.

For Rayleigh distribution the PDF is given by [7],

$$f(I) = \frac{1}{\sigma_R^2} \exp\left\{\frac{I}{2\sigma_R^2}\right\}, \ \mathrm{I} > 0 \quad (2)$$

3.2 Gamma–Gamma Distribution

For the Gamma–Gamma distributed channel the PDF of the intensity fluctuation is given by [7]

$$f(L) = \frac{2(\alpha\beta)^{\frac{\alpha+\beta}{2}}}{\tau(\alpha)\tau(\beta)} L^{\frac{\alpha+\beta}{2}-1} K_{(\alpha-\beta)}(2\sqrt{\alpha\beta L}), \ \mathrm{L} > 0 \quad (3)$$

L is the intensity of the signal, $K_{(\alpha-\beta)}$ is the modified Bessel function of the second order, which denotes the scintillation of the uniform plane waves and during the time of zero-inner scale, it will be [7, 9],

$$\alpha = \frac{1}{\exp\left[\frac{0.49\sigma_R^2}{\left(1+1.1\sigma_R^{\frac{12}{5}}\right)^{\frac{7}{6}}}\right] - 1} \quad (4)$$

$$\beta = \frac{1}{\exp\left[\frac{0.51\sigma_R^2}{\left(1+0.69\sigma_R^{\frac{12}{5}}\right)^{\frac{5}{6}}}\right] - 1} \quad (5)$$

3.3 *Lognormal Distributions*

In the log-normal model, the log intensity of the laser light traverses the medium is normally distributed with an average value of $-\sigma^2/2$. The probability density function of the received irradiance is given by [10–12],

$$\mathrm{f(I)} = \frac{1}{2\pi\sigma_R^{2^{1/2}} I} \exp\left\{ - \frac{\left(\ln\left(\frac{I}{I_0}\right) + \sigma_R^2/2\right)^2}{2\sigma_R^2} \right\} \mathrm{I} > 0 \quad (6)$$

where, I_0 is the signal irradiance in the absence of scintillation, I represents the destination irradiance.

4 Results and Discussion

The probability density function (pdf) and irradiance for various channel models have been plotted. Figures 2 and 3 show the pdf and irradiance for Gamma–Gamma and Rayleigh channel models. It is observed that that the Gamma–Gamma model

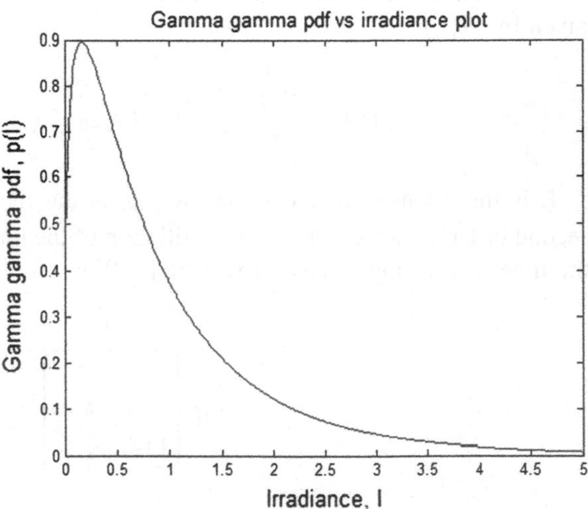

Fig. 2 Pdf and irradiance for Gamma–Gamma channel

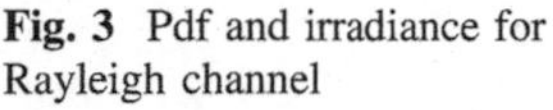

Fig. 3 Pdf and irradiance for Rayleigh channel

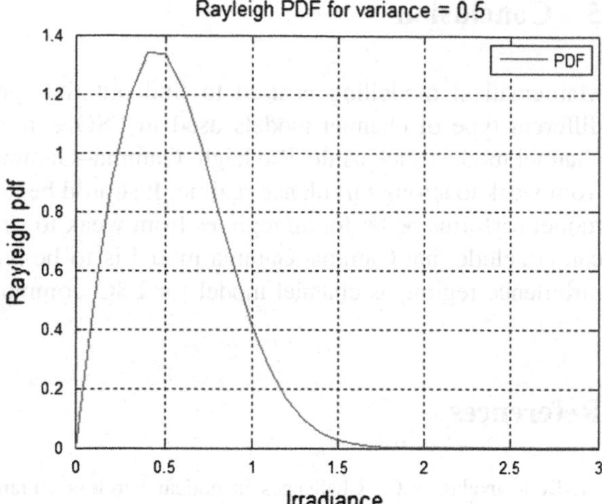

performs better for all regimes from weak to strong turbulence region. Figure 4 shows the output response of the multipath Gamma–Gamma modeled channel for the input PPM signal.

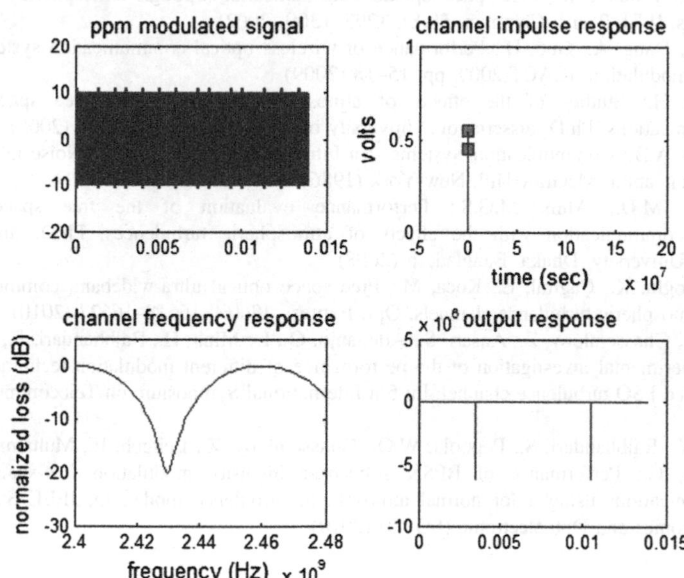

Fig. 4 Response of the multipath Gamma–Gamma modeled channel for the input PPM signal

5 Conclusion

Mathematical modelling is used to find out the symbol error probability under different type of channel models used in FSO communication for the turbulence channel models such as the Rayleigh, Gamma–Gamma model distribution are valid from weak to strong turbulence regime. It should be noted that the Gamma–Gamma model performs better for all regimes from weak to strong turbulence region. So we can conclude that Gamma–Gamma model is to be preferred under weak to strong turbulence regime as channel model for FSO communication systems.

References

1. Boucouvalas, A.C.: Challenges in optical wireless communications. Opt. Photonics News **16**(5), 36–39 (2005)
2. Abdullah, M.F.L., Bong, S.W.: Adaptive differential amplitude pulse-position modulation technique for optical wireless communication channels based on fuzzy logic. IET Commun. **8**(4), 427–432 (2014)
3. Akiba, M., Walkmori, K., Kodate, K., Ito, S.: Measurement and simulation of the effect of snow fall on free space optical propagation. Appl. Opt. **47**(31), 5736–5742 (2008)
4. Henniger, H., Wilfert, O.: An introduction to free-space optical communications. Radio Eng. **19**(2), 203–230 (2010)
5. Zhu, X., Kahn, J.M.: Free-space optical communication through atmospheric turbulence channels. IEEE Trans. Commun. **50**(8), 1293–1300 (2002)
6. Gao, S., Dang, A., Guo, H.: Performance of wireless optical communication systems using DPSK modulation. ICACT-2009, pp. 15–18 (2009)
7. Yuksel, H.: Studies of the effects of atmospheric turbulence on free space optical communications. Ph.D. dissertation, University of Maryland, College Park (2005)
8. Carlson, A.B.: Communication Systems: An Introduction to Signals and Noise in Electrical Communication. McGraw-Hill, New York (1986)
9. Hossen, M.D., Alim, M.G.S.: Performance evaluation of the free space optical (FSO) communication with the effects of atmospheric turbulences. Ph.D. dissertation, BRAC University, Dhaka, Bangladesh (2008)
10. Davaslıoğlu, K., Çağıral, E., Koca, M.: Free space optical ultra-wideband communications over atmospheric turbulence channels. Opt. Express **18**(16), 16618–16627 (2010)
11. Ijaz, M., Ghassemlooy, Z., Ansari, S., Adebanjo, O., Le Minh, H., Rajbhandari, S., Gholami, A.: Experimental investigation of the performance of different modulation techniques under controlled FSO turbulence channel. In: 5th International Symposium on Telecommunications (2010)
12. Tang, X., Rajbhandari, S., Popoola, W.O., Ghassemlooy, Z., Leitgeb, E., Muhammad, S.S., Kandus, G.: Performance of BPSK subcarrier intensity modulation free-space optical communications using a log-normal atmospheric turbulence model. In: IEEE Symposium on Photonics and Optoelectronic (SOPO) (2010)

Wavelet Based RTL-SDR Real Time Signal Denoising in GNU Radio

U. Reshma, H.B. Barathi Ganesh, J. Jyothi, R. Gandhiraj and K.P. Soman

Abstract Noise removal is considered to be an efficacious step in processing any kind of data. Here the proposed model deals with removal of noise from aperiodic and piecewise constant signals by utilizing wavelet transform, which is being realized in GNU Radio platform. We have also dealt with the replacement of Universal Software Radio Peripheral with RTL-SDR for a low cost Radio Frequency Receiver system without any compromise in its efficiency. Wavelet analyzes noise level separately at each wavelet scale in time-scale domain and adapts the denoising algorithm especially for aperiodic and piecewise constant signals. GNU Radio companion serves well in analysis and synthesis of real time signals.

Keywords Signal denoising · Wavelets · Continuous wavelet transform · Multirate signal processing · GNU radio companion

U. Reshma (✉) · H.B. Barathi Ganesh · J. Jyothi · K.P. Soman
Centre for Excellence in Computational Engineering and Networking,
Amrita Vishwa Vidyapeetham, Coimbatore 641112, Tamil Nadu, India
e-mail: reshma.anata@gmail.com

H.B. Barathi Ganesh
e-mail: barathiganesh.hb@gmail.com

J. Jyothi
e-mail: jyothijammigumpula@gmail.com

K.P. Soman
e-mail: kp_soman@amrita.edu

R. Gandhiraj
Department of Electronics and Communication Engineering,
Amrita Vishwa Vidyapeetham, Coimbatore 641112, Tamil Nadu, India
e-mail: r_gandhiraj@cb.amrita.edu

A. Nagar et al. (eds.), *Proceedings of 3rd International Conference on Advanced Computing, Networking and Informatics*, Smart Innovation, Systems and Technologies 43, DOI 10.1007/978-81-322-2538-6_42

1 Introduction

Problem proposed is removal of noise from an aperiodic and PieceWise Constant (PWC) signal. These signals occur in many contexts like bioinformatics, speech signals, astrophysics, geophysics, molecular biosciences and digital imaginary. Preprocessing of above said data is an undeniable paramount in any vital means of communication in present epoch. Each data has its own attribute and are essentially not the same. While denoising these attributes are considered based on the data chosen and denoised accordingly. The data chosen here is a real time 1D signal which is affected by noise by itself during transmission, or on recording the same. The salient idea behind the pre-processing stage is to retain the default quality of the data chosen rather than the re-styled one. In recent years many noise reduction techniques have been developed to produce a commotion free signal from the noise signal. Some of the algorithms used are Least Mean Squares Filtering, Adaptive Line Enhancer, Spectral Subtraction, Wiener filter, subspace, and spatial temporal prediction and much more [1, 2]. In PWC signal sudden variation in signal amplitude pose a fundamental challenge for conventional linear methods, e.g. Fast Fourier Transform (FFT) based noise filtering. While doing the above methods, signals converge slowly that is magnitudes of Fourier coefficients reduce much slower with increase in frequency. All these algorithms have their own pros and cons like reduced recoverability of data, computational complexity, less efficiency in processing real time signal and so on, which led to propose a wavelet transform based denoising method which offers simplicity and efficiency. Smoothing refers to removing high frequency components and retaining the low frequency components. Wavelet based denoising is not a smoothing function [3]. Wavelet retains the original signal regardless of the frequency content of the signal. The point to be stressed is that most of these algorithms are analyzed in MATLAB environment but introducing the same in GNU Radio can be applied more effectively for a real time signal [4]. In most cases Universal Software Radio Peripheral (USRP) and GNU Radio clubs up together to develop a real time Software Defined Radio (SDR) system. Since the outcome of such a combination were exorbitant this system could not be used in labs, research institutes, hobbyists and other commercial areas. RTL-SDR was then utilized for the same purpose, but this could not provide the efficiency (8 bit Anaolog to Digital Converter) that USRP (Typically 12 bit ADC) could afford to. Hence this work proposes an idea to overcome the above problem by introducing wavelet transform based denoising a signal which will be a key term for replacing USRP with RTL-SDR.

2 Literature Survey

Various researchers have proposed wavelet based denoising on 1D and 2D signals based on traditional way of thresholding and wavelet shrinkage which are unavailing for signals with high frequency noise content. Most of these methods were simulated in MATLAB which are exorbitant for commercial and educational basis. Moreover processing of real time signals becomes ineffectual. So a suitable platform and adequate algorithm is required. Slavy G. Mihov, Ratcho M. Ivanov and Angel N. Popov have proposed a denoising technique using wavelet transform which was applied on a large dedicated database of reference speech signals contaminated with various noises in several SNRs [5]. S. Grace Chang, Bin Yu and Martin Vetterli proposed a spatially adaptive wavelet thresholding method for image denoising. This is based on context modeling, a common technique used in image compression to adapt the coder to changing image characteristics [6]. Sylvain Sardy, Paul Tseng and Andrew Bruce have proposed a robust wavelet oriented predictor by having an optimizer model. Which is a nontrivial optimization problem and solved by choosing the smoothing and robustness parameters [7]. Sylvain Durand, Jacques Froment proposes a traditional wavelet denoising method based on thresholding and to restore the denoised signal using a total variation minimization approach [8]. It was made sure that the restoration process does not progressively worse the context information in the signal that has to be considered as a remark in the denoising step [8]. By observing this, wavelet transform based real time signal denoising is proposed in GNU Radio.

3 Wavelet Transform in GNU Radio for Denoising

3.1 Noise

Data captured from the real world known as signals do not exist without noise. In order to recover the original signal, noise must be removed. This task is more complex when the signal is aperiodic or piecewise constant in nature. The linear method achieved by low pass filtering is effective if and only if the signal to be recovered is sufficiently smooth, but it is not enough since piecewise constant signals are likely to introduce Gibbs phenomena while doing low pass filtering. By using Fourier transform aperiodic and signal overlapping cannot be substantially recovered from the noise affected signal in time-frequency domain. Wavelet transform in time-scale domain best suits for effective recovery of noise free signal from the contaminated one. In recent years Software Defined Radio has become an emerging field where it uses software instead of hardware components with Digital Signal Processor (DSP) or Field Programmable Gate Array (FPGA). SDR results in hardware reconfiguration by means of updating adoptable software written in DSP or FPGA. Therefore, SDR is a preferable choice in developing a wireless system instead of a traditional hardware radio when reconfigurability is required. At receiver end when SDR is built using RTL-SDR it results in reduced reconstruction performance when compared to SDR with USRP. This will introduce the noise occurrence in the signal mainly on flat top region of the signal. Wavelet transform based denoising will be solution to the above problem.

3.2 Wavelet

Wavelets could be dealt only with a prior knowledge in function spaces. In vector space, the vectors are represented as orthogonal unit vectors called bases. Fourier series is an extension of this idea for functions which are generally called as signals. Function is quantity which varies with respect to one or more running parameter, usually time and space. On the other hand Fourier transform gets struck with time-frequency localization. The above scenarios could be overcome using Wavelets. Similar to vector, function has its orthogonality characteristics and these orthogonal functions spans the function space. Wavelet obeys the above property in time-scale domain. Signal manipulation and interrogation using wavelets were developed primarily for the signals which are aperiodic, noisy, intermittent and transient. Current applications of wavelets include climate analysis, financial time series analysis, heart monitoring, condition monitoring of rotating machinery, seismic signal denoising, denoising of astronomical images, crack surface characterization, characterization of turbulent intermittency, audio and video compression, compression of medical and thump impression records, fast solution of partial differential equations, computer graphics and so on [3]. Based on the application continuous or discrete wavelet transform could be used. Here denoising is done using a continuous wavelet transform. Wavelets are waves that don't last forever or in other words functions that are not predominant forever. It is significant which means exists only in particular range. Mathematically wavelet is represented as [3],

$$\psi_{a,b}(t) = \frac{1}{\sqrt{|a|}}\psi\left(\frac{t-b}{a}\right). \tag{1}$$

where, a and b are scaling and translation parameters which when varied gives new scaling and translation functions. The transform is performed by placing various scales of wavelet in various positions of the signal. This transforms the signal into another representation which will be more useful form, for analyzing the signals which is known as wavelet transform. By performing this in smooth and continuous manner it becomes a Continuous wavelet transform (CWT). Wavelet transform is defined as [3],

$$w(a,b) = \int f(t)\frac{1}{\sqrt{|a|}}\psi\left(\frac{t-b}{a}\right)dt. \tag{2}$$

Unlike Fourier transform, we have a variety of wavelets that are used for signal analysis. Choice of a particular wavelet depends on the type of application used which will be dealt in the next section.

3.3 Multirate Signal Processing

One way to perform wavelet transform is through the use of different prediction and multirate signal processing steps. Here the predictor is type of wavelet transform used and Multirate signal processing blocks are used to form the scaling functions and filter banks. The complexity of Digital Signal Processing applications are reduced by altering the sampling rate used at different stages there by providing cost effective Digital Signal Processing structures [9]. The most basic blocks in multirate signal processing are M-fold decimator and L-fold interpolator. Wavelet transform is performed through interconnection of decimators, interpolators and filters by considering the rules that governs its interconnection [10]. Decimation is the process by which the sampling rate is reduced by an integer D without resulting in aliasing. Performing decimation mathematically is expressed as [9],

$$\emptyset_D(m) = \begin{cases} 1, & m = Dm, D \in N^+, \\ 0, & otherwise \end{cases} \tag{3}$$

$$x_d[n] = x[Dn]. \tag{4}$$

The M-fold decimator only keeps one out of every M elements of x[n] and ignores the rest. Interpolation is the inverse of the decimation where it is used to increase the sampling rate of a signal without changing its spectral component. This process will reduce the time period between the samples of the signal. Interpolation via zero insertion between the original sample signals will be mathematically expressed as [9] (Fig. 1),

$$y(n) = \begin{cases} x(\frac{n}{L}), \; n = 0, \pm L, \pm 2L, \ldots. \\ 0, \quad otherwise \end{cases} \tag{5}$$

3.4 GNU Radio

GNU Radio as a Software Defined Radio, which operates in multimode and performs multiple functions of receiving, processing and transmitting both real and non-real signals. A typical application of multirate digital signal processing in GNU

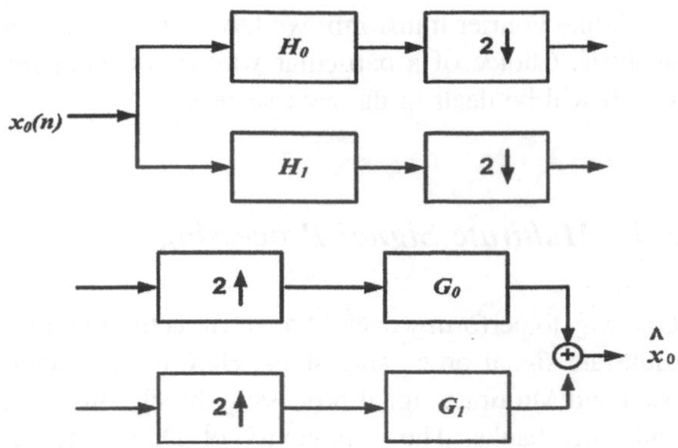

Fig. 1 Analyzer and synthesizer

Radio applications provides the tools like filter, decimator and and interpolator [11]. These tools could be used for performing wavelet transform in GNU Radio. Using available blocks in GNU Radio denoising the signal is effective by having wavelet transform. USRP is a range of SDR designed for development of transceiver system which could be controlled and used in GNU Radio platform. This set up results in an expensive system typically in the range of Lakhs. RTL-SDR is a substitute for USRP which generally has reduced efficiency when it comes to reconstruction of signals in receiver end. This typically ranges in few thousands and best suited for commercial purpose if modified for better efficiency which could be possibly achieved by inserting wavelet transform based denoising in receiver end.

4 Proposed Model

Before denoising a signal it is prerequisite for processing the signal i.e. information about signals maximum frequency and selection of wavelet. As mentioned in Sect. 3.2, it is necessary to select the required wavelet depending upon application for effective result and reconstruction. Miss selection of wavelets may lead to discontinuity in the signal. For example to work with the signals with flat tops Haar wavelets are relevant. Daubechies and Coiflet wavelet could be used for signals with sharp edges. Smooth signals use Symlet, Gaussian-Spline or Morlet wavelets. In this case the PWC signal are denoised using Haar wavelets. Speech signal and the signals which are received through RTL-SDR and USRP are denoised using Symlets. Maximum frequency information about a signal must be known in prior to avoid aliasing while applying scaling operation. Since scaling is done using a decimator sampling rate and maximum frequency information of a signal are taken

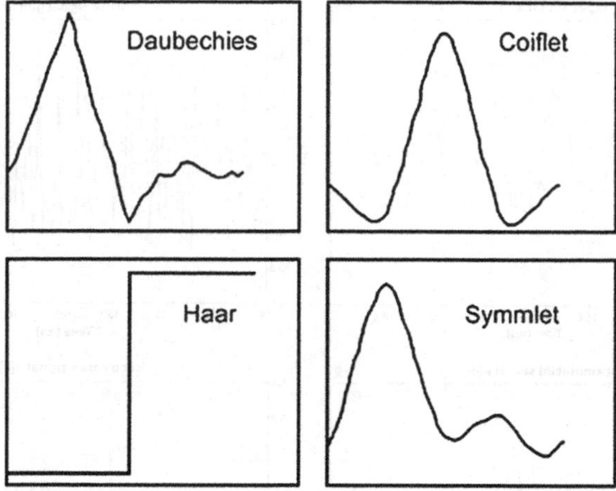

Fig. 2 Different wavelet functions

into consideration. The proposed model is shown below, Consider the following signal (Fig. 2),

$$e[n] = x[n] + g[n]. \quad (6)$$

To recover the original signal x[n] from contaminated signal e[n] the noise g[n] must be removed. The non-smooth signal undergoes wavelet analysis which is achieved by using decimator, low and high pass filters. Changing the tapping coefficients in decimator will result in a different wavelet functions and these are scaled by placing decimators in cascaded structure. The approximation order is K if all its wavelets have K vanishing moments [3]. That is,

$$\int t^m \psi(t)dt = 0,\ i = 0,1; 0 \leq m \geq K - 1. \quad (7)$$

In this case K = 4 when and sampling rate is 32 kHz. As the number of decimators gets increased new function spaces will be introduced and this could be mathematically represented as,

$$\ldots..V_0 \subset V \subset \ldots V_\infty \quad and \quad \ldots W_0 \perp W_1 \perp W_2 \ldots \quad (8)$$

where W_n is the complementary space of V_n. In the synthesis part thresholding of high frequency coefficients i.e. approximated coefficients will take place to smoothen the contaminated signal. Smoothing takes place while reconstructing

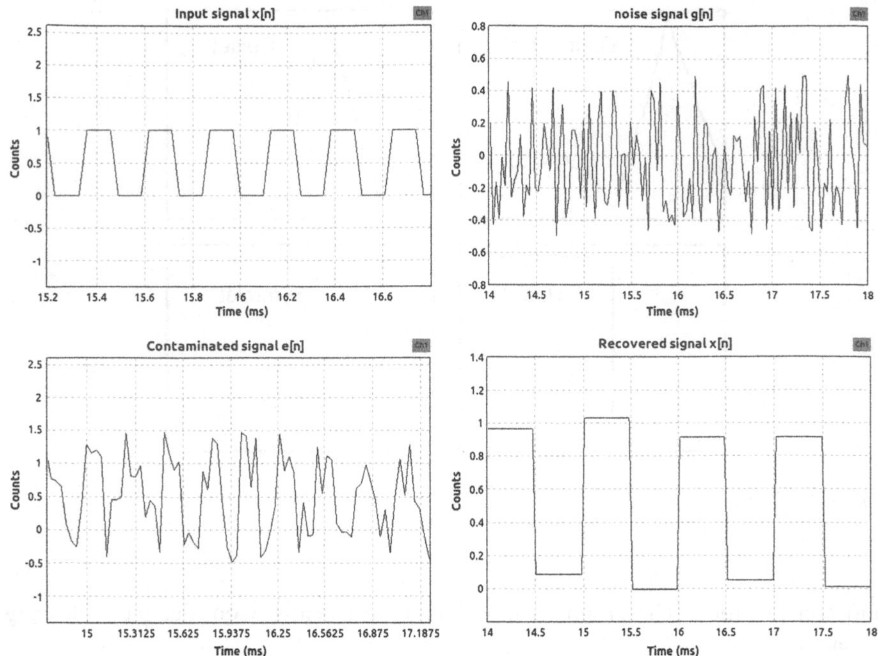

Fig. 3 Piecewise constant signal denoised in GNU Radio

signal from coarser scale to finer scale. In simple words we omit the coefficients required to reconstruct the high frequency components. For example,

$$V_j = V_{j-1} + W_{j-1} \quad and \quad V_j = V_{j-1} + CW_{j-1}. \tag{9}$$

where, C is the Thresholding factor.

5 Experimental Results

The test signals are transmitted using USRP (USRP N210-Carrier Frequency = 100 MHz) and received through RTL-SDR (RTL2832) and USRP. Further, denoising is done using wavelet transform in GNU Radio (Fig. 4). After denoising, the Mean Square Error (MSE) rate of these signals were calculated. Results showed that there was only small (1.38 dB) variation on receiving the same signal by means of RTL-SDR and USRP after denoising. Respective MSE rates were RTL-SDR

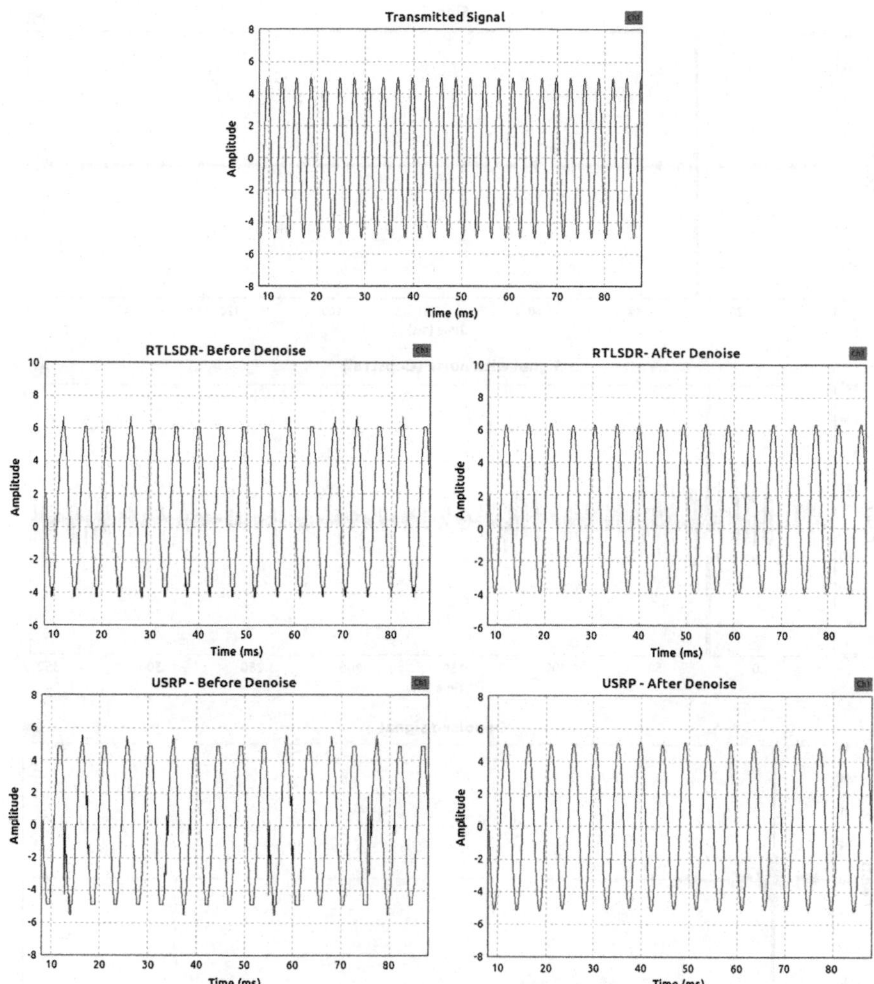

Fig. 4 Received RTL-SDR and USRP signals denoised in GNU Radio

4.997246 dB and USRP 4.997137 dB which shows that replacement of USRP with RTL-SDR shows almost no variation in system efficiency when wavelet transform based denoising applied in the receiver end for efficient signal reconstruction. Two more test signals [Speech signals (Figs. 5 and 6) and PWC signals (Fig. 3)] were included to check the performance of the proposed model. It showed great performance in denoising these signals.

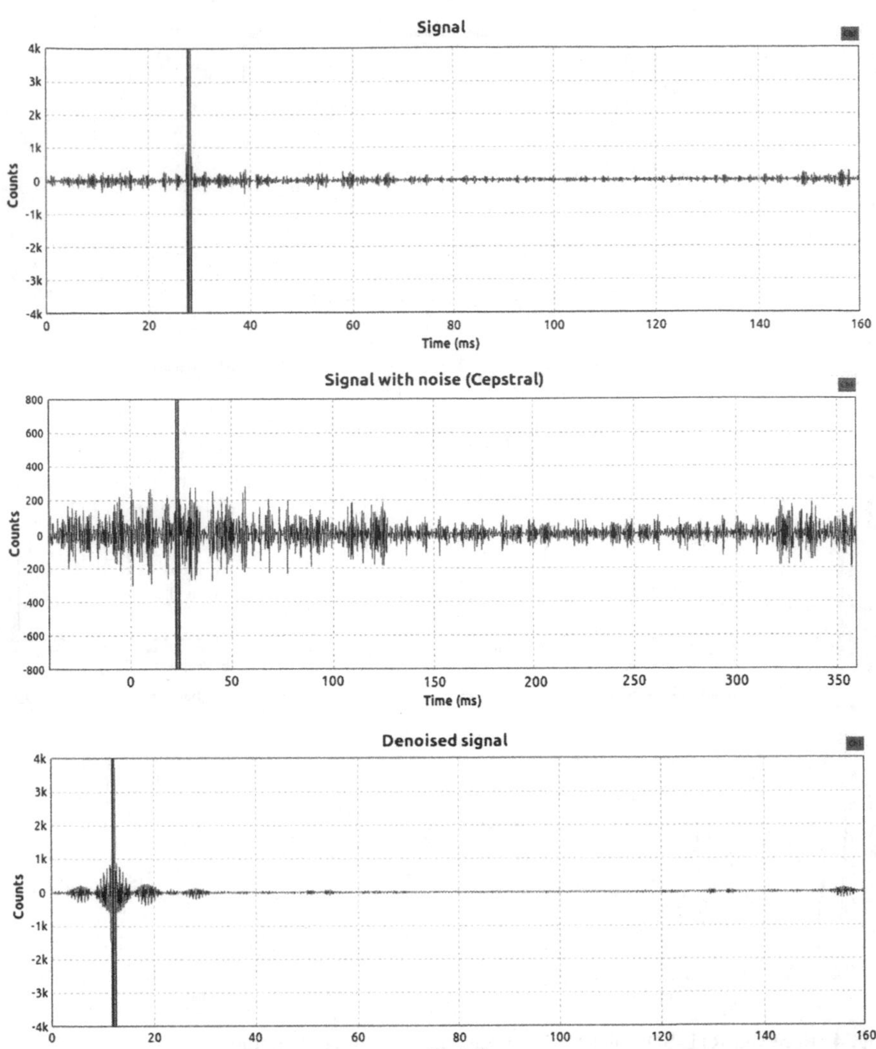

Fig. 5 Denoised Speech signal in Cepstral domain

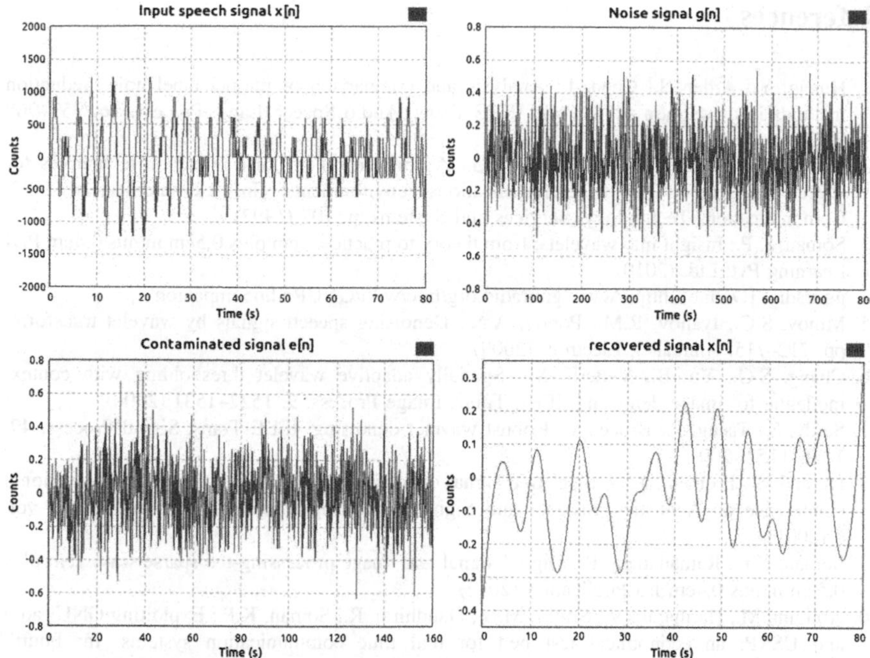

Fig. 6 Speech signal denoised in GNU radio

6 Conclusion

Various aperiodic and piecewise constant signals that were corrupted with noise was effectively retained by using wavelet transform based denoising in GNU Radio platform. Though there were methods which previously existed for the removal of noise from a 1-Dimensional signal in GNU Radio, this proposal proves to be simple and effective with less computational steps. There was only a small variation in mean square error rate (1.38 dB) on receiving the same signal by means of RTL-SDR and USRP after denoising. Hence a commercial cost effective SDR could be built by replacement of USRP with RTL-SDR at receiver end. As we have dealt with real time signals this method could be used in various applications like reception of NOAA satellite signal and FM signals. Denoising an image by receiving NOAA satellite signal may prove to be one such application for future development.

References

1. Huang, Y., Arden, J., Chen, J.: Analysis and comparison of multichannel noise reduction methods in a common framework. IEEE Trans. Audio Speech Lang. Process. **16**, 957–968 (2008)
2. Maher, A.G., King, R.W., Rathmell, J.G.: A comparison of noise reduction techniques for speech recognition in telecommunications environments. In: Communications' 92: Communications Technology, Services and Systems, p. 107 (1992)
3. Soman, K.P.: Insight into wavelets: from theory to practice. 1em plus 0.5em minus 0.4em PHI Learning Pvt. Ltd. (2010)
4. gnuradio. [Online]. http://www.gnuradio.org/trac/wiki/GNURadiocompanion
5. Mihov, S.G., Ivanov, R.M., Popov, A.N.: Denoising speech signals by wavelet transform, pp. 712–715. Annual, J. Electron. (2009)
6. Chang, S.G., Yu, B., Vetterli, M.: Spatially adaptive wavelet thresholding with context modeling for image denoising. IEEE Trans. Image Process. **9**, 1522–1531 (2000)
7. Sardy, S., Tseng, P., Bruce, A.: Robust wavelet denoising. IEEE Trans. Signal Process. **49**, 1146–1152 (2001)
8. Durand, S., Froment, J.: Artifact free signal denoising with wavelets. In: IEEE International Conference on Acoustics, Speech, and Signal Processing, Proceedings (ICASSP'01), vol. 6 (2001)
9. Soman, K.P., Ramanathan, R.: Digital signal and image processing-the sparse way. 1em plus 0.5em minus 0.4em Isa Publication (2012)
10. Abirami, M., Hariharan, V., Sruthi, M.B., Gandhiraj, R., Soman, K.P.: Exploiting GNU radio and USRP: an economical test bed for real time communication systems. In: Fourth International Conference on Computing, Communications and Networking Technologies (ICCCNT). IEEE (20130
11. Gandhiraj, R., Ram, R., Soman, K.P.: Analog and digital modulation toolkit for software defined radio. Procedia Eng. **30**, 1155–1162 (2012)

Part V
Emerging Techniques in Computing

A New Way for Combining Filter Feature Selection Methods

Waad Bouaguel and Mohamed Limam

Abstract This study investigates the issue of obtaining stable ranking from the fusion of the result of multiple filtering methods. Rank aggregation is the process of performing multiple runs of feature selection and then aggregating the results into a final ranked list. However, a fundamental question of is how to aggregate the individual results into a single robust ranked feature list. There are a number of available methods, ranging from simple to complex. Hence we present a new rank aggregation approach. The proposed approach is composed of two stages: in the first we evaluate he similarity and stability of single filtering methods then, in the second we aggregate the results of the stable ones. The obtained results on the Australian and German credit datasets using support vector machine and decision tree confirms that ensemble feature ranking have a major impact in the performance improvement.

Keywords Feature selection · Credit scoring

1 Introduction

The principal purpose of the dimensionality reduction process is, given a high dimensional dataset $\mathbf{x}_i$, $i \rightsquigarrow x_i = (x_i^1, x_i^2, \ldots, x_i^d)$ that describes a target variable Y_i using d features, to find the smallest pertinent set of features $\mathbf{X} = (X^1, X^2, \ldots, X^d)$, which represent the target variable as all the original set of features do [1, 2]. The process of feature selection in one of the most important task in the pre-analyse that not only consists in finding a reduced set of feature but also the choice of appropriate set based on their pertinence to the study [3].

W. Bouaguel (✉) · M. Limam
LARODEC, ISG, University of Tunis, Tunis, Tunisia
e-mail: bouaguelwaad@mailpost.tn

M. Limam
e-mail: mohamed.limam@isg.rnu.tn

M. Limam
Dhofar University, Salalah, Oman

A. Nagar et al. (eds.), *Proceedings of 3rd International Conference on Advanced Computing, Networking and Informatics*, Smart Innovation, Systems and Technologies 43, DOI 10.1007/978-81-322-2538-6_43

We consider the case of a financial dataset containing data about credit applicants. The class feature, is represented by the solvability level of an applicant, who can be credit worthy or not credit worthy. The class of solvability is assigned to each credit applicant by the credit mangers of the bank. Each customer is represented through a set of features that represent his current credit situation, the past credit history, duration of credits in months, behavior repayment of other loans, value of savings or stocks, stability in the employment, etc. [4].

In general, classification methods use collected information of each credit applicant to classify new ones. Feature reduction, in this context, is employed to find the minimal set of feature which can be used to represent the class of a new credit applicant.

Irrelevant and redundant features decrease the classification performance because they are usually mixed with the relevant ones witch confuse classification algorithms [5], feature selection is useful in this case in order to construct robust predictive models. Many feature selection methods are proposed in literature such as filter and wrapper methods [5]. Filter methods study the fundamental properties of each feature independently of the classifier [6]. In opposite to filters, wrappers use the classifier accuracy to evaluate the feature subsets [7]. Wrappers are the most accurate, but in this case accuracy comes with an exorbitant cost caused by repetitive evaluation [5].

According to [8] filter methods outperforms wrapper methods in many cases. However with the huge number of classical filter methods is difficult to identify which of the filter criteria would provide the best output for the experiments [9, 10]. Then, the best approach is to perform rank aggregation.

According to [11] rank aggregation improve the robustness of the individual feature selection methods such that optimal subsets can be reached [12]. Rank aggregation have many merits. However, an important number of different rank aggregation methods have been proposed in the literature witch make the choice difficult [11].

Thus, this paper discusses the major issues of filter approach and presents a new approach based on rank aggregation. Evaluations on a credit scoring problem demonstrate that the new feature selection approach is more robust and efficient.

This paper is organized as follows. Section 1 briefly reviews majors issues of filter feature selection and give the most famous filtering techniques. Section 3 describes our proposed approach. Experimental investigations and results on two datasets are given in Sect. 4. Finally, Sect. 5 provides conclusions.

2 Filter Framework and Rank Aggregation

In this section we will try to give an overall description of some of the most popular univariate filtering methods. Filter methods have many advantages but the most obvious ones are their computational efficiency and feasibility [13]. This advantage allows decision makers to create a complete picture of the available information by examining the data from different angles of various filtering approaches without

Table 1 Popular filter feature selection methods

Distance	Dependence	Information
Euclidean distance: Measure the root of square differences between features of a pair of instance	**Pearson**: Measure of linear dependence between two variables	**Mutual Information**: Measure the amount of information shared between two features
Relief: Measure the relevance of features according to how well their values separate the instances of the same and different classes that are near each other	**Chi-squared**: Measure the statistical independence of two events	**Information Gain**: Information gain but normalized by the entropy of an attribute. Addresses the problem of overestimating the features with multiple values

increasing the computational complexity and that what makes filter methods extremely effective in practice.

As discussed before, filters select relevant features regardless of the classification algorithm using a independent evaluation function. According to Dash [14], these independent evaluation functions may be grouped into four categories: distance, information, dependence and consistency, where the first three are the most used [15]. Each category have its own specificity and may have large number of filtering methods. Table 1 give the list of the most popular filter feature selection methods.

Filtering methods or further rankers choose one of the independent function discussed before to rank feature according to their relevance to the class label by giving a score to each feature. In general a high score indicates the presence of a pertinent and relevant feature and all features are sorted in decreasing order according to their scores [16]. Many ranks are available in the literature, making the choice difficult for a particular task [8]. According to [17, 18] there is no single best feature ranking method and can not chose the most appropriate filter, unless we evaluate all existing rankers, which is impossible to realise in most domains.

According to [18] rank aggregation which is an ensemble approach for filter feature selection that combine the results of different rankers, might produce a better and more stable ranking than individual rankings. Hence, in this work we investigate a new method combining several rankings.

3 Proposed Approach for Combining Filter Feature Selection Methods

3.1 *First Ranking for Single Filters and Stability Control*

Many studies show that the stability of an ensemble feature selection model is a curtail topic that influences the final result and the future classification [19]. According to [12] a stable feature selection method is preferred over unstable one in the construction of the final ensemble. Hence we begin by reducing statistical variations of each individual filter in order to retain just the stable ones. According to [19] the stability of each ranker may be quantified by its sensitivity to the variations in the training set. Hence we quantify the stability of each filter by the ranks they give to each feature on several iterations. Each filter was run 10 times for each dataset. In each run a feature ranking is obtained for each filtering method.

According to [19] the stability of a ranker can be measured using a measure of similarity for the ranking representation. Then, we use the Spearman footrule distance [20] as a simple way to compare two ordered lists. Spearman distance between two lists is defined as the sum overall the absolute differences between the ranks of all features from both lists [7]. According to [7, 20], As the Spearman value decreases as the similarity between the two lists increases. Then, the final stability score is the mean of similarity over all the lists for the evaluated filter. Once the final stability score is computed for each filter, we choose the stable ones for the next step. Hence we compare stability score with a threshold of 80 %. If the stability of a filter is less or equal to 80 % the selected filter is conserved for aggregation else it is considered as no stable and eliminated. Figure 1 illustrates the process of choosing the stable filters.

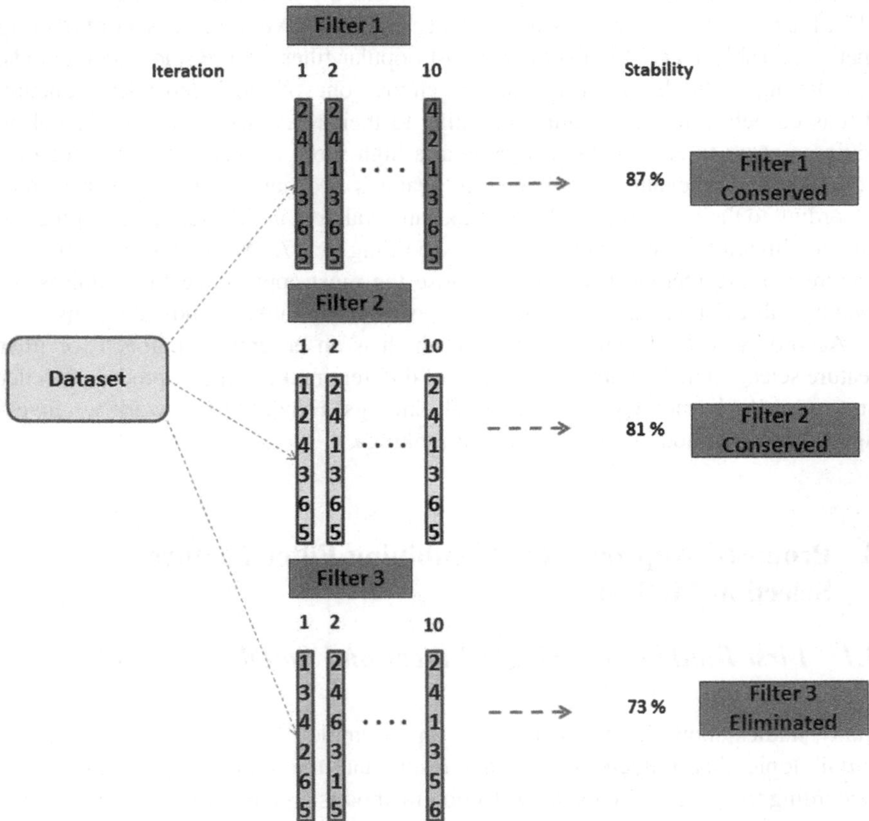

Fig. 1 The process of choosing the stable filters

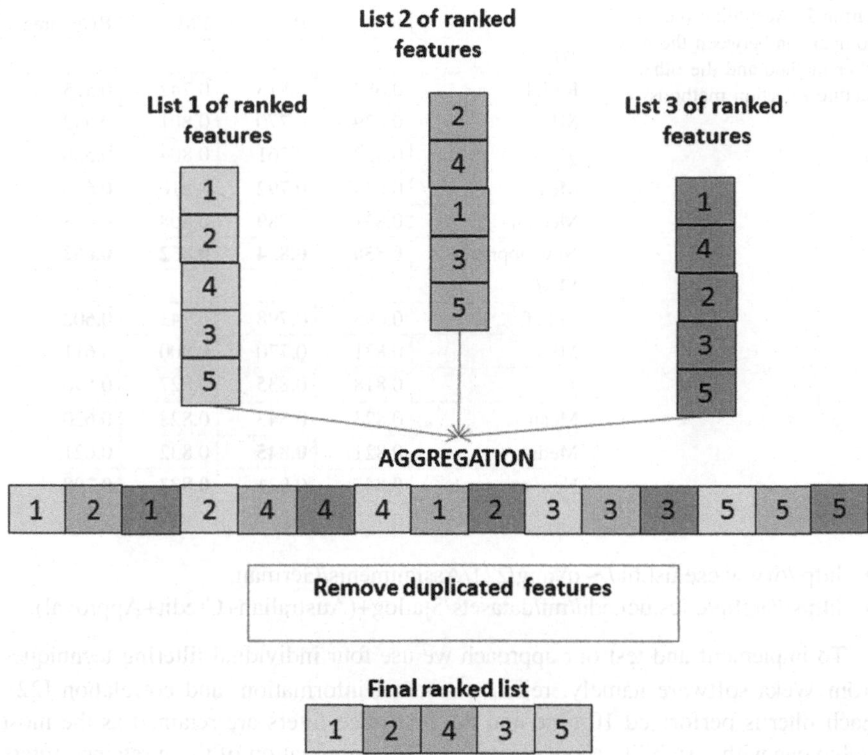

Fig. 2 The fusion process

3.2 *Merging Different Filter Methods*

Once the most stable filter are selected by the previous stage we move to their combination to provide a more robust result where the issue of selecting the appropriate filter is alleviated to some level [12].

Several rankers are independently applied to find different ranked lists of the same size. Then, these lists of features are merged by selecting feature by feature from each list, starting from the feature on the top of each list and so on [21]. Figure 2 illustrates the fusion process for an example of three filters.

4 Experimental Investigations

4.1 *Datasets Description and Performance Measures*

The adopted herein datasets used for evaluation are the Australian and German credit datasets from the UCI repository of machine learning, available on these links:

Table 2 Australian dataset: comparison between the new filter method and the other feature selection methods

	P	R	FM	ROC area
DT				
Relief	0.682	0.813	0.742	0.575
MI	0.829	0.770	0.801	0.542
χ^2	0.832	0.761	0.804	0.580
Mean	0.829	0.792	0.810	0.600
Median	0.831	0.789	0.808	0.613
New approach	**0.850**	**0.854**	**0.852**	**0.662**
SVM				
Relief	0.695	0.798	0.743	0.602
MI	0.831	0.770	0.800	0.611
χ^2	0.818	0.835	0.827	0.590
Mean	0.823	0.843	0.828	0.620
Median	0.821	**0.845**	0.832	0.621
New approach	**0.845**	0.821	**0.833**	**0.798**

- http://www.cse.ust.hk/ ~ qyang/221/Assignments/German.
- https://archive.ics.uci.edu/ml/datasets/Statlog+(Australian+Credit+Approval).

To implement and test our approach we use four individual filtering techniques from Weka software namely: relief, χ^2 mutual information, and correlation [22]. Each filter is performed 10 time and the first three filters are retained as the most stable one with a stability score over 80 %. The aggregation of these retained filters is performed with Spearman distances.

The obtained results by our proposed approach are compared to two well known rank aggregation techniques: mean, median [7, 23] and also compared to the results given by the individual feature selection methods. Decision trees DT and support vector machine SVM are used as classifiers to evaluate the obtained feature subsets.

The performance of our proposed method is evaluated using three performance measure from the information retrieval field [7]: precision (P), recall (R) and F-measure (FM) and results are presented in Tables 2 and 3.

We investigate the recall results for the set of feature selection methods. For the Australian dataset the best recalls are achieved by our approach for the DT and with the median aggregation for SVM classifier. For the German dataset Table 3 shows that the highest recall is achieved in two times by the new aggregation method with DT and SVM classifiers. We remark from Tables 2 and 3 that the results of aggregation techniques outperform the results of individual feature selection methods, this confirm our hypotheses that aggregation bring more robustness and stability to individual classifiers results. From Tables 2 and 3 we notice that for precision and F-measure the proposed approach always archives the best results.

Graphical tools can be also used as an evaluation criterion instead of a scalar criterion. In this section we use the area under the ROC curve to evaluate the effect of selected features on classification models. Hence, the best combination of

Table 3 German dataset: comparison between the new filter method and the other feature selection methods

	P	R	FM	ROC area
Decision tree				
Relief	0.682	0.555	0.669	0.631
MI	0.516	0.534	0.525	0.621
χ^2	0.737	0.477	0.579	0.600
Mean	0.750	0.542	0.612	0.682
Median	0.750	0.545	0.613	0.727
New approach	**0.782**	**0.601**	**0.689**	**0.725**
Support vector machine				
Relief	0.517	0.511	0.514	0.692
MI	0.603	0.534	0.566	0.701
χ^2	0.705	0.489	0.577	0.622
Mean	0.766	0.552	0.627	0.780
Median	0.756	0.560	0.643	0.781
New approach	**0.823**	**0.812**	**0.817**	**0.812**

features is the one that gives the highest area under the ROC curve will be considered as the most suitable for the classification task.

If we look in ROC area results' we notice from Tables 2 and 3 that proposed approach achieves the highest values with German dataset for both DT and SVM and respectively with the Australian dataset.

5 Conclusion

A new approach for rank aggregation in a feature selection context was presented in this study. We tried to implement a robust model for ranking based on ensemble feature selection. In a first part we investigated the stability of filtering methods then we conduct an aggregation on the most stable one. Results on two credit datasets show a remarkable improvement when using our new rank aggregation method compared to the individual rankers and other competitive aggregation methods taken as input. To simplify our work we used a simple similarity criterion, it would be better to study other similarity measure to compute the degree of stability of filtering methods.

References

1. Ben Brahim, A., Bouaguel, W., Limam, M.: Feature selection aggregation versus classifiers aggregation for several data dimensionalities. In: Proceedings of the International Conference on Control, Engineering & Information Technology (CEIT13) (2013)
2. Ben brahim, A., Bouaguel, W., Limam, M.: Combining feature selection and data classification using ensemble approaches: application to cancer diagnosis and credit scoring.

In: Francisr, T. (ed.) Case Studies in Intelligent Computing: Achievements and Trendss. CRC Press, Boca Raton (2013)
3. Fernandez, G.: Statistical data mining using SAS applications. In; Chapman & Hall/Crc: Data Mining and Knowledge Discovery. Taylor and Francis, Boca Raton (2010)
4. Forman, G.: BNS feature scaling: an improved representation over TF-IDF for SVM text classification. In: Proceedings of the 17th ACM Conference on Information and Knowledge Mining, pp. 263–270. ACM, New York, NY, USA (2008)
5. Rodriguez, I., Huerta, R., Elkan, C., Cruz, C.S.: Quadratic programming feature selection. J. Mach. Learn. Res. **11**(4), 1491–1516 (2010)
6. Saeys, Y., Inza, I.N., Larrañaga, P.: A review of feature selection techniques in bioinformatics. Bioinformatics **23**(19), 2507–2517 (2007)
7. Bouaguel, W., Bel Mufti, G., Limam, M.: A new feature selection technique applied to credit scoring data using a rank aggregation approach based on: optimization, genetic algorithm and similarity. In: Francisr, T. (ed.) Knowledge Discovery & Data Mining (KDDM) for Economic Development: Applications, Strategies and Techniques. CRC Press, Chicago (2014)
8. Wu, O., Zuo, H., Zhu, M., Hu, W., Gao, J., Wang, H.: Rank aggregation based text feature selection. In: Proceedings of the Web Intelligence, pp. 165–172. (2009)
9. Wang, C.M., Huang, W.F.: Evolutionary-based feature selection approaches with new criteria for data mining: a case study of credit approval data. Expert Syst. Appl. **36**(3), 5900–5908 (2009)
10. Bouaguel, W., Bel Mufti, G.: An improvement direction for filter selection techniques using information theory measures and quadratic optimization. Int. J. Adv. Res. Artif. Intell. **1**(5), 7–11 (2012)
11. Dittman, D.J., Khoshgoftaar, T.M., Wald, R., Napolitano, A.: Classification performance of rank aggregation techniques for ensemble gene selection. In: Boonthum-Denecke, C., Youngblood, G.M. (eds.) Proceedings of the International Conference of the Florida Artificial Intelligence Research Society (FLAIRS), AAAI Press, Coconut Grove (2013)
12. Saeys, Y., Abeel, T., Peer, Y.: Robust feature selection using ensemble feature selection techniques. In: Proceedings of the European conference on Machine Learning and Knowledge Discovery in Databases—Part II. ECML PKDD '08, pp. 313–325. Springer, Berlin, Heidelberg (2008)
13. Molina, L.C., Belanche, L., Nebot, A.: Feature selection algorithms: a survey and experimental evaluation. In: Proceedings of the IEEE International Conference on Data Mining, pp. 306–313. IEEE Computer Society (2002)
14. Dash, M., Liu, H.: Consistency-based search in feature selection. Artif. Intell. **151**(1–2), 155–176 (2003)
15. Krishnaiah, P., Kanal, L.: Preface. In: Krishnaiah, P., Kanal, L. (eds.) Classification Pattern Recognition and Reduction of Dimensionality. Handbook of Statistics, vol. 2, pp. v–ix. Elsevier (1982)
16. Guyon, I., Elisseeff, A.: An introduction to variable and feature selection. J. Mach. Learn. Res. **3**(9), 1157–1182 (2003)
17. Hastie, T., Tibshirani, R., Friedman, J.: The Elements of Statistical Learning. Springer Series in Statistics. Springer New York Inc, New York (2001)
18. Prati, R.C.: Combining feature ranking algorithms through rank aggregation. In: The 2012 International Joint Conference on Neural Networks (IJCNN), pp. 1–8. Brisbane, Australia, 10–15 June 2012
19. Kalousis, A., Prados, J., Hilario, M.: Stability of feature selection algorithms: a study on high-dimensional spaces. Knowl. Inf. Syst. **12**(1), 95–116 (2007)
20. Pihur, V., Datta, S., Datta, S.: RankAggreg, an R package for weighted rank aggregation. BMC Bioinform. **10**(1), 62–72 (2009)

21. Mak, M.W., Kung, S.Y.: Fusion of feature selection methods for pairwise scoring svm. Neurocomputing **71**(16–18), 3104–3113 (2008)
22. Bouckaert, R.R., Frank, E., Hall, M., Kirkby, R., Reutemann, P., Seewald, A., Scuse, D.: Weka manual (3.7.1) (2009)
23. Kolde, R., Laur, S., Adler, P., Vilo, J.: Robust rank aggregation for gene list integration and meta-analysis. Bioinformatics **28**(4), 573–580 (2012)

Design and Implementation of Interactive Speech Recognizing English Dictionary

Dipayan Dev and Pradipta Banerjee

Abstract Computer synthesis of natural speech is one of the major objectives for researchers as well as linguists. The technology furnishes plenty of functional applications for human-computer interaction. This paper describes the procedure and logic for implementation of a software application that uses the technology of conversion of Speech to Text and Text to Speech synthesis. Basically, its a software application which will recognize the word spoken by the user using the microphone and the corresponding meaning will be delivered through the computer speaker. We measured the performance of the application with different main memory size, processor clock speed, L1 and L2 cache line sizes. Further, this paper demonstrates the various factors on which the application shows the highest efficiency and providing the suitable environment for it. At the end, the paper describes the major use of the application and future work included for the same. This paper describes the procedure and logic for implementation of a software application that uses the technology of conversion of Speech to Text and Text to Speech synthesis. Basically, its a software application which will recognize the word spoken by the user using the microphone and the corresponding meaning will be delivered through the computer speaker. We measured the performance of the application with different main memory size, processor clock speed, L1 and L2 cache line sizes. Further, this paper demonstrates the various factors on which the application shows the highest efficiency and providing the suitable environment for it. At the end, the paper describes the major use of the application and future work included for the same.

Keywords Dictionary · Speech recognition · Speech synthesis · Software application · Phoneme

---

D. Dev (✉)
Department of Computer Science & Engineering, NIT Silchar, Silchar, India
e-mail: dev.dipayan16@gmail.com

P. Banerjee
Department of Information Technology, NIT Durgapur, Durgapur, India
e-mail: pradiptapb@gmail.com

A. Nagar et al. (eds.), *Proceedings of 3rd International Conference on Advanced Computing, Networking and Informatics*, Smart Innovation, Systems and Technologies 43, DOI 10.1007/978-81-322-2538-6_44

1 Introduction

Speech is the primary and effective way of communicating with Humans. Speech Technology found a great popularity in middle of 20th century due to few blockbuster movies of 1960s and 1970s. [1] For this it became an interest topic of research for the scientists from then onwards. In 2011, Apple created a speech technology enabled computer, named Knowledge Navigator, that explains the mechanism of Multimodal User Interface (MUI) and Speech User Interface (SUI). The computer also demonstrates the intelligence of voice-enable systems. These inventions show the inclinations of researchers towards building speech enabled applications.

As the time progresses, the field of speech processing application have found a vast diversity in day to day tasks of human being. There are many speech recognition systems available in the recent days. Few of them are IBMs viaVoice [2], Dragon Naturally Speaking [3], Phillips FreeSpeech and L&Hs Voice Xpress [4]. Though there are no such widely used speech recognition systems, due to the unreliable recognition capacity. From the last few years, scientists have put many efforts to perfect the recognition techniques. We have used Java Speech API, thats uses on hidden Markov Model (HMM), which is so far the most widely accepted technique for speech recognition.

This paper explains the implementation of a software application, Interactive Speech Recognition English Dictionary (ISRED) that deals with speech recognition as well as speech synthesis. The application is made in such a way that it is almost completely immune to noise distortion caused during the speech recognition phase. We have used Java Speech API to serve the recognition purpose, which uses the HMM, that is widely accepted now-a days. The task of this system is, it will recognize the user's spoken word using the microphone and it's corresponding meaning will be delivered by the through the speaker. The major challenges in this implementation are:

- Speech recognition
- Creating the database of a dictionary
- Connecting the database with the speech recognition engine
- Speech synthesis

Evaluation of best performance of ISRED, various tools is used, which is discussed in the later stage of the paper. Various factors, such that, main memory size, clock speed of the processor, L1 and L2 cache line size etc. are discussed to find out the best environment on which the application shows the optimum efficiency.

The remainder of this paper is organized as follows. We discuss the basics of speech recognition in Sect. 2. The Sect. 3 discusses speech synthesis part, In Sect. 4 we introduce the proposed flowchart of our work, algorithms and grammars used. The next part, Sect. 5 shows the step by step operation of the application UI. The next section deals with the performance evaluation of the application considering various factors. The conclusions are given in Sect. 7.

2 Speech Recognition

When we utter a speech, fundamentally, the speech gets converted to analog signal with the help of microphone. This signal is managed by the sound card installed in the computer, that converts the signal to digital form [5]. The whole speech remains in binary form in our system which is used by users through various programs.

An acoustic model is formed through capturing chucks of speech, then converting them into texts and corresponding statistical depiction that compose the words. These are operated by some speech recognition engine which recognizes the speech.

The acoustic model of speech-recognition software converts the speech into small speech elements, termed as phonemes [6, 7]. After that, the language is matched with the digital dictionary, that is stored in memory of the computer [5]. Based on the digital form, if any match found, the words get displayed on the screen. So, in general a speech recognition engine is composed of the following components:

- Language model or grammar
- Acoustic Model
- Decoder

In Fig. 1, the feature extraction phase comprised of three stages. The initial phase, acoustic front end or speech analysis executes spectral-temporal analysis of the signal and originates some unprocessed characteristics of the power spectrum. The second phase comprises an extension of feature vector, which is an amalgam of static and dynamic properties. The last phase of the whole process, though always not present, converts this extension of feature to more compressed and vigorous vectors, which is delivered to the recognizer as an input [8].

3 Speech Synthesis

The speech synthesis is one of the major challenges in making any kind of speech enabled software [9].

A speech synthesizer, usually called text-to-speech (TTS) technology transforms the text languages into digital speech.

3.1 Working of a TTS System

The TTS [10] process comprises of two main parts: a front end and a back end. The front end does the text analysis. In this phase, the unprocessed text is transmuted into proximate words. The method is termed as tokenization. The front end assigns

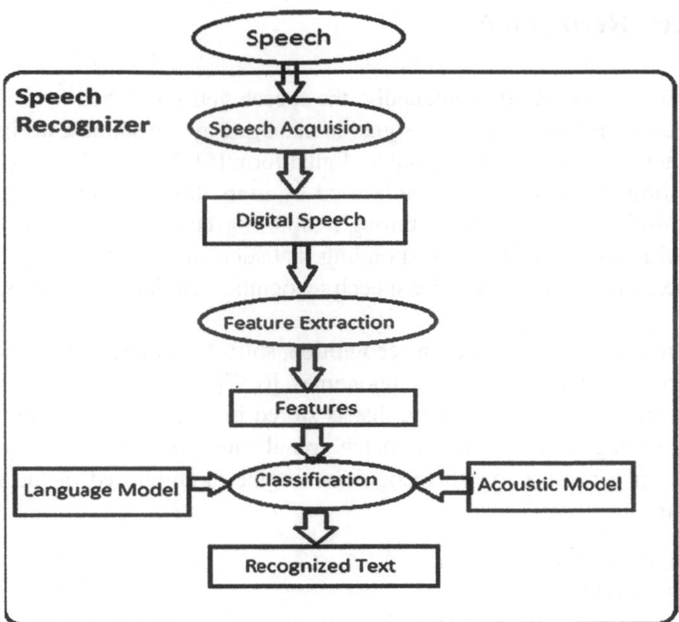

Fig. 1 Components of speech recognition engine

each word a phonetic transcription. Other task includes division of texts to prosodic units. This whole process of assignment of phonetic transcriptions into words is called text-to-phoneme conversion. The front end outputs the linguistic depiction, which consists of the prosody information as well as phonetic transformation. The synthesizer has a back end which is capable of speech waveform generation. At this point of time, the prosodic and phonetic information produces the acoustic output. Both these phases are termed as high and low level synthesis.

A simple model of whole approach is shown in Fig. 2. In our application, the input text is a spoken word by the user.

The string is then passed through a pre-processing stage and later on an analysis is performed to convert it into some phonetic depiction. At the last stage, low-level synthesizer comes to play, which generates the desired speech sound.

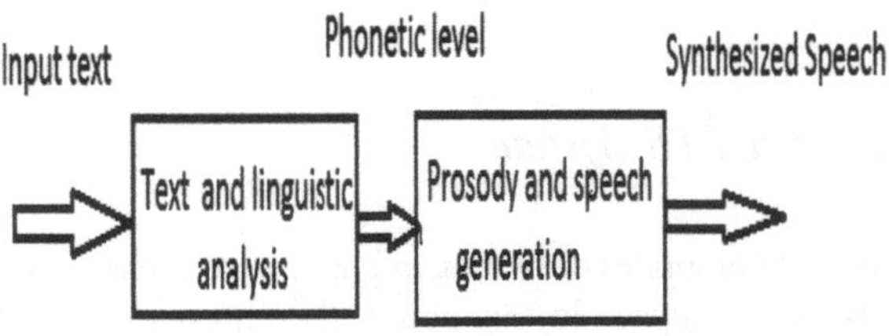

Fig. 2 A simple speech synthesizer working procedure

3.2 *Detailed Block Diagram of a Text-to-Speech Synthesizer*

Here the block diagram of the TTS engine is given showing the intermediate steps (Fig. 3).

4 Design Procedures

We have implemented and verified the whole application in J2SE platform. The user interface is built in Java-Swing platform. For the database part, the help of Windows Apache MySQL and PhP (WAMP) is taken [11].

4.1 *The Flowchart of the Implementation*

The checkpoints of the application ISRED is given in Fig. 4.

4.2 *Building the Dictionary for Acoustic Symbols*

To make the software more attractive to almost all users, the first thing to do is the flexibility in the pronunciation. Build the dictionary was the toughest thing to do in this project. For this, we had to judge the accent of different person, while doing the pronunciation of a particular word, as the proper word should be recognized by the speech engine for the correct recognition, which is the basis of this project. We trained the SRE with the thousand of words, having different kind of accents. This has made the application impregnable to any kind of noisy environment. In Fig. 5,

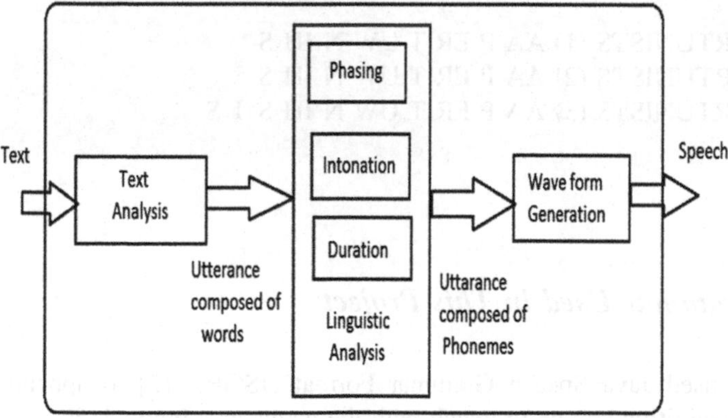

Fig. 3 Intermediate steps of a text-to-speech synthesizer

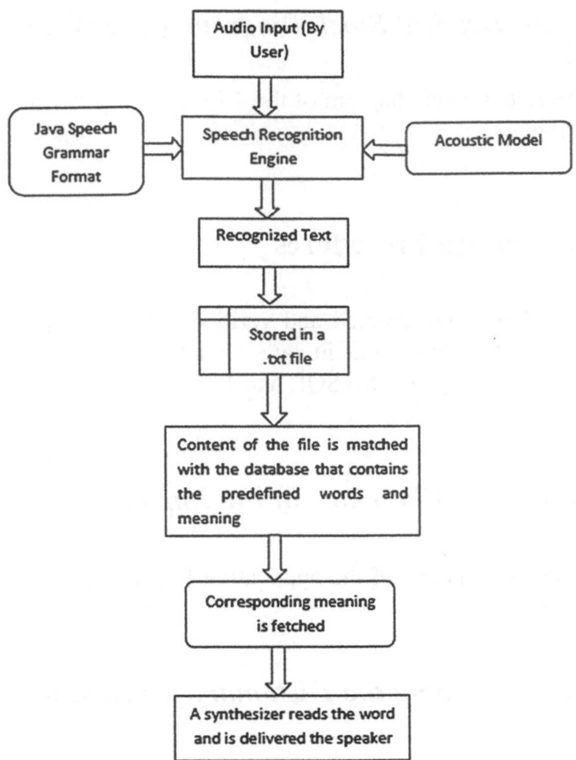

Fig. 4 Flowchart of the whole implementation

we have showed the syntax of the .DIC file with was the backbone for the application. Here we provided the syllable of word Opportunists in three different accents that are actually stored in the dictionary file .DIC.

During the speech recognition phase, SRE will match these words from the file with the spoken word in the digital form to get the desired result.

```
OPPORTUNISTS (1) AA P ER T UW N IH S
OPPORTUNISTS (2) AA P ER T UW N IH S S
OPPORTUNISTS (3) AA P ER T UW N IH S T S
```

4.3 Grammar Used in This Project

We have used Java Speech Grammar Format (JSGF) [12] to append with our speech recognition engine to match with the acoustic symbols [13].

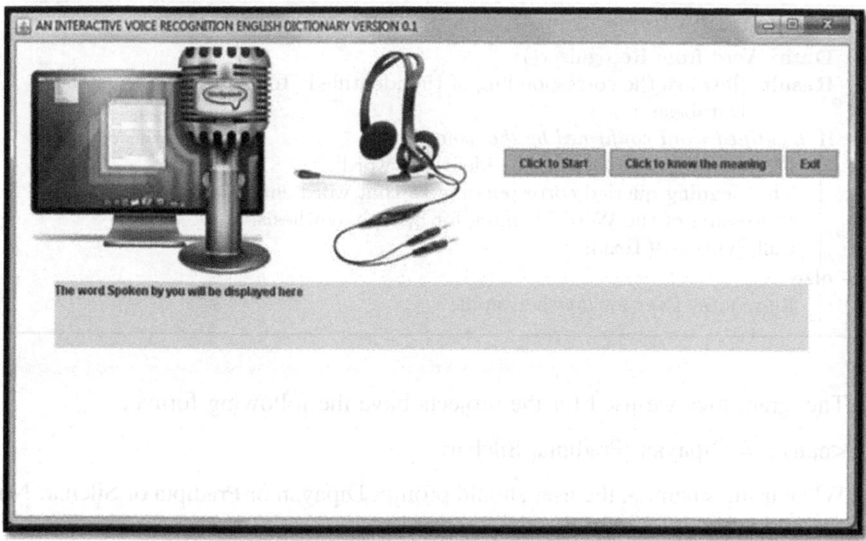

Fig. 5 The application is waiting for the user to utter some word

Following is an example to show the working of our grammar.

public <Welcome> = Hi;

public <completeWelcome> = <Welcome> Dipayan

This grammar file allows the system to identify two sentences. "Hi" and "Hi Dipayan".

Each grammar file has two components: The grammar header and the grammar body.

Algorithm 1: Recognizer()

```
Data: Input Speech through microphone
Result: Delivery of corresponding meaning
initialization;
User speaks through the microphone to send a word to the application.
The recognizer allocates its engine surface.
while true do
    if Audio Signal=Audible then
        Audio Processed;
        Processed Audio Parsed by the recognizer;
        All possible pronunciations checked as per our own Indian dialect
        recognition system;
        Parsed audio word deciphered to the most appropriate English word;
        Identified word displayed in text format;
        Text = matched word;
        Call Processor(Text);
        Processing of the meaning of the word, 'Text' for Audio synthesis;
        Identified word vocal output through Speakers after Speech Synthesis.
    else
        Error receiving proper signal;
        Re-request user for an decipherable audible input;
```

Algorithm 2: Processor(Text)

```
Data: Word from Recognizer()
Result: Retrieve the corresponding of the identified 'Text' meaning from the
        Database
if Identified word confirmed by the user then
    Dictionary is searched for the identified word;
    The meaning queried corresponding to that word entry;
    Processing of the Word Meaning for Speech synthesis;
    Call Synthesis(Text);
else
    Re-request User for another input;
```

The .gram files we used for the projects have the following format:

<name> = Dipayan |Pradipta| Silchar;

When using <name>, the user should prompt Dipayan or Pradipta or Silchar. No words other than these will be accepted.

There is a provision for Optional Grouping in JSGF: [ ]

public <welcome> = [Welcome] Dipayan;

Algorithm 3: Synthesis(Text)

```
Data: Text meaning of the word from Processor(Text)
Result: Synthesize it and pass it to Recognizer()
Customized speech engine configured;
Allocate the synthesizer; High speech quality selected;
Word parsed to identify its pronunciation;
Pronunciation deciphered voiced through the speech engine as per the set
  parameters;
Notify the engine on successful completion of the Synthesis;
```

When user wants to use <welcome>, he must say "Welcome Dipayan" or "Dipayan". As "Welcome" is defined inside the square bracket, it denotes an optional grouping. Therefore, the user may utter it but "Dipayan" is compulsory.

5 Experiments and Results

5.1 Before the Recognition

Figure 5 shows the initial phase of the application. Here after clicking the button click to start, the application starts recognizing voice from the user.

5.2 *Verifying the Word Recognized*

Figure 6 shows the next phase of the application. In this phase, the application displays the word that is recognized by it for verifying that with the user. If the displayed word matches with the word spoken, the user can press the button Click to know the meaning, to know to meaning of the word. Or else, he can press the button click to start again to start recognizing the correct word.

5.3 *Delivering the Output*

Figure 7 shows the last phase. In this phase, the application displays the meaning of the word, in the given text-area as well as delivers the meaning through the computer speaker also.

6 Performance Evaluation

ISRED is analyzed with three different approaches. Initially, the application is run on various platforms as to comprehend how ISRED behaves with changing environment. Speech inputs are collection and executed so as to regulate the quality of the samples. Later on, VTune, an Intel toolkit is used to gather the performance

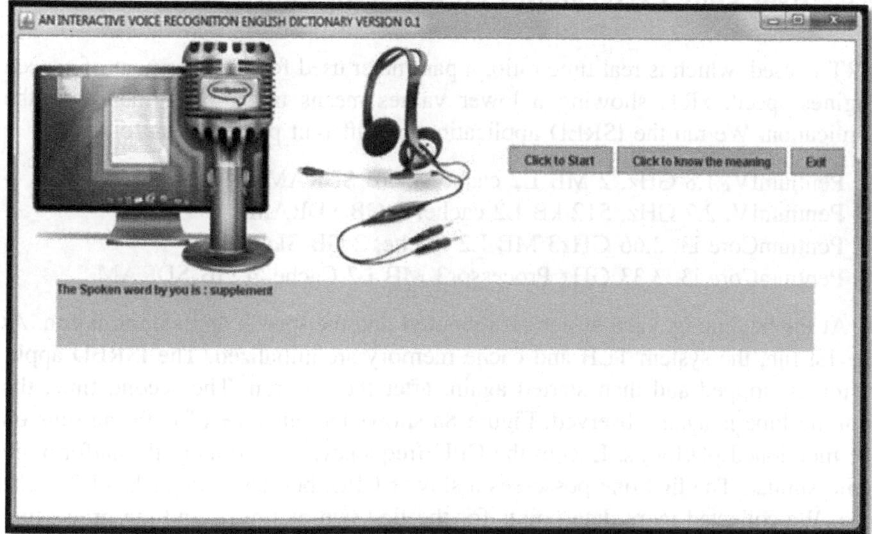

Fig. 6 The uttered word is recognized by the application and displays it the text-area

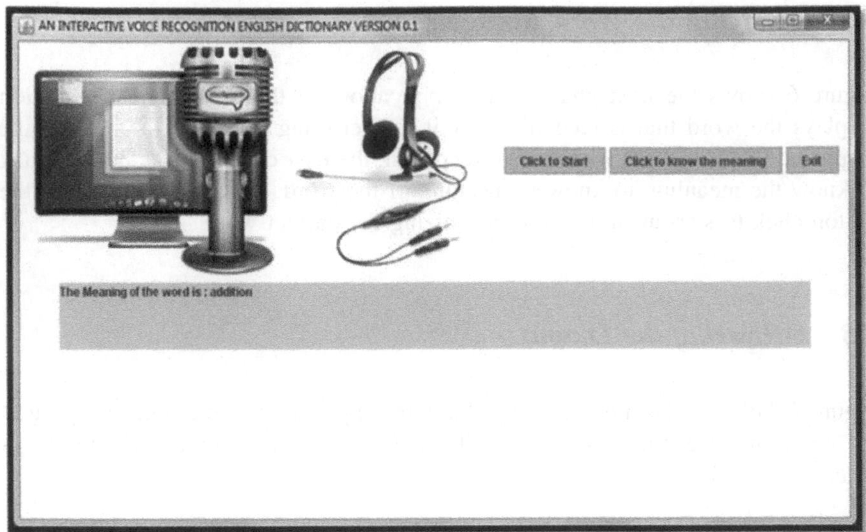

Fig. 7 The application displays the desired meaning of the word in the text-area along with delivering it through the speaker

counter values [14, 15]. A system simulator, SoftSDV as well as a simulator for cache memory is applied to observe the behavior of system memory in detail.

6.1 Real Time Performance on Different Platforms

XRT is used, which is real time ratio, a parameter used for measurement of speech engines speed. xRT, showing a lower values means better performance of the application. We ran the ISRED application on different platforms as follows:

1. PentiumIV, 1.8 GHz, 2 MB L2 cache, 4 GB SDRAM
2. PentiumIV, 2.2 GHz, 512 kB L2 cache, 2 GB SDRAM
3. PentiumCore i3, 2.66 GHz3 MB L2 Cache, 2 GB SDRAM
4. PentiumCore i3, 3.33 GHz Processor3 MB L2 Cache, 4 GB SDRAM

At the beginning, each system is rebooted and the speech application is run. At the 1st run, the system TLB and cache memory are initialized. The ISRED application is stopped and then started again, after the first run. The second time, the running time is again observed. Figure 8a shows the outcome of both the runs on the mentioned platforms. Except the CPU frequency, the 3rd and 4th platform are quite similar. The first one possesses a slower CPU, but has a larger level 2 cache size. We collected more data points for the first two platforms and the processing

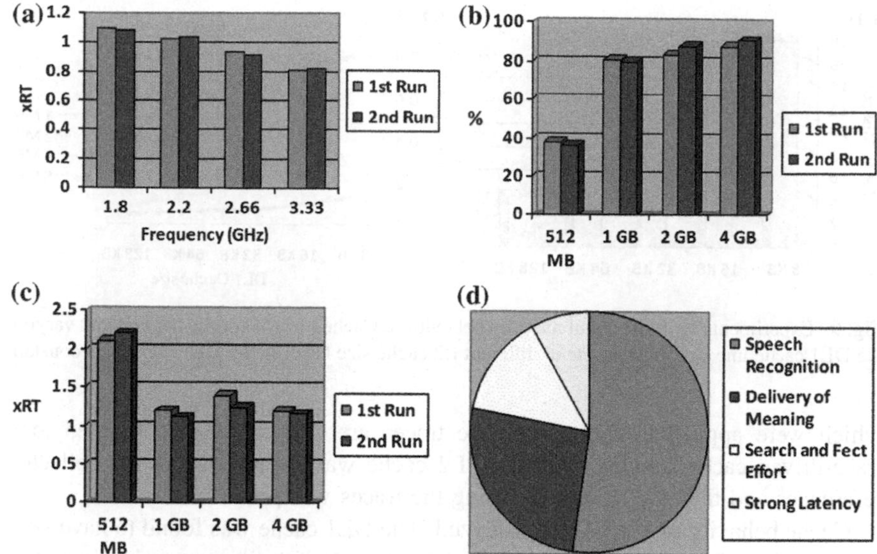

Fig. 8 Real time performance analysis. **a** Plot of xRT with 4 different processors. **b** CPU utilization of the application, with variation of RAM size. **c** Plot of xRT with various RAM size. **d** Time taken by each task to execute the application

speed is found to be nearly indistinguishable. The lack of CPU frequency is compensated by the larger L2 cache, though having a slower speed.

We extend our experiment with the testing of memory by altering the SDRAM of the system. This is done for the 1st platform from where the CPU utilization rate as well as real time rate is calculated. Figure 8b, c portrait the results. The CPU utilization and real time performance of the application give its best results after 1 GB RAM size. The efficiency seems extremely poor when the application is deployed in 512 MB. Therefore, a RAM of 1 GB is crucial to get the maximum efficiency of the application.

The task distribution of executing the ISRED is achieved from VTUNE. The result is shown is Fig. 8d, divided by major computation of the application. Figure suggests, the speech recognition part takes the longest amount of time while running the application.

6.2 More Detail Memory Behavior Study

Software simulator tools are used in our application to observe different experiments related to system's memory. We dealt with SiftSDVs CPU model and after initialization of SoftSDV we first ignore the initial 55 billion instructions. 370 million reference trace of L2 cache is collected from 1 billion instructions,

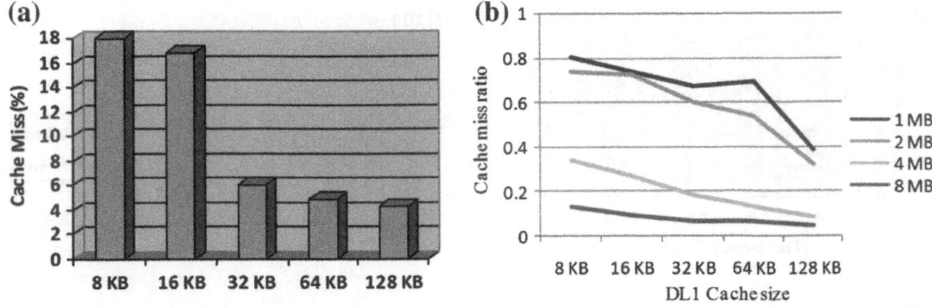

Fig. 9 Experiments for more detail memory behavior. **a** Cache miss % for the application, varying the DL1 cache line size. **b** Miss rate of different L2 cache size keeping the DL1 cache size constant

which were applied to the cache. The traces are collected and submitted to a trace-driven cache simulator as input. L2 cache was warmed using the collected references and then the simulation using the traces was performed.

Cache behavior of level 1 was analyzed. The DL1 cache was found to have very good locality. Figure 9a shows the cache line miss rate for different cache sizes. The cache sizes against the % of cache- miss is plotted in the figure. The plot shows that, during the process, ISRED shows poor cache miss ratio when the DL1 cache size is less than 32 kB and thereafter shows almost identical amount of cache miss. This high amount of cache miss with smaller DL1 cache size is due to the fact that, the application goes through various state change in the HMM stage as well as during the fetching of data from the Dictionary database.

Figure 9b portraits the cache miss ratio versus L2 cache size for fixed DL1 cache sizes. Figure shows a significant decrease in the miss rate 4 MB onward. This result is somewhat expected as three big data storage such as dictionary database, language model and acoustic model are used for managing the application. Moreover, ISRED uses a relative large database. In the process of searching the database, numerous words are visited one by one. So, having a big cache size is advisable which gives the best efficiency.

7 Conclusion and Future Work

The idea behind this project is to develop a compact digital dictionary that uses speech as a medium of input and the output will come as audio version. The main goal is to make interactive and user friendly software such that any person with even a little knowledge of computer can use this software and get the meaning of a desired word in no time.

ISRED application based on Java API which is platform independent. Several tools are used to explore its performance and various conclusions is suggested where ISRED shows best efficiency. Our study indicates that, this application requires a relative high clock speed processor to show its best performance. Since it uses a large data structure, it possesses excellent data spatial locality. But accessing the acoustic model and language model are quite random and hence shows very little temporal locality. Experiment also shows that, to achieve optimize performance, we should use L1 cache line of more than 32 kB and L2 cache line of more than 4 MB.

There are still some scopes of improvement left in this, like the application can be embedded on the chip and can be used with any mobile phones, tablet, pc etc.

References

1. Book on Speech Signals and Introduction to Speech Coding
2. IBM Desktop Via-Voice http://www-01.ibm.com/software/pervasive/viavoice.html
3. Dragon Speech Recognition Software. http://www.dragonsys.com
4. L&H Voice Xpress. http://l-h-voice-xpress-professional.software.informer.com/
5. Gu, H.-Y., Tseng, C-Y., Lee, L.-S.: Isolated-utterance speech recognition using hidden Markov models with bounded state durations. IEEE Trans. Signal Process. **39**(8), 1743–1752 (1991)
6. Yusnita, M. A., Paulraj, M.P., Yaacob, S., Bakar, S.A., Saidatul, A., Abdullah, A.N.: Phoneme-based or isolated-word modeling speech recognition system? An overview. 2011 IEEE 7th International Colloquium on Signal Processing and its Applications (CSPA), pp. 304–309, 4–6 Mar 2011
7. Jurafsky, D.: Speech Recognition and Synthesis: Acoustic Modeling, winter 2005
8. Agaram, K., Keckler, S.W., Burger, D.C.: A characterization of speech recognition on modern computer systems. In; 4th IEEE Workshop on Workload Characterization, at MICRO-34, Dec 2001
9. Tabet, Y., Boughazi, M.: Speech synthesis techniques. A survey. In; 2011 7th International Workshop on Systems, Signal Processing and Their Applications (WOSSPA)
10. O'Malley, M.H.; Berkeley Speech Technol., CA, USA. Text-to to-speech conversion technology, pp. 17–23 (1990)
11. http://www.wampserver.com/en
12. http://java.coe.psu.ac.th/Extension/JavaSpeech1.0/JSGF.pdf
13. Chow, Y.-L.; BNN Syst. & Technol. Corp., Cambridge, MA, USA, Roukis, S.: Speech understanding using a unification grammar, vol. 2, pp. 727–730 (1989)
14. Bhandarkar, D., Ding, J.: Performance characterization of PentiumÂ® Pro Processor. In; Proceedings of Symposium on High Performance Computer Architecture, San Antonio, 1–5 Feb 1997
15. Luo, Y., Cameron, K.W., Torrellas, J., Solihin, Y.: Performance modeling using hardware performance counters. In; Tutorial, HPCA-6, Toulouse, France Jan 2000

On Careful Selection of Initial Centers for K-means Algorithm

R. Jothi, Sraban Kumar Mohanty and Aparajita Ojha

Abstract K-means clustering algorithm is rich in literature and its success stems from simplicity and computational efficiency. The key limitation of K-means is that its convergence depends on the initial partition. Improper selection of initial centroids may lead to poor results. This paper proposes a method known as Deterministic Initialization using Constrained Recursive Bi-partitioning (DICRB) for the careful selection of initial centers. First, a set of probable centers are identified using recursive binary partitioning. Then, the initial centers for K-means algorithm are determined by applying a graph clustering on the probable centers. Experimental results demonstrate the efficacy and deterministic nature of the proposed method.

Keywords Clustering · K-means algorithm · Initialization · Bi-partitioning

1 Introduction

Clustering is the process of discovering natural grouping of objects so that objects within the same cluster are similar and objects from different clusters are dissimilar according to certain similarity measure. Various methods for clustering are broadly

---

R. Jothi (✉) · S.K. Mohanty · A. Ojha
Indian Institute of Information Technology, Design and Manufacturing Jabalpur, Jabalpur, Madhya Pradesh, India
e-mail: r.jothi@iiitdmj.ac.in
URL: http://www.iiitdmj.ac.in

S.K. Mohanty
e-mail: sraban@iiitdmj.ac.in
URL: http://www.iiitdmj.ac.in

A. Ojha
e-mail: aojha@iiitdmj.ac.in
URL: http://www.iiitdmj.ac.in

A. Nagar et al. (eds.), *Proceedings of 3rd International Conference on Advanced Computing, Networking and Informatics*, Smart Innovation, Systems and Technologies 43, DOI 10.1007/978-81-322-2538-6_45

classified into hierarchical and partitional methods [1, 2]. Hierarchical methods generate a nested grouping of objects in the form of dendrogram tree. Single-linkage and complete-linkage are the well known hierarchical clustering methods. The major drawback of these algorithms is their quadratic run time which poses a major problem for large datasets [3]. In contrast to hierarchical methods, partitional methods directly divide the set of objects into k groups without imposing the hierarchical structure [1].

K-means is a popular partitional clustering algorithm with the objective of minimizing the sum of squared error (SSE) which is defined as the sum of the squared distance between the cluster centers and the points in the cluster. K-means starts with k randomly chosen initial centers and assign each object in the dataset to a nearest center. Iteratively, the centers are recomputed and objects are reassigned to the nearest centers. Due to its simple implementation, K-means has been extensively used in various scientific applications. However, the results of K-means strongly depend on the choice of initial centers [1, 3, 4].

If the initial centers are improperly chosen, then the algorithm may converge to a local optimum. Number of approaches have been proposed to wisely choose the initial seeds for K-means [5–10].

A sampling based clustering solution for initial seed selection was suggested by Bradley and Fayyad [5]. The idea was to choose several samples from the given set of objects and applying K-means on each of the sample independently with random centers. The resulting centroids from each subcluster is the potential guess of the centers for the whole dataset. Likas et al. [8] proposed global K-means algorithm which incrementally chooses k cluster centers one at a time. Experimental results demonstrated that their method outperforms the K-means algorithm. But, this method is computationally expensive as it requires N execution of the K-means algorithm on the entire dataset, where N is the number of objects in the dataset [4].

Arthur and Vassilvitskii proposed an improved version of K-means known as K-means++ [7]. It chooses the first center c_1 randomly from the dataset and other centers c_i, $2 \leq i \leq k$ are chosen such that distance between c_i and the previously chosen centers is maximum. Both theoretically and experimentally they have shown that, K-means++ not only speed up the convergence of the clustering process but also yields a better clustering result than K-means.

Alternatively, may initialization methods were discussed by considering the principal dimensions for splitting the dataset [6, 9]. Ting and Jennifer [9] proposed two divisive hierarchical approaches, namely PCA-part method and Var-part method which identify the centers for K-means algorithm using k-splits along the better discriminant hyperplanes.

Erisoglu et al. [6] proposed a method which first defines the subspace of the dataset X along two main dimensions that best represent the spread of the dataset. Then, the data point with the longest distance from the centroid of X in the subspace is chosen as the first cluster center. The subsequent centers are chosen such that the center c_i has the maximum distance from the previously computed centers $c_1, \ldots, c_{i-1}$.

Min-Max algorithm proposed by Tzortzis and Likas [10] tackles initialization problem by implementing a weighted version of the K-means by assigning weights to the clusters in proportion to their variance. This weighting scheme attains high quality partition by controlling the clusters with larger variance.

There are various methods for cluster initialization, a comparative study of which can be seen in [4]. However, many of the methods run in quadratic time [8, 11] which degrades the efficiency of K-means. In this paper, we propose an initialization method which runs in $O(n \lg n)$ time using constrained recursive bi-partitioning. The performance of the proposed algorithm has been demonstrated using experimental analysis.

The rest of the paper is organized as follows. The description of K-means algorithm is given in Sect. 2. The proposed method is explained in Sect. 3. The complexity of the proposed method is discussed in Sect. 4. The experimental analysis is shown in Sect. 5. The conclusion and future scope are given in Sect. 6.

2 K-Means Algorithm

Let $X = \{x_1, x_2, \ldots, x_n\}$ be the set of objects to be clustered into k groups, where k is the number of classes of objects, which must be known a priori. The objective of K-means is to find a k-partition of X such that the sum of squared error (SSE) criterion is minimized. Let $S = \{S_1, S_2, \ldots S_k\}$ be the set of partitions returned by K-means and let μ_i be the center of the partition s_i. The sum of squared error (SSE) is defined as follows [1].

$$SSE = \sum_{i=1}^{k} \sum_{x_j \epsilon S_i} d(x_j, \mu_i)^2. \tag{1}$$

where $d(\ldots)$ denotes the distance (dissimilarity). K-means algorithm starts with k arbitrary centers and iteratively recomputes the centers and reassigns the points to the nearest centers. If there is no change in centers, then the algorithm stops. The K-means algorithm is described as follows.

Algorithm 1 *K-means Algorithm [2]*

Input: *Dataset* X.
Output: k *clusters of* X.

1. *Randomly choose* k *centers* μ_i, $1 \leq i \leq k$.
2. *For each object* x_j *in* X, *compute distance between* x_j *and* μ_i, $1 \leq i \leq k$.
3. *Assign* x_j *to the partition* s_i, *such that the distance between* x_j *and* μ_i *is minimum.*
4. *Recompute centers; repeat steps 2 and 3 if there is change in centers. Else stop.*

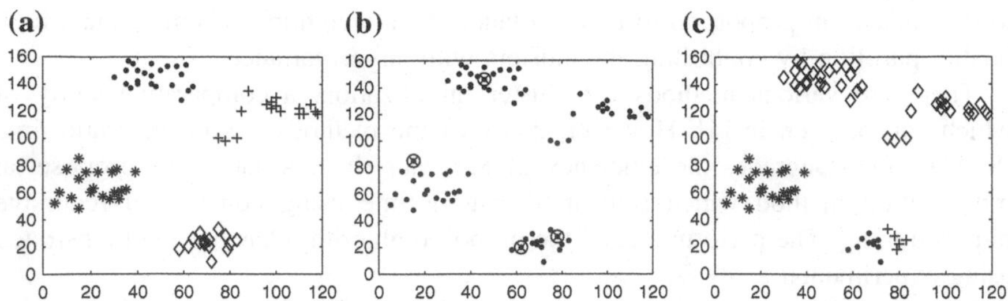

Fig. 1 K-means converging to a local optimum with random center initialization. **a** A given dataset with four clusters. **b** Randomly chosen initial centers. **c** The result of K-means with centers chosen in (**b**)

As the initial centers for K-means partition are chosen randomly, two or more centers may collide in a nearby region. As a consequence, the points are forced to be assigned to one of the nearest centers, leading to poor results. This is illustrated in Fig. 1. It is clear from the figure that, K-means gets trapped with local optimum if the initial centers are not chosen properly.

3 Proposed Method

The proposed method is a two-step process. In the first step, it identifies a set of probable centers by dividing the dataset X into k' partitions. During the second step, the actual centers are determined by grouping the probable centers into k subsets, where center of each subset is considered as one of the k centers for K-means algorithm.

3.1 *Identifying Probable Centers*

K-means algorithm tries to assign the points to a cluster based on a nearest center. Representing a cluster using a single prototype (center) may not capture the intrinsic nature of clusters [12]. Motivated from [12], we identify a pool of $k', k' \gg k$ most likely points from the dataset, such that the spread of each cluster in the dataset is well represented by a subset of these points. We call these points as *probable centers* denoted by $Y = \{y_1, y_2, \ldots, y_{k'}\}$. In order to identify a set of probable centers, this paper proposes an algorithm known as Constrained Bi-partitioning algorithm.

Constrained Bi-partitioning Algorithm Generally a Bi-partitioning method recursively split the dataset into a binary tree of partitions [13]. It starts from the root node that contains the given dataset X and split the node into two subsets X_1

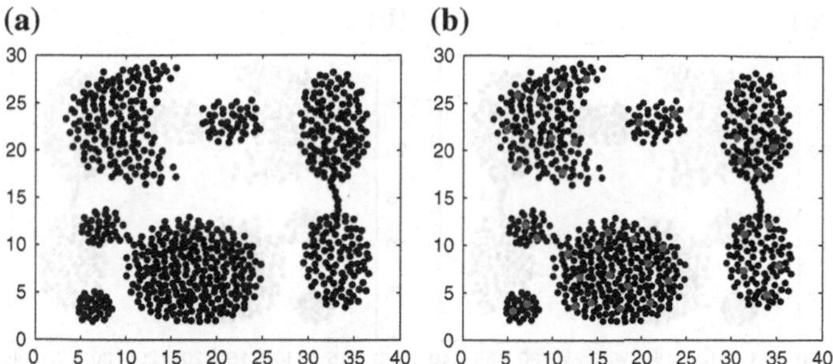

Fig. 2 The probable centers. **a** The dataset. **b** The probable centers are marked in *red dots* (color figure online)

and X_2 based on certain partitioning criteria. The recursive splitting stops once k partitions are identified.

This paper proposes an improved version of Binary partitioning known as Constrained Bi-partitioning algorithm. The basis of bipartitional criteria adapted in our proposed algorithm lies in choosing the two centers for splitting a node in the tree. Two centers p and q are chosen such that they are the farthest pair of points in the node in order to maximize the inter-cluster variance. While Bi-partitioning algorithm stops when k partitions are identified, the constrained algorithm continues the splitting process as long as the size of the subset to be partitioned is greater than $\sqrt{n}$. Thus, the number of partitions is not preset. The leaf nodes, the subsets which cannot be partitioned further, are stored in the set of partitions $S = \{S_1, S_2, \ldots, S_{k'}\}$. The center of each partition S_i in S is considered to be a probable center y_i.

Figure 2 shows an example of a dataset and its probable centers. It is obvious from the figure that, each cluster is covered by a subset of probable centers. The actual centers of the K-means algorithm can be identified by merging these probable centers into k groups.

3.2 Computation of k Initial Centers from k' Probable Centers

Once the set of probable centers $Y = \{y_1, y_2, \ldots, y_{k'}\}$ are recognized, next we need to identify the k disjoint subsets of Y. The probable centers must be grouped such that the closeness between the centers within a subset is high as compared to the closeness between the centers of different subsets. This in turn maximizes the inter-cluster separation of the final clusters produced by the K-means algorithm.

The relative inter-connectivity of the probable centers intuitively expresses the neighboring nature of the subsets and the breaks in the connectivity gives a clue on the separation of actual clusters in the dataset. In order to identify such breaks, we

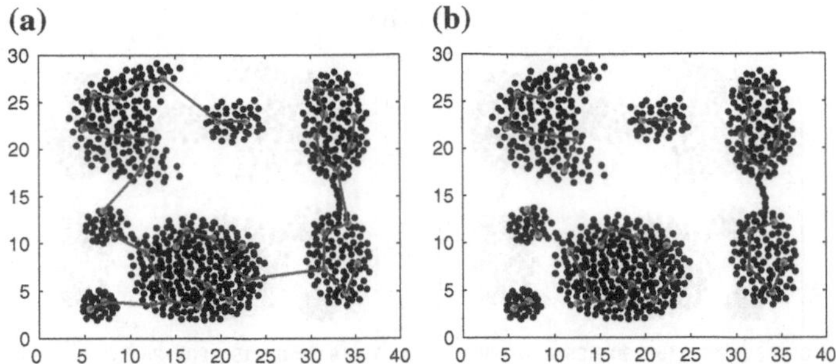

Fig. 3 MST-based partitioning of probable centers for the dataset shown in Fig. 2a. **a** The MST of probable centers. **b** k disjoint subsets of probable centers

employ Minimum Spanning Tree (MST)-based representation of probable centers, as MST of a set of points can be used to reflect the similarity of the points with their neighborhood [14]. Simply removing $k-1$ longest edges from the MST results in k disjoint subsets of nearest centers, such that each subset would represent a cluster. This is illustrated in Fig. 3.

As each $y_i \in Y$ may or may not belong to the dataset X, we choose a best representative point $r_i \in X$ from each S_i such that r_i is closest to y_i. Let $R = \{r_1, r_2, \ldots, r_{k'}\}$ denotes the set of best representative points identified in the above manner. Prim's algorithm on R generates MST T_1, removing $k-1$ longest edges from T_1 yields k clusters of best representative points. Finally, the actual centers for K-means algorithm are computed from the center of the k clusters. The proposed method DICRB is summarized in Algorithm 2. The result of K-means initialized with the proposed method is demonstrated in Fig. 4.

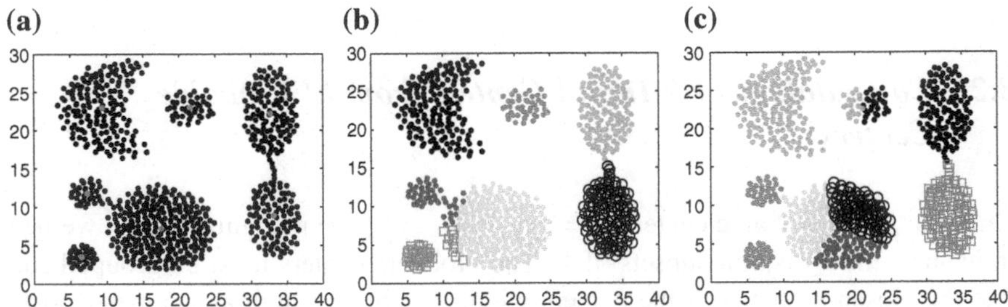

Fig. 4 Result of K-means initialized with proposed method. **a** The centers identified by proposed method marked in *red dots*. **b** The final clusters identified by K-means initialized with proposed method. **c** Result of K-means with random initialization (color figure online)

Algorithm 2 *Deterministic Initialization using Constrained Recursive Bi-partitioning (DICRB) Algorithm*

Input: *Dataset* X.
Output: k *initial centers for K-means algorithm.*

1. *Let* S *be the set of partitions and initialize* $S = \phi$.
2. *Initialize the node to be splitted* $X' = X$.
3. *Repeat*
   3.1 *if size of* $X' > \sqrt{n}$
      3.1.1 *Find the center* μ *of the node* X'.
      3.1.2 *Find a point* $o \epsilon X'$ *such that* $d(o, \mu)$ *is minimum.*
      3.1.3 *Choose two centers* $p \epsilon X'$ *and* $q \epsilon X'$ *such that* p *is farthest point from* o *and* q *is farthest point from* p.
      3.1.4 *Split the node* X' *into two nodes* X_p *and* X_q *according to centers* p *and* q.
      3.1.5 *Recursively apply bi-partitioning on* X_p *and* X_q.
   3.2 *else* $S = \{S \cup X'\}$
4. *Until there is no node to split*
5. *Identify a set of probable centers* $Y = \{y_1, y_2, \cdots, y_{k'}\}$, *where* y_i *is the center of the subset* s_i.
6. *Build a set of best representative points* (R) *by choosing points closer to each probable center* $y_i \epsilon Y$.
7. *Construct MST* T_1 *of* R.
8. *Remove* $k - 1$ *longest edges from* T_1 *to get* k *clusters.*
9. *Center from each of the* k *clusters corresponds to actual center for K-means algorithm.*

4 Complexity of the Proposed Method

The complexity of the DICRB method is analyzed as follows. The steps 1–4 take $O(n \lg n)$ time to construct binary partitioning tree. Step-5 takes $O(n)$ time to identify the probable centers from each partition. Similarly $O(n)$ time is needed to find the best representative point set R in step-6. As the size of the set R is $O(\sqrt{n})$, Prim's algorithm in step-7 takes $O(n)$ and clustering in step-8 takes $O(\sqrt{n})$ complexity. Hence, the overall time complexity to identify the initial cluster centers for K-means algorithm is $O(n \lg n)$.

5 Experimental Results

The proposed method of initialization is compared against random initialization based on the number of iterations (*I*) required for convergence, *SSE* and Adjusted Rand Index (ARand) [15] of clusters after the convergence. The tests are conducted on four synthetic and four real datasets. The K-means algorithm with randomly chosen initial centers and K-means algorithm with proposed method of initialization are run for 100 times on the identical datasets and the average value of I, SSE and ARand are observed. We also report the maximum and minimum number of iterations taken by both the methods.

The synthetic datasets DS1, DS2, DS3 and DS4 used in our experiments are chosen such that each dataset would represent different kind of clustering problem. The details of these datasets are given in Table 1 and are shown in Fig. 5. The results of K-means on these datasets are shown in Tables 2 and 3. It is evident from the results provided in the Tables 2 and 3 that, K-means initialized with proposed method performs better in terms of constant number iterations and improved cluster quality with respect to internal as well as external quality measures.

We have observed the results of proposed method also on few real datasets taken from UCI Machine Learning Repository (http://archive.ics.uci.edu/ml/). The details of these datasets can be seen in Table 1. Our proposed initialization method maintains the stable results from K-means according to the number of iterations, minimized error and improved cluster separation as compared to K-means with random centers. This is evident from the results shown in Tables 4 and 5.

Table 1 Details of the datasets. No. of instances (n), no. of dimensions (d), no. of clusters (k)

2-Dimensional synthetic datasets			
Dataset	n		k
DS1	399		6
DS2	788		7
DS3	600		15
DS4	300		2
Real datasets			
Dataset	n	d	k
Iris	150	5	3
Ruspini	75	2	4
WDBC	569	32	2
Thyroid	215	5	6

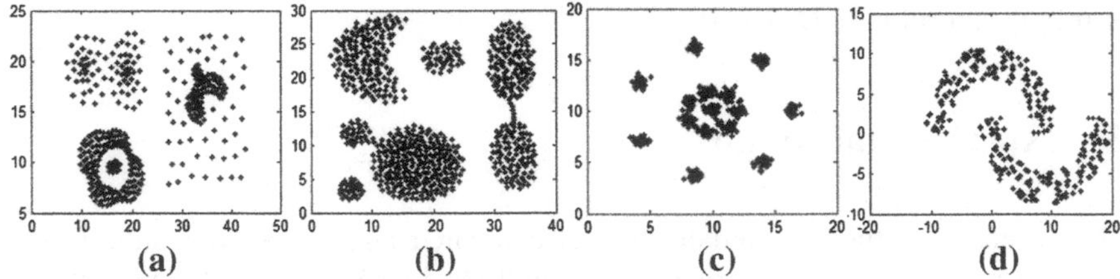

Fig. 5 Synthetic datasets used for experimental study

Table 2 Comparison of initialization methods according to number of iterations (I) on synthetic datasets

Dataset	Method	Avg (I)	Max (I)	Min (I)
DS1	Random	13	39	5
	Proposed	**9**	**9**	**9**
DS2	Random	16	33	6
	Proposed	**10**	**10**	**10**
DS3	Random	1	20	4
	Proposed	**6**	**6**	**6**
DS4	Random	8	13	4
	Proposed	**6**	**6**	**6**

Table 3 Comparison of initialization methods according to SSE and adjusted rand index of clusters on synthetic datasets

Dataset	Method	SSE	Adjusted rand
DS1	Random	4801.04	0.6936
	Proposed	**4733.10**	**0.8361**
DS2	Random	12322.38	0.7967
	Proposed	**11514.35**	**0.9277**
DS3	Random	1001.47	0.8485
	Proposed	**186.57**	**0.9091**
DS4	Random	11885.37	0.6229
	Proposed	**11876.01**	**0.6536**

Table 4 Comparison of initialization methods according to number of iterations (I) on real datasets

Dataset	Method	Avg (I)	Max (I)	Min (I)
Iris	Random	9	15	3
	Proposed	**6**	**6**	**6**
Ruspini	Random	4	9	2
	Proposed	**2**	**2**	**2**
WDBC	Random	10	14	8
	Proposed	**8**	**8**	**8**
Thyroid	Random	9	15	3
	Proposed	**3**	**3**	**3**

Table 5 Comparison of initialization methods according to SSE and adjusted rand index of clusters on real datasets

Dataset	Method	SSE	Adjusted rand
Iris	Random	106.35	0.8749
	Proposed	**87.31**	**0.9799**
Ruspini	Random	29385.95	0.7360
	Proposed	**13269.65**	**1.000**
WDBC	Random	11689.49	0.6972
	Proposed	**11640.71**	**0.7988**
Thyroid	Random	528.28	0.5608
	Proposed	**502.11**	**0.5907**

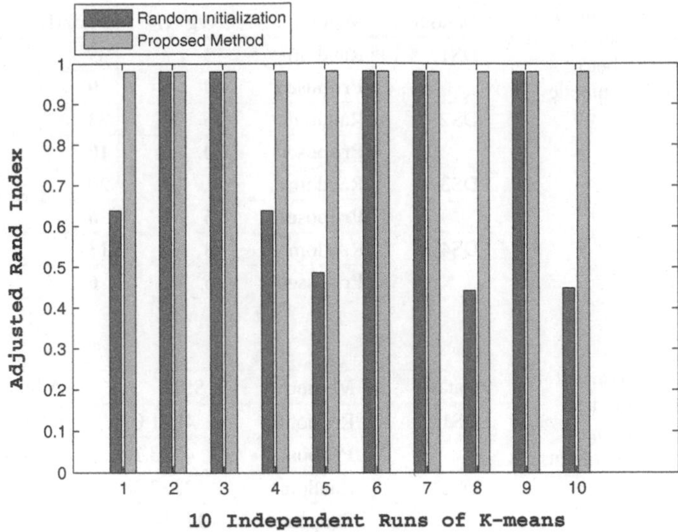

Fig. 6 Adjusted rand index of random and proposed method on iris dataset in each run

With random initialization method, the initial centers are arbitrary in each run and thus the final clustering results are not deterministic. With the proposed method, the initial centers are chosen from a pool of probable candidate centers which always produce deterministic clustering results. The Fig. 6 shows the Adjusted Rand value obtained from random initialization and proposed method on Iris dataset for 10 runs of K-means. As can be seen from the figure, the Adjusted Rand value for random initialization method changes with every run where as it remains constant in the proposed method, showing the deterministic nature of our algorithm.

6 Conclusion

This paper proposed an initialization method for K-means algorithm using constrained recursive bi-partitioning. The efficiency of the proposed method was demonstrated through experiments on different synthetic and real datasets. While the clustering results of K-means algorithm with random initialization is unstable, our proposed method of initialization produces deterministic results. As a future work, we will carry out an extensive analysis of DICRB method on more datasets.

References

1. Jain, A.K., Murty, M.N., Flynn, P.J.: Data clustering: a review. ACM Comput. Surv. **31**(3), 264–323 (1999)
2. Han, J., Kamber, M.: Data Mining, Southeast Asia Edition: Concepts and Techniques. Morgan Kaufmann, Los Altos (2006)
3. Xu, R., Wunsch, D., et al.: Survey of clustering algorithms. IEEE Trans. Neural Networks **16** (3), 645–678 (2005)
4. Celebi, M.E., Kingravi, H.A., Vela, P.A.: A comparative study of efficient initialization methods for the k-means clustering algorithm. Expert Syst. Appl. **40**(1), 200–210 (2013)
5. Bradley, P.S., Fayyad, U.M.: Refining initial points for k-means clustering. ICML **98**, 91–99 (1998)
6. Erisoglu, M., Calis, N., Sakallioglu, S.: A new algorithm for initial cluster centers in k-means algorithm. Pattern Recogn. Lett. **32**(14), 1701–1705 (2011)
7. Arthur, D., Vassilvitskii, S.: K-means++: the advantages of careful seeding. In: Proceedings of the Eighteenth Annual ACM-SIAM Symposium on Discrete Algorithms, pp. 1027–1035 (2007)
8. Likas, A., Vlassis, N., Verbeek, J.J.: The global k-means clustering algorithm. Pattern Recogn. **36**(2), 451–461 (2003)
9. Ting, S., Jennifer, D.G.: In search of deterministic methods for initializing k-means and gaussian mixture clustering. Intell. Data Anal. **11**(4), 319–338 (2007)
10. Tzortzis, G., Likas, A.: The minmax k-means clustering algorithm. Pattern Recogn. **47**(7), 2505–2516 (2014)
11. Cao, F., Liang, J., Jiang, G.: An initialization method for the k-means algorithm using neighborhood model. Comput. Math. Appl. **58**(3), 474–483 (2009)
12. Liu, M., Jiang, X., Kot, A.C.: A multi-prototype clustering algorithm. Pattern Recogn. **42**(5), 689–698 (2009)
13. Chavent, M., Lechevallier, Y., Briant, O.: DIVCLUS-T: a monothetic divisive hierarchical clustering method. Comput. Stat. Data Anal. **52**(2), 687–701 (2007)
14. Zahn, C.T.: Graph-theoretical methods for detecting and describing gestalt clusters. IEEE Trans. Comput. **100**(1), 68–86 (1971)
15. Halkidi, M., Batistakis, Y., Vazirgiannis, M.: On clustering validation techniques. J. Intell. Inf. Syst. **17**(2), 107–145 (2001)

Rule Based Machine Translation: English to Malayalam: A Survey

V.C. Aasha and Amal Ganesh

Abstract In this modern era, it is a necessity that people communicate with other parts of world without any language barriers. Machine translation system can be very useful tool in this context. The main purpose of a machine translation system is to convert text of one natural language to a different natural language without losing its meaning. The translation of text can be done in several ways mainly, rule based and corpus based—each of which uses various approaches to obtain correct translation. In this paper, we focus on the various machine translation approaches with a core focus on English to Malayalam translation using Rule Based approach.

Keywords English to malayalam machine translation · RBMT · Transfer rules

1 Introduction

Machine translation is a field in Natural Language Processing that converts text in one source language to a different target language. The ultimate aim of a translation system is to achieve correct language translation with minimum human intervention. Some of these systems use linguistic information of both source and target language, while some others translate on the basis of mathematical probabilities. The former are called Rule based machine translation (RBMT) systems and the latter are Statistical machine translation (SMT) systems. The conversion of languages can be either in single direction—from source to target language or bidirectional between the language pairs. Bilingual translation systems are designed for exactly two languages whereas the systems capable of producing translations for

V.C. Aasha · A. Ganesh (✉)
Department of Computer Science & Engineering, Vidya Academy of Science & Technology, Thalakkottukara P. O, Thrissur, Kerala, India
e-mail: amal.ganesh@vidyaacademy.ac.in

V.C. Aasha
e-mail: aashavc@gmail.com

A. Nagar et al. (eds.), *Proceedings of 3rd International Conference on Advanced Computing, Networking and Informatics*, Smart Innovation, Systems and Technologies 43, DOI 10.1007/978-81-322-2538-6_46

more than two languages are called multilingual systems. These systems are multilingual in the aspect that they translate from one source language to multiple target languages as well as to-and-fro between the languages. There exist different approaches [1] to machine translation mainly: Rule based approach and Corpus based approach.

1.1 Rule Based Machine Translation (RBMT)

One of the earliest translation method developed is the RBMT system. It uses linguistic rules and language order for its conversion to destination language. This information includes knowledge about syntax, semantics and morphology of each language respectively in addition to the information obtained from dictionary. The RBMT systems follow various approaches for translation namely; Direct, Transfer and Interlingua approach. In all of these approaches, the rules are an important factor during various stages of translation. The much discussed diagram of machine translation is given in Fig. 1. The figure shows the translation of source text to target text through different RBMT approaches [1, 2].

a. Direct approach
Direct approach of Machine Translation (MT) directly translates the input language into output language with the help of a dictionary. In a Direct MT system, each new language pairs require creation of new system. Also, it may produce satisfactory results only for simple sentences. The performance of such systems relies upon the accurate entries of words found in dictionary [3].

b. Interlingua approach
The Interlingua approach of translation involves translating the source sentence to an intermediary state called interlingua. The interlingua representation is then translated to the target language in order to produce meaningful translation.

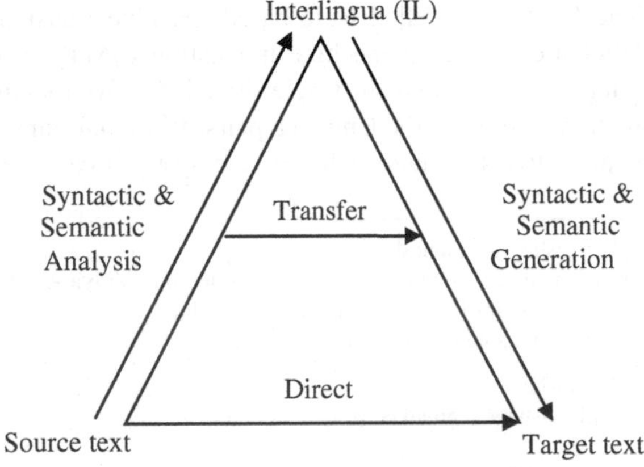

Fig. 1 RBMT approaches

Since this intermediate state is not language dependent, it can be used for any languages. Thus Interlingua approach can be followed for multilingual translation by translating the same interlingua to multiple languages.

c. Transfer approach

In Transfer approach the translation is based on transfer rules. This approach represents the text to a less language dependent representation rather than the independent representation of Interlingua approach. The three phases of translation are: Analysis stage, Transfer stage and Generation [2]. The source text is converted to a source dependent representation with the help of parser. It is then transferred to target—language dependent representation using transfer rules. The transfer rules are applied at each step for ensuring correct translation. The necessary morphological inflections for the sentences are added in the third phase.

The development of RBMT systems involve much effort to coordinate between various stages of translation. Even then, RBMT systems are guaranteed to produce nearly correct output always as it is based on syntax and semantics of both languages, and not on any probability measures. The accuracy of output can be enhanced if the translation is limited to a particular domain. The quality of translation can be further increased by pre-processing the input sentence. RBMT systems designed for various domains can be integrated to a single system. Such systems could correctly translate sentences irrespective of the domain. In [4] they use Moses, an open source decoder to integrate multiple RBMT systems to hybrid system.

1.2 Corpus Based Machine Translation

The complexity of creating RBMT system paved way for developing other MT approaches. The more recent approaches [1, 3] of translation are Corpus based machine translation (CMT) and Hybrid based translation (HBT). Corpus based MT require enormous amount of corpus data to translate the text. The two major classifications of CMT are SMT [Statistical Machine Translation] and EBMT [Example Based Machine Translation]. Hybrid approach is a relatively new advancement which utilises the properties of both Rule-Based and Corpus-Based approaches.

1.2.1 Statistical Machine Translation

SMT makes use of corpus and calculates the order of words in both the source—destination languages using mathematical statistical probability. i.e., SMT is based on the probability distribution P (t/s), which indicates the probability to convert a source sentence ‘s’ to a target sentence ‘t’ [3]. SMT addresses ambiguity problems and works well for unrestricted collection of text. Statistical machine translations are further categorised as word based and phrase based translation.

1.2.2 Example Based Machine Translation

The EBMT systems use the sample sentences stored in the database for translation of new sentence. The system stores large amount of translated texts in its corpora for reference. This prior information increases the chance of correctly translating new sentences of same pattern. When a new sentence has to be translated it is checked with existing corpus to find a matching pattern. Once the matching pattern is found, then the input sentence is translated accordingly with help of database. It is likely to get correct translation, if the corpus is sufficient with varying sentence patterns. The corpus of the system can be made rich by adding new translations to the database and reusing them afterwards.

1.2.3 Hybrid-Based Translation

The Hybrid approach of MT utilises properties of RBMT and SMT and is proven to have better efficiency. Some Hybrid systems use a rule-based approach followed by correction of output using statistical information. On the other hand, in some systems statistical pre-processing is done followed by correction using transfer rules. Hybrid systems are more powerful as they are based on both rules and statistics.

2 Related Works

Since our research concentrates on translation of language from source to target language, the related works are classified as MT works in India and English to Malayalam MT Systems. This classification helps us to better understand the tools and technologies used in various MT systems.

2.1 Machine Translation Works in India

Machine translation works in India started in 1991 although it began during 1960's in the world. Google translate, Bing-translate, Systran etc. are some famous statistical machine translation systems available worldwide. The works in machine translation from English to Dravidian languages had started a few years back. Due to the morphological complexities involved, not much interest has been shown in developing computational resources for Malayalam language. Thus the linguistic works in this language are limited and inadequate.

Machine translation from English to the Indian languages and among various Indian languages is dynamically in progress. *AnglaBharati* is one of the early MT system financed by TDIL Programme [5] and developed by Indian Institute of Technology (IIT)-Kanpur. AnglaBharati MT System [phase I (1991) and II (2004)] is for conversion of English to Assamese, Bangla, Hindi, Malayalam, Nepali, Punjabi, Telugu, and Urdu languages. Since they use Interlingua approach of RBMT, a single analysis on English language is sufficient to develop an interlingua. This intermediate structure is translated to different Indian language output. *Anubharati* [phase I (1995) and II (2004)] developed by IIT-Kanpur translates Hindi to any other Indian languages, using example-based approach. *Anusaaraka* (1995) [6], designed by IIT-Kanpur along with University of Hyderabad translates from Punjabi, Bengali, Telugu, Kannada, Marathi and English into Hindi using concepts of 'Paninian Grammar'. The RBMT approaches for English—Malayalam language pair limit only to a few systems. One of them has been done for six worded sentences [7] and another for simple sentences in general domain [8]. Better results of machine translation can be achieved if the transfer rules cover all possible sentence pattern, and also with the help of an improved bilingual dictionary. Since creating a machine translator system with such specifications is a tedious task, most translators limit the translation to a domain.

2.2 English to Malayalam Machine Translation

The growth of English to Malayalam machine translation systems began with the development of AnglaMalayalam. Analysis of some of the English to Malayalam Machine Translation systems are shown in Table 1.

Table 1 Analysis of English–Malayalam machine translation systems

MT system	Method	Year	Domain	Evaluation criteria
AnglaMalayalam [18]	Interlingua	2008	Health, tourism	Accuracy 75 %
English to Malayalam [7]	Rule based	2009	General	Works up to 6 word simple sentences
English to Malayalam [19]	Statistical	2009	General	BLEU score 16.10
English to Malayalam [8]	Rule based	2011	General	Accuracy: 53.63 %
English to Malayalam [13]	Transfer based	2012	Simple sentences	Word error rate: 0.429 F-measure: 0.57
English to Malayalam [20]	Hybrid based	2013	History	BLEU score 69.33

In addition to these, an Example based Malayalam to English MT system [9] claims quality results for 75 % of their test sets. Another rule based MT system translates sentences from English to Malayalam and Hindi [10] with F-mean of 0.74. This system was designed for translating sentences from administration domain.

2.3 Working Procedure of RBMT Systems

The Machine translation involves various pre-processing steps of NLP like tokenisation, stemming, POS [Part of Speech] tagging, and the rest. There are various NLP tools available like NLTK [11], openNLP etc. to do preprocessing tasks. The working procedure of a general rule based machine translation using transfer approach is as described. The input sentence is tokenised and parsed with help of parser. The source parse tree is converted to target parse tree using transfer rules. The transfer rules are formulated according to the target language structure. Then words from target parse tree are mapped with a bilingual dictionary to obtain the word meaning in output language. Later, the morphological inflections are added to the words for obtaining meaningful text. In [12] they use OAK parser to obtain the POS tag and parse tree of English sentence. In [7, 8, 13] Stanford parser offer POS tagging, parse tree and provide dependency information of the input sentence. The dependency information indicates how words in a sentence are related to each other. The input text is tokenised and a tag corresponding to each word is assigned by the parser. The Stanford parser [14] uses Penn Tree Tagset for this. Utilising this information as well as the transfer rules, the source parse tree is reordered to target parse tree. Word reordering is done according to the transfer rules and each sentence follows definite rule patterns. The tag of the word in source language is represented on left side and right side tags represent target language words. The rules are devised in such a way that English language maintains Subject—Verb—Object order whereas the target language (Indian) adhere Subject—Object—Verb order. In [15] the correct word order of Marathi language was obtained upon post-order traversals on the parse tree. It used a named entity tagger to identify person names, location etc. Also it had special features for solving word sense disambiguation of prepositions. In [13] the nouns, pronouns and verbs are morphologically generated with the help of transformation rules. A FST based Malayalam morphological synthesizer is used for machine translation in [8]. Also an SVM based English to Malayalam transliterator is used for translating named entities.

The major concern in machine translation is the ambiguities in the meaning of words based on the context. Malayalam language has diverse morphological variations and it uses suffixes for words, compared to prefixes in English language. Such issues in translation can be resolved by considering the semantic information about the text. The meaning of words suitable to the context can be then selected. In

Malayalam there are seven case markers for nouns alone. Similarly the inflections for verbs, adverbs and adjectives have to be added to get perfect translation.

In short, the working of RBMT system for English to Malayalam language using transfer based approach can be described as follows. After the POS tagging of English sentence by parser, they are reordered according to transfer rules. The transfer rules represent the rules required for reordering sentences to target language format. The words from reordered target parse tree are mapped with English to Malayalam dictionary. The dictionary provides the Malayalam meaning which can be substituted to respective English counterparts. The words that are not in dictionary like named entities, place names etc. are transliterated. The necessary morphological inflections are added to the Malayalam sentence to generate meaningful translation [16, 17]. The efficiency of the MT system can be raised by restraining the translation to a particular domain. Thereby the dictionary contains more words specific to the domain. This makes the system more productive by providing correct word meanings and thereby to get more accurate translations.

3 Conclusion

Attempts to build machine translation systems for various languages are in progress. But a perfect system with its output much close as that of human translator has not yet achieved. In order to achieve reasonable translation quality, one challenge is to create a corpus or dictionary with enormous number of corpora consisting of words or sentences pertaining to the target language. On the other hand, RBMT systems take time to generate transfer rules for all sentence structures. The area of interest here is a study on rule based MT system using transfer approach for translation from English to Malayalam language. The parser generates a parse tree for the input sentence. The source parse tree is reordered with the help of transfer rules to obtain target parse tree. The words are then mapped with bilingual dictionary and transliterator to obtain Malayalam words. The final output sentence in Malayalam is represented with proper suffixes attached. The translation accurateness of RBMT systems can be increased by making it domain specific. Thus it could cover all words of that particular domain and the grammar rules. The multiple RBMT systems of different domains can be combined to provide robust English to Malayalam translation system. i.e., our ultimate aim is to minimize the limitation of RBMT system by incorporating as many rules so as to make the system error prone free.

References

1. Tripathi, S., Sarkhel, J.K.: Approaches to machine translation. Ann. Libr. Inf. Stud. **57**, 388–393 (2010)
2. Daniel, J., Martin, J.H.: Speech and Language Processing: An Introduction to Natural Language Processing, Speech Recognition, and Computational Linguistics. 2nd edn. Prentice-Hall (2009)

3. Antony, P.J.: Machine translation approaches and survey for Indian languages. Comput. Linguist. Chin. Lang. Process. **18**, 47–78 (2013)
4. Eisele, A., Federmann, C., Saint-Amand, H., Jellinghaus, M., Herrmann, T., Chen, Y.: Using moses to integrate multiple rule-based machine translation engines into a hybrid system. In: Proceedings of the Third Workshop on Statistical Machine Translation, pp. 179–182. Columbus Ohio, USA (2008)
5. Technology Development for Indian Languages, Government of India: http://tdil.mit.gov.in. Accessed 1 Dec 2014
6. Bharati, A., Chaitanya, V., Kulkarni, A.P., Sangal, R.: Anusaaraka: machine translation in stages. In: Vivek, A. (ed.) Quarterly in Artificial Intelligence, vol. 10, no. 3. NCST, Mumbai (1997)
7. Rajan, R., Sivan, R., Ravindran, R., Soman, K.P.: Rule based machine translation from English to Malayalam. In: IEEE International Conference on Advances in Computing, Control and Telecommunication Technologies, pp. 439–441 (2009)
8. Harshwardhan, R.: Rule based machine translation system for English to Malayalam language. Dissertation, CEN, Amrita Vishwa Vidyapeetham, Coimbatore (2011)
9. Anju, E.S., Manoj Kumar, K.V.: Malayalam to English machine translation: an EBMT system. IOSR J. Eng. **4**(1), 18–23 (2014)
10. Meera, M., Sony, P.: Multilingual machine translation with semantic and disambiguation. In: Fourth International Conference on Advances in Computing and Communications (ICACC), pp. 223–226. IEEE (2014)
11. Bird, S., Klein, E., Loper, E.: Natural Language Processing with Python. O'Reilly Media Inc, CA (2009)
12. Alawneh, M.F., Sembok, T.M.: Rule-based and example-based machine translation from English to Arabic. In: Sixth International Conference on Bio-Inspired Computing: Theories and Applications. IEEE Computer Society, pp. 343–347 (2011)
13. Nair, A.T., Idicula, S.M.: Syntactic based machine translation from English to Malayalam. In: IEEE International Conference on Data Science and Engineering (ICDSE), pp. 198–202 (2012)
14. Stanford parser: http://nlp.stanford.edu/software/lex-parser.shtml. Accessed 1 Dec 2014
15. Garje, G.V., Kharate, G.K., Harshad, K.: Transmuter: an approach to rule-based English to Marathi machine translation. Int. J. Comput. Appl. **98**(21), 33–37 (2014)
16. Nair, L.R., Peter, D., Renjith, S., Ravindran, P.: A system for syntactic structure transfer from Malayalam to English. In: Proceedings of International Journal of Computer Applications, International Conference on Recent Advances and Future Trends in Information Technology (2012)
17. Nair, L.R., Peter, S.D.: Development of a rule based learning system for splitting compound words in Malayalam language. In: Recent Advances in Intelligent Computational Systems (RAICS), pp. 751–755. IEEE, 22–24 Sept (2011)
18. Centre for Development of Advanced Computing. CDAC Annual Report 2007–2008: http://cdac.in/index.aspx?id=pdf_AnnualReport_08_09. Accessed 1 Dec 2014
19. Rahul, C., Dinunath, K., Ravindran, R., Soman, K.P.: Rule based reordering and morphological processing for English-Malayalam statistical machine translation. In: IEEE International Conference on Advances in Computing, Control, and Telecommunication Technologies, pp. 458–460 (2009)
20. Nithya, B., Joseph, S.: A hybrid English to English to Malayalam machine translation. IOSR J. Eng. **81**(8), 11–15 (2013)

Time Series Forecasting Through a Dynamic Weighted Ensemble Approach

Ratnadip Adhikari and Ghanshyam Verma

Abstract Time series forecasting has crucial significance in almost every practical domain. From past few decades, there is an ever-increasing research interest on fruitfully combining forecasts from multiple models. The existing combination methods are mostly based on time-invariant combining weights. This paper proposes a dynamic ensemble approach that updates the weights after each new forecast. The weight of each component model is changed on the basis of its past and current forecasting performances. Empirical analysis with real time series shows that the proposed method has substantially improved the forecasting accuracy. In addition, it has also outperformed each component model as well as various existing static weighted ensemble schemes.

Keywords Time series forecasting · Forecasts combination · Changing weights · Forecasting accuracy

1 Introduction

A time series is a sequential collection of observations, recorded at specific intervals of time. Time series forecasting is the estimation of unknown future data through a mathematical model, constructed with available observations [1]. Obtaining reasonably precise forecasts is a major research concern that led to the development of various forecasting models in literature [2, 3]. However, there are generally a

R. Adhikari (✉) · G. Verma
Department of Computer Science & Engineering, The LNM Institute of Information Technology, 302031 Jaipur, India
e-mail: adhikari.ratan@gmail.com

G. Verma
e-mail: ghanshyam.verma8@gmail.com

A. Nagar et al. (eds.), *Proceedings of 3rd International Conference on Advanced Computing, Networking and Informatics*, Smart Innovation, Systems and Technologies 43, DOI 10.1007/978-81-322-2538-6_47

number of obstructing issues, of which some are related to the time series itself, whereas others are related to the particular model. As such, no single model alone can achieve uniformly best forecasts for a specific category of time series [4]. On the other hand, extensive research evidences show that combining forecasts from multiple conceptually different models is a very prolific alternative to using only a single model [3, 5, 6]. Together with mitigating the model selection risk and substantially improving the forecasting accuracy, this approach also often outperforms each component model.

Fascinated by the strength of forecasts ensemble, a large amount of works has been carried out in this direction over the past few decades [3, 5, 7]. As a result, several forecasts combination approaches have been evolved in literature. These include simple statistical ensembles, e.g. *simple average*, *trimmed mean*, *Winsorized mean*, *median* [8] as well as more advanced combination methods. These approaches are mostly based on forming a weighted linear combination of component forecasts, where the final combining weights are fixed for each model. However, instead of using constant weights throughout, it can be potentially beneficial to update the weights after specific intervals of time. But, this topic has received considerably limited research attention so far. Some notable works on combining forecasts through time-dependent weights are those of Deutsch et al. [9], Fiordaliso [10], and Zou and Yang [11]. Each of them has demonstrated that combining with time-varying weights can be advantageous over conventional static weighted linear combinations methods.

This paper proposes a forecasts combination approach that updates the weights after generation of each new forecast. As such, the component weights vary with each time index, thereby making the combination scheme dynamic. The associated weight of each constituent model is updated on the basis of its past and current forecasting performances. The proposed ensemble is constructed with nine individual models and is tested on four real time series datasets. The forecasting performances of the proposed method is compared with those of the component models as well as various other static weighted linear combination methods, in terms of two well-known absolute error measures.

The rest of the paper is organized as follows. Section 2 briefly describes the forecasts combination methodology and Sect. 3 presents the proposed dynamic ensemble approach. Section 4 reports the details of empirical works and finally Sect. 5 concludes the paper.

2 Combination of Multiple Forecasts

A linear combination is the most common method of combining forecasts. For the dataset $\mathbf{Y} = [y_1, y_2, \ldots, y_N]^{\mathrm{T}}$ and its n forecasts $\hat{\mathbf{Y}}^{(i)} = \left[\hat{y}_1^{(i)}, \hat{y}_2^{(i)}, \ldots, \hat{y}_N^{(i)}\right]^{\mathrm{T}}$ $(i = 1, 2, \ldots, n)$, the forecasts from a linear combination are given by:

$$\hat{y}_k = w_1\hat{y}_k^{(1)} + w_2\hat{y}_k^{(2)} + \ldots + w_n\hat{y}_k^{(n)} \tag{1}$$
$$\forall k = 1, 2, \ldots, N.$$

In (1), w_i is the weight to the ith forecasting model $\forall i = 1, 2, \ldots, n$. Usually, the combining weights are assumed to be *nonnegative*, i.e. $w_i \geq 0\, \forall i$ and *unbiased*, i.e. $\sum_{i=1}^{n} w_i = 1$ [2, 12]. Some widely popular methods of linearly combining forecasts are briefly discussed below.

- The *simple average* is the easiest combination method that assigns an equal weight to each component forecast, i.e. $w_i = 1/n\, \forall i = 1, 2, \ldots, n$. Due to its simplicity, excellent accuracy, and impartiality, simple average is often a top priority choice in forecasts combination [2, 8].
- The *trimmed mean*, *Winsorized mean*, and *median* are other robust alternatives to simple average. A trimmed and Winsorized mean both forms a simple average by fetching an equal number of α smallest and largest forecasts and either completely discarding them or setting them equal to the $(\alpha + 1)$th smallest and largest forecasts, respectively [8]. The simple average and median are both special cases of trimmed mean, corresponding to no trimming and full trimming, respectively.
- In an *Error-based* (*EB*) combination, the weight to each model is assumed to be inversely proportional to its in-sample forecasting error, so that:

$$w_i = e_i^{-1} / \sum_{i=1}^{n} e_i^{-1} \tag{2}$$
$$\forall i = 1, 2, \ldots, n.$$

  Here, e_i is an in-sample forecasting error of the ith model. This method is based on the rational ideology that a model with more error receives less weight and vice versa [2, 12].
- The *Ordinary Least Square* (*OLS*) method estimates the combining weights $w_i\, (i = 1, 2, \ldots, n)$ together with a constant w_0 through minimizing the *Sum of Squared Error* (*SSE*) of the combined forecast [12, 13]. As the actual dataset is unknown in advance, so in practice, the weights are estimated through minimizing an in-sample combined forecast SSE.
- The *outperformance* method, proposed by Bunn [14] determines a weight on the basis of the number of times the particular model performed best in the past in-sample forecasting trials.

As per the formulation (1), all the aforementioned methods assume static time-invariant weights. As stated in the outset, there has been very limited research on combining forecasts through time-varying weights.

3 The Proposed Dynamic Ensemble Approach

In this study, we modify the static time-invariant weighted linear combination scheme (1) to the following dynamic framework:

$$\hat{y}_k = w_{k,1}\hat{y}_k^{(1)} + w_{k,2}\hat{y}_k^{(2)} + \ldots + w_{k,n}\hat{y}_k^{(n)} \qquad (3)$$
$$\forall k = 1, 2, \ldots, N.$$

Here, $w_{k,i}$ is the weight to the ith model at the instance k. Now, the weight $w_{k,i}$ depends on both the model index i and the time index k. As usual, we assume nonnegativity and unbiasedness of the weights, i.e. $w_{k,i} \geq 0\ (\forall i = 1, 2, \ldots, n)$ and $\sum_{i=1}^{n} w_{k,i} = 1\ (\forall k = 1, 2, \ldots, N)$. We consider *Mean Squared Error* (*MSE*) and *Mean Absolute Percentage Error* (*MAPE*) to evaluate the forecasting performances of the component models [15]. These are defined below.

$$\text{MSE} = \frac{1}{N}\sum_{t=1}^{N}(y_t - \hat{y}_t)^2 \qquad (4)$$

$$\text{MAPE} = \frac{1}{N}\sum_{t=1}^{N}\left|\frac{y_t - \hat{y}_t}{y_t}\right| \times 100 \qquad (5)$$

In the proposed formulation, we update the weights $w_{k,i}$ after generation of each new forecast on the basis of the past and present forecasting performances of the component models. The past performance of each model is evaluated on an in-sample validation dataset, whereas the present performance is assessed through the out-of-sample forecasts, obtained till now. The requisite steps of the proposed ensemble framework are presented in Algorithm 1. The algorithm uses the standard matrix notations, popularized by Golub and Van Loan [16].

Algorithm 1 The proposed dynamic ensemble framework

Inputs: The in-sample dataset $\mathbf{Y}_{\text{in}} = [y_1, y_2, \ldots, y_{N_{\text{in}}}]^{\text{T}}$; the sizes N_{tr}, N_{vd}, N_{ts} of the training, validation, and testing sets, respectively, so that $N_{\text{tr}} + N_{\text{vd}} = N_{\text{in}}$ and $N_{\text{in}} + N_{\text{ts}} = N$ (size of the whole time series); the n forecasting models M_i $(i = 1, 2, \ldots, n)$.

Output: The combined forecast vector $\hat{\mathbf{Y}} = [\hat{y}_{N_{\text{in}}+1}, \hat{y}_{N_{\text{in}}+2}, \ldots, \hat{y}_{N_{\text{in}}+N_{\text{ts}}}]^{\text{T}}$.

Steps:

1. **for** $i = 1$ to n **do**
2. Fit the model M_i to the in-sample training dataset $\mathbf{Y}_{\text{tr}} = [y_1, y_2, \ldots, y_{N_{\text{tr}}}]^{\text{T}}$.
3. Use it to forecast the validation dataset $\mathbf{Y}_{\text{vd}} = [y_{N_{\text{tr}}+1}, y_{N_{\text{tr}}+2}, \ldots, y_{N_{\text{tr}}+N_{\text{vd}}}]^{\text{T}}$.
4. Obtain the in-sample forecast $\hat{\mathbf{Y}}_{\text{vd}}^{(i)} = \left[\hat{y}_{N_{\text{tr}}+1}^{(i)}, \hat{y}_{N_{\text{tr}}+2}^{(i)}, \ldots, \hat{y}_{N_{\text{tr}}+N_{\text{vd}}}^{(i)}\right]^{\text{T}}$. $//\hat{\mathbf{Y}}_{\text{vd}}^{(i)}$ *is the forecast of* $\mathbf{Y}_{\text{vd}}$ *through* M_i.
5. Using M_i forecast the testing dataset $\mathbf{Y}_{\text{ts}} = [y_{N_{\text{in}}+1}, y_{N_{\text{in}}+2}, \ldots, y_{N_{\text{in}}+N_{\text{ts}}}]^{\text{T}}$.
6. Obtain the forecast $\hat{\mathbf{Y}}_{\text{ts}}^{(i)} = \left[\hat{y}_{N_{\text{in}}+1}^{(i)}, \hat{y}_{N_{\text{in}}+2}^{(i)}, \ldots, \hat{y}_{N_{\text{in}}+N_{\text{ts}}}^{(i)}\right]^{\text{T}}$. $//\hat{\mathbf{Y}}_{\text{ts}}^{(i)}$ *is the forecast of* $\mathbf{Y}_{\text{ts}}$ *through* M_i.
7. **end for**
8. $\mathbf{w} = \mathbf{0}_{N_{\text{ts}} \times n}$; $\mathbf{F} = \left[\hat{\mathbf{Y}}_{\text{ts}}^{(1)} | \hat{\mathbf{Y}}_{\text{ts}}^{(2)} | \ldots | \hat{\mathbf{Y}}_{\text{ts}}^{(n)}\right]_{N_{\text{ts}} \times n}$
   $\mathbf{V} = \left[\hat{\mathbf{Y}}_{\text{vd}}^{(1)} | \hat{\mathbf{Y}}_{\text{vd}}^{(2)} | \ldots | \hat{\mathbf{Y}}_{\text{vd}}^{(n)}\right]_{N_{\text{vd}} \times n}$; $\hat{\mathbf{Y}} = [\,]$
   //$\mathbf{w}$, $\mathbf{F}$, *and* $\mathbf{V}$ *are respectively the weight, forecast, and validation matrices.*
   //$\mathbf{w}$ *and* $\hat{\mathbf{Y}}$ *are initialized to the zero matrix and empty vector, respectively.*
9. Set $w(1, :) = 1/n$ //*Set the initial weights as* $w_{1,i} = 1/n$ $(\forall i = 1, 2, \ldots, n)$.
10. **for** $k = 1$ to N_{ts} **do**
11. $\hat{\mathbf{Y}} = \left[\hat{\mathbf{Y}} | \mathbf{F}(k, :)\, w(k, :)'\right]$; $\mathbf{Y}^* = \left[\mathbf{Y}_{\text{vd}} | \hat{\mathbf{Y}}\right]$
    //*Find* $\hat{y}_k = \mathbf{F}(k, :)\, w(k, :)' = \sum_{i=1}^{n} \hat{y}_{N_{\text{in}}+1}^{(i)} w_{k,i}$ *through* (3) *and append it to* $\hat{\mathbf{Y}}$.
12. Set $i = 1$
13. **while** ($i \le n$ **and** $k < N_{\text{ts}}$) **do**
14. $\mathbf{Y}_i^* = \left[\mathbf{V}(:, i)' | \mathbf{F}(1:k, i)'\right]'$
15. $w(k+1, i) = 1/\mathit{error}(\mathbf{Y}^*, \mathbf{Y}_i^*)$ //*Weight is the inverse absolute error between* $\mathbf{Y}^*$ *and* $\mathbf{Y}_i^*$.
16. $i = i + 1$
17. **end while**
18. $w(k+1, :) = w(k+1, :)/\text{sum}(w(k+1, :))$ //*Make the weights unbiased.*
19. **end for**

The steps 14 and 15 of Algorithm 1 implements our proposed weight assignment and renewal mechanism. The successive forecasting errors has a key role in weights estimation. In this work, MSE is used as the absolute error function.

4 Empirical Results and Discussions

The effectiveness of the proposed dynamic ensemble approach is evaluated on four real time series datasets with nine component forecasting models. These nine models are selected from the following widely popular classes: *random walk* [17], *Box-Jenkins* [18], *Artificial Neural Network* (*ANN*) [17, 19], and *Support Vector Machine* (*SVM*) [20, 21]. The random walk and Box-Jenkins models assume linear data generation function, whereas, ANN and SVM have recognized success records in modeling nonlinear structure in a time series [17, 21]. The subclass Box-Jenkins models, viz. *Autoregressive Integrated Moving Average* (*ARIMA*) and *Seasonal ARIMA* (*SARIMA*) [15] are used in this work. These two models are commonly expressed as $\mathrm{ARIMA}\,(p,d,q)$ and $\mathrm{SARIMA}\,(p,d,q)\times(P,D,Q)^s$, respectively [15, 17]. The parameters (p,P), (d,D), and (q,Q) respectively signify the *auto-regressive, degree of differencing*, and *moving average* processes, "*s*" being the *period of seasonality*. We have considered three well-known classes of ANN models, viz. *Feedforward ANN* (*FANN*) [17], *Elman ANN* (*EANN*) [2, 22], and *Generalized Regression Neural Network* (*GRNN*) [4]. Both iterative (ITER) and direct (DIR) approaches [23] are adopted for implementing each ANN model. Finally, the *Least Square SVM* (*LS-SVM*) framework, pioneered by Suykens and Vandewalle [21] is considered for SVM models.

The experiments in this study are carried out on MATLAB. The default ARIMA class of the Econometric toolbox, the neural network toolbox [24], and the LS-SVMlab toolbox [25] are respectively used for fitting the Box-Jenkins, ANN, and SVM models. The appropriate number of input (i) and hidden (h) nodes of an $i\times h\times o$ ANN are selected through in-sample training-validation trials. The number of output (o) nodes is 1 for iterative approach and it is equal to the testing dataset size for direct approach. Each FANN is trained for 2000 epochs through the *Levenberg Marquardt* (*LM*) backpropagation training algorithm, using the corresponding MATLAB function *trainlm* [24]. Similarly, each EANN is trained through the *traingdx* function, based on gradient descent with momentum and adaptive learning [24]. For SVM, the *Radial Basis Function* (*RBF*) kernel is used. The optimal SVM regularization constant (C) and RBF kernel parameter (σ) are estimated from the ranges $\left[10^{-5},\,10^{5}\right]$ and $[2^{-10},\,2^{10}]$, respectively through tenfold crossvalidation technique.

The following four real time series datasets are used in this study. (1) **Emission:** contains monthly records of CO_2 emission in *parts per million* (*ppm*), Mauna Loa, US from January, 1965 to December, 1980 [26], (2) **Inflation:** contains monthly inflation in consumer price index numbers of industrial workers of India from 31 August, 1969 to 30 September, 2014 [27], (3) **Google:** contains monthly Google stock prices from 19 August, 2004 to 24 February, 2014 [28], (4) **Facebook:** contains the daily closing stock prices of Facebook from 18 May, 2012 to 3 September, 2014 [28]. The first one is a monthly seasonal time series, whereas the rest three are non-stationary, nonseasonal series. Prior to fitting the forecasting models, the observations of each time series are normalized to lie in the [0, 1] interval through the following transformation:

$$y_t^{(\text{new})} = \frac{y_t - y^{(\min)}}{y^{(\max)} - y^{(\min)}} \qquad (6)$$

$$\forall t = 1, 2, \ldots, N_{\text{tr}}.$$

In (6), $\mathbf{Y} = [y_1, y_2, \ldots, y_{N_{\text{tr}}}]^{\text{T}}$, $\mathbf{Y}^{(\text{new})} = \left[y_1^{(\text{new})}, y_2^{(\text{new})}, \ldots, y_{N_{\text{tr}}}^{(\text{new})}\right]^{\text{T}}$ are the training dataset and its corresponding normalized set, respectively, $y^{(\min)}$, $y^{(\max)}$ being respectively the minimum and maximum values of $\mathbf{Y}$. The forecasts of each model are again transformed back to the original scale. All the necessary modeling information of the four time series is presented in Table 1 and their respective plots are depicted in Fig. 1.

In the present work, we have compared the forecasting accuracy of our proposed dynamic ensemble method with that of the nine component models as well as seven other widely recognized static linear combination schemes. The forecasting error of each method is measured in terms of MSE and MAPE. The effectiveness of the proposed approach depends a lot on the choice of the in-sample validation dataset. As such, to have a reasonable similarity between the in-sample validation and out-of-sample testing datasets, we have considered them to be of equal length, i.e. $N_{\text{vd}} = N_{\text{ts}}$. Table 2 presents the obtained forecasting results through all methods for the four time series datasets. The results of the proposed ensemble method is shown in bold scripts.

From Table 2, we can see that the proposed dynamic weighted ensemble approach has achieved overall lowest forecasting errors among all methods in terms of both MSE and MAPE. From the obtained results, it can be said that the approach of using time-variant weights is a fruitful alternative to the common static weighted

Table 1 Information of the time series datasets

Information	Emission	Inflation	Google	Facebook
Size (total, testing)	(192, 32)	(542, 100)	(498, 100)	(576, 92)
Box-Jenkins	SARIMA	ARIMA	ARIMA	ARIMA
	$(0,1,1) \times (0,1,1)^{12}$	$(2,1,0)$	$(2,1,0)$	$(1,0,0)$
ITER-FANN	$14 \times 10 \times 1$	$2 \times 1 \times 1$	$5 \times 9 \times 1$	$9 \times 11 \times 1$
DIR-FANN	$8 \times 14 \times 32$	$5 \times 3 \times 100$	$4 \times 10 \times 100$	$2 \times 2 \times 92$
ITER-EANN	$14 \times 14 \times 1$	$9 \times 14 \times 1$	$12 \times 14 \times 1$	$13 \times 15 \times 1$
DIR-EANN	$10 \times 17 \times 32$	$5 \times 10 \times 100$	$3 \times 7 \times 100$	$3 \times 11 \times 92$
ITER-GRNN	$10 \times 1 \times 1$	$13 \times 11 \times 1$	$5 \times 9 \times 1$	$3 \times 1 \times 1$
DIR-GRNN	$16 \times 3 \times 32$	$3 \times 11 \times 100$	$4 \times 7 \times 100$	$2 \times 1 \times 92$
LS-SVM (C, σ)	(3.47, 3.843)	(0.457, 7.366)	(0.0054, 0.0713)	(4984.848, 0.527)

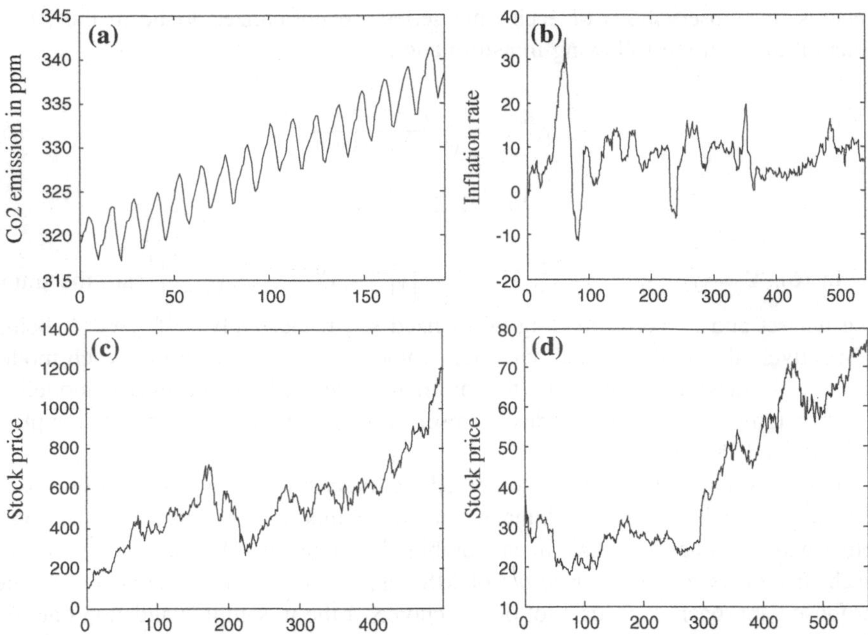

Fig. 1 Plots of the time series. **a** Emission. **b** Inflation. **c** Google. **d** Facebook

Table 2 Obtained forecasting results through all methods

Forecasting errors	Emission		Inflation		Google[a]		Facebook	
	MSE	MAPE	MSE	MAPE	MSE	MAPE	MSE	MAPE
Random walk	0.221	0.102	3.54	19.28	15.755	1.097	4.349	2.582
Box-Jenkins	0.451	0.163	9.117	30.763	14.908	1.051	4.671	2.785
ITER-FANN	0.132	0.0763	10.083	32.661	21.228	1.316	5.358	2.789
DIR-FANN	0.295	0.117	6.401	24.871	21.317	1.252	6.301	2.971
ITER-EANN	0.117	0.0721	9.885	27.115	18.916	1.059	5.772	2.951
DIR-EANN	0.331	0.131	8.224	28.376	22.332	1.22	5.43	2.794
ITER-GRNN	0.434	0.141	9.656	28.986	26.922	1.472	4.736	2.649
DIR-GRNN	0.353	0.126	6.527	24.365	16.029	1.111	6.899	3.344
LS-SVM	0.177	0.0867	5.592	20.174	23.028	1.413	5.93	2.944
Simple average	0.16	0.0814	2.172	13.838	4.388	0.588	1.678	1.518
Trimmed mean[b]	0.168	0.0834	2.322	14.164	4.2	0.584	1.886	1.639
Winsorized mean	0.166	0.0835	2.336	14.282	4.036	0.576	1.907	1.629
Median	0.176	0.0859	2.529	14.787	5.075	0.631	2.056	1.799
EB[c]	0.138	0.074	2.363	14.256	5.262	0.649	1.747	1.528
OLS	0.19	0.101	2.789	14.247	6.261	0.685	1.73	1.524
Outperformance	0.137	0.0746	2.259	14.902	7.742	0.789	2.142	1.723
Proposed	**0.132**	**0.0719**	**2.15**	**13.751**	**3.912**	**0.559**	**1.599**	**1.474**

[a] Original *MSE* obtained MSE × 10−4

[b] 20 % trimming is used

[c] The weight to each model is proportional to in-sample MSE

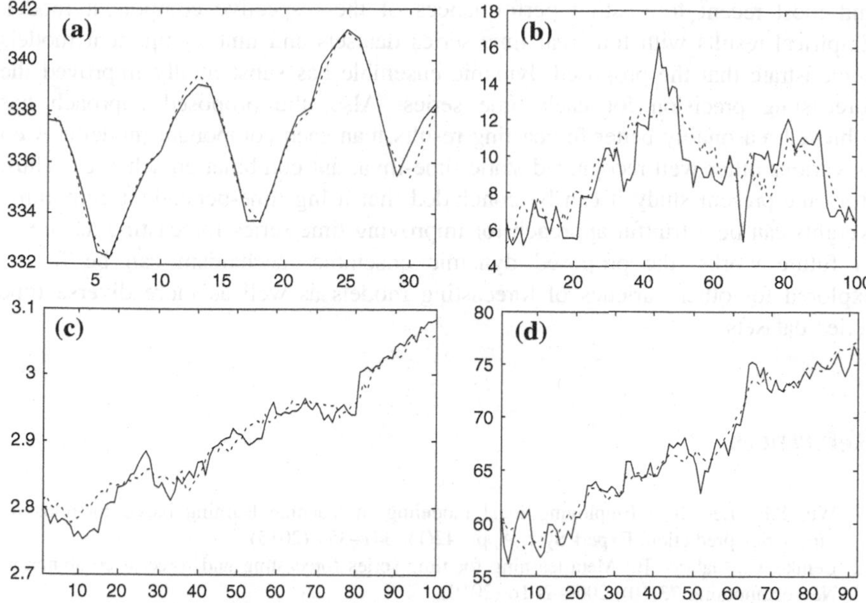

Fig. 2 Plots of the actual testing dataset and its forecast through the proposed method for: **a** Emission. **b** Inflation. **c** Google. **d** Facebook

combination methods. In Fig. 2, we depict the actual testing dataset and its forecast through the proposed dynamic ensemble method for each time series. The solid and dotted lines respectively represent the actual and predicted series. The closeness between the testing and predicted series is clearly evident in each subplot of Fig. 2.

5 Conclusions

Obtaining reasonably precise forecasts for a time series has indispensable importance in almost every branch of science and engineering. In spite of several existing forecasting models, none can be guaranteed to provide uniformly best results. On the other hand, extensive research works in this domain have demonstrated that combining forecasts from multiple models substantially improves the forecasting accuracy as well as often outperforms each component model. Most of the forecasts combination techniques in literature assigns static time-invariant combining weights to the component models.

This paper proposes a dynamic ensemble approach that updates the combining weights after obtaining each new forecast. Unlike the static ensemble methods, a combining weight in the present formulation depends on both the particular model as well as the time index. Each time, an weight is updated on the basis of the past and most recent forecasting performances of the respective component model. Empirical results with four real time series datasets and nine component models demonstrate that the proposed dynamic ensemble has substantially improved the forecasting precision for each time series. Also, the proposed approach has achieved reasonably better forecasting results than each component model as well as various other well recognized static time-invariant combination schemes. Thus, from the present study it can be concluded that using time-dependent combining weights can be a fruitful approach for improving time series forecasting accuracy. In future works, the proposed dynamic ensemble mechanism can be further explored for other varieties of forecasting models as well as more diverse time series datasets.

References

1. Wu, S.F., Lee, S.J.: Employing local modeling in machine learning based methods for time-series prediction. Expert Syst. Appl. **42**(1), 341–354 (2015)
2. Lemke, C., Gabrys, B.: Meta-learning for time series forecasting and forecast combination. Neurocomputing **73**(10), 2006–2016 (2010)
3. de Gooijer, J.G., Hyndman, R.J.: 25 years of time series forecasting. Int. J. Forecast. **22**(3), 443–473 (2006)
4. Gheyas, I.A., Smith, L.S.: A novel neural network ensemble architecture for time series forecasting. Neurocomputing **74**(18), 3855–3864 (2011)
5. Clemen, R.T.: Combining forecasts: a review and annotated bibliography. Int. J. Forecast. **5** (4), 559–583 (1989)
6. Andrawis, R.R., Atiya, A.F., El-Shishiny, H.: Forecast combinations of computational intelligence and linear models for the NN5 time series forecasting competition. Int. J. Forecast. **27**(3), 672–688 (2011)
7. De Menezes, L.M., Bunn, D.W., Taylor, J.W.: Review of guidelines for the use of combined forecasts. Eur. J. Oper. Res. **120**(1), 190–204 (2000)
8. Jose, V.R.R., Winkler, R.L.: Simple robust averages of forecasts: some empirical results. Int. J. Forecast. **24**(1), 163–169 (2008)
9. Deutsch, M., Granger, C.W.J., Teräsvirta, T.: The combination of forecasts using changing weights. Int. J. Forecast. **10**(1), 47–57 (1994)
10. Fiordaliso, A.: A nonlinear forecasts combination method based on Takagi-Sugeno fuzzy systems. Int. J. Forecast. **14**(3), 367–379 (1998)
11. Zou, H., Yang, Y.: Combining time series models for forecasting. Int. J. Forecast. **20**(1), 69–84 (2004)
12. Adhikari, R., Agrawal, R.K.: Performance evaluation of weights selection schemes for linear combination of multiple forecasts. Artif. Intell. Rev. **42**(4), 1–20 (2012)
13. Granger, C.W.J., Ramanathan, R.: Improved methods of combining forecasts. J. Forecast. **3**(2), 197–204 (1984)
14. Bunn, D.W.: A Bayesian approach to the linear combination of forecasts. Oper. R. Q. **26**(2), 325–329 (1975)
15. Hamzaçebi, C.: Improving artificial neural networks' performance in seasonal time series forecasting. Inf. Sci. **178**(23), 4550–4559 (2008)
16. Golub, G.H., Van Loan, C.F.: Matrix computations, vol. 3, 3rd edn. The John Hopkins University Press, Baltimore, USA (2012)

17. Zhang, G.P.: Time series forecasting using a hybrid arima and neural network model. Neurocomputing **50**, 159–175 (2003)
18. Box, G.E.P., Jenkins, G.M., Reinsel, G.C.: Time series analysis: forecasting and control. Prentice-Hall, Englewood Cliffs (1994)
19. Adhikari, R., Agrawal, R.K.: A combination of artificial neural network and random walk models for financial time series forecasting. Neural Comput. Appl. **24**(6), 1441–1449 (2014)
20. Vapnik, V.: The Nature of Statistical Learning Theory. Springer-Verlag, New York (1995)
21. Suykens, J.A.K., Vandewalle, J.: Least squares support vector machines classifiers. Neural Process. Lett. **9**(3), 293–300 (1999)
22. Zhao, J., Zhu, X., Wang, W., Liu, Y.: Extended kalman filter-based elman networks for industrial time series prediction with GPU acceleration. Neurocomputing **118**, 215–224 (2013)
23. Hamzaçebi, C., Akay, D., Kutay, F.: Comparison of direct and iterative artificial neural network forecast approaches in multi-periodic time series forecasting. Expert Syst. Appl. **36**(2), 3839–3844 (2009)
24. Demuth, H., Beale, M., Hagan, M.: Neural Network Toolbox User's Guide. The MathWorks, Natic (2010)
25. Pelckmans, K., Suykens, J.A., Van Gestel, T., De Brabanter, J., Lukas, L., Hamers, B., De Moor, B., Vandewalle, J.: LS-SVMlab toolbox user's guide. Pattern Recogn. Lett. **24**, 659–675 (2003)
26. Data market (2014): http://datamarket.com/
27. Open Govt. data platform, India (2014): http://data.gov.in
28. Yahoo! Finance (2014): http://finance.yahoo.com

A Novel Graph Clustering Algorithm Based on Structural Attribute Neighborhood Similarity (SANS)

M. Parimala and Daphne Lopez

Abstract Graph Clustering techniques are widely used in detecting densely connected graphs from a graph network. Traditional Algorithms focus only on topological structure but mostly ignore heterogeneous vertex properties. In this paper we propose a novel graph clustering algorithm, Structural Attribute Neighbourhood Similarity (SANS) algorithm, provides an efficient trade-off between both topological and attribute similarities. First, the algorithm partitions the graph based on structural similarity, secondly the degree of contribution of vertex attributes with the vertex in the partition is evaluated and clustered. An extensive experimental result proves the effectiveness of SANS cluster with the other conventional algorithms.

Keywords Structural attribute similarity · Graph clustering · Neighborhood

1 Introduction

Clustering plays a vital and indispensable role in grouping them into set of clusters. The main purpose of clustering is to group similar objects in same cluster and dissimilar objects in different cluster. Clustering process is applied in various domains [1, 2] such as business and financial data, biological data, health data and so on. Graph Structure is a well-studied expressive model [3] to study the relationship among the data and visualize the data in graph structure. Clustering in graph [4] deals with grouping of similar vertices based on connectivity and neighborhood. Clustering in graph involves in various scientific [5] and medical applications such as community detection in social network [6, 7] functional related

M. Parimala (✉) · D. Lopez
School of Information Technology & Engineering, VIT University, Vellore 632014, India
e-mail: parimala.m@vit.ac.in

D. Lopez
e-mail: daphnelopez@vit.ac.in

A. Nagar et al. (eds.), *Proceedings of 3rd International Conference on Advanced Computing, Networking and Informatics*, Smart Innovation, Systems and Technologies 43, DOI 10.1007/978-81-322-2538-6_48

protein modules in protein-protein interaction network [8] and co-author relationship in citation network. Basically graph clustering is classified under three categories. Structural Graph Clustering [9, 10] is based on the connectivity and common neighbours between two vertices but it ignores to focus on the properties of the vertices. Recently, with the increase of rich information available, vertices in graphs are associated with the number of attributes that describes behavior and properties of the vertices. Attributed Graph Clustering [11, 12] is similar to data clustering where the objects with similar properties are grouped but it fails to represent the relationship of the objects. Structural Attribute Graph Clustering [13] groups the objects based on both structure and attributes. The resultant cluster balances the structure and homogeneous properties of vertices. A best clustering algorithm would generate clusters that has cohesive intra-cluster similarity and low inter-cluster similarity by balancing both the structural and attribute similarities. The proposed method focuses on balancing both the similarities.

The rest of this paper is organized as follows. Section 2, discusses the literature work in the graph clustering field and the general idea of SANS algorithm is described in Sect. 3. Section 4 presents the experimental result and analysis. Finally, Sect. 5 summarizes the work.

2 Related Work

Most of the existing graph clustering algorithms focuses only on structural similarity [14] that avoids the attribute information. Normalized cut [15], Modularity [9], and Structural density [10] are some of the measures used to find the Structural Similarity. For example in social network, communities are detected based on relationship among the users but ignores the personal profile such as interest, gender and education. Considering only the attribute similarity [16] would lead to clusters with rich attribute information but it fails to identify the topological structure of the network. The concept of similarity score [11, 12] to find the similarity between the vertices of directed graphs is discussed by Vincent, 2004 [17]. When this similarity is applied to social network, communities are formed based on the personal profile but interaction among the users is not identified. Structural attributed graph [18, 19] plays a vital role in finding the communities based on their interaction as well as on their attribute information. Even though the existence of Structural attributed graph is inevitable, maintaining the tradeoff between the two types of information in the clustering process becomes a challenging task. Some of similarity measures used are distance model [20], probabilistic model [21] and neighborhood based model [13, 22].

3 Problem Definition

A directed graph (G) is represented as set of vertices (V) and set of edges (E) where n denotes the number of vertices present in the graph G. The proposed algorithm Structural Attribute Neighborhood Similarity Algorithm (SANS) partitions the graph into k clusters ($C_1 \ldots C_k$) based on the structural and attribute similarity. The Structural similarity is based on the Weight Index (WI) and attribute similarity is based on the Similarity matrix. The preliminary definition needed for the SANS algorithms is defined as follows.

Definition 1 (*Weight Index*) Given a n × n Weight Matrix (W), the Weight Index (WI) for each vertex v_i is the sum of weights of incoming and outgoing edges at v_i.

$$WI(v_i) = \sum_{j=1}^{n} W_{ij} + \sum_{j=1}^{n} W_{ji}$$

where W_{ij} denotes the edge weight between the vertices v_i and v_j.

Definition 2 (*Similarity Matrix*) Let m denotes the number of attributes for each vertex. The similarity between the v_i and v_j with respect to attribute is indicated as S_{ij}.

$$S_{ij} = \frac{1}{m} \sum\nolimits_{k=1}^{m} S_{ijm}$$

where,

$$S_{ijm} = \begin{cases} 0, \text{if } v_i \text{ and } v_j \text{ do not match for the } m^{th} \text{ attribute} \\ 1, \text{if } v_i \text{ and } v_j \text{ match for the } m^{th} \text{ attribute} \end{cases}$$

Properties

If any two objects v_i and v_j are similar then, it satisfies the following properties,

(i) Symmetry: $S(v_i, v_j) = S(v_j, v_i)$
(ii) Positivity: $0 \le S(v_i, v_j) \le 1 \, \forall \, v_i \, and \, v_j$
(iii) Reflexivity: $S(v_i, v_j) = 1 \; if \; i = j$

The following properties holds good if any two objects v_i and v_j are dissimilar,

(i) Symmetry: $D(v_i, v_j) = D(v_j, v_i)$
(ii) Positivity: $\mathrm{D}(v_i, v_j) \ge 0 \, \forall \, v_i \, and \, v_j$
(iii) Reflexivity: $\mathrm{D}(v_i, v_j) = 0 \; if \; i = j$

4 Algorithm

The proposed algorithm Structural Attribute Neighborhood Similarity (SANS) groups the object based on the attribute and structural similarity of the vertices. Weight Index (WI) value for each vertex is detected using the Weight matrix (W). The vertex with maximum weight index value is considered as centroid or Center. The number of centroids formed is the number of clusters (k) needed for the given dataset. Thus the SANS algorithm automatically detects the k value. The neighborhood vertices of the Center are grouped to that cluster. Then, each vertex in the cluster is compared with the vertices that are not in any of the cluster. The vertex is grouped with the cluster if the similarity value of the vertex is greater than threshold value. This process is iteratively done to generate k clusters. The SANS algorithm is proposed as follows.

SANS (Structural Attribute Neighborhood Similarity) Algorithm

Input: A directed graph G, set of vertices V and set of edges E.
Output: Number of *k* clusters (C_i.....C_k) formed with *k* centroids (*Center)*

Algorithm:
1. Calculate the Weight Index (WI) for vertices
2. Construct the n×n Similarity Matrix.
3. C_0=Ø; i=1
4. Repeat for each vertex v_i in V

(i) Select the vertex with highest weight index as centroid (*Center)* for the cluster C_i.

$$Center = \max_{v_i \in V - \cup_{r=0}^{i-1} C_r} WI(v_i)$$

(ii) Group the neighbourhood vertices of v_i to the C_i.

$$C_i = Center \cup N[v_i] \qquad \textit{// Structural similarity}$$

(iii) For each vertex in C_i

If the similarity value between the vertex in cluster and other vertices are greater than the threshold value, then it is grouped.

$$C_i = \left\{ v \in V - \bigcup_{r=0}^{i-1} C_i \text{ such that } S_{ij} \geq \textit{threshold} \right\} \text{//attribute similarity}$$

5. Until $V - \bigcup_{r=0}^{i} C_r \neq \{\phi\}$; i=i+1
6. Return *k* clusters {C_1......C_k}

5 Implementation

The data set for the implementation of the proposed algorithm is derived from DBLP Bibliography data focusing on three research areas as Graph Theory (GT), Data Mining (DM) and Graph Clustering (GC). Using this data a co-author graph with 200 nodes and their co-author relationships is built. Significant attributes that contribute meaning is the Primary topic and Count (Number of papers) for each author. Primary topic for each author is selected from one of their keywords in the paper. The algorithm is implemented and its performance is compared with the algorithms stated below:

(i) k-SNAP: Groups vertices with same attribute in the cluster. It ignores the topological structure of the graph that represents the co-author relationship.
(ii) S-Cluster: This algorithm considers Structure similarity but avoids the attribute similarity.
(iii) SA-Cluster: Both structural and attribute similarities are considered. In this algorithm the value for number of cluster (k) to be formed must be given by the user. Since the process is iterative, it is more expensive.
(iv) SANS Algorithm: The significance of this algorithm is it considers both structural and attributes similarities like SA cluster, but it automatically detects number of clusters to be formed and is less expensive.

The procedure is described as follows: This algorithm examines the DBLP dataset automatically, and calculates the number of clusters as k = 50. The k-SNAP clusters group the authors based on their primary topic and ignores the co-author relationship among them. The S-cluster method groups the authors who collaborate with each other but ignores the similar topic of research. Interestingly, SA-cluster and SANS algorithm cluster the dataset based on co-author relationship and attribute similarity of the author. The cluster formed by these two methods contain authors who have close co-author relationship and who work on the same topic but never collaborate. For example, *Xuding Zhu* and *Loannis G. Jollis* are experts on Graph theory but they have never collaborated. As a result, S-cluster assigns these two authors into two different clusters, since they are not reachable from each other and due to the lack of co-author connectivity. Whereas SA-cluster and SANS cluster assigns these two authors in the same cluster. Similar cases can be found in other cluster in different area of research. Even though SA-cluster and SANS methods maintains a tradeoff between the structural and attribute similarity, the proposed algorithm (SANS) outperforms SA-cluster in the following ways.

(i) The number of cluster (k) to be formed is automatically detected by the proposed algorithm.
(ii) The optimal nodes are selected based on the frequency of the vertex. For example, in DBLP dataset the author with more number of papers are selected for the clustering process which is inferred from the *Count* attribute for each author.

(iii) Since the clustering is based on neighborhood distance, SANS algorithm is faster than SA-Cluster which uses random walk distance to group the author.
(iv) SA-Cluster is more expensive than the SANS cluster, as it calculates the random walk distance iteratively.

The quality of the k cluster $\{V_i\}_{i=1}^{k}$ is measured using density function.

$$density\left(\{V_i\}_{i=1}^{k}\right) = \sum_{i=1}^{k} \frac{\left|\left\{(v_i, v_j) \middle| v_i, v_j \in V_i, (v_i, v_j) \in E\right\}\right|}{|E|}$$

Figure 1 shows the density comparison of four algorithms on DBLP dataset when the cluster number k = 50. The density value of SA-Cluster and S-Cluster are close. They remain around 0.4 (Table 1) or above even when k is increasing. The density value of proposed method SANS has a high density of around 0.5 when compared to the above two algorithms. On the other hand, k-SNAP has a very low density value because it groups the graph without considering the connectivity. We also compare the efficiency of different clustering algorithms. Figure 2 exhibits the runtime of various methods on DBLP dataset. As we can observe, k-SNAP is more efficient as it cluster the nodes based on the similarity of the attributes. SA-Cluster is slower than the S-Cluster, as it iteratively computes the random walk distance for each iteration, whereas the other method computes only once. The proposed clustering algorithm (SANS), group the objects based on the neighborhood which is slower than k-SNAP but faster than the S-Cluster and SA-Cluster.

Fig. 1 Quality comparison of cluster in DBLP dataset

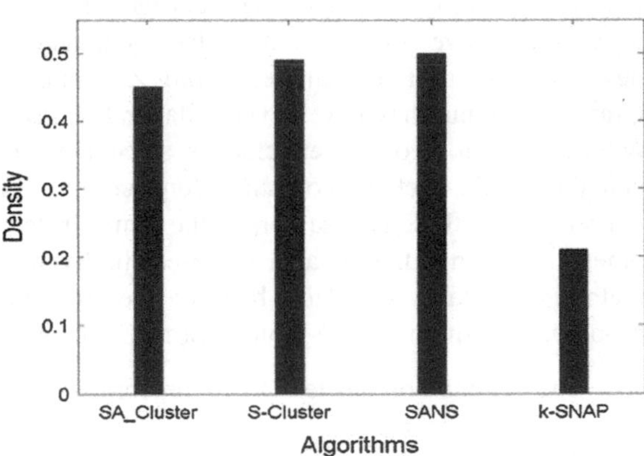

Table 1 Cluster quality in each iteration

Iter#	k-SNAP	S-Cluster	SA-Cluster	SANS
1	0.1	0.45	0.42	0.50
2	0.13	0.47	0.41	0.52
3	0.19	0.40	0.38	0.45
4	0.20	0.39	0.45	0.49
5	–	–	0.40	–

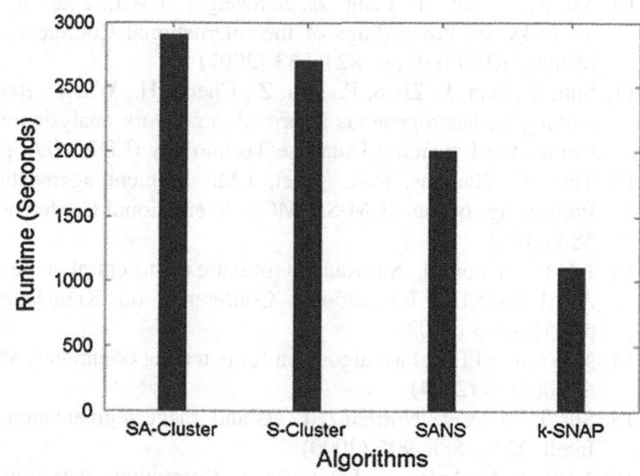

Fig. 2 Cluster efficiency

6 Conclusion

This paper introduces a new graph clustering algorithm based on structural and attribute similarities. The proposed algorithm partitions the graph into k clusters which have both high intra-cluster similarity and homogeneous attribute values. Neighborhood distance is used to measure the structural similarity to find the vertex connectivity whereas similarity index is used to measure the attribute similarity to find the vertex closeness. Quality of the clusters is validated using density function. Experimental results on the real dataset prove to be a balanced method between attribute and structural similarity.

References

1. Xu, R., Wunsch, D.: Survey of clustering algorithms. IEEE Trans. Neural Networks **16**(3), 645–678 (2005)
2. Ng, R., Han, J.: Efficient and effective clustering method for spatial data mining. In: Proceedings of the International Conference Very Large Data Bases (VLDB'94), pp. 144–155 (1994)
3. Lin, G., Bie, Y., Wang, G., Lei, M.: A novel clustering algorithm based on graph matching. J. Software **8**(4) (2013)

4. Fortunato, S.: Community detection in graphs. Physics Reports (2010)
5. Balsubramanan, R., Cohen, W.W.: Block-Ida: jointly modelling entity-annotated text and entity-entity links. In: SDM (2011)
6. McAuley, J., Leskovec, J.: Learning to discover social circles in ego network. In: NIPS (2009)
7. Ahn, Y., Bagrow, J.P., Lehmann, S.: Link communities reveal multi-scale complexity in networks. Nature **466**, 761 (2010)
8. Girvan, M., Newman, M.: Community structure in social and biological networks. PNAS **99** (12), 7821–7826 (2002)
9. Newman, M.E.J., Girvan, M.: Finding and evaluating community structure in networks. Phys. Rev. E **69**, 026113 (2004)
10. Xu, X., Yuruk, N., Feng, Z., Schweiger, T.A.J.: Scan: a structural clustering algorithm for networks. In: Proceedings of the International Conference Knowledge Discovery and Data Mining (KDD'07), pp. 824–833 (2007)
11. Sun, Y., Han, J., Zhao, P., Yin, Z., Cheng, H., Wu, T.: Rankclus: integrating clustering with ranking for heterogeneous information network analysis. In: Proceedings of the International Conference Extending Database Technology (EDBT'09), pp. 565–576 (2009)
12. Tian, Y., Hankins, R.A., Patel, J.M.: Efficient aggregation for graph summarization. In: Proceedings of the ACM-SIGMOD International Conference Management of Data, pp. 567–580 (2008)
13. Jeh, G., Widom, J.: SimRank: a measure of structural-context similarity. In: Proceedings of the ACM SIGKDD International Conference on Knowledge Discovery and Data Mining, pp. 538–543 (2002)
14. Newman, M.E.J.: Fast algorithm for detecting community structure in networks. Phys. Rev. E. **69**, 066133 (2004)
15. Shi, J., Malik, J.: Normalized cuts and image segmentation. IEEE Trans. Pattern Anal. Mach. Intell. **22**(8), 888–905 (2000)
16. Yang, J., McAuley, J., Leskovec, J.: Community detection in networks with node attributes. In: ICDM (2013)
17. Blondel, V.D., Van Dooren, P.: The measure of similarity between graph vertices, with applications to synonym extraction and web searching. J. Siam Rev. **46**(4), 647–666 (2004)
18. Zhou, Y., Cheng, H., Yu, J.X.: Clustering large attributed graphs: a balance between structural and attribute similarities. TKDD **5**(2), 12 (2011)
19. Zhou, Y., Cheng, H., Yu, J.X.: Graph clustering based on structural/attribute similarities. VLDB **2**(1), 718–729 (2009)
20. Zhou, Y., Cheng, H., Yu, J.X.: Clustering large attributed graphs: an efficient incremental approach. In: ICDM. pp. 689–698 (2010)
21. Xu, Z., Ke, Y., Wang, Y., Cheng, H., Cheng, J.: A model-based approach to attributed graph clustering. In: ACM-SIGMOD (2012)
22. Gleich, D.F., Seshadhri, C.: Neighborhoods are good communities. In: KDD (2012)

Change-Point Detection in Enterprise Attack Surface for Network Hardening

Ghanshyam S. Bopche and Babu M. Mehtre

Abstract Applications of change-point detection typically originate from the perspective of enterprise network security and network monitoring. With the ever-increasing size and complexity of enterprise networks and application portfolios, network attack surface keeps changing. This change in an attack surface is detected by identifying increase or decrease in the number of vulnerabilities at network level. Vulnerabilities when exploited successfully, either provide an entry point to an adversary into the enterprise network or can be used as a milestone for staging multi-stage attacks. In this paper, we have proposed an approach for change-point detection in an enterprise network attack surface. In this approach, a sequence of static attack graphs are generated for dynamic (time varying) enterprise network, and successive graphs in a sequence are compared for their dissimilarity for change-point detection. We have presented a small case study to demonstrate the efficacy and applicability of the proposed approach in capturing a change in network attack surface. Initial results show that our approach is capable of capturing the newly introduced vulnerabilities into the network and is able to differentiate these vulnerabilities for efficient network hardening.

Keywords Attack surface · Attack graph · Change-point detection · Network security and protection · Security metric · Similarity measures

G.S. Bopche (✉) · B.M. Mehtre
Center for Information Assurance & Management (CIAM),
Institute for Development and Research in Banking Technology (IDRBT),
Castle Hills, Masab Tank 500057, Hyderabad, India
e-mail: ghanshyambopche.mca@gmail.com

B.M. Mehtre
e-mail: mehtre@gmail.com

G.S. Bopche
School of Computer and Information Sciences (SCIS),
University of Hyderabad, Gachibowli 500046, Hyderabad, India

A. Nagar et al. (eds.), *Proceedings of 3rd International Conference on Advanced Computing, Networking and Informatics*, Smart Innovation, Systems and Technologies 43, DOI 10.1007/978-81-322-2538-6_49

1 Introduction

With the constant evolution of TCP/IP protocols and hence enterprise networks, there may be new plausible security risks as new servers, workstations, or services are added to the network. These network assets are increasingly vulnerable and always a soft target of sophisticated cyber attacks from potential adversaries such as hackers, corporate competitors, disgruntled employees, government agencies, etc. In order to keep the security posture of an enterprise network up to date, enterprise networks need to be scanned and hardened regularly.

Present day vulnerability scanners such as, *Nmap*, *Nessus*, *Retina*, *GFI LanGuard*, etc. identify vulnerabilities in isolation, i.e., vulnerabilities local to a system only. For an enterprise network of reasonable size, vulnerability scanners can report a large number of vulnerabilities. From the defender's standpoint, an administrator has to identify the vulnerabilities, which really matters in securing enterprise network. In other words, administrator has to identify vulnerabilities that allow an adversary to enter the network or the vulnerabilities that can be used as a milestone for staging multistage attacks against the enterprise network.

An attack graph- a "multihost, multistage" attack detection technique derives all possible attack paths available to an adversary. In order to do this, a network model description such as vulnerability information, network configuration, and network reachability information is used. The cause-consequence relationship between the extant vulnerabilities is taken into account to draw multistage correlated attack scenarios (that are used by an adversary to get incremental access to enterprise critical resources). Understanding such relationship is vital for optimal placement of the security countermeasures and hence for efficient network hardening. Vulnerability scanners are not capable of finding such correlation among the vulnerabilities. Further, prioritization of vulnerabilities for efficient network hardening based on a scanning report alone is highly impossible and not a viable solution in terms of time, and effort.

In this paper, we have used an attack graph-based approach for the detection and prioritization of vulnerabilities, which really plays a key role during network compromise. We have taken a snapshot of a network at time t and generated an attack graph. It is obvious that, it shows the vulnerabilities and their dependency on each other, which may lead to network compromise when exploited successfully. The graph-assisted metrics proposed in literature, for example shortest path metric [1], the number of paths metric [2], mean of path lengths [3], and others [4, 5] can be used to identify attack paths/attack scenarios of special interest. Even though administrator is aware about these attack paths and causal vulnerabilities, she cannot patch/fix all because of the various constraints. The constraint may be countermeasure cost, limited security budget, unavailability of patch/workarounds, patch time, etc. Unpatched vulnerabilities will straightaway appear in the attack graph generated for the same network over time Δt called sampling interval. The administrator already knows external causes of these vulnerabilities. She has to worry about the new vulnerabilities that will be introduced into this attack graph/enterprise network over time Δt. For an enterprise network of reasonable size, an attack graph is of enormous size. Searching newly introduced vulnerabilities manually in this graph is not feasible in real time. Again, identifying the most relevant vulnerabilities for efficient network hardening is also poses great difficulty/challenge. This challenge motivates our work. We used graph similarity-based approach to solve above said problem. We have compared the attack graphs generated at time t and $t + \Delta t$ for their similarity (may not be structural) in terms of common nodes and edges and by applying some heuristics we have identified the most relevant among the newly introduced vulnerabilities.

The remainder of this paper is organized as follows: Sect. 2 presents an attack graph model. Section 3 discusses our method of the change-point detection in terms of number of newly introduced vulnerabilities in a successive attack graph. A case study is presented in Sect. 4 to demonstrate the usefulness of our approach. Finally, Sect. 5 concludes the paper.

2 Preliminaries

An attack graph for the enterprise network is a directed graph representing prior knowledge about vulnerabilities, their dependencies, and network connectivity [6]. In this paper, we have generated labeled, goal-oriented attack graphs for the enterprise network. Exploits and security conditions are the two types of nodes in the attack graph. Here, exploit represents an adversary action on the network host in order to take advantage of extant vulnerability. Security conditions represent properties of system or network relevant for successful execution of an exploit. The existence of a host vulnerability, network reachability, and trust relationship between two hosts are the kind of security conditions required for successful exploitation of vulnerability on a remote host.

Exploits and conditions are connected by directed edges. No two exploits or two security conditions are directly connected. Directed edge from security condition to an exploit represent the *require relation*. It means, an exploit cannot be executed until all the security conditions have been satisfied. An edge from an exploit to a security condition indicates the *imply relation* [7]. It means, successful execution of an exploit will create few more conditions. Such newly created security conditions, i.e., post-conditions act as a pre-condition for other exploits. With the perception of an attack graph discussed above, Wang et al. [8] formally defined an attack graph as an exploit-dependency graph as follows:

Definition 1 (*Attack Graph Model*) Given a set of exploits e, a set of conditions c, a require relation $R_r \subseteq c \times e$, and an imply relation $R_i \subseteq e \times c$, an attack graph G is the directed graph $G(e \cup c,\ R_r \cup R_i)$, where $(e \cup c)$ is the vertex set and $(R_r \cup R_i)$ is the edge set [8].

As evident from the Definition 1, an attack graph G is a directed bipartite graph with two disjoint sets of vertices namely, exploit and condition. The edge set consist of two types of edges, namely, *require edge* and *imply edge*. As an important feature of an attack graph, require relation, i.e., R_r should be always conjunctive, whereas the imply relation, i.e., R_i should be always disjunctive [8]. The conjunctive nature of the conditions implies an exploit cannot be executed until all of its preconditions have been satisfied. An imply edge should identify those conditions which can be generated after the successful execution of an exploit. Introduction of a new vulnerability in an enterprise network is shown in an attack graph by the exploit, it's pre-conditions and post-condition.

3 Change-Point Detection

Change-point detection in the attack surface of an enterprise network can be determined by representing a given network, observed at time t and $t + \Delta t$, by attack graphs G and G', respectively. An attack surface represents, the set of ways an adversary can compromise an enterprise network [9]. Δt an arbitrary sampling interval depends on the time required for gathering vulnerability information, network configuration details and construction of an attack graph. It is an important parameter in security monitoring, defines how often an attack graph is constructed and governs the type of vulnerability (attack) can be detected. For a given window of time W and for a given enterprise network, attack graphs are generated (at discrete instants of time depending on Δt). This leads to the sequence of an attack graphs. Then the consecutive attack graphs in a sequence are analyzed for the detection of change in attack surface.

The algorithm presented in this paper works on labeled attack graphs. Let L_V and L_E denote the finite set of vertex and edge labels in an attack graph, respectively.

Definition 2 An attack graph G is a four-tuple $G = (V, E, \rho, \mu)$, where

- V is a finite set of vertices i.e., $V = e \cup c$.
- $E \subseteq V \times V$ is a set of Edges i.e., $E = R_r \cup R_i$
- $\rho : V \rightarrow L_V$ is a function assigning labels to the vertices
- $\mu : E \rightarrow L_E$ is a function assigning labels to the edges

Here, L_V and L_E represents the set of symbolic labels uniquely identifying each node and each edge, respectively in an attack graph G. Application of a minimum spanning tree algorithm (MST) on both G and G', gives the nodes and edges present in each graph. Once nodes and edges with respect to each input attack graph are identified, following three cases are considered:

1. $G' \backslash G$: nodes and edges unique to G' only, i.e., $G' - G$. These nodes represent newly introduced vulnerabilit(y)ies in the network.

2. $G \backslash G'$: nodes and edges unique to G only, i.e., $G - G'$. These nodes and edges belong to the vulnerabilities that are already patched.
3. $G \cap G'$: nodes and edges that are common to both G and G'. Common nodes and edges appeared because of the persistence of some vulnerabilities in an enterprise network both at time t and $t + \Delta t$. These vulnerabilities and their external causes (i.e., preconditions) are already known to the administrator but because of some constraints like limited security budget, unavailability of patches etc. those vulnerabilities are unpatched.

Case 1 is of special interest to an administrator since it gives information about the newly introduced vulnerabilities into the network. Our goal is to analyze these vulnerabilities for their use by an adversary in multistage attacks. Case 1 gives the node set N consisting of exploits, pre-conditions, post-conditions, i.e., $N = e \cup c$ and edge set M consisting of require relation and imply relation, i.e., $R_r \cup R_i$ with respect to the newly introduced vulnerabilities. The edge set M gives more information about the vulnerability dependency and hence is of special interest.

For each exploit there should be two or more pre-conditions need to be satisfied conjunctively. It means second node/vertex in two or three require relations (i.e., edges in set M), is common. This common vertex represents an exploit and the node preceding to it in those edges represent the pre-conditions required for its exploitation. Removal of any one of these pre-conditions can stop an exploit from executing. If the exploit is the first vertex in one or more imply relations (edges), then the second vertex of those edges represent the post-condition of an exploit. This post-condition may act as a precondition for other exploits. An approach of identifying and differentiating new vulnerabilities introduced in an enterprise network during the time period Δt is given in algorithm 1.

4 Case Study

In this section, a case study is presented to detect the vulnerability change in an attack surface of an enterprise network by means of dissimilarity between consecutive attack graphs generated for the same enterprise network at time t and $t + \Delta t$. From the context of an input attack graphs, it is shown that the obtained results provide unique security relevant information, which will enhance the security administrator's ability in hardening network security more efficiently.

A network similar to [10] has been considered as the test network. Topology of the test network is given in Fig. 1. There are 4 hosts in the network viz. $Host_0$ (H_0), $Host_1$ (H_1), $Host_2$ (H_2) and $Host_3$ (H_3). The description of each host is given below:

- H_0: a Web Server (Windows NT 4.0)
- H_1: a Windows Domain Server (Windows 2000 SP1)
- H_2: a Client (Windows XP Pro SP2)
- H_3: a Linux Server (Red Hat 7.0)

Algorithm 1 Change-point Detection in Attack surface

Input:
$G \rightarrow$ an attack graph for an enterprise network at time t
$G' \rightarrow$ an attack graph for the same network at time $t + \Delta t$
$\Delta t \rightarrow$ a sampling interval
Output:
$Exploit \subseteq (V' - V) \rightarrow$ is a set of new exploits/ vulnerabilities added to the G over the time interval Δt
$Initial \subseteq Exploit \rightarrow$ is a set of new exploits that can initiate new attack paths.
$Middle \subseteq Exploit \rightarrow$ is a set of new exploits that can be part of attack paths but does not initiate a new attack path
$PreCondn[i] \rightarrow$ pre-conditions of an exploit $i \in Exploit$

$\langle V, E \rangle \leftarrow$ MST (G) ▷ Apply minimum spanning tree algorithm (MST) on G to identify all the unique nodes and edges in G
$\langle V', E' \rangle \leftarrow$ MST (G') ▷ Apply MST on G' to identify all the unique nodes and edges in G'
$\langle (V' - V), (E' - E) \rangle \leftarrow G' \backslash G$ ▷ Compute $(G' - G)$. It gives vertices, i.e., $(V' - V)$ and edges, i.e., $(E' - E)$ belong to G' only
if $(G' \backslash G = \phi)$ **then**
    Print: "no new vulnerability introduced into the enterprise network"
else
    for all $v \in (V' - V)$ **do**
        if $(v == j)$ for two or more $(i, j) \in (E' - E)$ **then**
            $Exploit \leftarrow v$ ▷ v is an exploit
            $PreCondn \leftarrow i$ ▷ i is the pre-condition for an exploit v
        end if
    end for
    for all $e' \in Exploit$ **do**
        if $(e' == i)$ for one or two $(i, j) \in (E' - E)$ **then**
            $PostCondn \leftarrow j$ ▷ j is the post-condition of an exploit e'
        end if
    end for
    for all $(i, j) \in (E' - E)$ **do**
        if $i, j \in (Exploit \cup PostCondn)$ **then**
            $E'' = (i, j)$
        end if
    end for
    for all $i \in Exploit$ in G'' **do** ▷ G'' is constructed out of E''. G'' is a bipartite graph and may have one or more connected components
        find indegree $\delta(i)$
        if $(\delta(i) == 0)$ **then**
            $Initial \leftarrow i$ ▷ i is the first exploit in an attack sequence
        else
            $Middle \leftarrow i$ ▷ i is the middle exploit in an attack paths
        end if
    end for
    for all $i \in Initial$ **do**
        Find $PreCondn[i]$ ▷ Identify pre-conditions of an exploit, which can initiate a new attack sequence
    end for
    for all $i \in Middle$ **do**
        Find $PreCondn[i]$ ▷ Identify pre-conditions of an exploit, which does not initiate a new attack sequence but can be part of other attack sequence
    end for
end if

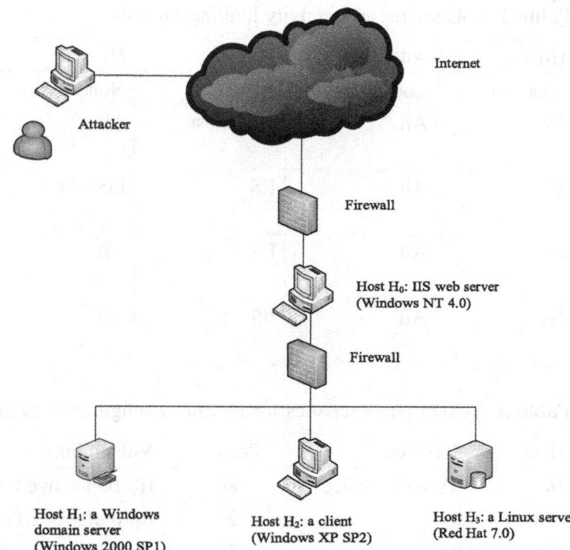

Fig. 1 Network configuration [10]

Here host H_3 is the target machine for an attacker and *MySQL* is the critical resource running over it. The attacker is an entity with malicious intent from outside the internal network. Here the attacker's intention is to gain root-level privileges on host H_3. The job of a firewalls is to separate internal network from the Internet and the connectivity-limiting firewall policies for the network configuration are given in Table 2. Tables 1 and 3 shows the system characteristics for the hosts available in the network at time t and $t + \Delta t$, respectively. Such kind of data is available in public vulnerability databases viz. *National Vulnerability Database* (*NVD*), *Bugtraq*, *Open Source Vulnerability Database* (*OSVDB*) etc. Here external firewall allows any external host to only access services running on host H_0. Connections to all other services/ports available on other hosts are blocked. Host's within the internal network are allowed to connect to only those ports specified by the

Table 1 System characteristics for network configuration at time t [10]

Host	Services	Ports	Vulnerabilities	CVE IDs
H_0	IIS web service	80	IIS buffer overflow	CVE-2010-2370
H_1	ssh	22	ssh buffer overflow	CVE-2002-1359
	rsh	514	rsh login	CVE-1999-0180
H_2	rsh	514	rsh login	CVE-1999-0180
H_3	LICQ	5190	LICQ-remote-to-user	CVE-2001-0439
	Squid proxy	80	squid-port-scan	CVE-2001-1030
	MySQL DB	3306	local-setuid-bof	CVE-2006-3368

Table 2 Policies for connectivity-limiting firewall

Host	Attacker	H_0	H_1	H_2	H_3
Attacker	Localhost	All	None	None	None
H_0	All	Localhost	All	All	Squid LICQ
H_1	All	IIS	Localhost	All	Squid LICQ
H_2	All	IIS	All	Localhost	Squid LICQ
H_3	All	IIS	All	All	Localhost

Table 3 System characteristics for network configuration at time $t + \Delta t$ [10]

Host	Services	Ports	Vulnerabilities	CVE IDs
H_0	IIS web service	80	IIS buffer overflow	CVE-2010-2370
	ftp	21	ftp buffer overflow	CVE-2009-3023
H_1	ftp	21	ftp rhost overwrite	CVE-2008-1396
	ssh	22	ssh buffer overflow	CVE-2002-1359
	rsh	514	rsh login	CVE-1999-0180
H_2	Netbios-ssn	139	Netbios-ssn nullsession	CVE-2003-0661
	rsh	514	rsh login	CVE-1999-0180
H_3	LICQ	5190	LICQ-remote-to-user	CVE-2001-0439
	Squid proxy	80	squid-port-scan	CVE-2001-1030
	MySQL DB	3306	local-setuid-bof	CVE-2006-3368

connectivity limiting firewall policies as shown in Table 2. In Table 2, *All* specifies that source host may connect to any port on a destination host in order to have access to the services running on those ports. *None* indicates that source host is prevented from having access to any port on the destination host [10].

An attack graph for the network configuration at time t and $t + \Delta t$ is shown in Fig. 2. These graphs for the same enterprise network at different instant of time (for the sampling interval Δt) are generated using model checking-based tool called SGPlan [11, 12]. Existing vulnerabilities in the test network are logically combined to generate different attack scenario, which in turn represents different attack paths. Attack graph is generated by collapsing several such attack paths for the same initial and goal condition. As shown in Fig. 2, nodes in both the attack graphs represent an exploit, required pre-conditions and implied post-conditions. Exploits are shown by a circle and named by alphabets.

As evident from the Fig. 2, number of attack paths for an attack graph at time t are only 2, whereas it is 16 for an attack graph at time $t + \Delta t$. It is because of an increase in the number of vulnerable services in the enterprise network within the sampling interval Δt. Vulnerabilities appeared in an attack graph (i.e., in an

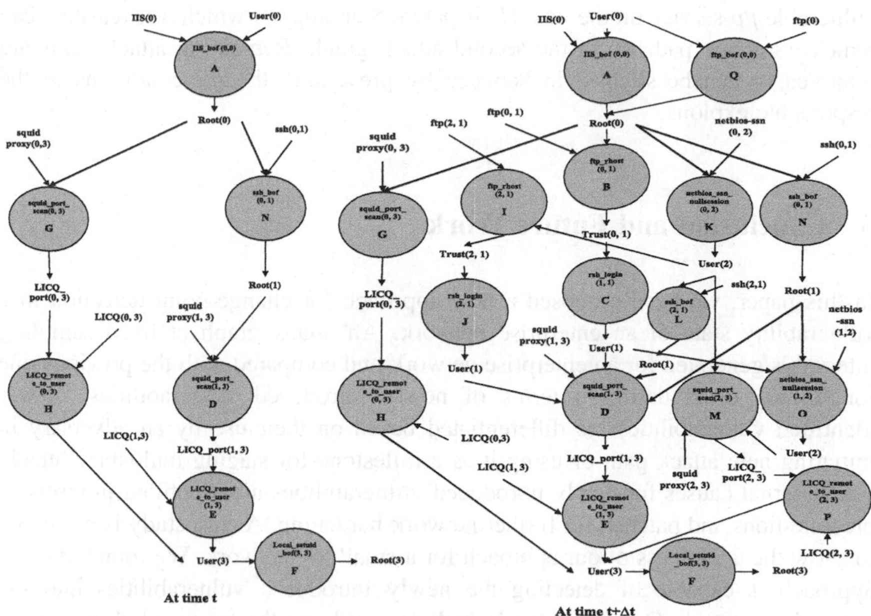

Fig. 2 Attack graph for the network configuration at time t and $t + \Delta t$

enterprise network) at time t are not patched and hence will appear straightaway in an attack graph at time $t + \Delta t$. The external causes for these unpatched vulnerabilities are already well known to the administrator. She has to worry about the new vulnerabilities introduced in the network.

Change point detection algorithm 1 proposed in Sect. 3 detects the new vulnerabilities appeared in attack graph at time $t + \Delta t$. It gives an exploit set $Exploit = \{B, C, I, J, K, L, M, O, P, Q\}$ and respective post-condition set *PostCondn*. From these two sets, we have derived a bipartite graph representing chains of exploits separated by post-conditions. Indegree, i.e., δ for each exploit in this graph is calculated. Exploit with indegree zero, i.e., $\delta = 0$, are responsible for initiating new attack paths. Remaining exploits with indegree value other than zero can be used by an attacker as a milestone for staging multi-stage attacks. From the defender's standpoint, the administrator must decide on which exploits to be focused on for efficient network hardening. Accordingly, pre-conditions must be identified and disabled using one or more security countermeasures to remove attack path or break an attack path in between in order to stop an adversary from reaching target machine. Algorithm 1 successfully identifies those preconditions and assist administrator in hardening network. In our case study, exploit Q provide a new way to an adversary in defeating network security and remaining exploits, i.e., $B, C, I, J, K, L, M, O, P$ can be used by an adversary as a milestone. If vulnerable *ftp* service on the host H_0 is patched or stopped whichever feasible, can remove 8 attack paths from the second attack graph. Remaining attack paths are removed or can be stopped in between by preventing the pre-conditions of the responsible exploits.

5 Conclusion and Future Work

In this paper, we have proposed a new approach for change-point detection in a vulnerability state of an enterprise network. An attack graph at fixed sampling interval is generated for an enterprise network and compared with the previous one for finding dissimilarity in terms of newly introduced vulnerabilities. Newly identified vulnerabilities are differentiated based on their use by an adversary in initiating new attack path or using it as a milestone for staging multistage attack. The external causes for newly introduced vulnerabilities are identified in terms of preconditions and patched for further network hardening. A case study is presented to show the usefulness of our approach for a small toy network. We found that our approach is capable of detecting the newly introduced vulnerabilities into the enterprise network. Our future work, includes applying the proposed change point detection algorithm for more sophisticated/complex enterprise network.

References

1. Phillips, C., Swiler, L.: A graph-based system for network vulnerability analysis. In: Proceedings of the 1998 Workshop on New Security Paradigms (NSPW '98), pp. 71–79. ACM, New York (1998)
2. Ortalo, R., Deswarte, Y., Kaaniche, M.: Experimenting with quantitative evaluation tools for monitoring operational security. IEEE Trans. Softw. Eng. **25**, 633–650 (1999)
3. Li, W., Vaughn, B.: Cluster security research involving the modeling of network exploitations using exploitation graphs. In: Sixth IEEE International Symposium on Cluster Computing and the Grid (CCGRID '06), vol. 2, pp. 26–26. IEEE Computer Society, Washington. May 2006
4. Idika, N., Bhargava, B.: Extending attack graph-based security metrics and aggregating their application. IEEE Trans. Dependable Secur. Comput. **9**(1), 75–78 (2012)
5. Noel, S., Jajodia, S.: Metrics suite for network attack graph analytics. In: 9th Annual Cyber and Information Security Research Conference (CISR '14), pp. 5–8. ACM, Oak Ridge National Laboratory, Tennessee, New York, April 2014
6. Wang, L., Singhal, A., Jajodia, S.: Measuring the overall security of network configurations using attack graphs. In: 21st Annual IFIP WG 11.3 Working Conference on Data and Applications Security, vol. 4602, pp. 98–112. Springer-Verlag, Berlin, Heidelberg (2007)
7. Wang, L., Liu, A., Jajodia, S.: An efficient and unified approach to correlating, hypothesizing, and predicting intrusion alerts. In: Proceedings of the 10th European Conference on Research in Computer Security (ESORICS'05), pp. 247–266. Springer-Verlag, Berlin, Heidelberg (2005)
8. Wang, L., Noel, S., Jajodia, S.: Minimum-cost network hardening using attack graphs. J. Comput. Commun. **29**(18), 3812–3824 (2006)

9. Sun, K., Jajodia, S.: Protecting enterprise networks through attack surface expansion. In: Proceedings of the 2014 Workshop on Cyber Security Analytics, Intelligence and Automation (SafeConfig '14). pp. 29–32. ACM, New York (2014)
10. Ghosh, N., Ghosh, S.K.: An approach for security assessment of network configurations using attack graph. In: First International Conference on Networks and Communications 2009, (NETCOM'09), pp. 283–288, Dec 2009
11. SGPlan 5: http://wah.cse.cuhk.edu.hk/wah/programs/SGPlan/
12. Ghosh, N., Ghosh, S.K.: A planner-based approach to generate and analyze minimal attack graph. J. Appl. Intell. **36**(2), 369–390 (2012)

Efficient Quality of Multimedia Experience Using Perceptual Quality Metrics

Rahul Gaurav and Hasan F. Ates

Abstract Although there are very few metrics which correlate well with the Human Visual System (HVS), most of them do not. We extensively studied about the Video Quality Assessment (VQA) of 2D svideos for enhancing the Quality of Experience (QoE) using better Quality of Service (QoS). We propose a solution which helps us to find a high correlation between HVS and the objective metrics using Perceptual Quality Metrics (PQM). The motive behind this work is to introduce an objective metric that is adequate to predict the Mean Opinion Score (MOS) of distorted video sequences based on the Full Reference (FR) method.

Keywords Quality assessment · Subjective testing · Objective testing · Structural similarity · QoE · PQM

1 Introduction

Up until now, the most precise and valued way of assessment of the quality of a video is the evaluation using subjects in the form of human participants [1]. As involving human subjects in such applications is laborious hence this leads to a need of a highly robust system which is able to assess the quality effectively without introducing any human observers. Few things can easily be deduced from literature reviews that the focus has been on the Quality of Service (QoS) rather than the Quality of Experience (QoE). The former term tries to objectively quantify the services handed over by the vendor and has nothing to do with the view point of the audience but it is more relevant to the media. While the latter speaks about the

R. Gaurav (✉)
Telecom SudParis, ARTEMIS Department, Evry, France
e-mail: rahulgaurav.kist@gmail.com

H.F. Ates
Department of Electrical-Electronics Engineering, ISIK University,
Sile, Istanbul, Turkey

A. Nagar et al. (eds.), *Proceedings of 3rd International Conference on Advanced Computing, Networking and Informatics*, Smart Innovation, Systems and Technologies 43, DOI 10.1007/978-81-322-2538-6_50

subjective measure of a person's experience. So in order to gauge the performance of the quality assessment, Mean Opinion Score (MOS) comes into play, which is the subjective quality measurement carefully done by using human subjects as observers and helps us to correlate with the obtained objective scores. Clearly, there is a need of a versatile QoS model which complies with the QoE in the best possible way. Our paper proposes one solution to this issue. We worked on such objective metrics which performs better than the state-of-art models and mimics the HVS. To a great extent, our work is inspired by the Perceptual Quality Metric (PQM) for dealing with 3D video datasets [2]. A robust objective algorithm has been proposed namely Perceptual Quality Metric for 2D (PQM2D) using the ideas from the above mentioned work. The aim of this work is to show better results for 2D VQA and outperform the various popular state-of-art metrics like Peak Signal to Noise Ratio (PSNR), Structural Similarity (SSIM) Index and Multi Scale SSIM (MS-SSIM) Index. For the verification phase, series of subjective experiments were performed to demonstrate the level of correlation between objective metrics and the user scores obtained by Subjective Evaluation using human observers, keeping in mind the standards set by the International Telecommunication Union (ITU) [1].

2 Quality of Experience Experimentation

2.1 Introduction

The QoE methods are essentially used to gauge the performance of multimedia or television systems with the help of responses obtained from observers who view the system under test [3]. With the help of this experiment, we will be able to find the MOS of the various video sequences under consideration[4–6].

2.2 General Viewing Conditions

The Table 1 gives us a short overview of the laboratory conditions and some of the details about the display of our system.

2.3 Source Sequences

The videos were obtained from the Laboratory for Image and Video Engineering (LIVE) at The University of Texas at Austin [7]. In our experiments, we used nine reference videos in the test session. Figure 1 shows the histogram of PSNR

Table 1 Laboratory conditions

Parameters	Settings
Peak luminance of the screen	150 cd/m^2
Other room illumination	Low
Height of image on screen (H)	11 cm
Viewing distance	88 cm

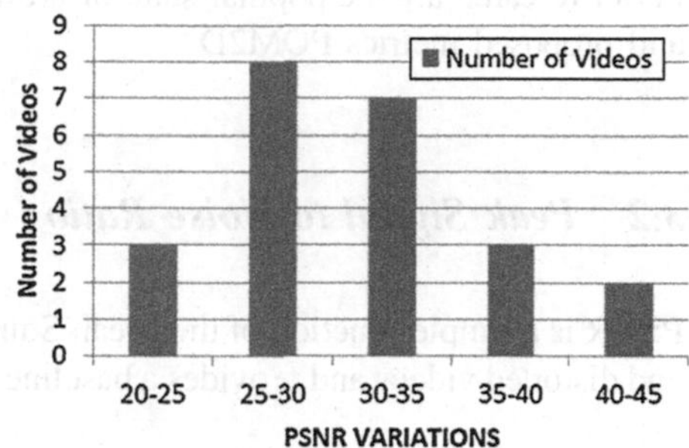

Fig. 1 PSNR range of video sequences

variations for the selected set of source sequences. Test cases were carefully selected so that the maximum range of PSNR is covered to get more reliable results.

2.4 Test Sequences

For the test sequence cases, we used four types of distortions namely wireless distortion, IP network distortion, H.264 compression and MPEG-2 compression.

2.5 Subjective Testing Design

The test methodology used is known as Double Stimulus Impairment Scale (DSIS) [1]. A carefully selected playlist was prepared by the authors, comprising of 24 videos in total, 9 reference samples in total with various kinds of distorted counterparts.

2.6 Observer Selection and Training

Most of the subjects who took part in our research were non-expert undergraduate students from the department of psychology of the ISIK University, Turkey. Each video was rated by 16 subjects in total with the help of a program formulated by the Graphics and Media Lab Video Group in Russia [8].

3 Quality of Service Experimentation

3.1 Introduction

We carried out the FR based objective VQA simulation using MATLAB codes written by the authors for our selected set of videos. The objective algorithms used in our research are the popular state-of-art metrics like PSNR, SSIM and MS-SSIM and proposed metrics PQM2D.

3.2 Peak Signal to Noise Ratio

PSNR is a simple function of the Mean Squared Error (MSE) between the reference and distorted videos and provides a baseline for objective algorithm performance [9].

$$PSNR = 10\log_{10}\frac{255^2}{MSE} \quad (1)$$

3.3 SSIM

We applied the SSIM index frame-by-frame on the luminance component of the video and computed the overall SSIM index for the video as the average of the frame level quality scores. We used two kinds of algorithms for SSIMs namely SSIM-Gaussian (SSIMG) and SSIM Block (SSIMB). The former is the standard SSIM using conventional Gaussian way and in the latter, SSIM is computed on an 8×8 block basis, and the average SSIM for the whole frame is the average of block SSIMs [10].

$$SSIM(x,y) = \frac{(2\mu_x\mu_y + C_1)(2\sigma_{xy} + C_2)}{(\mu_x^2\mu_y^2 + C_1)(\sigma_x^2\sigma_y^2 + C_2)} \quad (2)$$

3.4 MS-SSIM

The fact which distinguishes MSSIM from SSIM is that this VQA algorithm evaluates multiple SSIM values at multiple resolutions. Although it does not lay stress on the luminance component in general, nonetheless we implemented it frame by frame to the luminance part and finally average value was computed. In defining MS-SSIM, luminance, contrast and structure comparison measures are computed at each scale as follows [11]:

$$l(x,y) = \frac{(2\mu_x\mu_y + C_1)}{(\mu_x^2 + \mu_y^2 C_1)} \tag{3}$$

$$c(x,y) = \frac{\sigma_x\sigma_y + C_2}{(\sigma_x^2 + \sigma_y^2 + C_2)} \tag{4}$$

$$s(x,y) = \frac{\sigma_{xy} + C_3}{(\sigma_x + \sigma_y + C_3)} \tag{5}$$

where C_1, C_2 and C_3 and are small constants given by the following relation are small constants given by the following relation

$$C_1 = (K_1L)^2, \tag{6}$$

$$C_2 = (K_2L)^2 \tag{7}$$

and

$$C_3 = \frac{C_2}{2} \tag{8}$$

Furthermore,

$$L = 255, K_1 < < 1 and K_2 < < 1. \tag{9}$$

Using the above equations, we compute the MS-SSIM values as follows

$$MS - SSIM(x,y) = [l_M(x,y)]^{\alpha_M} \Pi_{j=1}^{M} [c_j(x,y)]^{\beta_j} [s_j(x,y)]^{\gamma_j} \tag{10}$$

Similarly, we used two kinds of MS-SSIMs namely MS-SSIM Gaussian (MS-SSIMG) and MS-SSIM Block (MS-SSIMB) by making slight changes, that is, rather than using the Gaussian window in the former, we computed the SSIM level by level by using 8 × 8 block level at each resolution in the latter.

3.5 Proposed PQM2D Metrics

Using the ideas from [1, 12, 13] and rather than dealing with the 3D video components, our metrics assessed the quality of 2D video sequences extensively. The idea behind the formation of the new metrics is taken from the fact that the luminance value is an essential component that determines the quality of an image. On the contrary, chrominance is basically responsible for colour in the image. Thus, we can say that the luminance provides structure based information about the image rather than the colour of the various objects in the image. This method is based on the idea of finding the difference between luminance values in the test and impaired frames [13]. As it is obvious that there might be variations in the structure as and when the frames become distorted, there should be prominent deviations in the luminance values. Furthermore, these luminance deviations, when considered at a specific pixel coordinate of reference as well as the impaired frames, give us meaningful values. That means, greater the impairment in the structure of the processed frame at a certain pixel coordinate, greater is the luminance deviation from the reference frame, at that very point. The step by step algorithm implementation is given below.

1. Compute the pixel mean, variance and covariance of blocks

$$\mu(b_o), \mu(b_R), \sigma^2(b_o), \sigma^2(b_R), \sigma(b_o b_R). \tag{11}$$

2. Compute weighted distortion coefficient for each pixel in the block

$$\alpha(m,n) = \begin{cases} 0, \mu(b_o) << 1 \quad \text{and} \quad \mu(b_R) << 1 \\ 1, \mu(b_o) << 1 \quad \text{and} \quad \mu(b_R) > 1 \\ \min\left[\frac{(c_o(m,n) - c_R(m,n))^4}{\mu(b_o)^2}\right], \quad \text{else} \end{cases} \tag{12}$$

For contrast distortion in the block, define:

$$K(b_R) = 1 + \frac{\sigma^2(b_o) - (\sigma^2(b_R))^2 + 255}{(\sigma^2(b_o))^2 + (\sigma^2(b_R))^2 - 2(\sigma(b_o b_R))^2 + 255} \tag{13}$$

3. Perceptual distortion Metrics (PDM) in the whole block is defined as:

$$PDM(b_R) = \frac{K(b_R)}{64} \sum_{(m,n)\epsilon(b_R)} \alpha(m,n) \tag{14}$$

After PDM is computed for all blocks, total perceptual distortion in the frame is equal to weighted mean of block distortions:

$$PDM(c_R) = \frac{\sum_{(b_R)\epsilon(c_R)} w(b_R) PDM(b_R)}{\sum_{(b_R)\epsilon(c_R)} w(b_R)} \tag{15}$$

$$w(b_R) = \begin{cases} 1, \mu(b_o) = 0 \\ \frac{255}{\mu(b_o)}, else \end{cases} \tag{16}$$

Finally PQM2D is defined as follows:

$$PQM2D(c_R) = \begin{cases} 0, PQM2D(c_R) < 0 \\ 1 - PDM(c_R), else \end{cases} \tag{17}$$

Frame level PQM2Ds are averaged for the whole video. In order to obtain the overall objective score for a sequence, the scores of the frame level PQM are added up and divided by the number of frames in the sequence and we get the PQM2D score representing the overall quality score judging on a scale of measurement of 0 to 1 where 0 stands for the worst quality and 1 for the best. The main idea for the PQM is based on the fact that the HVS gives the quality by first measuring the errors in the luminance which in fact comprises of the structure in an image and also is quite less sensitive to the chrominance element of an image.

3.6 Simulation Results and Discussions

The various performance criterions applied on the metrics were monotonicity and accuracy, determined on the basis of Pearson Linear Correlation Coefficient (PLCC) and the Spearman Rank Order Correlation Coefficient (SROCC) respectively. The number of selected presentations for each distortion type were 7 for wireless, 5 for H264, 6 for IP and 6 for MPEG-2. In the scatter plots Fig. 2, we used statistical procedures of various regression types like exponential, linear, logarithmic, power for finding the best fitting trend lines in our data values in order to predict the accuracy of our results. The various equations of the best fitting trend lines are also shown in the plot. On the basis of square of correlation, we fitted the best trend lines and after comparing all its values, we found that the linear fit is the best for all our models. Clearly, PQM2D has the highest value and both the MS-SSIM values are lowest. Tables 2 and 3 show us the performance estimation of all the objective models with respect to the statistical measures of coefficients of the PLCC and the SROCC respectively for all selected video scenes and also individually for each of the four distortion types. It is clearly evident from the results of the metric PQM2D, with respect to PLCC and SROCC that it outperforms all the other objective models. Our tactfully organised digital video database taken from the LIVE database also testifies the drawbacks of PSNR and both MS-SSIM as it is substantially lower than most of the objective models. When we study the linear correlations based on distortion types, we see that PQM2D is mostly superior like in IP and MPEG2 distortion cases and close to the superior in case of wireless and H264 ones. Nevertheless both SSIM have shown their fairly efficient performance. For example SSIMG and SSIMB perform the best in wireless and H264 distortions respectively. However SSIMG and SSIMB perform poorly for the IP distortions, causing their overall performance to be lower than the PQM2D. Likewise MS-SSIM has shown inferior performance in most of the distortion types. Therefore it can be said that the PQM2D performs consistently well for all distortion types while other metrics fail for certain types of distortions. When we study the monotonicity of the model using the SROCC results, we still see that PQM2D has the highest overall correlation score. When distortion types are individually considered, correlation values of PQM2D are fairly close to the best one except for the wireless case where it performs sub optimally. Yet, for the full data, PQM2D

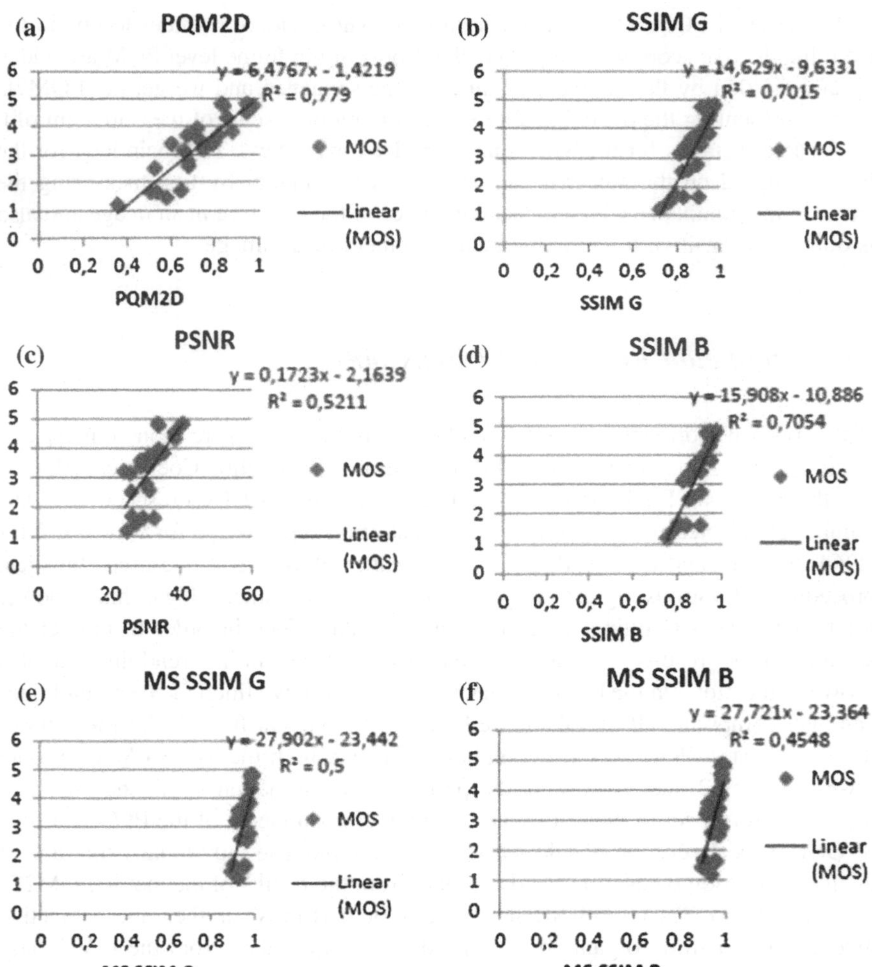

Fig. 2 Scatter plots of objective versus subjective model. **a** PQM2D. **b** SSIM G. **c** PSNR. **d** SSIM B. **e** MS SSIM G. **f** MS SSIM B

Table 2 Comparison of PLCC

Metrics	Wireless	H.264	IP	MPEG2	ALL
PQM2D	0.916	0.942	0.924	0.956	0.883
PSNR	0.537	0.915	0.713	0.918	0.722
SSIMG	0.928	0.940	0.609	0.930	0.838
SSIMB	0.904	0.950	0.622	0.920	0.840
MS-SSIMG	0.812	0.887	0.647	0.830	0.707
MS-SSIMB	0.789	0.868	0.654	0.800	0.674

Table 3 Comparison of SROCC

Metrics	Wireless	H.264	IP	MPEG2	ALL
PQM2D	0.786	0.90	0.771	0.843	0.885
PSNR	0.714	0.900	0.429	0.929	0.780
SSIMG	0.964	1.000	0.600	0.929	0.877
SSIMB	0.955	1.000	0.714	0.929	0.877
MS-SSIMG	0.893	1.000	0.647	0.829	0.746
MS-SSIMB	0.964	0.900	0.829	0.814	0.716

again has the highest SROCC scores among the tested metrics. The higher quality of performance of our metrics PQM2D is elucidated in both the correlation results as it is always slightly larger than SSIMG and SSIMB and also is fairly larger than MS-SSIMG and MS-SSIMB. Nevertheless, the SSIM results are apparently comparable to the best performing algorithm.

3.7 Conclusion and Future Work

The gist of our discussion is that the PQM2D is superior in performance and this gives us the perfect picture of our research theme that a robust objective algorithm, well-correlated with the human perceptual experience can provide us the best method to estimate the digital video quality. In other words, a well formed QoS can only be justified when the QoE has been obtained systematically. Through our objective metric discussions, the sensitiveness of HVS to the luminance component is clearly visible. Evidently, we came across several artefacts in our test videos arising due to different types of distortions in our experiment. Seemingly, it is hard to fathom that a single quality evaluation metrics can deal with all kinds of artefacts. In fact different quality metrics may be required to deal with different artefacts efficiently. The crux of the entire paper is that complexity of the HVS is still not much known and as we solve the complexity day by day, we can have more reliable and precise results for quality assessment. For future work and in order to enhance the MOS prediction models, other features of HVS can be stressed upon. Another possible enhancement could be made while dealing with the temporal features which are not employed in most of the QoS models. Presumably, incorporating both spatial as well as the temporal component into the QoE model could lead to a rather effective prediction of the QoE.

References

1. International Telecommunication Union-Radiocommunication Sector: Methodology for the subjective assessment of the quality of television pictures, Recommendation ITU-R BT, pp. 500–513 (2012)
2. Joveluro, P.: Perceptual quality estimation techniques for 2D and 3D videos. PhD Thesis, University of Surrey, UK (2011)

3. Seshadrinathan, K., Soundararajan, R., Bovik, A.C., Cormack, L.K.: A subjective study to evaluate video quality assessment algorithms. In: SPIE Proceedings Human Vision and Electronic Imaging (2010)
4. Video Quality Experts Group (VQEG): http://www.its.bldrdoc.gov/vqeg/vqeg-home.aspx
5. International Telecommunication Union-Radiocommunication Sector: Subjective video quality assessment methods for multimedia applications, Recommendation ITU-T P. 910 (1999)
6. Roque, L.: Quality evaluation of coded video. Masters Thesis, Universidade Tecnica de Lisboa, Portugal (2009)
7. LIVE, Image and Video Quality Assessment Database: http://live.ece.utexas.edu/research/quality/subjective.htm
8. Vatolin, D., Petrov, O.: MSU Perceptual Video Quality Tool 1.0 (Software for Subjective Video Quality Evaluation). Moscow State University, Russia (2006)
9. Winkler, S.: Digital Video Quality. Wiley, Chichester (2010)
10. Wang, Z., Bovik, A., Sheikh, H., Simoncelli, E.: Image quality assessment: from error visibility to structural similarity. IEEE Trans. Image Process. **13**(4), 600–612 (2004)
11. Seshadrinathan, K., et al.: Study of subjective and objective quality assessment of video, IEEE Trans. Image Process. (2010)
12. Amirshahi, S.A.: Towards a perceptual metric for video quality assessment. Masters Thesis report, Gjovik University College, Norway (2010)
13. Gaurav, R.: Video quality assessment using subjective and objective metrics. Masters Thesis, ISIK University, Turkey (2013)

Unsupervised Machine Learning Approach for Gene Expression Microarray Data Using Soft Computing Technique

Madhurima Rana, Prachi Vijayeeta, Utsav Kar, Madhabananda Das and B.S.P. Mishra

Abstract Machine learning is a burgeoning technology used for extractions of knowledge from an ocean of data. It has robust binding with optimization and artificial intelligence that delivers theory, methodologies and application domain to the field of statistics and computer science. Machine learning tasks are broadly classified into two groups namely supervised learning and unsupervised learning. The analysis of the unsupervised data requires thorough computational activities using different clustering algorithms. Microarray gene expression data are taken into consideration for cluster regulating genes from non-regulating genes. In our work optimization technique (Cat Swarm Optimization) is used to minimize the number of cluster by evaluating the Euclidean distance among the centroids. A comparative study is being carried out by clustering the regulating genes before optimization and after optimization. In our work Principal component analysis (PCA) is incorporated for dimensionality reduction of vast dataset to ensure qualitative cluster analysis.

Keywords Gene expression · Microarray data · Principal component analysis (PCA) · Hierarchical clustering (HC) · Cat swarm optimization (CSO)

M. Rana (✉) · P. Vijayeeta · U. Kar · M. Das · B.S.P. Mishra
KIIT University, Bhubaneswar, India
e-mail: madhurima.rana@gmail.com

P. Vijayeeta
e-mail: prachi.vijayeeta@gmail.com

U. Kar
e-mail: utsav.kar12@gmail.com

M. Das
e-mail: mndas_prof@kiit.ac.in

B.S.P. Mishra
e-mail: bsmishrafcs@kiit.ac.in

A. Nagar et al. (eds.), *Proceedings of 3rd International Conference on Advanced Computing, Networking and Informatics*, Smart Innovation, Systems and Technologies 43, DOI 10.1007/978-81-322-2538-6_51

1 Introduction

Clustering methods are used in the analysis of microarray data. It is a revolutionary step which is very useful in the identification of similar groups of samples or genes in an expression levels [1]. Microarray technologies are used to measure a vast number of gene expression level in different conditions, and produces a big scale of dataset.

It (clustering) is used for the identification of genes which have similar profile. Real world clustering problems are used for the analysis of gene expressions from microarray dataset. Various evolutionary algorithms have been put forward to optimize single objective of clusters [1]. Based on the similarity/dissimilarity measures among the individual objects, clustering are classified as either hard clustering or soft clustering. Hard clustering is based on classical set theory in which the clustering data are partitioned into specific number of subsets which are mutually exclusive. But in soft clustering an object can belong to more than one cluster simultaneously [2]. With the growing importance of computational intelligence in diverse application areas, soft computing is fast gaining popularity in the world of research [3]. In the early 1990, Latif A. Zadeh, a pioneer in this field, coined the term "Soft Computing which gives a holistic view of inexactness and non-determinism of computational problems that need to be solved. Researchers have categorized the soft computing based problem solving techniques into feasible components that are highly interactive over a given set of related domains. These components of soft computing comprises of Artificial Neural Network (ANN), Evolutionary search strategies such as Genetic Algorithm (GA), Ant Colony Optimization (ACO), Particle Swarm Optimization (PSO) [4] etc.

The expression levels of numerous genes using microarray data elucidate the hidden patters in a sample. Clustering plays a vital role in identifying similar patterns by revealing natural structures and grouping them. In order to reduce the cluster errors, cat swarm optimization [5] is being incorporated.

We have organized the entire paper into six sections. The first section consists of introduction where we have narrated the technologies in nutshell. Section 2 reveals the literature review where all related works are mentioned briefly. Section 3 describes the working process model of the entire work. Section 4 explains and discusses the experimental results which are represented by graphs. In Sect. 5, the work has been concluded. In the last future scope has been discussed.

2 State of the Art Processing Technique

The science of clustering has been one of the research challenges in the field of data mining. Clustering the expressible genes in an micro-array data and studying their quality under favourable conditions are one of the prime focus of Gibbons and Roth [6]. They have carried out a comparative study of gene clustering algorithms and have concluded that self-organized map is one of the best approach in case of higher number of clusters.

Lee et al. [7] have applied a heuristic global optimization method called Deterministic Annealing to the clustering method. This DA optimization method locally minimizes the cost function, subject to the constraint on given entropy by controlling the temperature factor. Mohsen et al. [8] have proposed a Bi-clustering algorithm to discover similar genes which have common behaviour. A Hybrid MOPSO (multi-objective Particle Swarm Optimization) algorithm is used for identifying bi-clustering in gene expression data. Rousseeuw et al. [9] have proposed for partitioning technique to represent clusters by the Silhouette based on the tightness and separation of objects. Jiang et al. [10] have surveyed on DNA microarray technology to observe of thousands of genes expression levels. Their work comprises of analysing different proximity measurement for gene expression data.

Dey et al. [11] have implemented PSO technique along with K-means to get a better accuracy of cluster. They have considered a data matrix where class labels are associated with the samples. Andreopoulos et al. [12] have performed extensive survey on different clustering algorithms and had carried out match for biomedical applications exclusively. They have evaluated the complexity of nearly forty clustering algorithms and demonstrated it in a tabular manner very lucidly. Dudoit et al. [13] have analysed various clustering compactness by evaluating different distance parameters and coefficients. They have revealed a generalized application of clustering based on the gene expression obtained from microarray data. Yin et al. [14] have taken a yeast dataset and experimented over it three different clustering algorithms (i.e. Hierarchical Clustering, Self-Organizing-Map) and compared their performance. They have also analysed similarity among them. The efficiency of different algorithms has also been evaluated.

Priscilla et al. [15] have proposed a two dimensional hierarchical clustering and observed that expressions of genes are increased effectively. Analysis of the results has showed that the proposed techniques give better f-measure, precision and recall value than the traditional approach. Liu et al. [16] have proposed two clustering methods over cat swarm optimization (i.e. CSOC (Cat Swarm Optimization Clustering) and KCSOC (K-harmonic means Cat Swarm Optimization Clustering)). They have taken six different datasets and showed the efficiency of the proposed algorithm. Sathishkumar et al. [17] have proposed a technique for reduction of dimensionality and clustering. For reduction LSDA (Locality Sensitive Discriminant Analysis) is used. As compared to traditional technique the proposed algorithm has given much higher accuracy and less time.

3 Working Process Model

The working process model comprises of Gene Expression Dataset, Clustering and Cat Swarm Optimization.

3.1 Gene Expression Data

Gene expression data consist of microarray dataset. A microarray is a 2-D array on a small solid glass surface that holds large amount of biological data. The gene microarray is a highly sophisticated chip used in the detection of malignant tissues, proteins, peptides and carbohydrates expansion and in many other scientific researches related to biotechnology.

The samples of genes which are put into the probes of a tray are experimented on the basis of some physical conditions. The feature corresponds to the expression of different genes. We have adopted a methodology to group similar types of genes. Then we have to optimize the cost function by reducing the number of clusters [18].

3.2 Clustering Technique

Clustering technique is applied when several instances are separated into groups. They apparently reflect mechanisms which works in the domain from the extracted instances based upon the similarity that exist among themselves [19]. Different mechanism is imposed to express the results of clusters depending upon their occurrence in a group. If the groups are exclusively identified then the instances will be in one group. If they are overlapping, then the instances may fall into several groups [20]. If they are probabilistic, then any instance belongs to each group with ascertain probability.

Hierarchical Clustering: This clustering resembles a hierarchy or tree like structure where we do not assume any particular type of clusters. It clusters data over different scales to create tree like structure. It is multilevel tree where one level cluster is joined with next. To improve the quality of hierarchical clustering firstly we need to perform careful analysis of object linkages at each hierarchical portioning. Secondly proper integration of micro-clusters needs to be ensured. Mostly hierarchical clustering methods are accomplished by use of proper metric (distance measure between observations pairs), and a linkage criterion that specifies the variation of sets as a function of the pairwise distances in the sets of observations.

Gene-Based Clustering: Gene-based clustering is a methodology that is adopted to form co-expressed genes together which indicate co-regulation and co-function and hence are responsible for causing diseases.

Significance of gene-based clustering:

The aim of gene expression data clustering is to express about data distribution and structure. This is being carried out by imposing hierarchical clustering algorithm technique.

Since the dataset is high voluminous so we have reduced the dataset using Principle Component Analysis (PCA). The PCA removes the noisy data and increases the simplicity of the useful information.

By using microarray data we can conveniently estimate the relationship between the gene clusters as well as between the genes within the same cluster. This is obtained by evaluating clustering algorithms and also by optimizing the distances between the centroids.

In this paper we have used Hierarchical Clustering method to produce a dendrogram without specifying the number of clusters. By applying agglomerative method we combine the two nearest clusters by calculating distance between clusters that represents the cluster shape. Between cluster distance specifies three measures i.e. average linkage, single linkage and complete linkage. But in our work we have taken into consideration average linkage of pair wise distances dendrogram.

3.3 Cat Swarm Optimization

Cat Swarm Optimization algorithm is designed by simulating the behaviour of the cats in two different modes [14]. Mode-1: In this first mode (i.e. Seeking Mode), the cat's resting situation, looking and seeking for moving the next position is being modelled. Mode-2: In this second mode (i.e. Tracing Mode), the modelling is done in such a way that it traces out only to the targeted ones.

It is consider that each and every cat is being placed in its own M-dimensional position. Let $V = \{V_1, V_2, \ldots, V_n\}$ be the velocities in the respective dimensions and let a suitable fitness function be taken into consideration so that the corresponding fitness value is computed to represent the location of the cat. A flag is used to ascertain when it is in tracing mode or seeking mode. The ultimate solution at the end would lead to one of the cat's best positions. Here CSO is mainly used with HC to reduce the number of cluster.

Steps of CSO:

Step 1: Initialization of number of cats (N).

Step 2: Distribution of cats arbitrarily over M-dimensional search area.

Step 3: Assignment of random values to the velocity of the cat within the limit of maximum velocity. Picking up a number of cats in a random manner and setting them into tracing mode as per the value of Mixture Ratio (MR) and setting the remaining cats to the seeking mode.

Step 4: Calculating the fitness function based on the position of the cat and compute the fitness value of each cat. Save the best cat position in the memory (x_{best}) that generates the global best solution.

Step 5: Then the cats are moved to other locations depending on their flag value. Check whether kth cat is in the seeking mode or tracing mode, if seeking mode then apply it to the process of seeking mode otherwise place it in the process of tracing mode.

4 Experimental Results and Discussion

The data constitutes of http://rana.lbl.gov/data/dlcl/dlcl_figureplus.txt, which is organized in a matrix format having a size of 13412 × 40. In Fig. 1 the work flow process model is stepwise briefly described. The contents of the matrix possessing the express able forms of gene are filtered to separate non-expressible genes that have no effect on the changes being encountered. We have removed the missing data from the dataset.

The gene filtering function is used to filter out those genes with small variance over time. Finally entropy filtering function removes all the profiles having low entropy [13]. Now, we have obtained a manageable list of genes by using PCA, the dimensionality of the dataset is being reduced to 40 × 40. A new hybrid technique (PCA+ CSO) is introduced here to reduce dimensionality of the dataset. In Table 1 the parameter setting of CSO is described.

After that clustering technique is applied to the reduced dataset using Matlab codes. The distance function is used to verify the cluster tree and calculate the pairwise distance between profile and another function is executed. Subsequently some hierarchical cluster function is used to obtain accurate number of cluster.

A dendrogram has been constructed to distinguish the arrangement of clusters graphically. In following Figs. 2 and 3 describes the dendrogram results.

In our paper the intra and inter cluster distances are obtained. Here, we have use Euclidean distance for comparing distances:

$$d(x, y) = \sqrt{\sum_{i=1}^{n} (yi - xi)_2}$$

where $x = (x_1, x_2, \ldots x_n)$ and $y = (y_1, y_2, \ldots y_n)$ are two points.

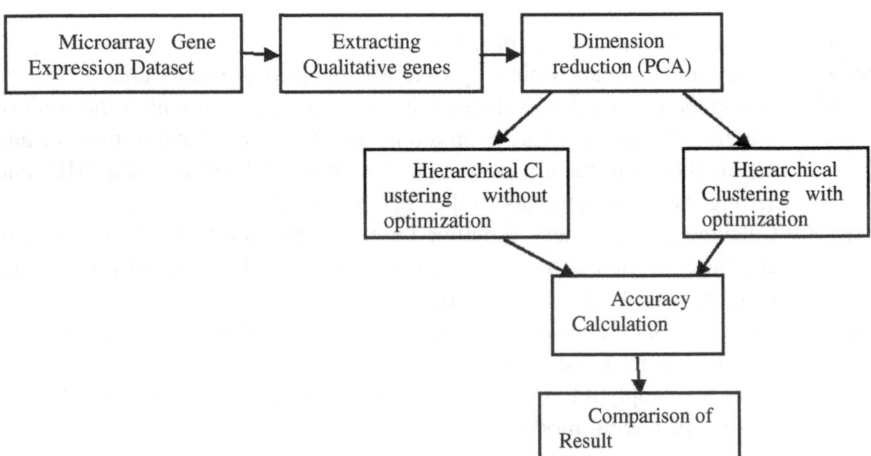

Fig. 1 Workflow diagram model

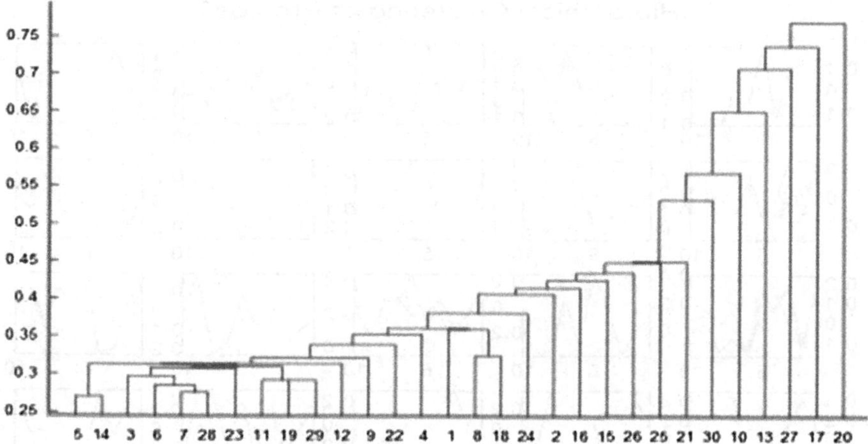

Fig. 2 Dendrogram before using optimization

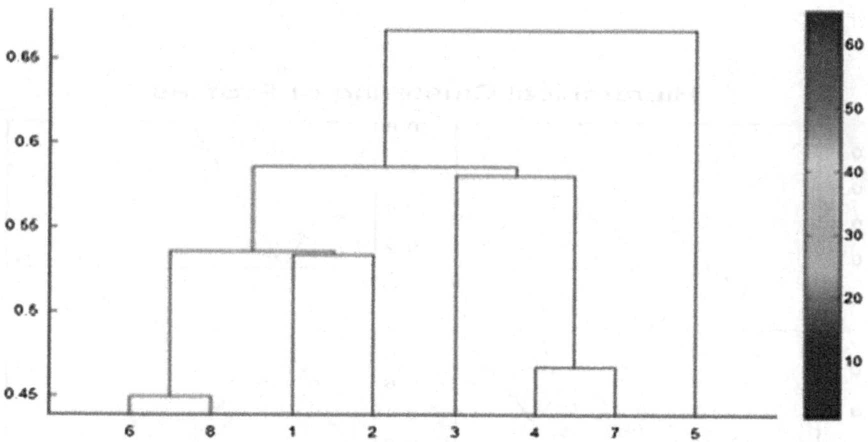

Fig. 3 Dendrogram after using optimization

Lastly we calculate natural clustering over the dataset and plotted it by using hierarchical clustering. Our experimental result clearly shows that after optimization the numbers of clusters are reduced. Here Figs. 4 and 5 show the difference. This led to smaller computation cost (Table 2).

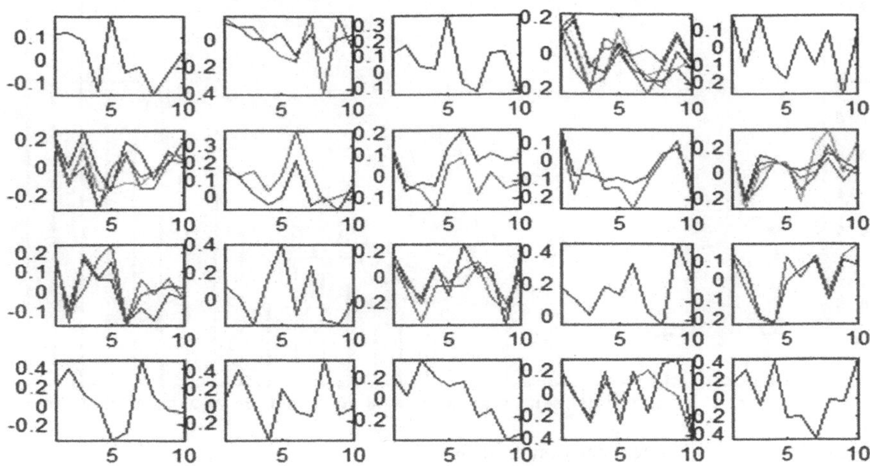

Fig. 4 Gene profile in the Hierarchical cluster

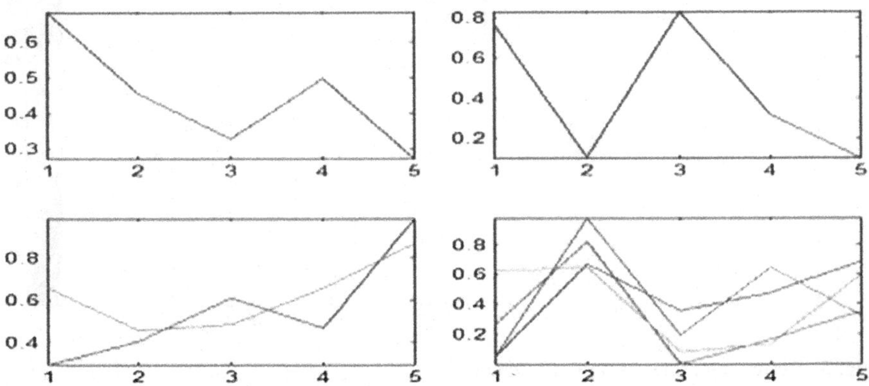

Fig. 5 Gene profile after CSO and HC

Table 1 Parameter values for CSO

Parameters	Values
SMP	3
SRD	0.2
CDC	0.2
MR	0.3

Table 2 Experimental results

Technology	Intra cluster distance	Inter cluster distance	No. of clusters
HC	0.8398	0.6354	20
CSO+HC	0.6797	0.7344	4

5 Conclusion

Although there are many searching strategies for evaluating an optimal solution, but we have applied cat swarm optimization (CSO) technique to optimize the dataset. The simulations carried out for dimension reducibility have resulted in a graph with a very feasible targeted output. On the basis of experimental result (Table 2) we can conclude that the number of clusters obtained after optimization is comparatively lesser than that obtained before optimization. The dendrogram analysis of hierarchical clustering clearly shows that how one cluster is inter-related with another one and lastly the gene profiling plot is the graphical representation of the expression of the different genes in each cluster. In our paper the experimental results clearly satisfy the desired aim.

6 Future Scopes

One of the limitations of hierarchical clustering is that once a step is done it can never be undone. So, it fails to correct erroneous decision. To overcome this problem, error detection can be done by other hybrid algorithm in the future. In near future number of clusters should be obtained by some statistical machine learning algorithms. And different clustering algorithm can also be used. More swarm based techniques and more clustering techniques will be used for better result and accuracy purpose. The number of cluster should be calculated experimentally by some techniques.

References

1. Ma, P.C.H., Chan, K.C.C., Xin, Y., Chiu, D.K.Y.: An evolutionary clustering algorithm for gene expression microarray data analysis. IEEE Trans. Evol. Comput. **10**(3), 296–314 (2006)
2. Witten, I.H., Frank, E., Hall, M.A.: Data Mining—Practical Machine Learning Tools and Techniques. Morgan Kaufmann (2005)
3. Thamaraiselvi, G., Kaliammal, A.: A data mining: concepts and techniques. SRELS J. Inform. Manage. **41**(4), 339–348 (2004)
4. Roy, S., Chakraborty, U.: Introduction to soft computing: NeuroFuzzy and Genetic Algorithms. Pearson Publication
5. Dudoit, S., Gentleman, R.: Cluster analysis in DNA microarray experiments. Bioconductor Short Course Winter (2002)

6. Gibbons, F.D., Roth, F.P.: Judging the quality of gene expression-based clustering methods using gene annotation. Genome Res. **12**(10), 1574–1581 (2002)
7. Deng, Y., Kayarat, D., Elasri, M.O., Brown, S.J.: Microarray data clustering using particle swarm optimization K-means algorithm. In: Proceedings 8th JCIS, pp. 1730–1734 (2005)
8. Lee, K.M., Chung, T.S., Kim, J.H.: Global optimization of clusters in gene expression data of DNA microarrays by deterministic annealing. Genom. Inform. **1**(1), 20–24 (2003)
9. Dudoit, S., Gentleman, R.: Cluster analysis in DNA microarray experiments. Bioconductor Short Course Winter (2002)
10. Rousseeuw, P.J.: Silhouettes: a graphical aid to the interpretation and validation of cluster analysis. J. Comput. Appl. Math. **20**, 53–65 (1987)
11. Jiang, D., Chun, T., Aidong, Z.: Cluster analysis for gene expression data: A survey. IEEE Trans. Knowled. Data Eng. **16**(11), 1370–1386 (2004)
12. Dey, L., Mukhopadhyay, A.: Microarray gene expression data clustering using PSO based K-means algorithm. UACEE Int. J. Comput. Sci. Appl. **1**(1), 232–236 (2009)
13. Andreopoulos, B., An, A., Wang, X., Schroeder, M.: A roadmap of clustering algorithms: finding a match for a biomedical application. Briefings Bioinform. **10**(3), 297–314 (2009)
14. Santosa, B., Ningrum, M.K.: Cat swarm optimization for clustering and pattern recognition. In: International Conference of Soft Computing SOCPAR'09, pp. 54–59. 20 (2009)
15. Yin, L., Huang, C.H., Ni, J.: Clustering of gene expression data: performance and similarity analysis. BMC Bioinform. (2006)
16. Priscilla, R., Swamynathan, S.: Efficient two dimensional clustering of microarray gene expression data by means of hybrid similarity measure. In Proceedings of the International Conference on Advances in Computing, Communications and Informatics, pp. 1047–1053. ACM (2012)
17. Santosa, B., Ningrum, M.K.: Cat swarm optimization for clustering. In: International Conference of in Soft Computing and Pattern Recognition, pp. 54–59 (2009)
18. Iassargir, M., Ahhmad, A.: A hybrid multi-objective PSO method discover biclusters in microarray data. Mohsen. Int. J. Comput. (2009)
19. Karaboga, D., Ozturk, C.: A novel clustering approach: artificial bee colony (ABC) algorithm. Appl. Soft Comput. 652–657 (2011)
20. Castellanos-Garzón, J.A., Diaz, F.: An evolutionary and visual framework for clustering of DNA microarray data. J. Integr. Bioinform. **10**, 232–232 (2012)

Phase Correlation Based Algorithm Using Fast Fourier Transform for Fingerprint Mosaicing

Satish H. Bhati and Umesh C. Pati

Abstract The fingerprint identification is a challenging task in criminal investigation due to less area of interest (ridges and valleys) in the fingerprint. In criminal incidences, the obtained fingerprints are often partial having less area of interest. Therefore, it is required to combine such partial fingerprints and make them entire such that it can be compared with stored fingerprint database for identification. The conventional phase correlation method is simple and fast, but the algorithm only works when the overlapping region is in the leftmost top corner in one of the two input images. However, it does not always happen in partial fingerprints obtained in forensic science. There are total six different possible positions of overlapping region in mosaiced fingerprint. The proposed algorithm solves the problem using the mirror image transformation of inputs and gives correct results for all possible positions of overlapping region.

Keywords Cross power spectrum · Fingerprint mosaicing · Fourier transform · Mirror image transformation · Ridges and valleys

1 Introduction

Fingerprint mosaicing is the similar to the image mosaicing. It stitches two or more than two fingerprints and combines them to create a large view of fingerprint. The fingerprint identification system compares the query fingerprint with stored fingerprint template database by matching minutiae points in fingerprints. The fingerprints obtained in forensic science during criminal investigation are mostly partial [1].

S.H. Bhati (✉) · U.C. Pati
Department of Electronics and Communication, National Institute of Technology, Rourkela 769008, Odisha, India
e-mail: bhatisatish4045@gmail.com

U.C. Pati
e-mail: ucpati@nitrkl.ac.in

A. Nagar et al. (eds.), *Proceedings of 3rd International Conference on Advanced Computing, Networking and Informatics*, Smart Innovation, Systems and Technologies 43, DOI 10.1007/978-81-322-2538-6_52

Therefore, sometimes it is difficult to match with stored fingerprints. These partial fingerprints can be combined using mosaicing process to increase the area of ridges and valleys in fingerprints, which results in more number of minutiae points. Hence, it improves the accuracy and robustness of identification system.

In the case of different types of images like landscape photos, wildlife photos, aerial photos, architectural photos and real life photography, it is easy to decide between two images whether images need to join side by side or up-down manner for mosaicing. But, it is too difficult in case of fingerprints because these only contain black and white curvature lines (ridges and valleys). It is not possible to predict the position (left or right or upper or down) of fingerprints in the mosaiced fingerprint.

Kuglin et al. [2] proposed the phase correlation method which has been proved to register images when there is only translation between two images. Reddy et al. [3] extended phase correlation technique to estimate rotation and scaling parameters along with translation parameter involved in two overlapped images. Zhang et al. [4] proposed a hybrid swipe fingerprint mosaicing scheme using the phase correlation method with singular value decomposition to find a non-integer translation shift. Tarar et al. [5] proposed algorithm using phase correlation method to mosaic adjacent fingerprints involving only translation parameter with respect to one another. The conventional phase correlation method proposed in [2–5] works only in the case when the overlapping region is in the leftmost top corner of one of two input images. The method also has the drawback that the change in the sequence of input images affects the output mosaiced image. Many other techniques have been proposed to mosaic two fingerprints in [6–9].

The proposed algorithm in this paper transforms the input fingerprints in such a way that these satisfy the condition of having overlapping region in the leftmost top corner of one input fingerprint. After that, the translation parameter is estimated correctly using conventional phase correlation method. Finally, two input fingerprints are stitched at estimated translation parameter. The correctness of the mosaiced fingerprint can be examined by observing ridges and valleys. In correct mosaiced fingerprint image, ridges and valleys alternate and flow in a locally constant direction [10].

The rest of paper is organized as follows. Section 2 describes the six different possible positions of overlapping region in input partial fingerprints. Section 3 describes the details of the proposed algorithm. Experimental results are discussed in Sect. 4. The paper is concluded in Sect. 5.

2 Six Possible Positions of Overlapping Region

The phase correlation method in [2–5] gives correct mosaiced image if and only if the second input image contains the overlapping region in the leftmost top corner of image. There can be six possible ways in which two fingerprints can be overlapped as shown in Fig. 1.

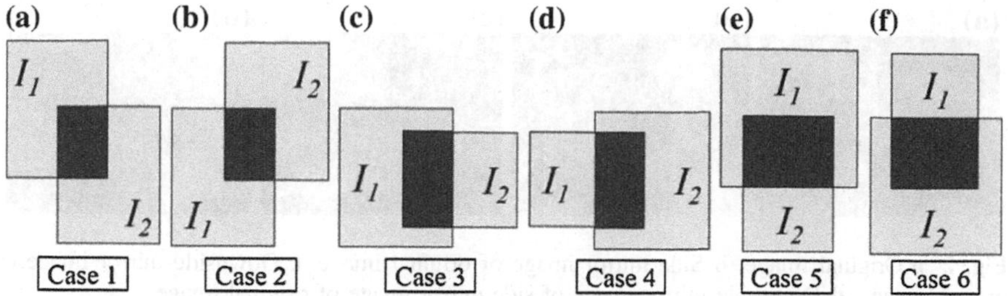

Fig. 1 Different types of possible ways in which two fingerprints can be overlapped

The conventional method works only for *Case 1*, *Case 3* and *Case 5* which are shown in Fig. 1a, c, e respectively. In these three cases, the leftmost top corner of second image contains the overlapping region. But, the method gives incorrect mosaiced fingerprint for remaining *Case 2*, *Case 4* and *Case 6* which are shown in Fig. 1b, d, f respectively.

3 Proposed Algorithm

The phase correlation method in [2–5] estimates the translation parameter at the position of the highest peak in the cross power spectrum of two images. The method always estimates translation parameter even though images do not have overlapping region. The estimated translation parameter is correct only when the overlapping region is in the leftmost corner of second image. The proposed algorithm transforms the both images such a way that the overlapping region goes to the leftmost top corner of one image.

3.1 Mirror Image Transformation

There can be three possible different mirror images of an image, which are as shown in Fig. 2.

As shown in Fig. 3a, it can be observed that *Case 1* and *Case 3* are the side mirror image of *Case 2* and *Case 4* respectively. Similarly, *Case 5* and *Case 1* are the downside mirror image of *Case 6* and *Case 2* respectively as shown in Fig. 3b. Thus, we can divide all cases into three categories. *Category 1* includes *Case 1*, *Case 3* and *Case 5*. These do not need to be converted to other cases by the mirror image transformation. *Category 2* includes *Case 2* and *Case 4*, which need to be converted into *Case 1* and *Case 3* respectively using side mirror image transformation. *Category 3* includes *Case 6* and *Case 2*. These need to be converted into *Case 5* and *Case 1* respectively using downside mirror image transformation.

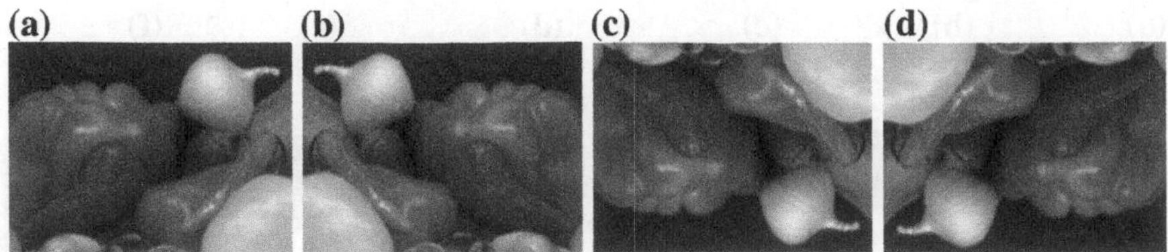

Fig. 2 **a** Original image. **b** Side mirror image of original image. **c** Downside mirror image of original image. **d** Downside mirror image of side mirror image of original image

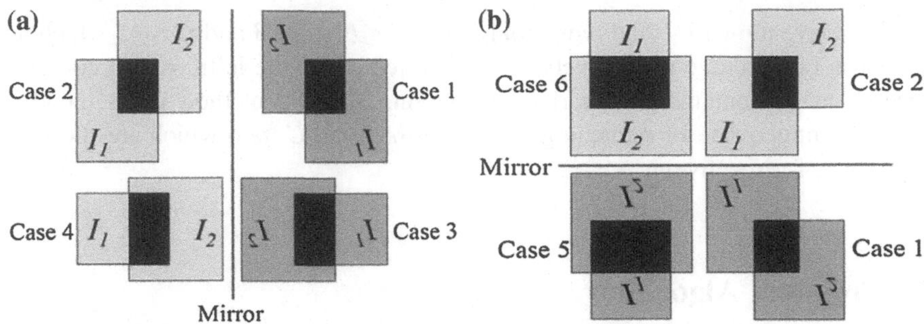

Fig. 3 **a** Side mirror image of *Case 2* and *Case 4*. **b** Downside mirror image of *Case 6* and *Case 2*

Our proposed method finds mirror image of both input fingerprints to convert *Category 2* and *Category 3* into *Category 1* and then implements the same steps as for *Category 1*.

3.2 Algorithm

The steps of the proposed phase correlation method are as follows:

1. Read two input fingerprint images I_1 and I_2.
2. Enhance both images I_1 and I_2 using adaptive histogram equalization method [11]. Let enhanced fingerprint images are E_1 and E_2 respectively.
3. Find the side or downside mirror images M_1 and M_2 of fingerprints E_1 and E_2 respectively.
4. Compute the Fourier transforms F_1 and F_2 of images M_1 and M_2 respectively.
5. Make both images of same size by zero padding. Let resized images are R_1 and R_2 respectively.
6. Compute the cross power spectrum of R_1 and R_2. Let P is the cross power spectrum which is given by,

$$P = \frac{R_1 \times R_2^*}{||R_1 \times R_2^*||} \quad (1)$$

where, * indicates complex conjugate value and || || indicates absolute value.

7. Compute the inverse Fourier transform p of cross power spectrum P.
8. Locate the peak pixel value in p. Let (t_x, t_y) is the position of peak. This position (t_x, t_y) is the translation parameter for fingerprint M_2.
9. Stitch both fingerprints by overlapping image M_2 on image M_1 at the translation parameter (t_x, t_y).
10. Apply the same mirror image transformation on mosaiced fingerprint as applied in step 3.

4 Experimental Results and Discussion

Figure 4 shows results using the conventional phase correlation method for *Case 1*. It can be observed in Fig. 4d that the position of overlapping region is in the leftmost top corner of second input fingerprint. Therefore, it does not need to be converted into any mirror image to mosaic fingerprint correctly as shown in Fig. 4f, g. Here, the continuous ridges and valleys in the both fingerprints through overlapping region prove the correctness of mosaiced fingerprint visually. Similarly, we can get mosaic fingerprint for *Case 3* and *Case 5* using the conventional method. Both cases also show correct mosaiced fingerprint without using the mirror image transformation.

Figure 5 shows results for *Case 2* using the conventional method. As shown in Fig. 5b, d, the overlapping region is not positioned in the leftmost corner in any of inputs. Therefore, if the conventional method is directly used to estimate translation parameter as shown in Fig. 5e, it estimates the incorrect translation parameter. As a result fingerprints are mosaiced incorrectly as shown in Fig. 5g, h. It can be

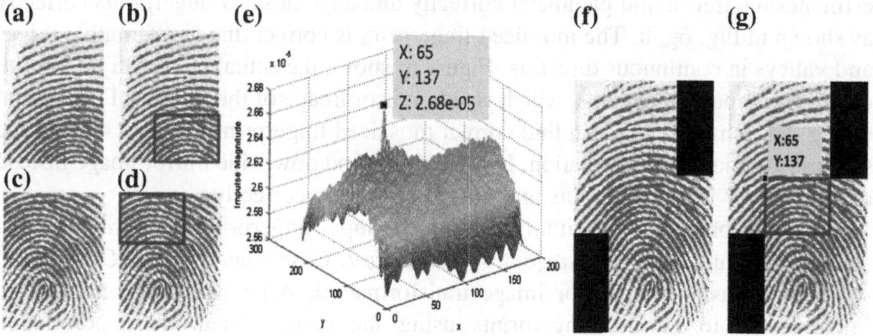

Fig. 4 **a**, **c** Input fingerprints. **b**, **d** Overlapping region in input fingerprints. **e** Peak in cross power spectrum. **f**, **g** Mosaiced fingerprint

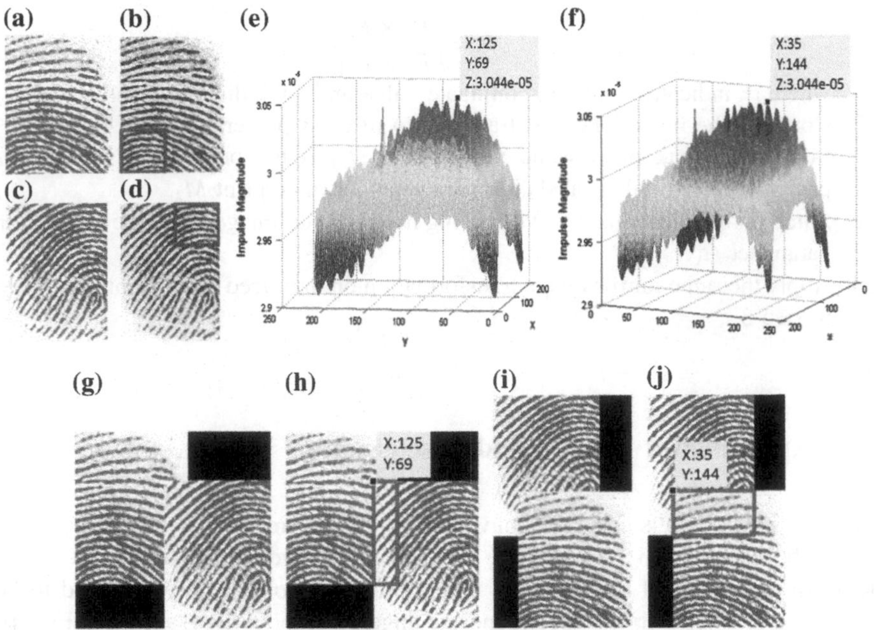

Fig. 5 **a**, **c** Input fingerprints. **b**, **d** Overlapping region. **e** Cross power spectrum. **f** Cross power spectrum after changing the sequence of inputs. **g**, **h** Mosaiced fingerprint. **i**, **j** Mosaiced fingerprint after changing the sequence of inputs

observed that the mosaiced fingerprint is not correct due to discontinuities of ridges and valleys through overlapping region. Figure 5f, i, j show results for the same input fingerprints but after changing the sequence of inputs. However, the method estimates incorrect translation parameter and as a result fingerprints are mosaiced incorrectly.

In our proposed algorithm, the conventional phase correlation method is applied to side mirror image of both inputs as shown in Fig. 6b, e instead of direct inputs. It estimates the translation parameter correctly and also mosaics fingerprints correctly as shown in Fig. 6g, h. The mosaiced fingerprint is correct due to alternative ridges and valleys in continuous direction. Figure 6i shows the actual mosaiced fingerprint of original input fingerprints, which is side mirror image of the mosaiced fingerprint of Fig. 6h. Similarly, we can find correct mosaiced fingerprint for *Case 4* using the side mirror image transformation. In *Case 6*, we find downside mirror image instead of side mirror image of inputs and after that mosaic correctly.

Thus, our proposed algorithm first converts input fingerprints of *Case 2*, *Case 4* (*Category 2*) and *Case 6* (*Category 3*) into *Case 1*, *Case 3* and *Case 5* (*Category 1*) respectively using the mirror image transformation. After that, similar steps are implemented to mosaic fingerprints using the conventional phase correlation method.

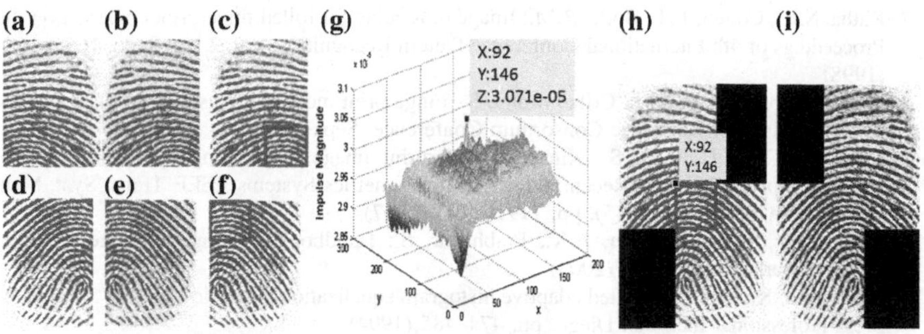

Fig. 6 **a**, **d** Input fingerprints. **b**, **e** Side mirror image of inputs. **c**, **f** Overlapping region in side mirror image of inputs. **g** Cross power spectrum of side mirror image of inputs. **h** Mosaiced fingerprint of side mirror image of inputs. **i** Original mosaiced fingerprint

5 Conclusion

The proposed algorithm mosaics two partial fingerprints having common overlapping region for all possible positions of overlapping region in the partial fingerprints. The algorithm uses the mirror image transformation to change the position of overlapping region in the leftmost top corner of one fingerprint. Due to this position, correct translation parameter is estimated using the conventional phase correlation method. This improves the robustness of the phase correlation method. We are investigating further to develop an algorithm which will be independent of the sequence of inputs and will mosaic the fingerprints involving rotation as well as scaling transformation.

References

1. Zhang, D., Qijun, Z., Nan, L.U.O., Guangming, L.: Partial fingerprint recognition. U.S. Patent 8,411,913, issued April 2, 2013
2. Kuglin, C.D., Hines, D.C.: The phase correlation image alignment method. In: Proceedings of IEEE International Conference on Cybernetics Society, New York, pp. 163–165 (1975)
3. Reddy, B.S., Chatterji, B.N.: An FFT-based technique for translation, rotation and scale-invariant image registration. IEEE Trans. Image Process 5(8), 1266–1271 (1996)
4. Zhang, Y.-L., Yang, J., Wu, H.: A hybrid swipe fingerprint mosaicing scheme. In: Proceedings of International Conference on Audio and Video-based Biometric Person Authentication (AVBPA), Rye Brook, New York, pp. 131–140, July 2005
5. Tarar, S., Kumar, E.: Fingerprint mosaicing algorithm to improve the performance of fingerprint matching system. In: Computer Science and Information Technology, Horizon Research Publication Corporation, vol. 2, no. 3, pp. 142–151, Feb 2014
6. Jain, A.K., Ross, A.: Fingerprint mosaicing. In: Proceedings of IEEE International Conference on Acoustic, Speech, and Signal Process, vol. 4, pp. 4064–4067, May 2002

7. Ratha, N.K., Conell, J.H., Bolle, R.M.: Image mosaicing for rolled fingerprint construction. In: Proceedings of 4th International Conference Pattern Recognition, vol. 2, no. 8, pp. 1651–1653 (1998)
8. Shah, S., Ross, A., Shah, J., Crihalmeanu, S.: Fingerprint mosaicking using thin plate splines. In: Proceedings of Biometric Consortium Conference, Sept 2005
9. Choi, K., Choi, H., Lee, S., Kim, J.: Fingerprint image mosaicking by recursive ridge mapping. Special Issue on Recent Advances in Biometrics Systems. IEEE Trans. Syst. Man, Cybern. Part B: Cybern. 37(5), pp. 1191–1203 (2007)
10. Maltoni, D., Maio, D., Jain, A.K., Prabhakar, S.: Handbook of Fingerprint Recognition. Springer-Verlag, New York (2003)
11. Zuiderveld, K.: Contrast limited adaptive histogram equalization. Graphic Gems IV. Academic Press Professional Inc., San Diego, pp. 474–485 (1994)

An LSB Substitution with Bit Inversion Steganography Method

Nadeem Akhtar

Abstract Several works have been proposed and implemented for image Steganography using LSB substitution. In this paper, a module based LSB substitution method is implemented, which is further improved by using a novel bit inversion technique. The substitution method hides the message data after compressing smoother areas of the image in a lossless way, resulting in fewer number of modified cover image pixels. After this, a bit inversion technique is applied. In the bit inversion technique, certain LSBs of pixels of cover image are changed if they occur with a particular pattern. In this way, less number of pixels is modified. So PSNR of stego-image is improved. For correct de-steganography, the bit patterns for which LSBs has inverted needs to be stored within the stego-image somewhere.

Keywords LSB substitution · Bit inversion · Steganography · PSNR · Image quality

1 Introduction

Image Steganography is a method of hiding secret data into images for transmission over the network. Since images are popular over the Internet. Steganography differs from cryptography which encrypts the data and transmits it without concealing the existence of data. Steganography provides secrecy of text or images to prevent them from attackers. For the Steganography to be imperceptible, the distortion in the stego-image must be as less as possible.

There are several methods for hiding data into images [1, 2]. The simplest method is LSB substitution method which replaces the LSBs of cover images with

N. Akhtar (✉)
Department of Computer Engineering, Zakir Husain College of Engineering & Technology, Aligarh Muslim University, Aligarh, India
e-mail: nadeemalakhtar@gmail.com

A. Nagar et al. (eds.), *Proceedings of 3rd International Conference on Advanced Computing, Networking and Informatics*, Smart Innovation, Systems and Technologies 43, DOI 10.1007/978-81-322-2538-6_53

the message bits. The advantages of Least-Significant-Bit (LSB) Steganography data embedding are that it is simple to understand, easy to implement, and it produces stego-image that is almost similar to cover image and its visual infidelity cannot be judged by naked eyes. Several Steganography methods based on LSB have been proposed and implemented [3–6].

A good technique of image Steganography aims at three aspects. First one is capacity (the maximum data that can be stored inside cover image). Second one is the imperceptibility (the visual quality of stego-image after data hiding) and the last is robustness [7]. The LSB based technique is good at imperceptibility but hidden data capacity is low because only one bit per pixel is used for data hiding. Simple LSB technique is also not robust because secret message can be retrieved very easily once it is detected that the image has some hidden secret data by retrieving the LSBs.

One of the earliest staganalytic methods is chi-square test [8], which performs statistical analysis to identify the message. By reducing the size of message, detection risk in this attack can also be reduced. In [9], authors have proposed technique known as RS staganalysis which can estimate message size efficiently when the message is embedded randomly. In [10], a powerful staganalysis method is proposed, called SPA, which uses sample pair analysis to detect the message length.

In this paper, a bit inversion technique is presented, which is a general scheme which can be combined with any Steganography method. Chen's Steganography method [11] is implemented and it is combined with bit inversion technique. Next section describes the method proposed. Section 3 describes the experiments and results. In Sect. 4, conclusion is discussed.

2 Implementation

Several data hiding schemes exist which exploits the repetitions of message data to improve the quality of stego-image [11–13].

2.1 Module Based LSB Substitution Method

In [11, 13], Steganography is implemented consisting of two phases- hiding phase and adjustment phase. In the hiding phase, each message image pixel is hidden into corresponding cover image to be a stego-image pixel. While hiding the data, next message image pixels are also considered to check if they are identical to the current pixel. If there are adjacent pixels repeating heavily, a flag (m) is stored to indicate the heavy repetitions along with the value (z) being repeated and number of times (r) the value is being repeated. The heavy repetition flag is taken as m if the range of the message image pixels is from 1 to m − 1. These three values m, z, r are stored in three cover pixels only instead of all r repetitions in case of heavy repetitions (r > 3). In this way, there are less number of pixels getting modified. Let t_i, p_i, s_i be the stego-image, cover image and message image pixels respectively. If there is no heavy repetition then message image pixel s_i is hidden into cover pixel p_i as (t_i=) $p_i - (p_i \bmod(m + 1)) + s_i$. If there is heavy repetition, then the values of m, z and r are stored as following

$$t_i = p_i - [p_i \bmod (m + 1)] + m$$
$$t_{i+1} = p_{i+1} - [p_{i+1} \bmod (m + 1)] + z$$
$$t_{i+2} = p_{i+2} - [p_{i+2} \bmod (m + 1)] + (r - 4)$$

After the first phase, the difference between the stego-image and cover image pixels is m. In the adjustment phase, the stego-image pixels are modified to make them closer to the corresponding cover image pixel. After the adjustment phase, the difference reduces to m/2.

Following transformations are made in the adjustment phase, where t(i) is the stego-image pixel after first phase and p(i) is the cover image pixel.

Case 1: $(\lfloor (m+1)/2 \rfloor < t_i - p_i \leq m \text{ and } t_i \geq m+1)\; t_i \leftarrow t_i - m - 1$
Case 2: $(-m \leq t_i - p_i < -\lfloor (m+1)/2 \rfloor \text{ and } t(i) \leq 254 - m) t_i \leftarrow t_i + m + 1$
Case 3: $(-\lfloor (m+1)/2 \rfloor \leq t_i - p_i \leq \lfloor (m+1)/2 \rfloor)\; t_i \leftarrow t_i$
Case 4: $(\lfloor (m+1)/2 \rfloor < t_i - p_i \leq m \text{ and } t_i \leq m) t_i \leftarrow t_i$
Case 5: $(-m \leq t_i - p_i < -\lfloor (m+1)/2 \rfloor \text{ and } t_i \geq 255 - m)\; t_i \leftarrow t_i$

2.2 *Bit Inversion*

A novel bit inversion method to improve the quality of final image is proposed. To understand the scheme, following examples are considered.

Four message bits 1011 are to be hidden into four cover image pixels 10001100, 10101101, 10101011 and 10101101. After plain LSB steganography, stego-image pixels are 1000110**1**, 1010110**0**, 10101011 and 10101101. Two pixels i.e. first and second of cover image have changed. Now, the second and third LSB of three cover image pixels are 0 and 1 respectively. For two of these three pixels, LSB has changed. If the LSB of these three pixels are inverted, cover image pixels will be 10001100, 10101101, 10101011 and 1010110**0**. Now, there is only one pixel of stego-image which differs from cover image i.e. the last one. Thus, the PSNR would increase improving the quality of stego-image. For correct de-steganography, there is need to store the fact that LSBs are inverted for those pixels in which second and third LSB are 0 and 1 respectively.

If two bits are considered, there are four (00, 10, 10, 11) possible combinations. For each of the combination of some two bits, stego-image is analyzed to find the number of pixels of first type i.e. whose LSB has changed and second type i.e.

whose has not changed. If the number of pixels of first type is greater than the number of second type pixels, LSB of first type pixels are inverted. In this way, less number of pixels of cover image would be modified.

2.3 Combination of LSB Substitution and Bit Inversion

The bit inversion method is a general method and can be combined with any steganography method. In this work, bit inversion method is combined with the Chen's steganography method explained in Sect. 2.1. The fourth least significant bit of the stego-image pixel is considered for bit inversion because its contribution to the PSNR among first four LSBs is maximum. The 5th and 6th bit of the stego-image pixel are considered for checking the bit pattern i.e. the 4th LSB of stego-image pixels is analyzed for the two-bit pattern of 5th and 6th bits. Only those pixels of stego-image are considered which are modified by the Chen's method because remaining pixels are already same as cover image pixels. So there is no need to consider them.

3 Results and Analysis

Two 512 * 512 cover images baboon and Peppers Fig. 1g, h; and six 256 * 512 message images Fig. 1a–f are used. All the images are taken from UCI image data repository [14]. Table 1 show the analysis and bit inversion decisions for the message image House for the cover images Baboon and Peppers. The stego-images generated by Chen's method are analyzed considering 5th and 6th LSB.

For the House image when embedded into Baboon, in Table 1, number of unchanged bits is much less than the number of changed bits for bit pattern 00, so inversion is performed. Number of unchanged bits is also less than the number of changed bits for bit pattern 11; inversion is also performed for this case too. Number of unchanged bits is not less than the number of changed bits for other bit patterns. Before inversion, the number of stego-pixels which differs from the cover image pixels is 103105. After inversion is performed, the number of stego-pixels which differs from cover image pixels is 75373. A pixel-benefit of 27732 pixels is achieved, increasing the PSNR by 025.

Table 2 shows the improvement in PSNR for all the six message images when they are embedded into baboon and peppers cover images.

In the bit inversion method, the choice of cover image for a message image depends on how much improvement in PSNR it provides. For example, Baboon is better choice for cover image for crods message image because it provides better improvement in PSNR. For Laser message image, JuliaSet is better choice for cover image.

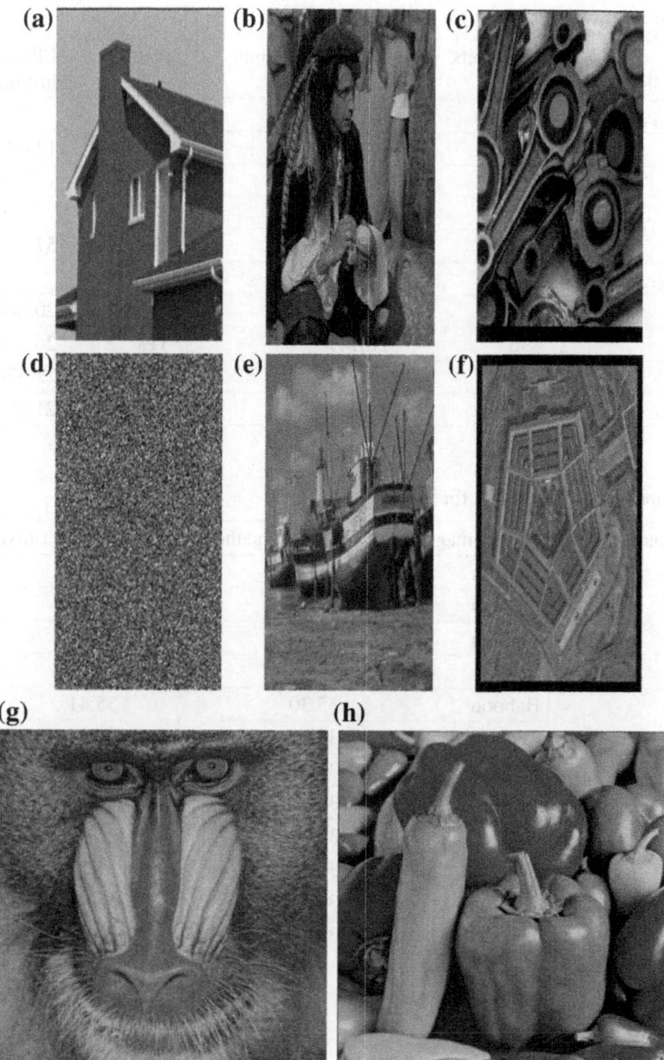

Fig. 1 Message images. **a** House. **b** Man. **c** Crods. **d** Random. **e** FishingBoat. **f** Pentagone cover images. **g** baboon. **h** Peppers

Table 1 Statistics and decisions for house image

Bit pattern (5th, 6th LSB)	Changed bits	Not changed bits	Invert	Changed bits in final image
Cover image: baboon				
0 0	31,932	17,473	Yes	17,473
0 1	0	0	No	0
1 0	0	0	No	0
1 1	71,173	57,900	Yes	57,900
Cover image: peppers				
0 0	22,562	20,608	Yes	20,608
0 1	27,896	26,147	Yes	26,147
1 0	22,570	18,182	Yes	22,570
1 1	21,192	19,321	Yes	21,192

Table 2 Improvement in PSNR for message images

Message image	Cover image	Chen's method	After bit inversion
House	Baboon	35.91	36.14
	Peppers	36.03	36.11
Man	Baboon	34.96	35.12
	Peppers	35.04	35.13
Crods	Baboon	35.30	35.41
	Peppers	35.32	35.41
Random	Baboon	34.21	34.40
	Peppers	34.34	34.43
FishingBoat	Baboon	35.34	35.49
	Peppers	35.44	35.53
Pentagone	Baboon	35.24	35.47
	Peppers	35.38	35.47

4 Conclusion

The proposed bit inversion schemes enhance the stego-image quality. The enhancement in PSNR is not bounded. The improvement in PSNR may be very large for some image as in the case of House image and for some other image, it may be small. For given a message image, a set of cover image can be considered and that cover image is selected for which the improvement is largest.

Although the third party could determine if the message bits are embedded using stegnalysis methods, he would have difficulty to recover it because some of the LSBs have been inverted; it will misguide the staganalysis process and make the message recovery difficult. The bit inversion method makes the Steganography better by improving its security and image quality.

References

1. Sencar, H.T., Ramkumar, M., Akansu, A.N.: Data hiding fundamentals and applications: content security in digital multimedia. Access Online via Elsevier (2004)
2. Cox, I., et al.: Digital watermarking and steganography. Morgan Kaufmann (2007)
3. Chan, C.K., Cheng, L.M.: Hiding data in image by simple LSB substitution. Pattern Recogn. **37**(3), 469–474 (2004)
4. Chang, C.-C., Tseng, H.W.: Data hiding in images by hybrid LSB substitution. In: International Conference on Multimedia and Ubiquitous Engineering. IEEE (2009)
5. Akhtar, N., Johri, P., Khan, S.: Enhancing the security and quality of LSB based image steganography. In: IEEE International Conference on Computational Intelligence and Computer Networks (CICN), Mathura, India, 27–29 Sept. 2013
6. Mohamed, M., Afari, F., Bamatraf, M.: Data hiding by LSB substitution using genetic optimal key-permutation. Int. Arab J. e-Technol. 2.1, 11–17 (2011)
7. Kessler, C.: Steganography: hiding data within data. An edited version of this paper with the title "Hiding Data in Data". Windows & .NET Magazine. http://www.garykessler.net/library/steganography.html (2001)
8. Westfeld, A., Pfitzmann, A.: Attacks on steganographic systems. In: 3rd International Workshop on Information Hiding (IHW 99), pp. 61–76 (1999)
9. Fridrich, J., Goljan, M., Du, R.: Detecting LSB Steganography in color and gray images. Magazine of IEEE Multimedia (Special Issue on Security), Oct.-Nov. 22–28 (2001)
10. Dumitrescu, S., Wu, X., Wang, Z.: Detection of LSB steganography via sample pair analysis. Springer LNCS **2578**, 355–372 (2003)
11. Chen, S.-K.: A module-based LSB substitution method with lossless secret data compression. Comput. Stand. Interf. **33**(4), 367–371 (2011)
12. Cheddad, A., et al.: Digital image steganography: survey and analysis of current methods. Signal Process. **90.3**, 727–752 (2010)
13. Akhtar, N., Bano, A., Islam, F.: An improved module based substitution steganography method. In: Fourth IEEE International Conference on Communication Systems and Network Technologies (CSNT), pp. 695–699 (2014)
14. The USC-SIPI Image Database http://sipi.usc.edu/database/

Image Quality Assessment with Structural Similarity Using Wavelet Families at Various Decompositions

Jayesh Deorao Ruikar, A.K. Sinha and Saurabh Chaudhury

Abstract Wavelet transform is one of the most active areas of research in the image processing. This paper gives analysis of a very well known objective image quality metric, so called Structural similarity, MSE and PSNR which measures visual quality between two images. This paper presents the joint scheme of wavelet transform with structural similarity for evaluating the quality of image automatically. In the first part of algorithm, each distorted as well as original image are decomposed into three levels and in second part, these coefficient are used to calculate the structural similarity index, MSE and PSNR. The predictive performance of image quality based on the wavelet families like db5, haar (db1), coif1 with one, two and three level of decomposition is figured out. The algorithm performance includes the correlation measurement like Pearson, Kendall, and Spearman correlation between the objective evaluations with subjective one.

Keywords Wavelet image quality · Objective image quality measurement · Subjective assessment

1 Introduction

In this period of time, objective image quality assessment is tending to become more active and wider area of research. Objective image quality researchers are trying to develop the computational model that predicts the quality of the given image correctly rather affected by the noise. According to availability of the original

J.D. Ruikar (✉) · A.K. Sinha · S. Chaudhury
National Institute of Technology, Silchar, India
e-mail: jayeshruikar@gmail.com

A.K. Sinha
e-mail: ashokesinha2001@yahoo.co.in

S. Chaudhury
e-mail: saurabh1971@gmail.com

A. Nagar et al. (eds.), *Proceedings of 3rd International Conference on Advanced Computing, Networking and Informatics*, Smart Innovation, Systems and Technologies 43, DOI 10.1007/978-81-322-2538-6_54

image, image quality assessment techniques are classified into three categories [1] (a) Full reference (b) No reference (c) Partial reference. Full reference methods [2–7] calculate the quality of the image by the difference between both the original and the processed image. No-reference approaches [8–11] assess quality of the image without knowing the original image but consideration of specific distortion and its causes in the image. Partial-reference approaches [12–15] estimate quality of the image in the absence of the original image but utilize some of the features in the image. In this paper, authors are working in full reference image quality assessment techniques especially wavelet decomposition based structural similarity index. According to the various aspects related to quantitative and qualitative analysis for image quality measurement, several useful algorithms are constructed, including the well-known structural similarity metric [2, 3]. A basic generalization that is accepted behind structural similarity is that, human visual system are much correlated and adapted to the structure of the image. Image decomposition with wavelet transform and application of SSIM quality metric analysis was illustrated in [16–18], however with haar, DB5 and COIF1 wavelet analysis with one, two and three level of decomposition analysis of images individual components was not performed. The proposed scheme is tested in terms of SSIM, PSNR and MSE with four different dataset.

2 Structural Similarity

The structural similarity (SSIM) is defined in [2, 3] for measuring the similarity between two images. Natural image signals are highly structured [19] that means the signal samples have strong dependencies among themselves especially if they are close in space. This is extraction of structural information of image as human traditional method of error visibility separation from distorted images [2]. The structures of the objects from the picture or image are not depending upon illumination [2]. Therefore, to investigate the structural information in an image, separation of the influence of the illumination is necessary. Let x = $\{x_i | i = 1, 2, \ldots N.\}$ and y = $\{y_i | i = 1, 2, \ldots N\}$ are the two image signals, x—signal is reference image and y—signal is distorted image. Structural similarity measurement is a function of three units mainly: luminance, contrast and the structure. Luminance of original image and distorted image is calculated by calculating the mean intensity of both the image. The formula for calculating the mean intensity is $\mu_x = \frac{1}{N}\sum_{i=1}^{N} x_i$ and $\mu_y = \frac{1}{M}\sum_{i=1}^{M} y_i$. Image contrast can be estimated from standard deviation and compared the σ_x and σ_y by the formula $\sigma_x = \sqrt{\left(\frac{1}{N-1}\right)\sum_{i=1}^{N} (x_i - \mu_x)^2}$, $\sigma_y = \sqrt{\left(\frac{1}{M-1}\right)\sum_{i=1}^{M} (y_i - \mu_y)^2}$. Then the signal is normalize and the structural comparison is estimated by the formula $(x - \mu_x)/\sigma_x$ and $(y - \mu_y)/\sigma_y$. So, generally similarity measure [2] can be defined $SSIM(x, y) = f(l(x, y), c(x, y), s(x, y))$.

Here, $l(x,y) = \frac{2\mu_x\mu_y + C_1}{\mu_x^2 + \mu_y^2 + C_1}$, $c(x,y) = \frac{2\sigma_x\sigma_y + C_2}{\sigma_x^2 + \sigma_y^2 + C_2}$ and $s(x,y) = \frac{\sigma_{xy} + C_3}{\sigma_x\sigma_y + C_3}$ where, C1, C2, and C3 are small constant; $C_1 = (K_1L)^2$, $C_2 = (K_2L)^2$ and $C_3 = C_2/2$ respectively. In this equation, L is the dynamic range of pixel values (L = 255 for 8 bit image), $K1, K_2 << 1$ are scalar constant. Combining and simplifying the above equation arrives at SSIM index [2]: $SSIM(x,y) = \frac{(2\mu_x\mu_y + C_1)(2\sigma_{xy} + C_2)}{(\mu_x^2 + \mu_y^2 + C_1)(\sigma_x^2 + \sigma_y^2 + C_2)}$.

The similarity measure must satisfy the following conditions. (1) Symmetry: $s(x,y) = s(y,x)$. (2) Boundedness: $s(x,y) \leq 1$. (3) Unique maximum: $s(x,y) = 1$ if and only if x = y.

SSIM index can be used as complimentary method to the traditional approaches of estimation of image quality such as Mean Square Error (MSE), Peak Signal to Noise Ratio (PSNR). Following are the formula of them $MSE = \frac{1}{MN}\sum_{i=1}^{M}\sum_{j=1}^{N} (X - Y)^2$ and $PSNR = 10\log_{10}\frac{L^2}{MSE}$.

These approaches are based on the pixel by pixel calculation that's why they fail in predicting the correct image quality as shown in [20, 21]. In [22], authors demonstrated how the MSE and PSNR values are changes non-linearly with linear change in quality factor. In [23], effect of different filtering techniques on structural similarity has been evaluated.

3 Experimental Study and Results

In this brief, Daubechies-5 (db5) and Haar wavelet are used from wavelet family having properties like asymmetric, orthogonal, biorthogonal and thirdly Coiflets-1 (coif1) having the properties like near symmetric, orthogonal, biorthogonal is selected. Four database are used in this study are TID [24], CSIQ [25], LIVE [26], and IVC [27] and are comprises of the subjective rating for various calculation and validation. The step by step for evaluation of the proposed algorithm is as follows.

(1) Read the original and distorted image convert it to gray: (here in this paper images from TID, CSIQ, IVC and LIVE databases are considered)
(2) Calculate the SSIM, MSE and PSNR for original and distorted images
(3) Decompose the image for 2-D Haar wavelet transform at level 1, level 2, and level 3.
(4) Obtain the quality index such as SSIM, MSE, and PSNR in the LL band at each level as the luminance and the contrast information are available in the LL band.
(5) Use available subjective assessment result given in the database and Correlate the objective image quality with subjective assessment result (In this paper, Pearson, Kendall, and Spearman correlation is used)
(6) Repeat the steps 2–5 for Daubechies-5 (db5) and Coiflets-1 (coif1) for decomposition at 1 level, 2 level, and 3 level

Table 1 LCC, SROCC and KROCC correlation for CSIQ, IVC and LIVE database

	Correlation	SSIM	SSSI-I	SSIM-II	SSIM-III	MSE	MSE-I	MSE-II	MSE-III	PSNR	PSNR-I	PSNR-II	PSNR-III
CSIQ database (866 images are considered)													
DB5	PCC	0.7736	0.8079	0.7269	0.5694	0.6154	0.3783	0.2920	0.2169	0.6829	0.7615	0.7441	0.7055
	KROCC	0.6872	0.6827	0.6145	0.5372	0.5472	0.6436	0.6128	0.5603	0.5472	0.6436	0.6128	0.563
	LCC	**0.8685**	**0.8689**	0.8105	0.7348	**0.7445**	0.8125	0.7817	0.7346	**0.7445**	0.8125	0.7817	0.5603
HAAR	PCC	0.7736	0.8056	0.7217	0.5477	0.6154	0.3615	0.2702	0.1952	0.6829	0.7585	0.7388	0.6995
	KROCC	0.6872	0.6813	0.6220	0.5598	0.5472	0.6461	0.6157	0.5612	0.5472	0.6461	0.6157	0.5612
	LCC	**0.8685**	0.8677	**0.8167**	**0.7539**	**0.7445**	0.8128	0.7808	0.7308	**0.7445**	**0.8128**	0.7808	0.5612
COIF1	PCC	0.7736	0.8055	0.7212	0.5483	0.6154	0.3720	0.2876	0.2184	0.6829	0.7615	0.7424	**0.7075**
	KROCC	0.6872	0.6819	0.6146	0.5418	0.5472	0.6453	0.6134	0.5641	0.5472	0.6453	0.6134	0.5641
	LCC	0.8685	0.8681	0.8107	0.7369	**0.7445**	**0.8128**	**0.7818**	**0.7376**	**0.7445**	0.8128	**0.7818**	0.5641
IVC database (185 images are considered)													
DB5	PCC	0.8092	0.8481	0.8174	0.7068	0.5102	0.6591	0.6124	0.5576	0.6705	0.8219	0.8697	0.8464
	KROCC	0.7223	0.7236	0.6945	0.5964	0.5218	0.6520	0.6896	0.6688	0.5218	0.6520	0.6896	0.6688
	LCC	**0.9018**	0.9021	**0.8822**	**0.798**	**0.6884**	0.8377	0.8815	0.8621	0.6884	0.8377	0.8815	0.6688
HAAR	PCC	0.8092	0.8605	0.8152	0.6510	0.5102	0.6902	0.6798	0.6863	0.6705	0.8611	0.8973	**0.8961**
	KROCC	0.7223	0.7314	0.6777	0.5610	0.5218	0.6920	0.7185	0.7192	0.5218	0.6920	0.7185	0.7192
	LCC	**0.9018**	**0.9105**	0.8722	0.7691	0.6884	**0.8719**	**0.9017**	**0.9011**	**0.6884**	**0.8719**	**0.9017**	0.7192
COIF1	PCC	0.8092	0.8546	0.8018	0.6553	0.5102	0.6724	0.6212	0.5393	0.6705	0.8376	0.8687	0.8328
	KROCC	0.7223	0.7266	0.6650	0.5541	0.5218	0.6667	0.6892	0.6562	0.5218	0.6667	0.6892	0.6562
	LCC	**0.9018**	0.9058	0.8609	0.7603	0.6884	0.8515	0.8804	0.8520	0.6884	0.8515	0.8804	0.6562
LIVE database (779 images are considered)													
DB5	PCC	0.42084	0.476	0.3973	0.2868	0.458	0.17	0.16	0.1336	0.563	0.586	0.5853	0.5623
	KROCC	0.48614	0.484	0.4537	0.4193	0.405	0.441	0.443	0.4255	0.405	0.441	0.4429	0.4255
	LCC	**0.63067**	0.631	0.6114	0.5786	0.603	0.606	0.581	**0.6072**	0.552	0.603	0.6064	0.4255
HAAR	PCC	0.42084	0.472	0.403	0.2911	0.458	0.158	0.142	0.1086	0.563	0.587	0.6001	**0.591**
	KROCC	0.48614	0.489	0.47	0.4532	0.405	0.446	0.459	0.4631	0.405	0.446	0.4587	0.4631

(continued)

Table 1 (continued)

	Correlation	SSIM	SSSI-I	SSIM-II	SSIM-III	MSE	MSE-I	MSE-II	MSE-III	PSNR	PSNR-I	PSNR-II	PSNR-III
	LCC	**0.63067**	**0.632**	**0.62**	**0.6103**	**0.603**	**0.623**	**0.624**	0.603	0.552	**0.607**	**0.6228**	0.4631
COIF1	PCC	0.42084	0.472	0.4046	0.2822	0.458	0.166	0.161	0.1429	0.563	0.584	0.59	0.5596
	KROCC	0.48614	0.48	0.463	0.4343	0.405	0.44	0.44	0.4218	0.405	0.44	0.4439	0.4218
	LCC	**0.63067**	0.627	0.6142	0.5936	0.603	0.609	0.576	0.6035	0.552	0.603	0.6092	0.4218

Table 2 LCC, SROCC and KROCC correlation for TID database

	Correlation	SSIM	SSSI-I	SSIM-II	SSIM-III	MSE	MSE-I	MSE-II	MSE-III	PSNR	PSNR-I	PSNR-II	PSNR-III
FOR TID database (1700 images are considered)													
DB5	PCC	0.7401	0.7422	0.7083	0.5691	0.1113	0.1904	0.0765	0.0039	0.2209	0.5393	0.4903	0.4247
	KROCC	0.5768	0.5458	0.5259	0.4352	0.2665	0.4569	0.4121	0.3565	0.2396	0.4569	0.4121	0.3565
	LCC	**0.7749**	0.7437	0.7275	**0.6121**	**0.3723**	**0.5665**	0.5112	0.4483	0.3356	**0.5665**	0.5112	0.3565
HAAR	PCC	0.7401	0.7410	0.7156	0.5415	0.1113	0.1659	0.0379	0.0487	0.2209	0.5309	0.4823	0.3722
	KROCC	0.5768	0.5498	0.5315	0.3968	0.2665	0.4569	0.4255	0.3419	0.2396	0.4569	0.4255	0.3419
	LCC	**0.7749**	**0.7475**	**0.7319**	0.5556	**0.3723**	0.5630	0.5112	0.4149	0.3356	0.5630	0.5112	0.3419
COIF	PCC	0.7401	0.7412	0.7142	0.5498	0.1113	0.1851	0.0747	0.0165	0.2209	0.5360	0.4920	**0.4302**
	KROCC	0.5768	0.5457	0.5280	0.4376	0.2665	0.4555	0.4168	0.3600	0.2396	0.4555	0.4168	0.3600
	LCC	**0.7749**	0.7443	0.7296	0.6116	**0.3723**	0.5643	**0.5137**	**0.4522**	**0.3356**	0.5643	**0.5137**	0.3600

(7) Find out the result which is correctly predicting the image quality according to the subjective assessment.

The proposed algorithm was implemented using MATLAB R2009a on a HCL personal computer with 2 GB RAM, Intel(R) Core(TM)2 Duo CPU, 32 bit windows 7 ultimate operating system. The most common measures of predictive performance are the Pearson correlation coefficient (LCC), Spearman rank-order correlation coefficient (SROCC) and Kendall order correlation coefficient (KROCC). The results of CSIQ, IVC, LIVE and TID databases for correlations like Pearson, Spearman and Kendall on given algorithm are presented in Tables 1 and 2 respectively.

The bold entries in Tables 1 and 2 shows best overall performance for the SSIM, MSE and PSNR without and with the decomposed images with one two and three level of decomposition.

4 Conclusions

Image visual quality metrics play a very important role in various real time applications. In this paper, Authors analyze haar, DB5 and COIF1 wavelet with one, two and three level of decomposition of images for four different dataset. The performance of given methods shows a anticipated result with IVC and CSIQ dataset as it has been correlates well with subjective score.

References

1. Wang, Z., Bovik, A.C.: Modern image quality assessment. In: Synthesis Lectures on Image, Video, and Multimedia Processing, vol. 2 (1) (2006)
2. Wang, Z., Bovik, A.C., Sheikh, H.R., Simoncelli, E.P.: Image quality assessment: from error visibility to structural similarity. IEEE Trans. Image Process. **13**(4), 600–612 (2004)
3. Wang, Z., Simoncelli, E.P., Bovik, A.C.: Multi-scale structural similarity for image quality assessment. Invited Paper, IEEE Asilomar Conference on Signals, Systems and Computers (2003)
4. Yong, D., Shaoze, W., Dong, Z.: Full-reference image quality assessment using statistical local correlation. Electron. Lett. **50**(2), 79–81 (2014)
5. Demirtas, A.M., Reibman, A.R., Jafarkhani, H.: Full-reference quality estimation for images with different spatial resolutions. IEEE Trans. Image Process. **23**(5), 2069–2080 (2014)
6. Sheikh, H.R., Bovik, A.C.: Image information and visual quality. IEEE Trans. Image Process. **15**(2), 430–444 (2006)
7. Sheikh, H.R., Sabir, M.F., Bovik, A.C.: A statistical evaluation of recent full reference image quality assessment algorithms. IEEE Trans. Image Process. **15**(11), 3440–3451 (2006)
8. Zaramensky, D.A., Priorov, A.L., Bekrenev, V.A., Soloviev, V.E.: No-reference quality assessment of wavelet-compressed images. In: IEEE EUROCON, pp. 1332–1337 (2009)
9. Ji, S., Qin, L., Erlebacher, G.: Hybrid no-reference natural image quality assessment of noisy, blurry, JPEG2000, and JPEG Images. IEEE Trans. Image Process. **20**(8), 2089–2098 (2011)

10. Ke, G., Guangtao, Z., Xiaokang, Y., Wenjun, Z., Longfei, L.: No-reference image quality assessment metric by combining free energy theory and structural degradation model. IEEE International Conference on Multimedia and Expo (ICME), pp. 1–6 (2013)
11. Golestaneh, S.A., Chandler, D.M.: No-reference quality assessment of jpeg images via a quality relevance map. IEEE Signal Process. Lett. **21**(2), 155–158 (2014)
12. Qiang, L., Zhou, W.: Reduced-reference image quality assessment using divisive normalization-based image representation. IEEE J. Selected Topics Signal Process. **3**(2), 202–211 (2009)
13. Rehman, A., Zhou, W.: Reduced-reference image quality assessment by structural similarity estimation. IEEE Trans. Image Process. **21**(8), 3378–3389 (2012)
14. Jinjian, W., Weisi, L., Guangming, S., Anmin, L.: Reduced-reference image quality assessment with visual information fidelity. IEEE Trans. Multimedia **15**(7), 1700–1705 (2013)
15. Soundararajan, R., Bovik, A.C.: RRED indices: reduced reference entropic differencing for image quality assessment. IEEE Trans. Image Process. **21**(2), 517–526 (2012)
16. Chun-Ling, Y., Wen-Rui, G., Lai-Man, P.: Discrete wavelet transform-based structural similarity for image quality assessment. In: 15th IEEE International Conference on Image Processing, pp. 377–380 (2008)
17. Guo-Li, J., Xiao-Ming, N., Hae-Young, B.: A full-reference image quality assessment algorithm based on haar wavelet transform. International Conference on Computer Science and Software Engineering, pp. 791–794 (2008)
18. Rezazadeh, S., Coulombe, S.: A novel discrete wavelet transform framework for full reference image quality assessment. SIViP **7**(3), 559–573 (2013)
19. Wang, Z., Ligang, L., Bovik, A.C.: Video quality assessment based on structural distortion measurement. Sig. Process. Image Commun. **19**(2), 121–132 (2004)
20. Zhou, W., Bovik, A.C.: A universal image quality index. Signal Processing Lett. IEEE, **9**(3), 81–84 (2002)
21. Zhou, W., Bovik, A.C.: Mean squared error: Love it or leave it? A new look at signal fidelity measures. IEEE Signal Proc. Magaz. **26**(1), 98–117 (2009)
22. Ruikar, J.D., Sinha, A.K., Chaudhury, S.: Review of image enhancement techniques. In: International Conference on Information Technology in Signal and Image Processing—ITSIP 2013 (2013)
23. Ruikar, J.D., Sinha, A.K., Chaudhury, S.: Structural similarity and correlation based filtering for image quality assessment. In: IEEE International Conference on Communication and Signal Processing—ICCSP' 14 (2014)
24. Ponomarenko, N.: Tampere Image Database 2008 version 1.0, (2008). http://www.ponomarenko.info/tid2008.htm
25. Chandler, D.M.: CSIQ Image Database (2010). http://vision.okstate.edu/?loc=csiq
26. Sheikh, H.R.: LIVE Image Quality Assessment Database, Release 2, (2005). http://live.ece.utexas.edu/research/quality/subjective.htm
27. Le Callet, P.: Subjective quality assessment IRCCyN/IVC Database, 2005. http://www2.irccyn.ec-nantes.fr/ivcdb/

Part VI
Applications of Informatics

Modelling the Gap Acceptance Behavior of Drivers of Two-Wheelers at Unsignalized Intersection in Case of Heterogeneous Traffic Using ANFIS

Harsh Jigish Amin and Akhilesh Kumar Maurya

Abstract The gap acceptance concept is an important theory in the estimation of capacity and delay of the specific moment at unsignalized junctions. Most of analyzes have been carried in advanced countries where traffic form is uniform, and laws of priorities, as well as lane disciplines, are willingly followed. However, in India, priority laws are less honored which consequently create more conflicts at intersections. Modeling of such behavior is complex as it influenced by various traffic features and vehicles' as well as drivers' characteristics. A fuzzy model has been broadly accepted theory to investigate similar circumstances. This article defines the utilization of ANFIS to model the crossing performance of through movement vehicles at the four-legged uncontrolled median separated intersection, placed in a semi-urban region of Ahmedabad in the province of Gujarat. Video footage method was implemented, and five video cameras had been employed concurrently to collect the various movements and motorists', as well as vehicles' characteristics. An ANFIS model has been developed to estimate the possibilities of acceptance and rejections by drivers of two-wheelers for a particular gap or lag size. Seven input and one output parameters, i.e. the decision of the drivers are considered. Eleven different diverse combination of variables is employed to construct eleven different models and to observe the impact of various attributes on the correct prediction of specific model. 70 % observations are found to prepare the models and residual 30 % is considered for validating the models. The forecasting capability of the model has been matched with those experiential data set and has displayed good ability of replicating the experiential behavior. The forecast by ANFIS model ranges roughly between 77 and 90 %. The models introduced in this study can be implemented in the dynamic evaluation of crossing behavior of drivers.

H.J. Amin (✉) · A.K. Maurya
Department of Civil Engineering, Indian Institute of Technology Guwahati, Guwahati 781039, Assam, India
e-mail: aminharshj@gmail.com

A.K. Maurya
e-mail: akmaurya@gmail.com

A. Nagar et al. (eds.), *Proceedings of 3rd International Conference on Advanced Computing, Networking and Informatics*, Smart Innovation, Systems and Technologies 43, DOI 10.1007/978-81-322-2538-6_55

Keywords Adaptive neuro-fuzzy inference system · ANFIS · Critical gap · Uncontrolled intersection · Neuro-fuzzy · Gap acceptance behavior

1 Introduction

Gap acceptance behavior is a decision to accept or reject a gap of a specific length, and it is an outcome of decision method of a human understanding which incorporates an assessment of the set of explanatory characteristics. Previous analysis of gap acceptance behavior has admitted that drivers' performance depends on drivers' features [1] vehicles' features [2], gap size, waiting time, occupancy, approaching speed [2], etc. Considering a more practical way, this article attempts to model gap acceptance behavior by analyzing the set of critical attributes that affect the behavior of the driver. The evaluation of the set of descriptive attributes in the decision process is influenced by vagueness and uncertainty. Fuzzy logic performs a vital part in the human capability to deliver a decision in the circumstances of uncertainty and imprecision as shown in Fig. 1. For the existing research, a Neuro-fuzzy approach is utilized to establish a model of crossing performance of the driver at TWSC intersection. MATLAB is employed to develop an ANFIS, which contributes an optimization system to obtain variables in the fuzzy system that finest fit the data. A network-type arrangement is equivalent to a neural network [3].

Fuzzy tools and ANFIS have been positively utilized in numerous engineering disciplines, like, walker performance [4], volume assessment, accident avoidance assistance system [5], traffic signal controller at remote intersection [6], and highway impact investigation [7]. Effort correlated to utilization of fuzzy model for gap taking performance at TWSC junction has been achieved by Rossi et al. [8], and Rossi and Meneguzzer [9]. Amin et al. [3] evaluated the critical gap parameter for pedestrian using Raff model and constructed a gap acceptance behavioral model for pedestrian using ANFIS for two four-legged unsignalized junctions of India. They examined lag/gap time, type of approaching vehicle and age as well as sex of the pedestrian to obtain more real outcome, and results indicated that parameter age is most effective parameter. Rossi and Meneguzzer [9] demonstrated the crossing behavior at TWSC junctions utilizing a Neuro-Fuzzy system. In their effort, the input parameters considered are (i) whether it is lag or gap (ii) their length in seconds, and (iii) kind of maneuvering; the outcome variable is the driver's choice to take or discard the specific length of gap or lag. In addition Rossi et al. utilized the data from driving the simulator to produce gap taking fuzzy model. Ottomanelli et al. [4] made model based on the decision of pedestrians during crossing the midblock section using ANFIS and concluded that ANFIS model permit to replicate human approximate reasoning, presenting substantial and useful outcomes within soft computing environment with requiring a little data.

This paper is arranged in six sections, involving this section. Section 2 presents the procedure for data collection and extraction. The complete structure of ANFIS

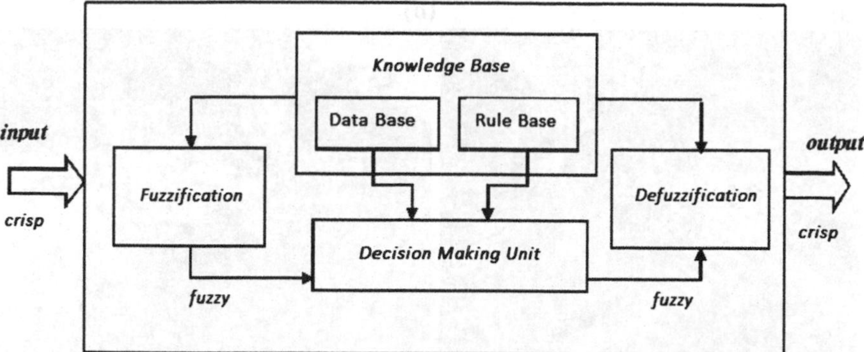

Fig. 1 Structure of fuzzy inference systems

is explained in Sect. 3. Sections 4 and 5 addresses the development and validation of the model respectively, using Neuro-fuzzy approach. Section 6 concludes the article by showing relevant conclusions and pointing the areas for further research.

2 Data Collection

This paper presents a study of crossing behavior of the driver at an uncontrolled median separated junction. In order to examine the crossing behavior, the four-legged uncontrolled crossing located in a semi-urban region of Ahmedabad in the province of Gujarat was picked. This intersection was found in plain topography with sufficient vision for every movements. The total traffic volume on the main road was in the range of 1300–2400 vph and on secondary stream was in the range of 700–1400 vph. Experiment was conducted throughout the peak hours of a typical weekday (9–12 A.M). Video footage method was implemented for data gathering. Five video camera had been employed simultaneously to collect the various attributes like vehicle arrival rate/time, accepted and rejected gap/lag time, speed and conflicting vehicle types, waiting time of minor stream vehicle at stopover, driver's generation as well as sex, occupancy, etc. Out of five, three video camera were set on the terrace of the building situated near junction for gathering the various vehicles movement on different approaches of this intersection, accepted and rejected gap/lag time, speed, and conflicting vehicle types, waiting time, etc. And two camera placed on shoulder and median of the secondary road with using 5 feet long tripod to gather the drivers' features. The captured cassette was played again on a big display monitor to extract and analyze the various necessary data. In this study, only 2-wheelers with through movements from minor road were considered. Gap data were extracted using software with an accuracy 0.01th of a second.

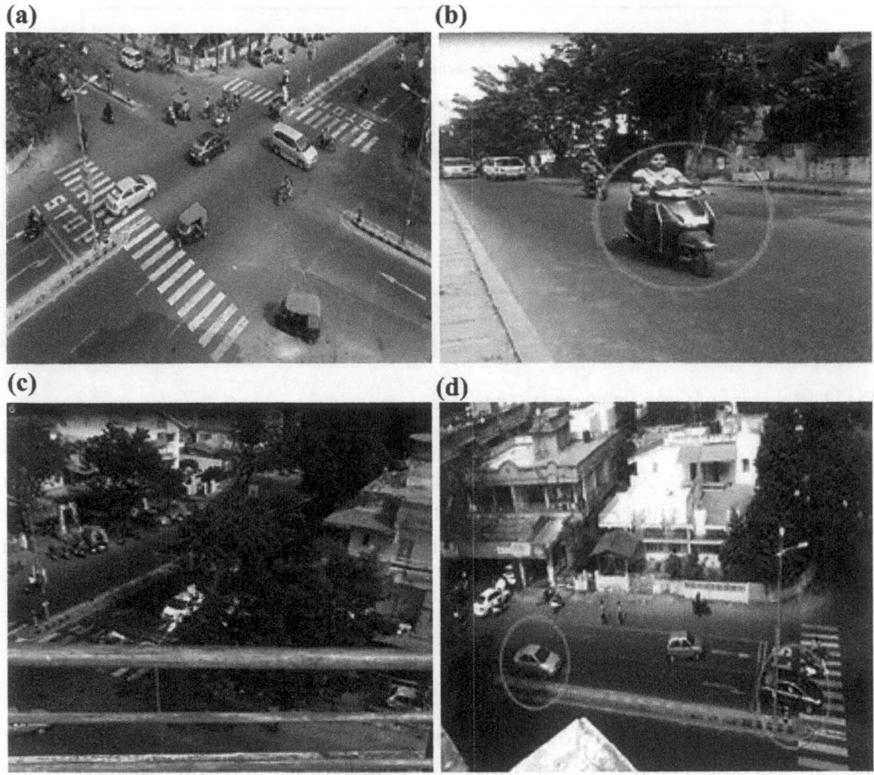

Fig. 2 **a–d** Videography of selected site obtained by various video cameras

Figure 2a–d represent the videography by four different cameras to obtain various parameters on the selected site. Using this procedure, 2364 minor through movements were extracted.

3 ANFIS Structure

An ANFIS utilizes two fuzzy logic and neural network programs. If these two programs are united, they may qualitatively and quantitatively generate a decent outcome that will incorporate either calculative capacities of a neural network or fuzzy intellect. This alteration makes it feasible to blend the benefits of neural network and fuzzy logic. A network generated by this system can use exceptional training algorithms that neural networks have at their disposal to achieve the parameters that would not have been imaginable in fuzzy logic tools. The arrangement of ANFIS is organized in two parts as similar as a Fuzzy system; first

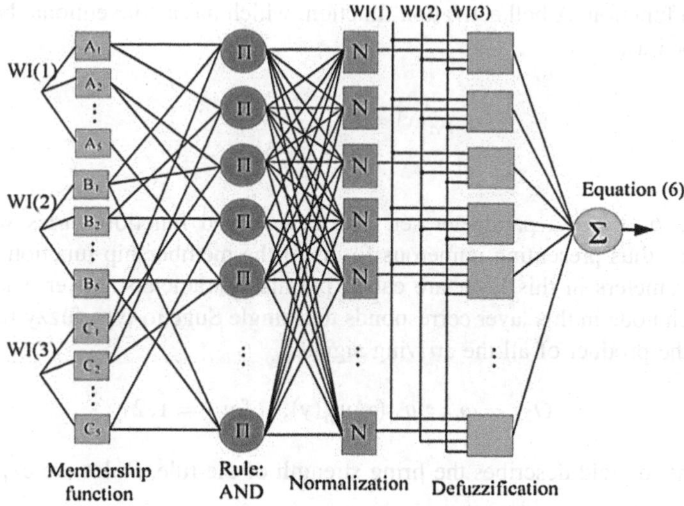

Fig. 3 Structure of ANFIS

is introductory or antecedent part, and the second is concluding parts that are attached jointly by a set of rules. We can identify six separate levels in the formation of ANFIS, which makes it in the form of multi-layer network. A type of this network, which is revealed in six different layers, is shown in Fig. 3. This structure comprises two inputs (x and y) and one output (Z) that is correlated with the subsequent rules:

Rule 1 If (x is A_1) and (y is B_1) then $Z_1 = p_1x + q_1y + r_1$
Rule 2 If (x is A_2) and (y is B_2) then $Z_2 = p_2x + q_2y + r_2$

where A and B are the input, Z is the output and p, q and r is a set of logical parameters of the rule. If we consider the outcome of each section of the ANFIS system as (*i*th node output in *j*th layer) then we can demonstrate the different layer functions of this system as follows.

Level 1 is designated as input level. In this level, neurons transfer the input signals (identified as crisp) to following level. Level 2 is the fuzzification level where all node i in this layer is an adaptive node with a node function,

$$O_{1,i} = \mu_{Ai}(x), \quad \text{for } i = 1, 2$$
$$O_{1,i} = \mu_{Bi}(y), \quad \text{for } i = 1, 2 \tag{1}$$

where μ_{Ai} is the membership grade of x in Ai is a fuzzy set and μ_{Bi} is the membership of y in B_i fuzzy set. In this model, fuzzification nodes have a bell activation function. A bell activation function, which has a conventional bell shape, is described as,

$$\mu_A(x) = \frac{1}{1 + \left|\frac{x - c_i}{a_i}\right|^{2b}} \tag{2}$$

where (a_i, b_i, c_i) is a parameter set. The bell-shaped function varies with these parameters, thus presenting numerous forms of the membership function for fuzzy set A. Parameters in this layer are called premise parameters. Layer 3 is the rule layer. Each node in this layer corresponds to a single Sugeno-type fuzzy rule whose result is the product of all the arriving signals:

$$O_{2,i} = \omega_i = \mu_{Ai}(x)\mu_{Bi}(y), \quad \text{for } i = 1, 2 \tag{3}$$

Each node yield describes the firing strength of the rule. Rules are explained as beneath:

If Gap/Lag is lower and approaching vehicle is Bus then likelihood of acceptance is 0

If Gap/Lag is medium and oncoming vehicle is Bus then likelihood of acceptance is 1

…and so on.

Layer 4 is the normalized layer. Every node in this layer is a fixed node tagged as N. The ith node determines the ratio of the ith rule's firing power to the sum of all rules' firing power:

$$O_{3,i} = \varpi_i = \frac{\omega_i}{\omega_1 + \omega_2}, \quad \text{for } i = 1, 2. \tag{4}$$

The products of this layer are termed normalized firing strengths. Level 5 is the defuzzification level. Defuzzification is the manner in which the weighted resultant value is estimated for a specified rule. Every node i in this level is an adaptive node with a node function:

$$O_{4,i} = \varpi_i f_i = \varpi_i(p_i x + q_i y + r_i) \tag{5}$$

where ANFIS is a normalized firing strength from layer 4 and (pi, qi, ri) is the parameter set of this node. These parameters are called resultant parameters. A single summation node interprets layer 6. Which calculates the overall ANFIS outcome as the summation of all defuzzification nodes:

$$O_{5,i} = \sum_i \varpi_i f_i = \frac{\sum_i \omega_i f_i}{\sum_i \omega_i} \tag{6}$$

Concisely, third layer performs the fuzzy AND of the ancestor part of the fuzzy rules, the fourth layer normalizes the membership functions (MF's), the fifth layer delivers the following part of the fuzzy rules, and lastly the last layer calculates the outcome of fuzzy model by aggregating the outcomes of layer fourth [3, 10, 11].

4 ANFIS Model Development

An ANFIS has been developed using Fuzzy Logic toolbox in MATLAB. Seven input parameters, i.e. lag/gap size, drivers' generation; driver sexes, occupancy, waiting time, oncoming vehicle speed and their vehicle type and one output variable i.e. choice of the drivers are considered. Size of lag/gap (sec) is distinguished by a set of three linguistic fuzzy values viz; lower, medium and higher. In the same way, the attribute oncoming vehicle speed and attribute waiting time are divided by three linguistic fuzzy values viz; light, moderate and high. Remaining four input variables are taken as crisp variables, and they need the similar approach as a fuzzy variable so as to present them in Nero-fuzzy system. It was not possible to get the genuine age of the driver; hence, it is obtained by only visual inspections and considered only those drivers who drive the motorized two-wheelers. Generation of the driver is taken as crisp parameter, 1—Young, 2—Moderate, 3—Elderly driver. Additional inputs are occupancy (1, and more than 1) and category of oncoming vehicle (motorized two-wheelers-1, three wheelers-2, car-3 and bus/truck-4).

Eleven diverse combination of attributes is employed to construct eleven diverse models and to observe the impact of various attributes on the correct prediction of a particular model. ANFIS uses a similar type of network structure as using by neural-network. Input membership function and linked attribute used in mapping of input and output membership function and linked attribute used in the mapping of output to represent the input/output map. Construction of ANFIS in MATLAB utilizing Fuzzy Logic Toolbox involves three stages; (1) developing Sugeno type Fuzzy Inference System (FIS), (2) training practice with a designated set of input and output data and (3) testing with independent data i.e. validation of the replicas. Utilization of rules is very important part of modeling. Fuzzy rules are a combination of linguistic statements that illustrate how the FIS should deliver a decision relating classifying an input or regulating the production. Two or more rules can't serving the similar outcome membership function, hence, the number of outcome membership functions ought to be in similar number of rules. Trapezium and Triangle shape are generally used by understanding of observed data. It depends on the variable associated with that function, and it can improve through a learning process. The estimation of parameters, associated with membership function, (or their modification) is simplified by a gradient vector, which contributes a measure of how strong the FIS is modeling the input/output data for an assigned set of parameters. Once the gradient vector is received, any of numerous optimization techniques can be utilized in order to modify the parameters so as to diminish few error measure. An ANFIS utilizes either back propagation or a mixture of least squares calculation and back propagation for parameter evaluation. The training data helps in automatically modifications and adjustment of associated parameter. This training method is utilized to diminish the lapse between the actually perceived outcome and predicted outcome. Generally 70 %, obtained data are used to calibrate the models. Error Tolerance, which is correlated to the size of an error, is used to generate a training stopping pattern. An error was fixed to 0 as their behavior is unexpected and unexplained especially for this situations. Noticed how the testing error declines up to a specific position in the training, and then it rises. This rise describes the spot of the model overfitting. ANFIS accepts the model parameters correlated with the least checking error. Least error is found at 40th epoch; hence, it was set to 40.

5 Validation of Model

This kind of demonstrating works correctly if the training data served to ANFIS for training (evaluating) membership function parameters is wholly demonstrative of the structures of the data that the trained FIS is proposed to replicas. This is not the situation, however, in a few situations, data is gotten by noisy estimations, and the training data may not illustrative of entire the characteristics of the data those will be represented in the replicas. This is the circumstances where model approval becomes an integral factor. The testing lapse (root mean square error RMSE) for each model after confirming with independent data is given in Table 1. Eleven different models have been made by using eleven different combinations of variable as shown in Table 1. By observing Table 1, Model 1 has least testing laps, however difference between training and testing lapse values is least for model 2; hence, it might be said that the Model 2 has ideal error values. Since the real choice data are denoted in binary state, i.e. 1-crossing and 0-waiting, the outcome of ANFIS model is a degree of preference for acceptance decisions. Practical application of this choice creating model needs conversion of model outcome values, in numerical conditions to be utilized in driver's gap acceptance behavior simulators. Due to this circumstances, ANFIS model results have been standardized bringing preference values in the range of 0–1. Then, to relate the outcomes of ANFIS model with measured data, it is presumed that if preference values larger than or equal to 0.50 the driver elect to accept the lag/gap. 70 % observations are found to prepare the models and residual 30 % is considered for validating the models. By seeing prediction of models presented in Fig. 4, it can be said that Model 1 and Model 2 are presenting a strong forecast level with measured data in comparison to other models. However, models containing the variable speed of the conflicting vehicle show very low forecast level may because of noisy data. By examining the predicted level of the Model 2, it can be concluded that the age of the driver is a chief variable in crossing behavior of driver. Additionally the parameter combinations of Model 4, Model 6 and Model 9 is indicating fair predictions.

Table 1 Comparisons between various models of two wheelers with their training and testing errors

Model no.	Input for ANFIS model	Number of rules	Training error	Testing error	Difference between errors
1	Gap/lag time (s)	3	0.37256	0.36333	0.00923
2	Gap/lag time (s), drivers' gender	6	0.36162	0.36435	0.00273
3	Gap/lag time (s), occupancy	6	0.3601	0.37197	0.01187
4	Gap/lag time (s), drivers' age	9	0.3474	0.3184	0.029
5	Gap/lag time (s), speed of the confl. vehicle	9	0.40518	0.45146	0.04628
6	Gap/lag time (s), confl. veh type	12	0.37521	0.33671	0.0385
7	Gap/lag time (s), waiting time	9	0.39514	0.42954	0.0344
8	Gap/lag time, gender, occupancy	12	0.39365	0.42854	0.03489
9	Gap/lag time, drivers' gender and age	18	0.38141	0.40621	0.0248
10	Gap/lag time, occupancy, drivers' age	18	0.40662	0.43234	0.02572
11	Gap/lag time, confl. veh type, gender of the driver	24	0.41214	0.44547	0.0333

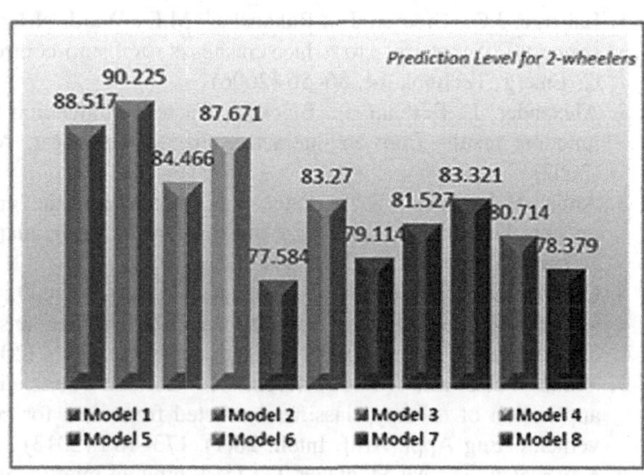

Fig. 4 Prediction level for two-wheelers

6 Conclusion

Modelling of gap accepting behavior is tough as it impacted by traffic features and vehicles' as well as drivers' features. Fuzzy model has been broadly accepted model to investigate similar conditions. This article explains the use of Adaptive Neuro-Fuzzy.

Inference System to model the gap taking performance of through movement vehicles at the four-legged uncontrolled junction, located in semi-urban area of Ahmedabad in the province of Gujarat. An ANFIS is utilized to compute the likelihood of accepting a specified gap/lag based on diverse drivers and traffic attributes. The model includes variably relating to gap/lag size, driver age; driver genders, occupancy, waiting time, approaching speed and conflicting vehicle type. Training forecasting accuracy by all replicas is better than 75 %. ANFIS models also forecasted well for the 30 % data reserved for validation. Eleven different models with a diverse combination of variables developed and checked with independent data. The forecasting abilities of models have been matched with those experiential data set and has exposed healthy capabilities of replicating the experiential performance. As a result, it can be inferred that the age of the driver played an important role in gap acceptance process for motorized two-wheelers and the variable approaching speed is not found much effective.

ANFIS model can be induced to different gap acceptance circumstances like right turning from the secondary road, right turning from the main road and left turning from the secondary road at four-legged uncontrolled junction with considering various parameters. This model can be a possible addition to microscopic traffic simulators. Hence, this will be the future scope of this research.

References

1. Laberge, J.C., Creaser, J.I., Rakauskas, M.E., Ward, N.J.: Design of an intersection decision support (IDS) interface to reduce crashes at rural stop-controlled intersection. Transp. Res. Part C: Emerg. Technol. **14**, 36–56 (2006)
2. Alexander, J., Barham, P., Black, I.: Factors influencing the probability of an incident at a junction: results from an interactive driving simulator. Accid. Anal. Prev. **34**(6), 779–792 (2002)
3. Amin, H.J., Desai, R.N., Patel, P.S.: Modelling the crossing behavior of pedestrian at uncontrolled intersection in case of mixed traffic using adaptive neuro fuzzy inference system. J. Traffic Logistic Eng. **2**(4), 263–270 (2014)
4. Ottomanelli, M., Caggiani, L., Iannucci, G., Sassanelli, D.: An adaptive neuro-fuzzy inference system for simulation of pedestrians behaviour at unsignalized roadway crossings. In: Softcomputing in Industrial Applications. Netherlands (2010)
5. Valdés-Vela, M., Toledo-Moreo, R., Terroso-Sáenz, F., Zamora-Izquierdo, M.A.: An application of a fuzzy classifier extracted from data for collision avoidance support in road vehicles. Eng Appl. Artif. Intell. **26**(1), 173–183 (2013)
6. Keyarsalan, M., Ali Montazer, G.: Designing an intelligent ontological system for traffic light control in isolated intersections. Eng. Appl. Artif. Intell. **24**(8), 1328–1339 (2011)
7. Jang, J.S.R., Sun, C.T., Mizutani, E.: Neuro-fuzzy and Soft Computing-A Computational Approach to Learning and Machine Intelligence. Prentice Hall, NJ (1997)
8. Rossi, R., Massimiliano, G., Gregorio, G., Claudio, M.: Comparative analysis of random utility models and fuzzy logic models for representing gap-acceptance behavior using data from driving simulator experiments. Procedia-Soc Behav. Sci. Elsevier **54**, 834–844 (2012)
9. Rossi, R., Meneguzzer, C.: The effect of crisp variables on fuzzy models of gap-acceptance behaviour. In: Proceedings of the 13th Mini-EURO Conference: Handling Uncertainty in the Analysis of Traffic and Transportation Systems, pp. 240–246 (2002)

10. Ghomsheh, V.S., Shoorehdeli, M.A., Teshnehlab, M.: Training anfis structure with modified pso algorithm. In: Proceeding of the 15th Mediterranean Conference on Control & Automation, Athens–Greece (2007)
11. Mehrabi, M., Pesteei, S.M.: An adaptive neuro-fuzzy inference system (anfis) modelling of oil retention in a carbon dioxide air-conditioning system. In: International Refrigeration and Air Conditioning Conference, Iran (2010)

Optimizing the Objective Measure of Speech Quality in Monaural Speech Separation

M. Dharmalingam, M.C. John Wiselin and R. Rajavel

Abstract Monaural speech separation based on computational auditory scene analysis (CASA) is a challenging problem in the field of signal processing. The Ideal Binary Mask (IBM) proposed by DeLiang Wang and colleague is considered as the benchmark in CASA. However, it introduces objectionable distortions called musical noise and moreover, the perceived speech quality is very poor at low SNR conditions. The main reason for the degradation of speech quality is binary masking, in which some part of speech is discarded during synthesis. In order to address this musical noise problem in IBM and improve the speech quality, this work proposes a new soft mask as the goal of CASA. The performance of the proposed soft mask is evaluated using perceptual evaluation of speech quality (PESQ). The IEEE speech corpus and NOISEX92 noises are used to conduct the experiment. The experimental results indicate the superior performance of the proposed soft mask as compared to the traditional IBM in the context of monaural speech separation.

Keywords Monaural speech separation · Computational auditory scene analysis · Perceptual evaluation of speech quality · Ideal binary mask · Optimum soft mask

M. Dharmalingam (✉)
PRIST University, Thanjavur, Tamilnadu, India
e-mail: dharmalingamrandd@gmail.com

M.C. John Wiselin
Department of EEE, Travancore Engineering College, Kollam, Kerala, India
e-mail: dr.wiselin16@gmail.com

R. Rajavel
Department of ECE, SSN College of Engineering, Chennai, India
e-mail: rajavelr@ssn.edu.in

A. Nagar et al. (eds.), *Proceedings of 3rd International Conference on Advanced Computing, Networking and Informatics*, Smart Innovation, Systems and Technologies 43, DOI 10.1007/978-81-322-2538-6_56

1 Introduction

In day to day life, the human auditory system receives number of sounds, in which some sounds may be useful and others are not. The speech separation problem in digital signal processing is to separate out the target speech signal from the other unwanted interferences. Human auditory system handle this complex source separation problem easily, whereas it is very difficult for the machine to perform the same as human beings. However, the acoustic interferences (music, telephone ringing, passing a car and other people speaking/shouting etc.) in a natural environment is often unavoidable. Reducing the impact of acoustic interference on the speech signal may be useful in a number of applications, such as voice communication, speaker identification, digital content management, teleconferencing system and digital hearing aids. Several approaches have been proposed in the last two decades for monaural speech separation, such as speech enhancement, blind source separation (BSS), model-based and feature-based CASA.

Speech enhancement approaches [1] utilize the statistical properties of the signal to separate the speech under stationary noisy conditions. Blind source separation is an another signal processing technique for speech separation. BSS can be done using independent component analysis (ICA) [2], spacial filtering, and nonnegative matrix factorization (NMF) [3]. BSS technique fails to separate the target speech signal effectively in the case where trained basis functions of two speech source overlaps [4]. Computational auditory scene analysis (CASA) is the most successful technique among these approaches to monaural speech separation. It aims to achieve human performance in auditory scene analysis [ASA] [5] by using one or two microphone recording of the acoustic scene generally called acoustic mixture [6]. CASA based speech separation techniques can be divided into model-based and feature-based techniques [7]. The model-based CASA techniques use trained models of the speaker to separate the speech signal [8]. Feature-based technique transform the observed signal into a relevant feature space and then it is segmented into cells, which are grouped into two main streams based on cues [7]. Generally in CASA based speech separation system, the noisy speech signal will be decomposed into various T-F units to decide whether a particular T-F unit should be designated as target or interference. After T-F decomposition, a separation algorithm will used to estimate the binary T-F mask based on the signal and noise energy. This binary mask is used in the synthesis process to convert the T-F representation into target speech and background noise. DeLiang Wang suggested that the IBM can be considered as a computational goal of CASA [9]. It is basically a matrix of binary numbers which is set to one when the target speech energy exceeds the interference energy in the T-F unit and zero otherwise. Even though, IBM is the optimal binary mask [3], it introduces objectionable distortions, called musical noise. It is mainly due to the repeated narrow-frequency-band switching [6] and moreover the perceived quality of binary-masked speech is poor. In order to address this musical noise problems in IBM, this work propose a genetic algorithm based optimal soft mask (GA-OSM) as the goal of CASA. The rest of the paper is organized in the following manner. The next section provides an overview of IBM and its shortcomings. Section 3 presents the proposed GA-optimum soft mask as the computational goal of CASA. Systematic evaluations and experimental results are provided in Sect. 4. Finally, Sect. 5 summarizes the research work and gives the direction for future extension.

2 The Ideal Binary Mask and Its Shortcomings

DeLiang Wang proposed ideal binary mask as a computational goal in CASA algorithms [9]. The IBM is a two-dimensional matrix of binary numbers and it is determined as in [9] by comparing the power of target signal to the power of masker (interfering) signal for each T-F unit obtained using Gammatone filter bank [10].

$$IBM(t,f) = \begin{cases} 1, & \text{if } s(t,f) - n(t,f) > LC \\ 0 & \text{otherwise} \end{cases} \quad (1)$$

where $s(t,f)$ is the power of the speech signal, $n(t,f)$ is the power of the noise signal, at time t and frequency f respectively and LC is the local SNR criterion [9]. Only the T-F units with local SNR exceeding LC are assigned the binary value 1 in the binary mask and others are assigned zero [10]. In CASA, an LC value of 0 dB is commonly used, since it shows higher speech intelligibility even at low SNR levels (−5 dB, −10 dB) [11]. In IBM based speech separation, T-F units with binary value 1 are retained, and with value 0 are discarded. The region with binary value 0 is generally interpreted as the deep artificial gap and it is being discarded during synthesis and produce musical noise at the output.

3 Proposed Genetic Algorithm Based Optimum Soft Mask

Research results show the musical noise arising from binary mask can be reduced by using soft mask [12]. However, the choice of soft mask should be made carefully such that it does degrade the quality of the speech signal [6]. Cao et al. [13] has proposed a kind of soft mask by filling the artificial gaps with un-modulated broadband noise. The un-modulated broadband noise shallows the areas of artificial gaps in the time-frequency domain of the IBM processed speech mixture and improves the speech quality. In this work, rather than adding additional broadband noise, the T-F units with local SNR less than LC are filled with certain amount of unvoiced speech to enhance the speech quality. Here, a simple question comes, how much amount of unvoiced speech can be added to get better speech quality?. This motivates to use the Genetic algorithm to find the optimum amount of unvoiced speech to be added to improve the speech quality. The schematic of the proposed GA-optimum soft mask based speech separation system is shown in Fig. 1.

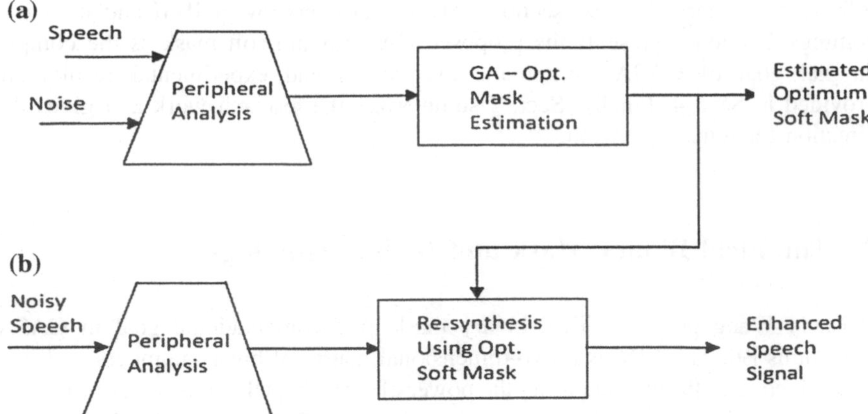

Fig. 1 Proposed GA-optimum soft mask based speech separation system

The input speech signal is first decomposed into various T-F units by a bank of 128 Gammatone filters, with their center frequencies equally distributed on the equivalent rectangular bandwidth (ERB) rate scale from 50 to 4000 Hz. The impulse response of the gammatone filter is given as [14]

$$g_{f_c}(t) = At^{N-1}\exp[-2\pi b(f_c)]\cos(2\pi f_c t + \phi)u(t) \tag{2}$$

where A is the equal loudness based gain, $N = 4$ is the order of the filter, b is the equivalent rectangular bandwidth, f_c is the center frequency of the filter, ϕ is the phase, and $u(t)$ is the step function. In each band, the filtered output is divided into a time frame of 20 ms with 10 ms overlapping between consecutive frames. As a result of this process, the input speech is decomposed into a two-dimensional time-frequency representation $s(t,f)$. Similarly, the noise signal also decomposed into a two-dimensional time-frequency representation $n(t,f)$. The proposed soft mask is defined as

$$GA - OSM(t,f) = \begin{cases} 1, & \text{if } s(t,f) - n(t,f) > LC \\ x & \text{otherwise} \end{cases} \tag{3}$$

where $GA - OSM(t,f)$ is the optimum soft mask. The GA is used here to find the optimum value of x which improves the speech quality. The GA frame work as similar as in [15] to find the optimum value of x is explained as follows:

Step-1 : Initialization: Generate a random initial population N_{POP} of size $[N \times 1]$ for the best value of x in Eq. (3). Where $N = 20$, i.e. N_{POP} = [20 × 1] matrix of chromosomes.

Step-2 : Fitness Evaluation: Fitness of all the solutions $x_1, x_2, x_3, \ldots\ldots x_N$ in the population N_{POP} is evaluated. The steps for evaluating the fitness of a solution are given below:

Step-2a: Determine the signal power at time t and frequency f and denote it as $s(t,f)$.

Step-2b: Determine the noise power at time t and frequency f and denote it as $n(t,f)$.

Step-2c: Compute the optimum soft mask $GA - OSM(t,f)$ as

$$GA - OSM(t,f) = \begin{cases} 1, & \text{if } s(t,f) - n(t,f) > LC \\ x_i & \text{otherwise} \end{cases} \tag{4}$$

where $x_i \in x_1, x_2, x_3, \ldots\ldots x_N$

Step-2d: Synthesize the speech signal using the computed soft mask as defined in step 2c.

Step-2e: The PESQ (fitness value) is calculated.

Step-2f: The steps 2c–2e are repeated for all solutions in the population.

Step-3: Updating Population: The populations are updated via mating and mutation procedure of Genetic algorithm.

Step-4: Convergence: Repeat steps 2–3 until an acceptable solution is reached or number of iteration is exceeded. At this point the algorithm should be stopped.

The final solution of this GA algorithm gives the best value of x in Eq. (3) and hence the optimum soft mask. This estimated optimum soft mask is used in the online speech separation stage to resynthesize the speech signal.

4 Performance Evaluations and Experimental Results

4.1 Experimental Database and Evaluation Criteria

The clean speech signals are taken from the IEEE corpus [16] and noise signals are taken from the Noisex92 database [17]. To generate noisy signals, clean speech signals are mixed with the babble and factory noises at different SNRs. The performance of the proposed optimum soft mask and IBM is assessed by using PESQ value, since PESQ measure is the one recommended by ITU-T for speech quality assessment [1, 18].

4.2 Performance Evaluation of GA-OSM Versus IBM

The clean speech signal and the babble noise are used to estimate the values of x in Eq. (3) and hence the optimum soft mask. Later, the clean speech and noise signals are manually mixed at SNRs in the range of −5 to 10 dB. The IBM computed using Eq. (2) and GA—optimum soft mask using Eq. (3) are applied to the mixture signals after T-F decomposition by the Gammatone analysis filter bank. Finally, the IBM and the GA-OSM weighted responses are processed by the Gammatone synthesis filterbank to yield an enhanced speech signal.

Figures 2, 3 and 4 show the PESQ value obtained by processing mixture signals using the proposed GA-OSM and IBM for the babble and factory noise respectively. As it can be seen from the figure, the GA-OSM processed signal has the best speech quality, compared to the IBM processed noisy signals. The average PESQ improvement for the speech signal "The sky that morning was clear and bright blue" [16] with babble noise is 0.4253 and with factory noise is 0.3871. The PESQ improvement for the speech signal "A large size in stocking is hard to sell" with babble noise is 0.4551 and with factory noise is 0.3673. Similarly, for the speech

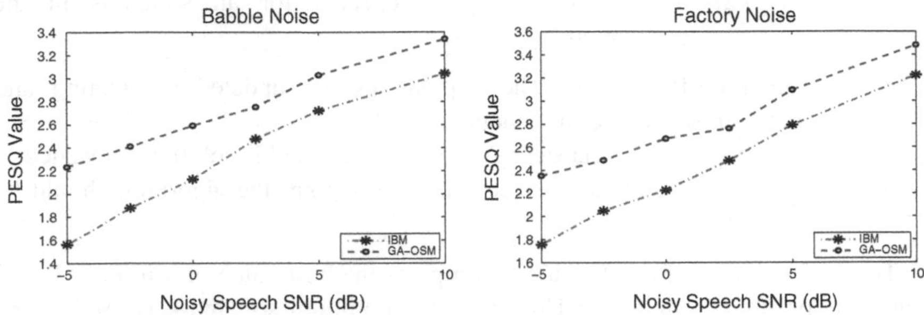

Fig. 2 Average PESQ values obtained by IBM and GA-OSM for the sentence "The sky that morning was clear and bright blue" [16] at different input SNRs and noise types

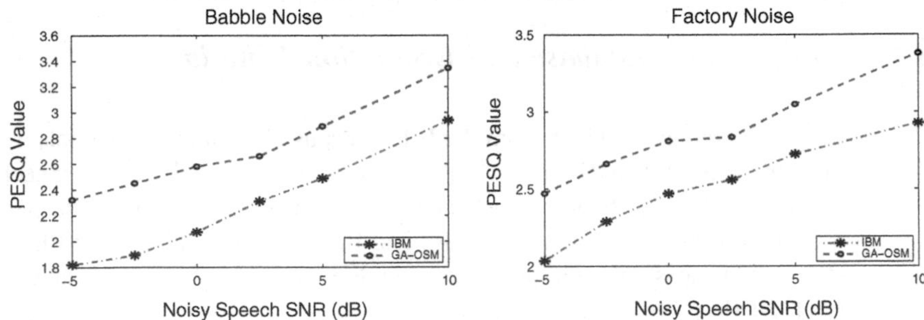

Fig. 3 Average PESQ values obtained by IBM and GA-OSM for the sentence "A large size in stocking is hard to sell" [16] at different input SNRs and noise types

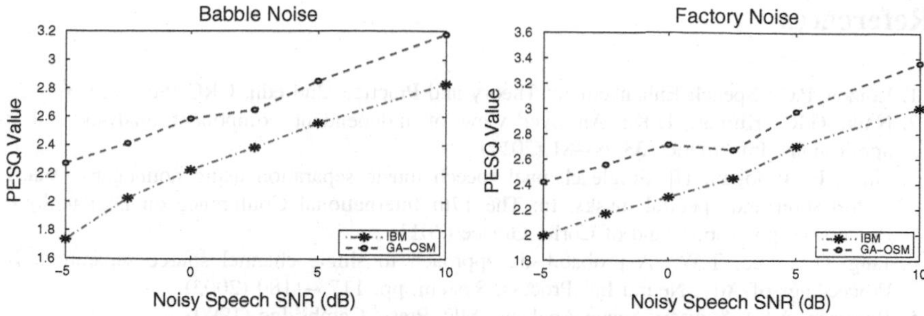

Fig. 4 Average PESQ values obtained by IBM and GA-OSM for the sentence "Sunday is the best part of the week" [16] at different input SNRs and noise types

signal "Sunday is the best part of the week" [16] the PESQ improvement with babble noise is 0.3649 and with factory noise is 0.3476. Moreover, the proposed GA-OSM mask estimation method use the clean speech signal "The sky that morning was clear and bright blue" [16] with babble noise to estimate the optimum soft mask using GA. This estimated soft mask is later used to separate and evaluate the performance of the system with other type of noise knows as factory noise. The results in Figs. 2, 3 and 4 illustrate the generalization ability of the proposed GA-OSM mask estimation approach to unseen speech and noises.

5 Summary and Future Work

Monaural speech separation is one of the challenging problem in the field of signal processing. Various approaches had been proposed including speech enhancement, Wiener filtering, noise tracking, BSS, CASA and so on. CASA based speech separation is the best among these techniques and sets the IBM as the computational goal. However, it introduces objectionable distortions called musical noise and degrades the quality of speech signal. In order to address this musical noise problem in IBM, this work proposed a genetic algorithm based optimal soft mask (GA-OSM) as the goal of CASA. The PESQ measure is used to examine the quality of the speech signal in the framework of monaural speech separation system. The experimental results in Fig. 2, 3 and 4 show the superior performance of the optimal soft mask (GA-OSM) as compared to the traditional IBM based speech separation system. However, the proposed GA-OSM estimation algorithm in its current form requires the prior knowledge of the clean speech and noise signals. This is one of the limitations of the current proposed algorithm. Further investigation of estimating the soft mask without the prior knowledge of speech and noise signal for monaural speech separation is in progress.

References

1. Loizou, P.C.: Speech Enhancement: Theory and Practice, 2nd edn, CRC Press (2013)
2. Naik, G.R., Kumar, D.K.: An over view of independent component analysis and its applications. Informatica **35**, 63–81 (2011)
3. Grais, E., Erdogan, H.: Single channel speech music separation using nonnegative matrix factorization and spectral masks. In: The 17th International Conference on Digital Signal Processing, pp. 1–6. Island of Corfu, Greece (2011)
4. Jang, G.J., Lee, T.W.: A probabilistic approach to single channel source separation. In: Proceedings of Adv. Neural Inf. Process. System, pp. 1173–1180 (2003)
5. Bregman, A.S.: Auditory Scene Analysis. MIT Press, Cambridge (1990)
6. Christopher, H., Toby, S., Tim B.: On the Ideal Ratio Mask as the Goal of Computational Auditory Scene Analysis. In: Naik, G. R., Wang, W. (eds) Blind Source Separation Advances in Theory, Algorithms and Applications. Signals and Communication Technology, pp. 369–393. Springer-Verlag, Heidelberg (2014)
7. Radfar, M.H., Dansereau, R.M., Chan, W.Y.: Monaural speech separation based on gain adapted minimum mean square error estimation. J. Sign. Process Syst. **61**, 21–37 (2010)
8. Mowlaee, P., Saeidi, R., Martin, R.: Model-driven speech enhancement for multisource reverberant environment. In: Theis, F., Cichocki, A., Yeredor, A., Zibulevsky, M. (eds.) Latent Variable Analysis and Signal Separation. Lecture Notes in Computer Science, vol. 7191, pp. 454–461. Springer-Verlag, Heidelberg (2012)
9. Wang, D.: On ideal binary mask as the computational goal of auditory scene analysis. In: Divenyi, P. (ed.) Speech Separation by Human and Machines, pp. 181–197. Kluwer Academic, Norwell (2005)
10. Geravanchizadeh, M., Ahmadnia, R.: Monaural Speech Enhancement Based on Multi-threshold Masking. In: Naik, G. R., Wang, W. (eds) Blind Source Separation Advances in Theory, Algorithms and Applications. Signals and Communication Technology, pp. 369–393. Springer-Verlag, Heidelberg (2014)
11. Li, N., Loizou, P.C.: Factors influencing intelligibility of ideal binary-masked speech: implications for noise reduction. J. Acoust. Soc. Am. **123**(3), 1673–1682 (2008)
12. Araki, S., Sawada, H., Mukai, R. Makino, S.: Blind sparse source separation with spatially smoothed time-frequency masking. In: International Workshop on Acoustic, Echo and Noise Control, Paris (2006)
13. Cao, S., Li, L., Wu, X.: Improvement of intelligibility of ideal binary-masked noisy speech by adding background noise. J. Acoust. Soc. Am. **129**, 2227–2236 (2011)
14. Patterson R.D., Nimmo-Smith, I., Holdsworth J.: Rice P : An Efficient Auditory Filter bank Based on the Gammatone Function. Report No. 2341, MRC Applied Psychology Unit, Cambridge (1985)
15. Rajavel, R., Sathidevi, P.S.: A new GA optimised reliability ratio based integration weight estimation scheme for decision fusion audio-visual speech recognition. Int. J. Sig. Imaging Syst. Eng. **4**(2), 123–131 (2011)
16. Rothauser, E.H., Chapman, W.D., Guttman, N., Hecker, M.H.L., Nordby, K.S., Silbiger, H.R., Urbanek, G.E., Weinstock, M.: Ieee recommended practice for speech quality measurements. IEEE Trans. Audio Electro Acoust. **17**, 225–246 (1969)
17. Noisex-92, http://www.speech.cs.cmu.edu/comp.speech/Section1/Data/noisex.html
18. ITU-T: Perceptual Evaluation of Speech Quality (PESQ): An Objective Method for End-to-End Speech Quality Assessment of Narrow-Band Telephone Networks and Speech Codecs, Series P: Telephone Transmission Quality Recommendation P.862, ITU, 1.4. (2001)

Automated Segmentation Scheme Based on Probabilistic Method and Active Contour Model for Breast Cancer Detection

Biswajit Biswas, Ritamshirsa Choudhuri and Kashi Nath Dey

Abstract Mammography is one of the renowned techniques for detection of breast cancer in medical domain. The detection rate and accuracy of breast cancer in mammogram depend on the accuracy of image segmentation and the quality of mammogram images. Most of existing mammogram detection techniques suffer from exact continuous boundary detection and estimate amount of affected area. We propose an algorithm for the detection of deformities in mammographic images that using Gaussian probabilistic approach with Maximum likelihood estimation (MLE), statistical measures for the classify of image region, post processing by morphological operations and Freeman Chain Codes for contour detection. For these detected areas of abnormalities, compactness are evaluated on segmented mammographic images. The validation of the proposed method is established by using mammogram images from different databases. From experimental results of the proposed method we can claim the superiority over other usual methods.

Keywords Mammographic images · Segmentation · Breast cancer

1 Introduction and Literature Review

In image analysis, segmentation is the partitioning method for digital image to break into multiple segments. The purpose of segmentation is to simplify or change the demonstration of an image that is more significant for image analysis. Image

B. Biswas (✉) · R. Choudhuri · K.N. Dey
Department of Computer Science and Engineering, University of Calcutta, Kolkata, India
e-mail: biswajit.cu.08@gmail.com

R. Choudhuri
e-mail: ritam.shirsa@gmail.com

K.N. Dey
e-mail: kndey55@gmail.com

A. Nagar et al. (eds.), *Proceedings of 3rd International Conference on Advanced Computing, Networking and Informatics*, Smart Innovation, Systems and Technologies 43, DOI 10.1007/978-81-322-2538-6_57

segmentation is potentially used to locate all the objects and boundaries in images [1]. The performance of an object detection and classification scheme should be consistent and perfect segmentation for the interest of object image [2]. Most important application of image segmentation and classification in mammogram images is for diagnosis of early stage of breast cancer [3]. Mammography is crucial in viewing and diagnosis of breast cancer. So the perfection biopsy using computer-aided methods are advantageous for the detection and classification. We concentrate on image processing for segmentation of breast cancer and estimate the presence of suspicious area based on the image information [4].

There are a number of methods devised towards segmenting human anatomical structures such as variants of active contours [5] and active shape models [6]. The major disadvantages of these methods are they need manual control point initialization and require predefined control [5, 6]. In recent times geometric active contours were established based on theory of curve evolution and geometric flows with level set method [7, 8]. These techniques can detect only the outer boundaries of objects defined by the gradient with well defined structured image. K-means clustering (KM), Fuzzy C means (FCM) clustering are also common for the segmentation of suitable anatomical structures [9]. Markov random field (MRF) based approaches are used for segmentation [10] but uses prerequisite information for segmenting an anatomical structures. Another approach for segmentation is spectral graph clustering method [11], which usually uses for MRI image segmentation [9]. It is a well known robust segmentation method and is commonly used for segmentation applications [2, 4]. A notably different method is proposed in [12]. Commonly a useful image segmentation method is watershed segmentation method that requires exact information about selected pixel set from detected regions [13, 14].

Here we use a probabilistic based segmentation algorithm for the mammogram image to detect breast cancer. This consists of parameter estimation, determine probability distribution of intensity for every pixel in the image, then threshold selection and threshold image creation, morphological processing on segmented region and the active contour model for segmentation of mammogram images. We present a statistical classification method to choose characteristics that satisfies classification condition without from misclassification on observation model [2]. The proposed segmentation approach is applied to various breast cancer images and the results are compared to watershed segmentation method from the related research works. Key steps of proposed method are as follows:

- Find the probability distributions of intensity in image with Gaussian model and estimate Maximum likelihood estimation (MLE) for parameters (Mean and variance).
- Using probabilistic model is prepared a classification method for pixel classification.
- Use morphological operation and Boundary Chain Codes (BBC) for defined active contour map of images.

The organization of the paper is as follows: Sect. 2 deals with background concepts in Probabilistic model, parameter estimation, pixel classification, morphology operations on image, Boundary Chain Codes (BBC) for active contour model and the proposed algorithm has been explained. Section 3 covers the results analysis followed by conclusion on Sect. 4.

2 Proposed Methodology

The lists the various steps of the detection algorithm which are emphasized in detail in the following subsections. The block diagram illustrated in Fig. 1.

2.1 Parameters Estimation and Probability Distribution

The first step of the proposed technique is to estimate the probability density of intensity distribution of each pixel in the input gray scale image $\mathbf{I}(M, N)$ where M, N are number of rows and columns of image $\mathbf{I}$. This task is done by the use of Gaussian probability density function $\mathcal{I}(\mu^*, \sigma^*2)$ of image $\mathbf{I}$ based on Maximum likelihood estimation of parameters. Here parameters are sample mean μ^* and variance σ^*2. Let $x_{i,j}$ denotes the intensity of the pixel at (i, j) in the image $\mathbf{I}(\mathbf{i} \leq \mathbf{M}\,\mathrm{and}\,\mathbf{j} \leq \mathbf{N})$. The value of θ is unknown for the probability density function (PDF) $f_{\mathbf{I}}(x, \theta)$ of image $\mathbf{I}$. We estimate θ based on the sample pixel $\mathbf{x} = x_{i,j}$ where $x_{i,j} \in \mathbf{I}$. We define the following function of θ, with $\mathbf{x}$ for discrete random variable $\mathbf{x} \in \mathbf{I}$, which is called the likelihood function [12]:

$$\mathbf{L}_{\mathbf{x}}(\theta) = p_I(\mathbf{x}, \theta)$$

where any value of θ that maximizes the likelihood function is known as a maximum-likelihood estimate (MLE) and denoted as $\hat{\theta}$:

$$\hat{\theta} = \arg\max_{\theta} L_{\mathbf{x}}(\theta) \quad (1)$$

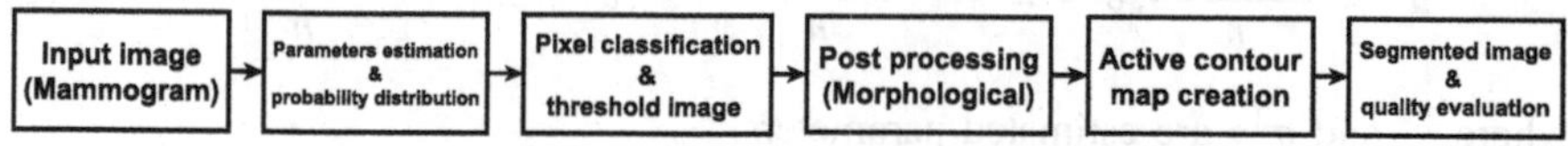

Fig. 1 Main steps of the proposed segmentation scheme for cancer detection

If the likelihood function is differentiable with respect to its parameter, a necessary condition for an MLE to satisfy is

$$\nabla_{\theta} L_{\mathbf{x}}(\theta) = 0, \text{ i.e.}, \frac{dL_x(\theta)}{d_m} = 0, m = 1, 2, \cdots, M \tag{2}$$

where the ∇_{θ} is the differential operator (1). The logarithmic function is a monotonically increasing and differentiable function, θ that defined the conditions (1) as

$$\frac{\partial \log L_{\mathbf{x}}(\theta)}{\partial \theta_m} = 0, m = 1, 2, \cdots, M \tag{3}$$

We select the solution that gives the largest value of $L_{\mathbf{x}}(\theta)$. Where the function $\log L_{\mathbf{x}}(\theta)$ is called the log-likelihood function, and its partial derivative with respect to θ is called the score function, denoted as $s(\mathbf{x}; \theta)$:

$$s(\mathbf{x}; \theta) \triangleq \left(\frac{\partial \log L_{\mathbf{x}}(\theta)}{\partial \theta_1}, \frac{\partial \log L_{\mathbf{x}}(\theta)}{\partial \theta_2}, \ldots, \frac{\partial \log L_{\mathbf{x}}(\theta)}{\partial \theta_m} \right) \tag{4}$$

The score function specifies the rate at which $\log L_{\mathbf{x}}(\theta)$ changes as θ varies.

Let **n** independent samples are taken from a common Gaussian normal distribution $\mathcal{I}(\mu, \sigma^2)$, where both mean and variance are unknown, and we find an MLE of these parameters based on $\mathbf{x} \in I(i,j)$. By setting $\theta = (\mu, \sigma^2)$, we have the likelihood function

$$L_{\mathbf{x}}(\theta) = \mathrm{I}(\mathbf{x}, \theta) = \frac{1}{(2\pi\sigma^2)^{\frac{n}{2}}} \exp\left[-\frac{\sum_{i=1}^{n} (\mathrm{x}_i - \mu)^2}{2\sigma^2} \right] \tag{5}$$

Then from expression (5) using Eq. (4), yields

$$\frac{1}{2\sigma^2} \sum_{i=1}^{n} (\mathbf{x}_i - \mu) = -n\sigma + \frac{1}{\sigma^3} \sum_{i=1}^{n} (\mathbf{x}_i - \mu)^2 = 0$$

from which we have

$$\mu^* = \frac{\sum_{i=1}^{n} \mathbf{x}_i}{n}, \sigma^2 * = \frac{\sum_{i=1}^{n} (\mathbf{x}_i - \mu^*)^2}{n}, \sigma^2 * \simeq \frac{\sum_{i=1}^{n} (\mathbf{x}_i^2 - \mu^{*2})}{n} \tag{6}$$

where μ^* and σ^2* are estimated parameters.

We used these estimated parameters, viz such as mean μ^* and variance σ^*2 in normal Gaussian PDF $\mathcal{I}(\mu, \sigma^2)$ to estimate probability density of every pixel $\mathbf{x}$ of the given image $\mathbf{I}(\mathbf{x} \in \mathbf{I})$. The probability of intensities of image $\mathbf{I}$ are used to pixel classification (background and foreground) and image segmentation for next process. The main necessary steps of parameters estimation is given as follows 1.

Algorithm 1 : Parameters estimation

Input: Gray scale image **I**;
Parameters: Gaussian PDF, unknown parameter μ and σ;
Output: Probabilistic image image **P**;
Step 1: Read gray image image **S**;
Step 2: Estimate meanμ and standard deviation σ using equations (6) on image **I** successively;
Step 3: Evaluate the probability image **P** using equation (5);

2.2 *Pixel Classification and Threshold Image*

In order to separating a pixel as an object to background and foreground classes, we label the classes π_1 and π_2 where π_1, π_2 are background and foreground class. We assume that if any pixel belong to class π_1 then that pixel assigned by 0 and 1 for class π_2. The pixels are classified on the value of probability $\mathbf{p}$ associated the random variables $\mathbf{x} \in \mathbf{I(i,j)}$. Thus the background class will be the population of $\mathbf{x}$ values for class π_1 and foreground class as population of $\mathbf{x}$ for class π_2. We represents these two population by probability density function $\mathbf{g_1(x)} \in \pi_1$ and $\mathbf{g_2(x)} \in \pi_2$ respectively. Let Ω is collection of all observations of $\mathbf{x}$. We define the sets $\mathbf{R_1} = \mathbf{x}|\mathbf{x} \in \pi_1$ and $\mathbf{R_2} = \mathbf{x}|\mathbf{x} \in \pi_2$ clearly, $\mathbf{R_2} = \Omega - \mathbf{R_1}$.

To classify a pixel object in class π_2 with respect class π_1, we define conditional case, that is

$$\mathbf{P(X \in R_2|\pi_1)} = \int_{R_2} f_2(\mathbf{x})\partial x$$

and similarly classify a pixel object in class π_1 with respect class π_2

$$\mathbf{P(X \in R_1|\pi_2)} = \int_{R_1} f_1(\mathbf{x})\partial x$$

However an object pixel can be belong to either $\mathbf{R_1}$ or $\mathbf{R_2}$ with basis of a threshold. We estimate threshold parameter η from probability matrix and constructed threshold image based on following rules:

$$\mathbf{R_1} : \sum \frac{\mathbf{g_1(x)}}{\mathbf{g_2(x)}} < \eta|\mathbf{x} \in \pi_1 = 0, \mathbf{R_2} : \sum \frac{\mathbf{g_1(x)}}{\mathbf{g_2(x)}} \geq \eta|\mathbf{x} \in \pi_2 = 1 \qquad (7)$$

where η is threshold value defined as $\eta = 1 - \frac{1}{2}\sqrt{\frac{\sigma^*}{\mu^*}}$ with estimated parameters σ^*, μ^*. The overall steps for construction of threshold image represented as follows 2:

Algorithm 2 : Threshold image construction

Input: probability image P;
Parameters: threshold η;
Output: Binary or threshold image $\mathbf{B}$;
Step 1: Read probability image $\mathbf{S}$;
Step 2: Estimating threshold η;
Step 3: Classifying pixel object using (7), (7);
Step 4: Construct threshold image;

2.3 Morphological Image Processing (Post Processing)

The shape based image processing operations known as morphology. The elementary operations of morphology are dilation and erosion. In dilation pixels are added to the boundaries of objects in an image. On the other hand, in erosion pixels are removed from object boundaries [15].

Algorithm 3 : Morphological operation

Input: Binary image B and morphological filter $\mathbf{S}$;
Parameters: Structure elements with size 3×3;
Output: Binary morphological image $\mathbf{M}$;
Step 1: Read binary image $\mathbf{S}$ and morphological filter $\mathbf{T}$;
Step 2: Applying erode and dilate operator (8), (9) on image B in order;
Step 3: Construct binary morphological image $\mathbf{M}$;

The binary dilation of a given binary image $\mathbf{X}$ by structure element $\mathbf{Y}$, denoted $\mathbf{X} \oplus \mathbf{Y}$, is defined as the set operation:

$$\mathbf{X} \oplus \mathbf{Y} = \{p|(\widehat{\mathbf{Y}}_p \cap \mathbf{X}) \neq \phi\} \tag{8}$$

where $\widehat{\mathbf{X}}_p$ is the reflection of the structuring element $\mathbf{X}$ and it is the set of pixel locations $\mathbf{p}$, where the reflected structuring element overlaps with foreground pixels in $\mathbf{Y}$ when translated to $\mathbf{p}$. Similarly, the binary erosion of $\mathbf{X}$ by $\mathbf{Y}$, denoted $\mathbf{X} \ominus \mathbf{Y}$, is defined as the set operation

$$\mathbf{X} \ominus \mathbf{Y} = \{p|(\widehat{\mathbf{Y}_\mathbf{p}} \subseteq \mathbf{X}\} \tag{9}$$

where, it is the set of pixel locations $\mathbf{p}$, where the structuring element translated to location $\mathbf{p}$ overlaps only with foreground pixels in $\mathbf{X}$ [16].

The proposed scheme first applies erode operator on binary image to remove unwanted shapes of the object and dilate is used to operator for filled edge gap in object boundary respectively. Algorithmic steps are as follows 3:

2.4 Active Contour Map Generation (Boundary Chain Codes)

We use Freeman chain code method on binary image to find all contour hierarchy [15, 16]. The contour hierarchy is a polygonal like representation as a sequence of steps in one of eight directions; each step is designated by an integer from 0 to 7 [16]. The closed contour of a region $\mathcal{R}$, is described by the sequence of points $\mathbf{s}_{\mathcal{R}} = [\mathbf{p_0}, \mathbf{p_1}, \ldots, \mathbf{p_{M-1}}]$ with $\mathbf{p_k} = \langle \mathbf{x_k}, \mathbf{y_k} \rangle$. We create the elements of its chain code sequence $\mathbf{s'}_{\mathcal{R}} = [\mathbf{c'}_0, \mathbf{c'}_1, \ldots, \mathbf{c'}_{M-1}]$ by $\mathbf{c'_k} = \mathbf{CODE}(\Delta a_k, \Delta b_k)$ where

$$(\Delta a_k, \Delta b_k) = \begin{cases} (a_{k+1} - a_k, b_{k+1} - b_k), & \text{for } 0 \le k \le M - 1; \\ (a_0 - a_k, b_0 - b_k), & \text{if } k = M - 1; \end{cases}$$

and $\mathbf{CODE}(\Delta a_k, \Delta b_k)$ being defined by Table 1.

For each point on the contour, only the initial point is recorded. The remaining points are encoded by 3 bits. Hence eight directional values can be stored. Algorithmic steps are as follows 4 (Fig. 2):

Algorithm 4 : Chain code operation

Input: Binary image B and a vector of contours $\mathbf{C}$;
Output: Contour image $\mathbf{CI}$;
Step 1: Read binary image $\mathbf{S}$;
Step 2:Get chain coded from image $\mathbf{B}$ using $\mathbf{CODE}$;
Step 3:Draw all retrieved contour on image $\mathbf{B}$;
Step 4: Construct contour image $\mathbf{CI}$;

Table 1 An 8-connected neighborhood

Δa	1	1	0	−1	−1	−1	0	1
Δb	0	1	1	1	0	−1	−1	−1
CODE $(\Delta a_k, \Delta b_k)$	0	1	2	3	4	5	6	7

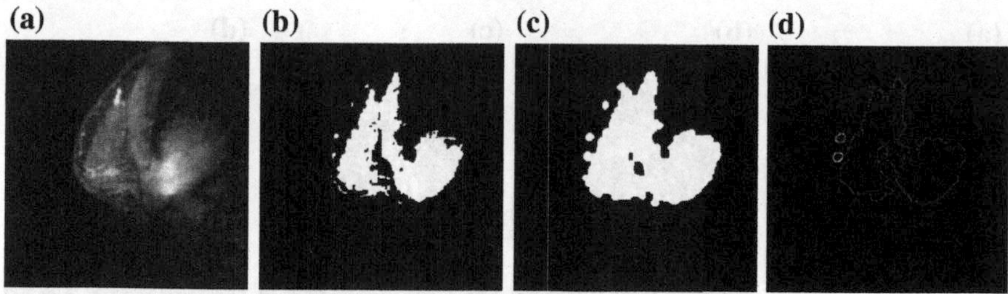

Fig. 2 Results of proposed approach: **a** source image, **b** segmented image, **c** post processing, **d** active contour map

3 Result and Analysis

In this paper all experiments are conducted on images of 30 women diagnosed with breast cancer [17]. The proposed algorithm detects the affected regions in an effective way and the resulting images are shown from Figs. 2, 3, 4, 5, 6 and 7. In each figures the first image is original mammogram, the second image is the threshold image, third is the image after applying morphology (post processing), and last one is the active contour map of cancer detected image. From the detected image, the affected area of breasts could be identified. In order to verify the

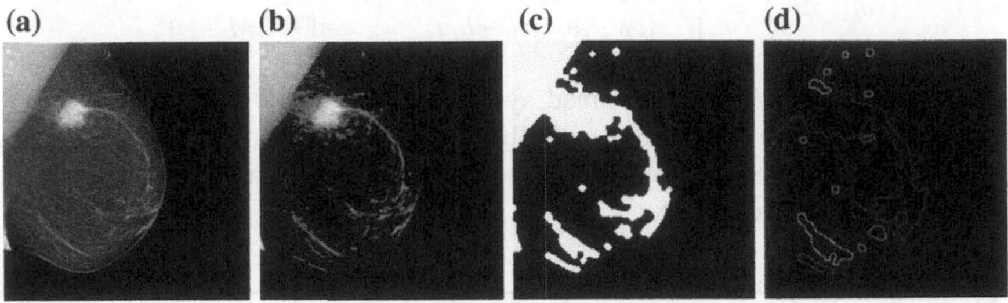

Fig. 3 Results of proposed approach: **a** source image, **b** segmented image, **c** post processing, **d** active contour map

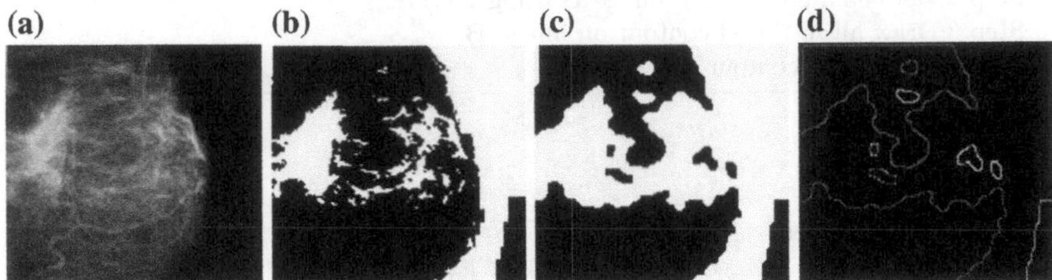

Fig. 4 Results of proposed approach: **a** source image, **b** segmented image, **c** post processing, **d** active contour map

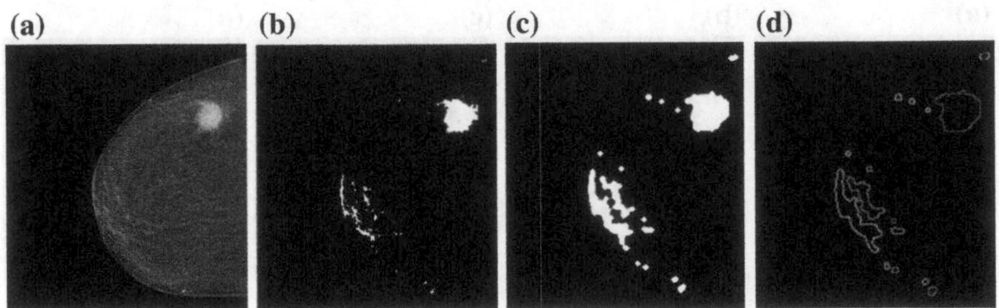

Fig. 5 Results of proposed approach: **a** source image, **b** segmented image, **c** post processing, **d** active contour map

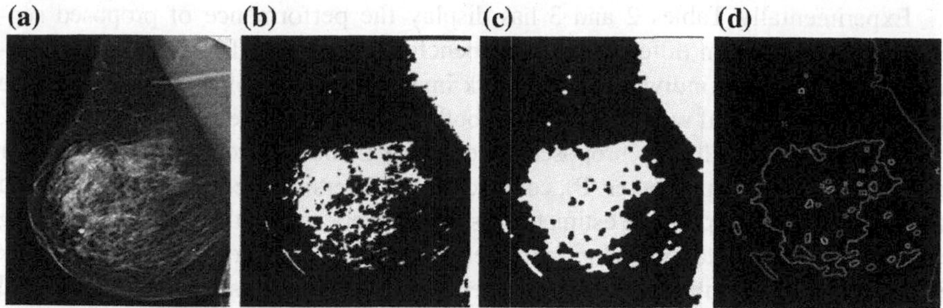

Fig. 6 Results of proposed approach: **a** source image, **b** segmented image, **c** post processing, **d** active contour map

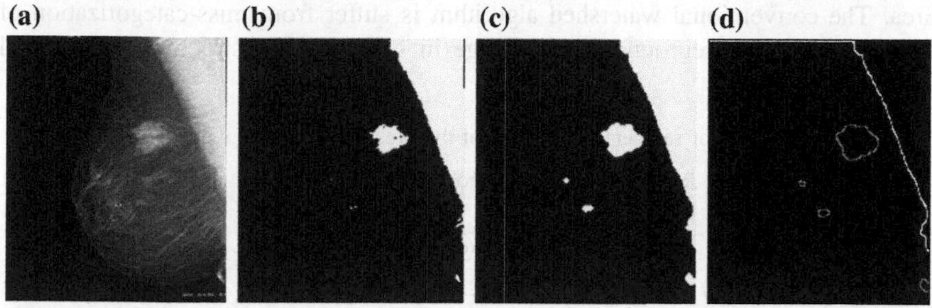

Fig. 7 Results of proposed approach: **a** source image, **b** segmented image, **c** post processing, **d** active contour map

effectiveness of segmentation methods, we used various region descriptors for performance accuracy. These region descriptors are as follows [15, 16]:

- Area of a region in the image plane $\mathbf{A(R)} = \sum_i \sum_j \mathbf{I(i,j)}$ where **I** image and **A (R)** the area is measured in pixels.
- The perimeter of an image is $\mathbf{P(R)} = \sum_i \sqrt{(\mathbf{x}_i - \mathbf{x}_{i-1})^2 + (\mathbf{y}_i - \mathbf{y}_{i-1})^2}$ where x_i and y_i are the co-ordinates of the *i*th pixel of region.
- Compactness—compactness is an often expressed measure of shape given by the ratio of perimeter to area. That is, $C(R) = \frac{4\pi A(R)}{P^2(R)}$
- Dispersion is the ratio of the maximum to the minimum radius of curve region. That is, $C(R) = \frac{\max\sqrt{(x_i - x_{i-1})^2 + (y_i - y_{i-1})^2}}{\min\sqrt{(x_i - x_{i-1})^2 + (y_i - y_{i-1})^2}}$ where x_i and y_i are the co-ordinates of the *i*th pixel of region.

The performance of the proposed segmentation method is tested on benchmark image database [17] and compared with the most popular segmentation algorithm watershed method [13]. From these experiments, we conclude that the proposed scheme in this work presents excellent and accurate segmentation results compared with the watershed algorithm.

Experimentally Tables 2 and 3 has display the performance of proposed segmentation method on different selective benchmark images. Table 2 presents estimation accuracy for individual benchmark image set. For comparative analysis the proposed method and watershed method both are applies on benchmark image sets. The simulation results for proposed scheme w.r.t. the different region descriptor evaluation area (A), perimeter (P), compactness, dispersion are shown in Table 2. In Fig. 8, we have compare the estimated results of proposed method with the existing and popular segmentation technique watershed segment method for the benchmark images (7). From Table 3, it can be observed that proposed segmentation method presents best values (bold faced) w.r.t statistical data of different region descriptor evaluation parameters of segmented output images.

These experiments clearly show that the proposed algorithm excels the limitations of the traditional method by poor structured region and detecting the exact area. The conventional watershed algorithm is suffer from miss-categorization of background and foreground object image in case of poor structured anatomical

Table 2 Assessments of segmentation results of the breast cancer image of proposed scheme

Contour Image	Regional shape descriptors			
	Perimeter	Area	Compactness	Dispersion
Figure 3d	99,978	56,939	0.00073	26.7809
Figure 4d	30,046	45,822	0.00068	67.2463
Figure 5d	174,960	15,515	0.00637	47.9097
Figure 6d	158,690	51,532	0.00026	5.9064
Figure 7d	30,046	45,822	0.00068	67.2463

Table 3 Assessments of regional description parameters of proposed and watershed segmentation scheme

Method	Image	Area	Perimeter	Compactness	Dispersion
Watershed	Figure 8	48,040	38,258	0.00142	42.3725
Proposed	Figure 8	**30,046**	**45,822**	**0.00068**	**67.2463**

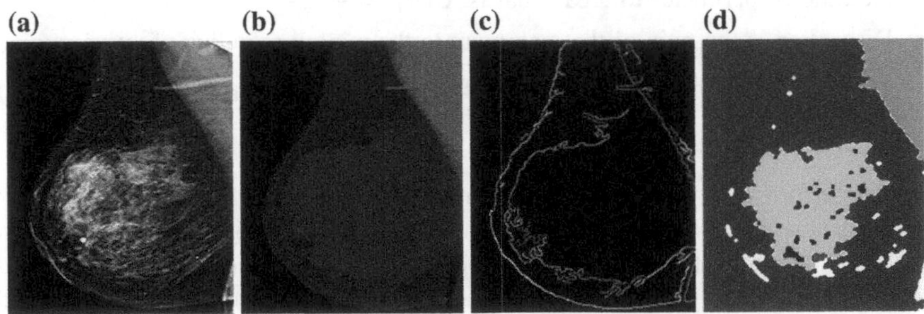

Fig. 8 Results of watershed approach: **a** source image, **b** segmented before watershed, **c** active contour map, **d** watershed image

image. Consequently, the shape and the size of segmented area are fairly accurate. In contrast, the proposed method gives a good partition of image objects without losing any discontinuous boundary. As we can observe in the result sets, the shape and the size of separating boundary of segmented region are detected correctly. In addition, our method gives the precisely continuous and smooth boundaries for image objects.

4 Conclusions

In this paper, we have conceptualized and simulated a probabilistic segmentation algorithm for breast cancer images. To detect abnormalities, the statistical parameters estimation and a classification method is used. All the parameters of the proposed model are estimated automatically from the images. In order to segment, morphological for operation are applied post processing and active contour map for smooth boundary of detected regions. Our method provides a good segmentation of affected cancer regions and smooth borders with continuous active contour map. The experimental results show that aforesaid parameters quantifies the breast abnormality.

References

1. W. Xu, S. Xia, et al., A Model based Algorithm for Mass Segmentation in Mammograms, IEEE, pp. 2543 - 2546, (2006)
2. Cheng, H.D., Shi, X.J., Min, R., Hu, L.M., Cai, X.P., Du, H.N.: Approaches for automated detection and classification of masses in mammograms. Pattern Recogn. **39**(4), 646–668 (2006)
3. Mudigonda, N.R., Rangayyan, R.M., Desautels, J.E.L.: Segmentation and classification of mammographic masses. SPIE Med. Imag. Image Process. **3979**, 55–67 (2000)
4. te Brake, G.M., Karssemeijer, N.: Single and multiscale detection of masses in digital mammograms. IEEE Trans. Med. Imag. **18**, 628–639 (1999)
5. Kass, M., Witkin, A., Terzopoulos, D.: Snakes: active contour models. Int. J. Comput. Vision **1**(4), 321–331 (1988)
6. Xu, C., Yezzi, A., Prince, J.: On the relationship between parametric and geometric active contours. Proc. Asilomar Conf. Sig. Syst. Comput. **1**, 483–489 (2000)
7. Vese, L.A., Chan, T.F.: A multiphase level set framework for image segmentation using the Mumford and Shah model. Int. J. Comput. Vision **50**(3), 271–293 (2002)
8. Brox, T., Weickert, J.: Level set based image segmentation with multiple regions. Pattern Recognit. Springer LNCS **3175**, 415–423 (2004)
9. Chen, W., Giger, M.L., Bick, U.: A fuzzy c-means (FCM)-based approach for computerized segmentation of breast lesions in dynamic contrast-enhanced MR images. Acad. Radiol. **13**(1), 63–72 (2006)
10. Li, H.D., Kallergi, M., Clarke, L.P., Jain, V.K., Clark, R.A.: Markov random field for tumor detection in digital mammography. IEEE Trans. Med. Imag. **14**, 565–576 (1995)
11. Shi, J., Malik, J.: Normalized cuts and image segmentation. IEEE Trans. Pattern Anal. Mach. Intell. **22**(8), 888–905 (2000)

12. Fieguth, P.: Statistical image processing and multidimensional modeling. In: Information Science and Statistics. Springer, New York (2011). ISBN:978-1-4419-7293-4
13. Hamarneh, G., Li, X.: Watershed segmentation using prior shape and appearance knowledge. Image Vis. Comput. **27**, 59–68 (2009)
14. Lezoray, O., Cardot, H.: Cooperation of color pixel classification schemes and color watershed: a study for microscopic images. IEEE Trans. Image Process. **11**(7), 783–789 (2002)
15. Nixon, M.S., AguadoFeature, A.S.: Extraction and Image Processing. British Library, UK (2002). ISBN:0-7506-5078-8
16. Burger, W., Burge, M.J.: Principles of Digital Image Processing Core Algorithms. Springer, London (2009). ISSN:1863-7310
17. http://marathon.csee.usf.edu/Mammography/Database.html

Disease Detection and Identification Using Sequence Data and Information Retrieval Methods

Sankranti Joshi, Pai M. Radhika and Pai M.M. Manohara

Abstract Current clinical methods base disease detection and identification heavily on the description of symptoms by the patient. This leads to inaccuracy because of the errors that may arise in the quantification of the symptoms and also does not give a complete idea about the presence of any particular disease. The prediction of cellular diseases is still more challenging; for we have no measure on the exact quantity, quality and extremeness. The typical symptoms for these diseases are visible at a later stage allowing the disease to silently progress. This paper provides an efficient and novel way of detection and identification of pancreatitis and breast cancer using a combination of sequence data and information retrieval algorithms to provide the most accurate result. The developed system maintains a knowledge base of the mutations of the diseases causing breast cancer and pancreatitis and thus uses techniques of protein sequence scoring and information retrieval for providing the best match of patient protein sequence with the mutations stored. The system has been tested with mutations available online and gives 98 % accurate results.

Keywords Levenshtein edit distance · Needleman Wunsch · Parallel programming · Sequence data · Disease detection

S. Joshi · P.M. Radhika (✉) · P.M.M. Manohara
Department of Information and Communication Technology,
Manipal Institute of Technology, Manipal University, Manipal, India
e-mail: radhika.pai@manipal.edu

S. Joshi
e-mail: sankrantijoshi@gmail.com

P.M.M. Manohara
e-mail: mmm.pai@manipal.edu

A. Nagar et al. (eds.), *Proceedings of 3rd International Conference on Advanced Computing, Networking and Informatics*, Smart Innovation, Systems and Technologies 43, DOI 10.1007/978-81-322-2538-6_58

1 Introduction

Health care is one the most booming industry where in lot of research and development is taking place. Lots of studies have been done on many interdisciplinary topics in lieu of achievement of high standards of treatment [1]. Though advanced technologies are being used successfully for diagnosis of many diseases, research is still being done, in case of cellular diseases. The identification and detection of these diseases is difficult owing to the fact that the current methods need to be applied to cellular level for detection of abnormalities, which many a times, are not clearly observed in various imaging tests [2]. Current clinical methods rely heavily on patient's description of discomfort. As almost all cellular diseases, in their onset stages, have symptoms common to many other diseases, they are difficult to identify early and are often over looked when based on patient's description. For example, in case of colon cancer, the cancer might be present for many years before the development of symptoms. The real symptoms related to any cellular diseases surface quite late and lead to progression of disease before they can be identified. Figure 1 provides the timeline for identification of the cellular diseases and development of symptoms.

Protein sequence data can easily reflect the changes in the human body and can be used to identify the presence of diseases in the body on the onset. This paper suggests the use of protein sequence data for detection and identification of diseases using a combination of sequencing and information retrieval methods. The data is taken in FASTA format.

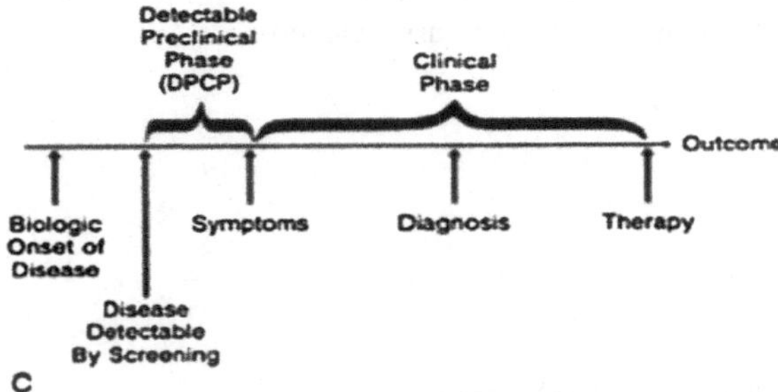

Fig. 1 Time line of a cellular disease (*Courtesy* https://onlinecourses.science.psu.edu/stat507/node/70)

2 Background Study

DNA sequencing software like Blast [3], Clustal [4] and Clustal W [5] have been developed to provide detailed analysis of human DNA. These work on the concept of parallelization of Smith-Waterman [6] and Needle-man Wunsch [7] algorithm and/or applying multiple sequence alignment for identifying patterns of similarity between various sequences and aligning them to identify the preserved patterns. Figure 1 identifies the shortcoming of current methods as the disease has already progressed by the time it becomes detectable through various imaging software. For this reason, the American Cancer Society (ACS) provides various guidelines on early detection of these diseases [8].

There are various technologies involved in development of systems to aid identification of such diseases and various other works have been done on sequence data to aid in early disease detection and identification. Neural networks and artificial intelligence are used extensively in most of the methods to provide the classification of various images. Data mining techniques have been used to predict the spread of lung cancer [9]. Genetic markers for understanding of leprosy, cancer, asthma and various automatic systems to monitor cardio health have been developed [10]. In case of working with DNA, large amount of sequence data is available for study and observation purposes. Apart from the famous human genome project [11], other notable works for the study of DNA and its involvement in diseases are DNA vaccines [12], DNA microarray techniques [13] and molecule markers [14]. Data Mining has also been a promising area of study in the development of methods for disease identification. Data mining has been used for the diagnosis and treatment of heart diseases [15]. A data mining approach has also been developed for diagnosis of coronary heart disease [16]. A review of data mining application in systems that have been developed to aid in the health care industry has been provided in [17]. Information Retrieval is now being used as a method to aid in diagnostics of various rare diseases [18]. Systems have also been developed for the measuring the accuracy of diagnostic systems using Information Retrieval [19].

In the proposed system, Levenshtein edit distance algorithm is used for the ease of mining relevant information in the form of mutations from protein sequences. It also increases the probability of finding the best possible match. Since Needleman Wunsch provides all possible matches it has been employed as the main algorithm for identification of all mutated protein sequences.

3 Proposed Study

3.1 Overview of the System

The current system provides a mechanism for identification of cellular diseases in pancreas or breast by detecting the presence of any mutations in the protein sequences of these organs. Minimalistic approach has been used to integrate the database and the system.

The data is passed through three scoring systems to provide the most accurate score and then the prediction is based on the symptoms associated with the patient. This system has three views.

1. The symptoms view where user selects disease identification through symptoms. This mode allows a user to select the symptoms the patient is having and provides information about diseases showing those symptoms. It also provides related probability of the diseases based on selection of symptoms.
2. The sequence selection takes the sequence data from the user and aligns it with all the sequences present in the database. This is followed by Levenshtein Edit Distance scoring of the input sequence. The sequences with the maximum score are classified into diseased or normal based on various mutations present. Thus after the final mutation match, the system provides a result based on the various mutations and the symptoms.
3. The final view provides information regarding performance of the system. This view has been used for comparison of time complexities in case of parallel and traditional computing.

The knowledge base contains various protein sequence mutations found in the onset of breast cancer and pancreatitis or those which lead to these diseases. These diseases were chosen due to the wide amount of study done on them so as to provide us with the exact mutations available. Some other mutations are also included based on their presence found in the patients of these diseases and based on some previous works done. Other general mutations found in almost all kinds of cancers are also included in the database. Along with this, it also takes into account the symptoms provided by the patient to solve the issues of various diseases which have similar early symptoms. Based on these two factors, it proposes a probability score for all the possible diseases.

The system takes the query in the form of sequence data or symptoms data and provides a match going through the various accuracy points. In case of a no match condition the system goes to the symptoms form and uses the symptoms to provide the top diseases.

3.2 Scoring Methodology

There are three checkpoints for scoring of the sequence before any final prediction can be made. This has been done to reduce the number of false positives and eliminate any false negatives, thereby increasing accuracy.

Needleman Wunsch. This algorithm has been used to provide scores for initial filtering of the sequences. Needleman Wunsch scoring is used to obtain all the possible matches. Since we do not want to find the preserved regions, we do not use the alignment step. Substantial decrease in run time has been obtained by the removal of alignment and trace back step. We also provide an alternative to gap penalty in the implementation since a large difference in length of sequences can lead to a high gap penalty increasing the false positives and negatives in context of the current system. The algorithm takes the sequences from the database and finds their alignment score with the input sequence data. Since it tries to find every possible match and gives the best possible match for all algorithms, it provides as a good means to filter the sequences to get the sequences with the maximum scores. The sequences with the maximum scores are then passed to Levenshtein algorithm.

Levenshtein. The levenshtein algorithm is a high quality string matching algorithm, which provides the least cost required to calculate the number of edit operation between two strings (in this case, sequences). It takes the input sequence and various mutations selected from the knowledge base based on scoring of Needleman Wunsch and enhance the score associated with sequences which have the minimum number of edits required to find a match. The system can also be implemented using back tracking, but due to the extensive number of insertions and deletions required for the process, it may require a large amount of time. We have used a recursive implementation approach for Levenshtein distance to provide the most effective match score by storing the prefix of the substring already matched and thus always focusing on the unmatched substring. This process also reduces the time of calculation. Though Levenshtein distance is a variation of Needleman Wunsch, the easy deletion mechanism instead of gap penalty provides a heuristic answer, i.e. the best possible approximation.

Based on the selection of symptoms and scores obtained from Levenshtein edit distance algorithm and Needleman Wunsch algorithm, the final results contain the diseases with their predicted probability of occurrence. In the absence of any mutations, the system provides an exhaustive list of symptoms, and their associated diseases.

3.3 Proposed Algorithm

Needleman Wunsch is a sequential alignment algorithm. It searches every substring for any two sequences and provides the best alignment. This value of each cell needs to be calculated using the values of previous cells surrounding it as seen in Fig. 2a. The presence of this dependency for each cell makes the algorithm to have the time complexity as O (nm) where n and m are the sizes of each sequence.

The system is implemented using parallel approach to reduce the running time. The modified approach calculates the values for the cells in a parallel manner and stores the value for previous columns in an array list for immediate availability. The parallel implementation has been done using threads and processors. Figure 2b shows the cells which are calculated in parallel.

The cells with the same numbers are calculated in parallel. This is possible for each cell; as the values of all previous cells are already calculated before reaching

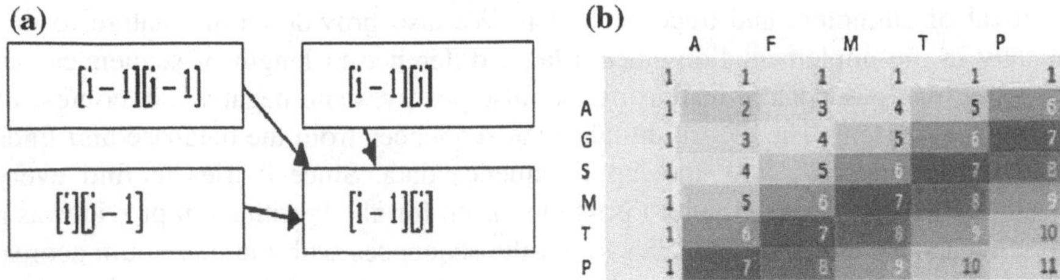

Fig. 2 **a** Calculation for each cell in Needleman Wunsch algorithm is dependent on the previous three cells, contributing to a runtime of O (nm). **b** Cells in similar color and having similar numbering are calculated simultaneously reducing the runtime to O (n + m) where n and m are length of sequences

the cell. This reduces the time of execution of the algorithm from $O\ (n * m)$ to $O\ (n + m)$ where n and m are the lengths of the two sequences.

3.4 Some Assumptions

The proposed method also assumes that a large collection of data is available for employing the parallel implementation of the system that has been developed. Hence, the data has been populated repeatedly for determining the speed of computations of the algorithms in the proposed system. The speed of threading has been determined while keeping all other processes idle and assuming the proposed software as a fully devoted system.

4 Results

4.1 Test Analysis and Parallel Implementation

In case of hyper threading with 10000, 10 threads are being invoked in the program and hence the time spent in allocation of data to the threads and switching is less because of parallel execution of threads. However, when we use the inbuilt Parallel class in C#, it invokes 4 threads owing to the quad processor CPU. There is a difference in time when working with Parallel class and hyper threading. In case with less number of sequences, even though the difference between the parallel and hyper-threading method is less, the difference between the hyper threading and sequential is substantial. This is because of the large size of each sequence.

While the computation time in case of parallel class in C# gives a constant value, the value of threading may depend on the various numbers of threads already running in the system and has found that 8 threads are optimum for 10,000 sequences.

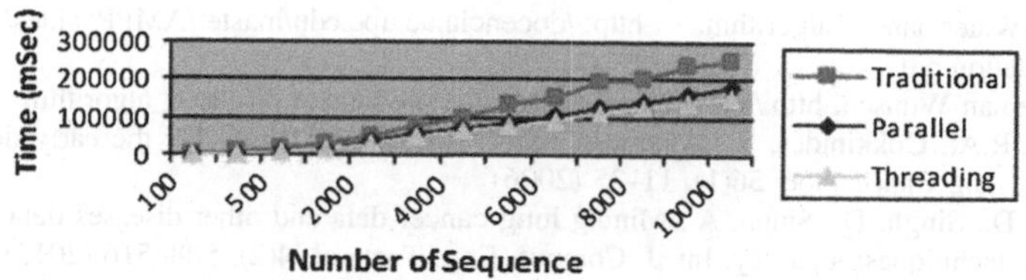

Fig. 3 Time required in milliseconds using Needleman Wunsch (NW) scoring algorithm for traditional and parallel approach (using CUDA) and using threads

Figure 3 shows the time analysis for both parallel and traditional Needleman Wunsch.

The time analysis for score matrix calculation of 10,000 sequences shows a drastic difference in the execution times of the traditional Needleman Wunsch algorithm and parallel Needleman Wunsch algorithm. Similar results have been observed in case of Levenshtein distance algorithm.

5 Conclusions and Future Work

This paper discusses an efficient method to identify and detect cellular diseases and makes use of protein sequence data and information retrieval methods. The developed system uses modified Needleman Wunsch algorithm and Levenstein's edit distance algorithms for scoring purposes. Parallel versions of these algorithms have been implemented using multithreading and system cores which reduces the time of computation considerably.

Future focus is to integrate the system with protein expression data for better diagnosis of the presence or absence of the diseases. The knowledge base can be developed as a distributed database network to get an availability of a wide variety of mutations.

References

1. http://gca.cropsci.illinois.edu/kaffe/tools1daf.html
2. Mayeus, R.: Biomarkers: potential use and limitations. NeuroRx*: J. Am. Soc. Exp. Neuro Therapeutics **2**(1), 182–188 (2004)
3. McClean, P.: Blast: basic local alignment search tool. http://www.ndsu.edu/pubweb/~mcclean/plsc411/Blast-explanation-lecture-and-overhead.pdf (2004)
4. Desomnd, H.G., Paul, S.M.: CLUSTAL: a package for performing multiple sequence alignment on microcomputer. Gene **73**(1), 237–244 (1990)
5. Thompson, J.D., Higgins, D.G., Gibson, T.J.: CLUSTAL W: improving the sensitivity of progressive multiple sequence alignment through sequence weighting, position-specific gap penalties and weight matrix choice. Nucleic Acid Res. **22** (22), 4673–4680 (1994)

6. Smith-Waterman algorithm. http://docencia.ac.upc.edu/master/AMPP/slides/ampp_sw_presentation.pdf
7. Needleman Wunsch. http://en.wikipedia.org/wiki/Needleman_Wunsch_algorithm
8. Smith, R.A., Cokkinides, V.: American cancer society guidelines for the early detection of cancer. Am. Cancer Soc. **56**(1), 11–25 (2006)
9. Parag, D., Singh, D., Singh, A.: Mining lung cancer data and other diseases data using data mining techniques: a survey. Int. J. Comput. Eng. Technol. **4**(2), 508–516 (2013)
10. Acharya, U.R., Sankaranarayanan, M., Nayak, J., Xiang, C., Tamura T.: Automatic identification of cardiac heath using modeling: a comparative study, Elsevier Inf. Sci. **178**(23), 4571–4582 (2008)
11. Human genome project. http://www.genome.gov/12011238
12. Saha, R., Killian, S., Donofrio, R.S.: DNA vaccines: a mini review. Recent Pat. DNA Gene Seq. **5**(2) (2011)
13. Majtán, T., Bukovska, G., Timko, J.: DNA microarray-techniques and applications in microbial systems. Folia Microbiol. **49**(6), 635–664 (2004)
14. Franzen, C., Müller, A.: Molecular techniques for detection, species differentiation, and phylogenetic analysis of microsporidia. Clin. Microbiol. Rev. **12**(2), 243 (1999)
15. Shouman, M., Turner, T., Stocker, R.: Using data mining techniques in heart disease diagnosis and treatment. In: Proceedings of International Conference on Electronics, Communications and Computers, vol. 6, no. 9, pp. 173–177. IEEE, Alexandria, March 2012
16. Alizadehsani, R., Habibi, J.: A data mining approach for diagnosis of coronary artery disease. Comput. Methods Progr. Biomed. **111**(1), 52–61 (2013)
17. Kaur, H., Wasan, S.K.: Empirical study on applications of data mining techniques in healthcare. J. Comput. **2**, 194–200 (2006)
18. Dragusin, R., Petcu, P.: Rare disease diagnosis as an information retrieval task. In: Proceeding in International Conference of Theoretical Information Retrieval, pp. 356–359. Springer, Berlin (2011)
19. Maity, A., Sivakumar, P., Rajasekhara Babu, M., Pradeep Reddy, Ch.: Performance evolution of heart sound information retrieval system in multi-core environment. IJCSIT **3**(3), 4404–4407 (2012)

Advance Teaching–Learning Based Optimization for Global Function Optimization

Anand Verma, Shikha Agrawal, Jitendra Agrawal and Sanjeev Sharma

Abstract Teaching–Learning based optimization (TLBO) is an evolutionary powerful algorithm in optimal solutions search space that is inspired from teaching learning phenomenon of a classroom. It is a novel population based algorithm with faster convergence speed and without any algorithm specific parameters. The present work proposes an improved version of TLBO called the Advance Teaching–Learning Based Optimization (ATLBO). In this algorithm introduced a new weight parameter for more accuracy and faster convergence rate. The effectiveness of the method is compare against original TLBO on many benchmark problems with different characteristics and shows the improvement in performance of ATLBO over traditional TLBO.

Keywords Global function optimization · Teaching-learning based optimization (TLBO) · Population based algorithms · Convergence speed

1 Introduction

To solve complex optimization problems such as mathematical optimization, multimodality, dimensionality and numeric optimization problems, several modern heuristic algorithms have been introduced. In a recent time a new optimization

---

A. Verma (✉) · J. Agrawal · S. Sharma
School of Information Technology, Rajiv Gandhi Proudyogiki Vishwavidyalaya, Bhopal, India
e-mail: Anandverma1291@gmail.com

J. Agrawal
e-mail: jitendra@rgtu.net

S. Sharma
e-mail: sanjeev@rgtu.net

S. Agrawal
University Institute of Technology, Rajiv Gandhi Proudyogiki Vishwavidyalaya, Bhopal, India
e-mail: shikha@rgtu.net

A. Nagar et al. (eds.), *Proceedings of 3rd International Conference on Advanced Computing, Networking and Informatics*, Smart Innovation, Systems and Technologies 43, DOI 10.1007/978-81-322-2538-6_59

technique Teaching–Learning Based Optimization (TLBO) gaining popularity for their reliability, accuracy and faster convergence rate. It has not required any algorithm-specific parameters so TLBO can also be called as an algorithm-specific parameter-less algorithm [1, 2]. TLBO is based on the effect of the influence of a teacher on the outcome of learners in a classroom and output is considered in terms of grades.

In this proposed work, includes a weight parameter 'W' in both phases and team leader concept with original TLBO that is known as Advance Teaching–Learning Based Optimization (ATLBO). The inclusive parameter gives a better convergence rate and accuracy. ATLBO compared with original TLBO for solving global function optimization problems and the performance characteristic are provided to show that ATLBO has better results than basic TLBO.

Rest of this paper is organized as follows. Section 2 presents the concept of basic TLBO. Section 3 describes the proposed Advance TLBO. Section 4 discusses results and comparison with basic TLBO. Section 5 gives conclusion. The detailed explanation of TLBO is given in next section.

2 Teaching Learning Based Optimization (TLBO)

Teaching–Learning Based Optimization (TLBO) is a newly introduced an efficient optimization algorithm, inspired by the teaching—learning process in the classroom. It is a population-based evolutionary computer algorithm that modeled on transferring knowledge in the classroom and use student result to proceed on global solution. TLBO does not need any specific parameters, it only requires common controlling parameters like population size and number of generations, so it is called parameter less optimization algorithm [3]. TLBO is divided into two phases: 'Teacher Phase' and 'Learner Phase'. In Teacher phase learners gain knowledge from teacher and then Learner also gain knowledge from their classmates by mutual interaction, group-discussion etc. [4–6] in 'Learning phase'.

2.1 *Initialization*

Following notation is used to describing TLBO

MAXIT: maximum number of allowable iterations.
N: number of learners in a class i.e. "class size";
D: number of courses offered to the learners;

The X is randomly initialized population by a search space bounded by matrix of N rows and D columns.

$$X_{new,i} = X_{min,i} + ri\left(X_{maxi} - X_{min,j}\right) \quad (1)$$

where ri is a uniformly distributed random variable within the range (0, 1), $X_{max,i}$ and $X_{min,j}$ represent the minimum and maximum parameter value [3, 7].

2.2 Teacher Phase

In human society teachers are seen as best learner and have more knowledge than learners. Teachers always tries to improve the knowledge of learners by which knowledge level of classroom in turn increase and help to get better grades and marks. But a teacher only improves the knowledge mean among students according to their ability and effectiveness. Merely it is unacceptable for a teacher that it can improve the knowledge mean of classroom towards specific degree [3, 8].

Let a teacher Ti is try to improve classroom mean Mi upward own knowledge level. Ti is denoted by "M_{new}" as new mean. The result is changed by the inequality between current [7, 9, 10] and new mean expressed by

$$\text{Difference\_Mean}_i = ri(M_{new} - T_f M_i) \quad (2)$$

where ri denotes for a random number in the range of [0, 1] and Tf is denominated as a teaching factor which determines the students mean value that is changed either select 1 or 2 with equivalent probability [9]

$$T_f = \text{round}[1 + \text{rand}(0, 1)\{2 - 1\}] \quad (3)$$

where termed rand denotes a random values between 0 and 1. This difference updates the current solution by following expression.

$$X_{new,i} = X_{old,i} + \text{Difference\_Mean}_i \quad (4)$$

2.3 Learner Phase

In Learner phase, Learners try to increases their knowledge from their classmates in the form of group discussion, mutual interaction and tutorial etc. A learner randomly interacts with each other's in the form of presentation, formal communication and group discussion, etc. If any learner has better knowledge than others, learners improve their knowledge form whom. For a population size Pn, learner improves their knowledge from following algorithm [7, 9, 10].

For i = 1 :Pn
Xi and Xj are two randomly selected learners
Where i<> j
if f (Xi) < f (Xj)
$X_{new,i}= X_{old,i}+$ ri (Xi – Xj)..5
Else
$X_{new,i}= X_{old,i}+$ ri(Xj – Xi)..6
End if
End for
Accept Xnew if gives a better function value.

2.4 Algorithm Termination

The algorithm is terminated when maximum iteration MAXIT is reached.

3 Proposed Advance Teaching–Learning Based Optimization

The proposed ATLBO is the improved version of traditional TLBO algorithm. In basic TLBO learn improves their knowledge by a single teacher or by interaction of other learners. In classroom environment some student gains knowledge during tutorial hours, group discussion or by self-motivation learning. TLBO algorithm tries to shift the mean of learners towards the teacher of class. In ATLBO algorithm a new weight factor W are included in teacher phase (in Eq. 1) of basic TLBO and in the part of learning phase (in Eqs. 5 and 6) includes a team leader concept that provides the global position of best learner. In learning phase of ATLBO previous best learner are act as a team leader, team leader are change during every iteration until result is not reached on global solution. Weight factor is linearly decreases with time.

$$\text{Weight factor}(w) = \text{Wmax} - \Delta w \tag{7}$$

$$\Delta\omega = [(\text{Wmax} - \text{Wmin}) * (\text{Tmax} - t)/\text{Tmax})] \tag{8}$$

where W_{max} and W_{min} are the maximum and minimum values of weight factor w, iteration is the current iteration number and max iteration is the maximum number of allowable iterations. W_{max} and W_{min} are selected to be 0.9 and 0.1, respectively.

So the improved teacher phase express following

$$X_{new,i} = w * X_{old,i} + ri(M_{new} - T_f M_i) \tag{9}$$

In contrast of Learning phase of basic TLBO, ATLBO improves its learning phase by concept of best learner that act as a team-leader, In every iteration previously best learner treat as a team-leader. A team-leader has always has the more knowledge than their classmate [9, 10].

Hence in the learning phase the new set of improved learns can be

For i = 1 : Pn
Xi and Xj are two randomly selected learners
Where I <> j
if f (Xi) < f (Xj)
$X_{new,i}$=W*$X_{old,i}$+ ri (Xi – Xj)+ri(Xbest-Xi)..............................10
Else
$X_{new,i}$= W*$X_{old,i}$+ ri(Xj – Xi)+ri(Xbest-Xi)..............................11
End if
End for

where r_i denotes for a random number in the range of 0–1.

Xbest is best learner act as team leader.

Accept X_{new} if gives a better function value.

4 Experimental Results

In this section, we have an exhaustive comparison of our proposed algorithm (ATLBO) with basic TLBO. For all the experiments population size was set to 20 and the maximum number fitness function evaluation was set to 100. Table 1 shown

Table 1 List of benchmark functions have been used in experiment

No	Function	n	C	Range	Formulation value (f_{min}) = 0
f1	Step	30	US	[−100, 100]	$f(x) = \sum_{i=1}^{n} (x_i + 0.5)^2$
f2	Sphere	30	US	[−100,100]	$f(x) = \sum_{i=1}^{n} (x_i)^2$
f3	Sum Squares	30	US	[−100, 100]	$f(x) = \sum_{i=1}^{n} (ix_i)^2$
f4	Schwefel 2.22	30	UN	[−10, 10]	$f(x) = \sum_{i=1}^{n} \lvert x_i \rvert + \Pi_{i=1}^{n} \lvert x_i \rvert$
f5	Bohachevsky1	2	MS	[−100,100]	$f(x) = X_1^2 + X_2^2 - 3\cos(3\Pi x_1) - 0.4\cos(4\Pi x_2) + 0.7$
f6	Bohachevsky2	2	MS	[−100, 100]	$f(x) = x_1^2 + x_2^2 - 3\cos(3\Pi x_1) \times \cos(4\Pi x_2) + 0.7$
f7	Bohachevsky3	2	MS	[−100, 100]	$f(x) = x_1 + x_2 - 3\cos(3\Pi x_1 + 4\Pi x_2) + 0.7$
f8	Rastrigin	30	MS	[−5.12, 5.12]	$f(x) = [x_1^2 - 10\cos(2\Pi x_i) + 10]$
f9	Griewank	30	MN	[−600, 600]	$f(x) = \frac{1}{4000} \sum_{i=1}^{n} (x_i)^2 - \Pi_i^n = 1\cos(xi/\sqrt{i}) + 1$

Table 2 Fitness evaluation of TLBO and ATLBO (mean and standard deviation)

No	Function	Mean (TLBO)	Sd (TLBO)	Mean (TLBO)	Sd (TLBO)
f1	Step	6.99E+01	64.99796	5.599362	13.34852
f2	Sphere	142.0809	174.0369	4.10E+00	7.557908
f3	Sum squares	591.9211	799.6606	4.53E+00	5.717949
f4	Schwefel 2.22	4.464452	3.663719	1.72E−01	0.224143
f5	Bohachevsky1	158.2136	155.8383	2.79E+00	3.477812
f6	Bohachevsky2	230.8089	254.7465	3.20E+00	4.672983
f7	Bohachevsky3	116.5454	178.0865	−2.4217	26.36527
f8	Rastrigin	5.06E+00	2.188272	6.32E−01	0.490737
f9	Griewank	0.023156	0.026999	1.48E−02	0.013463

below gives the description of all the nine benchmark functions that was used for experimentation.

4.1 Performance Metrics

To check the performance of proposed algorithm (ATLBO), Mean and standard deviation of different benchmarks is calculated for determining the quality of solution and the convergence rate is plotted and the resultant performance is compared to basic TLBO.

a. **Quality of Solution**: To determine the quality of solution, mean and standard deviation of 30 independent runs on nine different benchmark functions is calculated and recorded in Table 2. In all the cases ATLBO performs better than TLBO. To further verify the result, a two tailed t-test is performed and results are shown in Table 3. The result of t-test proves that the quality of ATLBO's solution is better than TLBO solution quality.
b. **Convergence Rate of Proposed Algorithm**: To determine the convergence rate, Both algorithms (ATLBO and TLBO) are tested on 10 independent sets with 100 iteration on 10 independent run over different benchmarks and draw graphs for comparing the convergence rate follows (Fig. 1):

Table 3 t-Test Result of benchmark functions

Function	f1	f2	f3	f4	f5	f6	f7	f8	f9
TLBO/ATLBO	Sign	Sign	Sign	Sign	Ex. Sign	Sign	Sign	Ex. Sign	Sign

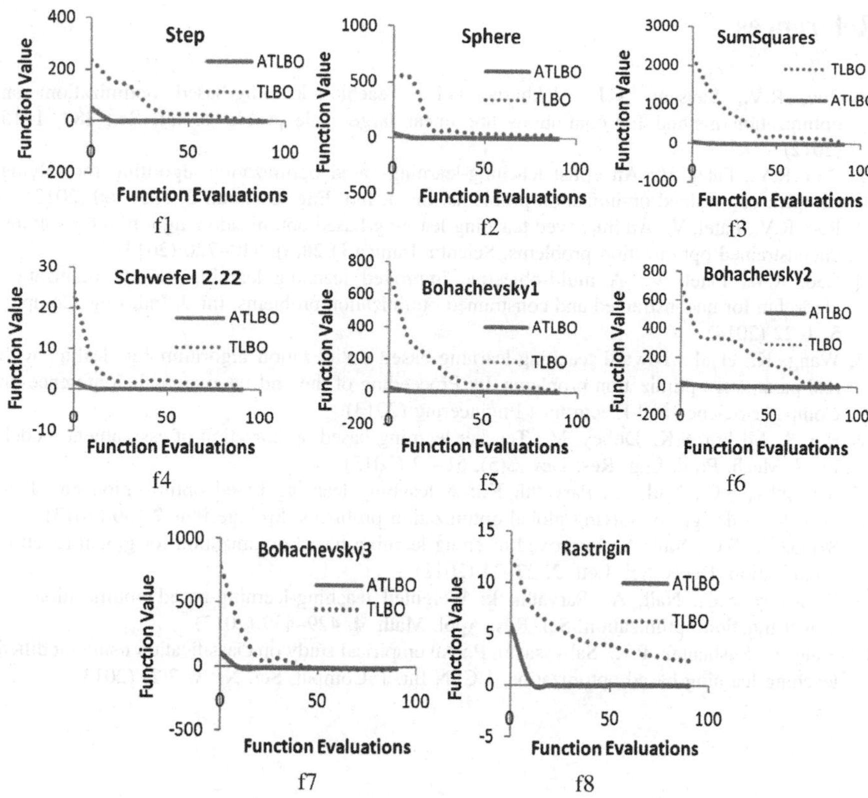

Fig. 1 Convergence curve for different benchmarks using ATLBO against TLBO

5 Conclusion and Further Research

A new algorithm Advance Teaching–Learning Based Optimization (ATLBO) is suggested. It is an improved version of basic Teacher Learning Based Optimization (TLBO). In ATLBO algorithm a new weight factor is introduced in both teaching and learning phase also with a new concept of team leader. Team leader which was the best learner in the previous phase will act as a teacher in learning phase. The ATLBO algorithm provides better results compared to basic TLBO algorithm on several benchmark functions. The proposed ATLBO algorithm is able to find the global optimum values more accurately with faster convergence rate. The experiment results have shown satisfactory performance of the ATLBO algorithm. In future performance of ATLBO may be enhanced with new parameter tuning strategy and adapts for multi-objective optimization and also its the practical application areas like clustering, data mining, design and optimization of communication, networks would be worth studying.

References

1. Rao, R.V., Savsani, V.J., Vakharia, D.P.: Teaching–learning-based optimization: an optimization method for continuous non-linear large scale problems. Inf. Sci. **183**, 1–15 (2012)
2. Rao, R.V., Patel, V.: An elitist teaching-learning-based optimization algorithm for solving complex constrained optimization problems. Int. J. Ind. Eng. Comput. **3**, 535–560 (2012)
3. Rao, R.V., Patel, V.: An improved teaching-learning-based optimization algorithm for solving unconstrained optimization problems. Scientia Iranica D **20**(3), 710–720 (2013)
4. Rao, R.V., Patel, V.: A multi-objective improved teaching–learning based optimization algorithm for unconstrained and constrained optimization problems. Int. J. Ind. Eng. Comput. **5**, 1–22 (2014)
5. Wang, K., et al.: Toward teaching-learning-based optimization algorithm for dealing with real-parameter optimization problems. In: Proceeding of the 2nd International Conference on Computer Science and Electronics Engineering (2013)
6. Rai, S., Mishra, S.K., Dubey, M.: Teacher learning based optimization of assignment model. Int. J. Mech. Prod. Eng. Res. Dev. **3**(5), 61–72 (2013)
7. Satapathy, S.C., Nail, A., Parvathi, K.,: A teaching learning based optimization based on orthogonal design for solving global optimization problems. SpringerPlus **2**,130 (2013)
8. Satapathy, S.C., Nail, A.: Improved teaching learning based optimization for global function optimization. Decis. Sci. Lett. **2**, 23–24 (2012)
9. Satapathy, S.C., Nail, A., Parvathi, k: Weighted teaching-learning-based optimization for global function optimization. Sci. Res. Appl. Math. **4**, 429–439 (2013)
10. Sahu, A., Sushantak, P.K., Sabyasachi, P.: An empirical study on classification using modified teaching learning based optimization. IJCSN Int. J. Comput. Sci. Netw. **2**(2) (2013)

Analysis of Online Product Purchase and Predicting Items for Co-purchase

Sohom Ghosh, Angan Mitra, Partha Basuchowdhuri and Sanjoy Kumar Saha

Abstract In recent years, online market places have become popular among the buyers. During this course of time, not only have they sustained the business model but also generated large amount of profit, turning it into a lucrative business model. In this paper, we take a look at a temporal dataset from one of the most successful online businesses to analyze the nature of the buying patterns of the users. Arguably, the most important purchase characteristic of such networks is follow-up purchase by a buyer, otherwise known as a co-purchase. In this paper, we also analyze the co-purchase patterns to build a knowledge-base to recommend potential co-purchase items for every item.

Keywords Viral marketing · Dynamic networks · Social networks · Recommendation system · Amazon co-purchase networks

1 Motivation and Related Works

Online market places are becoming largely popular with knowledge of internet among customers. Web-based marketplaces have been active in pursuing such customers with high success rate. For example, a large portion of the Black Friday

S. Ghosh · P. Basuchowdhuri (✉)
Department of Computer Science and Engineering, Heritage Institute of Technology, Chowbaga Road, Anandapur, Kolkata 700107, WB, India
e-mail: parthabasu.chowdhuri@heritageit.edu

S. Ghosh
e-mail: sohom1ghosh@gmail.com

A. Mitra · S.K. Saha
Department of Computer Science and Engineering, Jadavpur University, 188 Raja S. C. Mullik Road, Jadavpur, Kolkata 700032, WB, India
e-mail: anganmitra@outlook.com

S.K. Saha
e-mail: sks_ju@yahoo.co.in

A. Nagar et al. (eds.), *Proceedings of 3rd International Conference on Advanced Computing, Networking and Informatics*, Smart Innovation, Systems and Technologies 43, DOI 10.1007/978-81-322-2538-6_60

sales in USA or the Boxing Day sales in UK have gone online so that instead of waiting outside of the stores in cold for hours, the customers can sit in their home comfortably and still avail the deals by browsing through their websites. The online shopping stores have gone to great lengths to invest on high configuration servers that could seamlessly handle the requests thousands of buyers at the same time.

Such alarming growth in business in online market-places has drawn interest towards analyzing the customers' browsing or buying history. The customers' browsing or buying data is stored in the database and such data could be analyzed to observe patterns to understand the customers better. One such feature, popularly used by online market-places to recommend users for a potential sale, is the item co-purchase pattern of the customers. Customers often buy multiple products together, which may or may not be related. The frequency of a co-purchase may indicate the probability of the same co-purchase in future. The co-purchase patterns changes with time depending on many temporal factors. Understandably, this problem is a temporal problem and data from different time instances are needed to understand the dynamic behaviour of the customers.

In a study on Amazon co-purchase network Leskovec et al. [1], features of person-to-person recommendation in viral marketing were disclosed. For convincing a person to buy something, such recommendations were not very helpful. But they showed partitioning data based on certain characteristics enhances the viral marketing strategy. E-commerce demand has been explained in another paper using Amazon co-purchase network. Their claim is that item categories with flatter demand distribution is influenced more by the structure of the network [2]. A community detection method was suggested by Clauset et al. which took O ($md \log n$) time where, n, m are the number of nodes and edges respectively and d is the number of hierarchical divisions needed to reach the maximum modularity value. Overall goodness of detected communities is popularly measured by Modularity [3]. This method is popularly known as CNM [4] and they have used Amazon co-purchase network as a benchmark data to find the communities in the network. The communities detected by their algorithm has maximum modularity 0.745. But, these communities were so large that finding out patterns from them was not of much use. For instance, the largest community having about 100,000 nodes consisted of about one-fourth of the entire network. Luo et al. made a study of the local communities in Amazon co-purchase network and claimed that recommendation yields better results for digital media items than books [5]. 3 and 4 node motifs had been analysed in Amazon co purchase network [6]. It was found that frequent motifs did not contribute much in comprehending the behaviour of the network. Recent works on detection of frequent sub-graphs [7–10] has helped us in interpreting the dynamics of temporal network. J. Han et al. devised a method popularly known as FP-growth to find frequently occurring patterns in transaction databases [11]. C. Bron and J. Kerbosch formulated an algorithm to find out maximal cliques in an network [12]. Basuchowdhuri et al. studied the dynamics of communities in amazon copurchase network [13]. They have analysed the evolving product purchase patterns.

2 Problem Definition

Given a directed graph $G(V, E)$, where nodes are products of an e-commerce transaction database and directed edges represent the co-purchase relation, we analyze market dynamics, relation between some of the inherent properties of the graph, temporal human predilection, variation in number of reviews and build a recommendation system that would help in bolstering the revenue. The frequency of co-purchase although important, was not available in the dataset. Hence, we ignore the frequency of co-purchased items and use topological characteristics to build a recommender system.

3 Examining Features of the Co-purchase Network

We have examined the Amazon co-purchase network data in this paper. Unlike other social network datasets, temporal characteristics are present in this dataset, i.e., it comprises a network and its snapshots at four timestamps. The snapshots reveals the dynamism in the structure of the original network and enables us to study the evolving follow-up purchasing patterns. For the ease of comprehending, we represent this dynamic network as a set of time-stamp graphs G_0, G_1, G_2, G_3, where G_0 is the original network at t = 0 and the evolved versions of G_0, after one, two and three units of time are G_1, G_2 and G_3, respectively. We them tabulate them as follows in Table 1. In this table, LWCC and LSCC refers to largest weakly connected component and largest strongly connected component respectively. Considering all time stamps the number of bi-directed edges are 5,853,404 while that of distinct edges are 10,847,450. We subsequently look into a few features of the network. Firstly, we look into its extended reachability by looking at the nodes' 2-hop degree. The 2-hop degree distribution expresses the extent to which viral marketing would be able to boost up sales in the network. Next, we review two features of the network that gives us a knowledge of how good transitivity of co-purchase is maintained in the network. The last two features observed, express the distribution of review writing by the customers. We relate the review writing with buying and thereby project the buying frequency and its distribution.

Table 1 Snapshot graphs in amazon co-purchasing network

Graph	|V|	|E|	Size of LWCC	Size of LSCC	Global clustering co-efficient	Month, Year
G_0	262,111	1,234,877	262,111	241,761	0.4198	March 02, 2003
G_1	400,727	3,200,440	400,727	380,167	0.4022	March 12, 2003
G_2	410,236	3,356,824	410,236	3,255,816	0.4064	May 05, 2003
G_3	403,394	3,387,388	403,364	395,234	0.4177	June 01, 2003

3.1 2-Hop Degree Distribution of Items

Degree distribution of a typical social network is known to display a plot of scale-free nature. But, 2-hop degree of a node represents its ability to extend its reach beyond its immediate neighbors. Its significance lies in the realization of influence that a node exerts on its neighbors' neighbors. Such measures in sparse networks can be useful as average reachability between nodes is high. In sparse networks, the nodes with high 2-hop degree value are the central nodes, from which all the nodes can be reached quickly. Figure 1 shows the 2-hop degree distribution of Amazon co-purchase network in log-log scale. The distribution plot follows power law with a heavy tail.

3.2 Clustering Coefficient Distribution

Local clustering coefficient gives an essence of how good or bad a node can pulls others into formation of a dense network, a quasi-clique for example. Figure 2 shows the distribution of clustering coefficient in log-log scale. Nodes with high clustering coefficient (CC) has appeared more number of times than those with lower CC. This reveals the cliquishness of the network.

3.3 Triplet and Triangle Distributions

A triplet is a structure where a node spawns two children. A triplet provides with an opportunity of a triad closure which is the minimum unit of friendship. For each

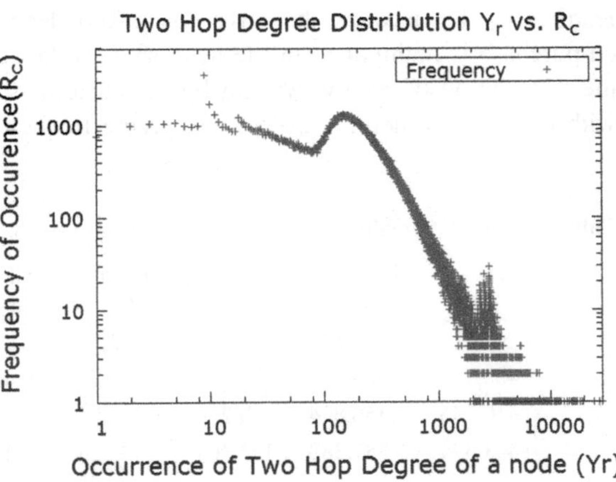

Fig. 1 2-hop degree distribution of amazon co-purchase network in log-log scale

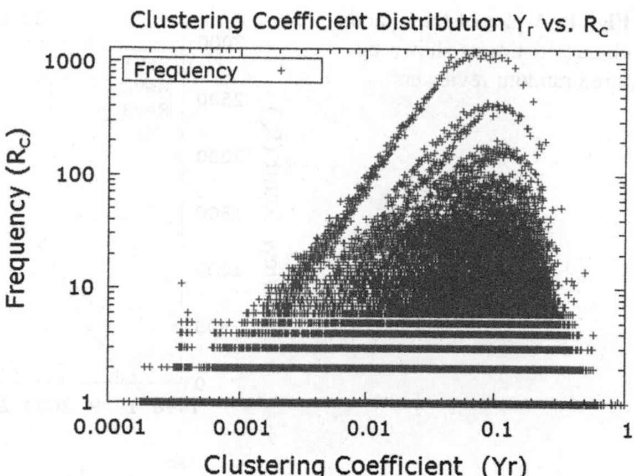

Fig. 2 Clustering coefficient distribution of amazon co-purchase network in log-log scale

node we have found out how many possible triplets exists in which it takes part. On the other hand, triangles depict closure of such triplets and is essentially the smallest structure displaying sense of community. Figure 3 shows triplet and triangle distribution in Amazon co-purchase network. It is a plot between occurrence of triplets and triangles in which a node is present and its frequency in log-log scale. Algorithm [1] states the way of discovering triangles from triplets.

3.4 Detection of Burst Mode

Burst mode is usually observed in human behavior and therefore is a characteristic of social networks. It shows a human predilection of being increasingly engrossed in a particular activity before loosing interest and settling down. A few plots below

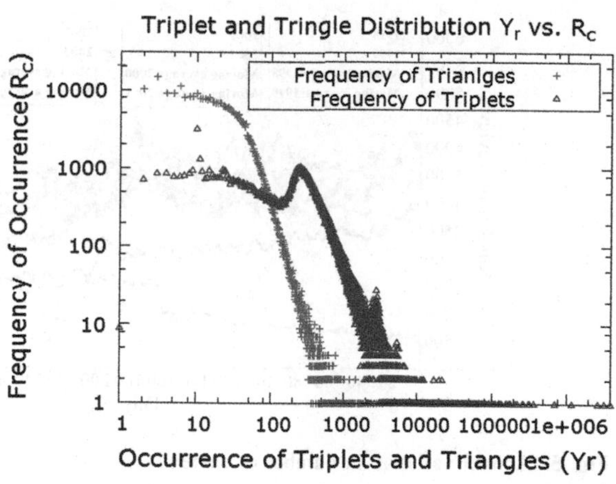

Fig. 3 Triangle and triplet distribution of amazon co-purchase network in log-log scale

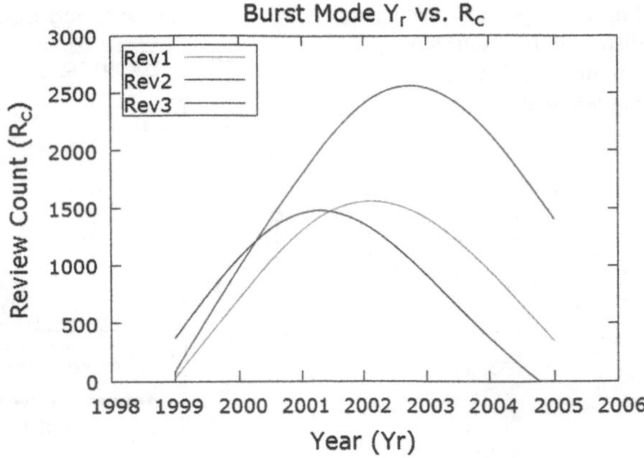

Fig. 4 A plot of burst mode characteristics as shown by three random reviewers

show how users started to review items with increasing vigor and then with a sharp descent coming down to a state of stability. Figure 4 shows the burst mode characteristics shown by three reviewers.

3.5 Annual Variation in Number of Reviews

During a year there are peak times when the sales are maximum. In the Amazon co-purchase network we have plotted the variation in number of reviews for 7 years and some common features came up. The graph begins with a peak because of the continued sales during the festive season. The graph ends with a peak which around the month of December. The number of reviews are predictably high due to the festive mood of Christmas and New Year. Another commonly occurring peak is near April which is time for Good Friday and Easter. Figure 5 gives us a glimpse of how the variation in number of reviews takes place over the years.

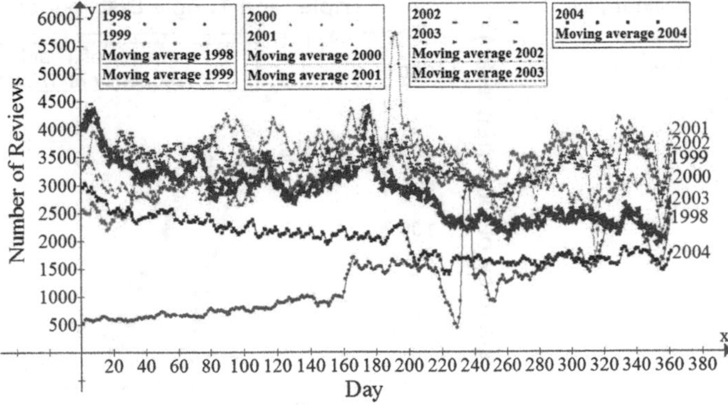

Fig. 5 Annual variation in number of reviews

Algorithm 1 Algorithm to Find all Possible Triangles from Triplets in a Graph

```
K contains the list of adjacency list of nodes
R contains the list of all possible triangles.Initialized to NULL
for all t in K do
  t is a node whose adjacency list is contained in K
  for all x in adjacency list of t  do
    P is initialized to a null set
    P= Set of Nodes such that P_i belongs to Adjacency list of x − t
    N= P ∩ Adjacency List of t
    for all triplet in N do
      if Transform(triplet) not in R then
        R=R+ Transform(triplet)
      end if
    end for
  end for
end for
Transform(triplet) {
return (x_1,x_2,x_3) such that x_1 < x_2 < x_3 }
```

4 Recommendation System

One of the ways to boost sales is providing ease to customers for finding those products which match their tastes and choices. A certain percentage of this comfort level is achieved by a good recommender system. We design a recommender system which partly uses content based filtering and collaborative filtering.

4.1 Assigning Weights to Edges

The Amazon co-purchase network graph is essentially unweighted. An attempt has been made to make it a weighted network such that meaningless data can be eliminated before going into further stages of recommendation. This also forms the first stage of recommendation as the edges with significant weights can now be recommended with each other.

$$\text{Weight}\,(a,b) = |A \cap B|$$

A Set of Reviewer ID's who reviewed Item *a*

B Set of Reviewer ID's who reviewed Item *b*

Figure 6 gives a distribution of hashed edge and its corresponding weight in log log scale. Here, the edges are hashed by assigning consecutive numbers after sorting them in decreasing order of their corresponding weights i.e. edge (156,632–156,634) having the highest weight (842) is hashed as 1, followed by edge

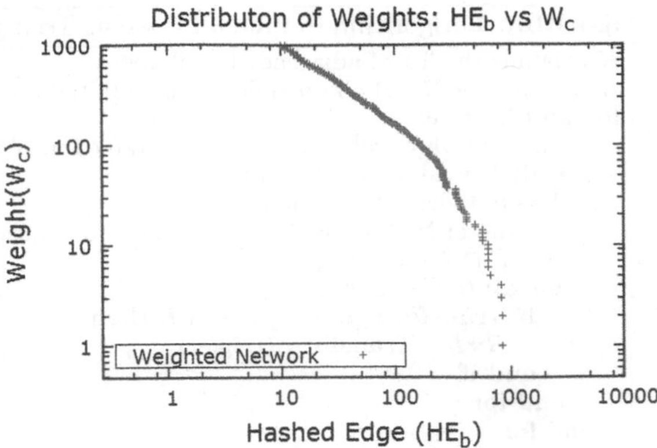

Fig. 6 A plot of hashed weights

(197,325–197,327) with weight (826) and so on. From the graph we have located the region where the graph attains steepest slope and thus the top 2000 edges are taken for recommendation. Using this method, we find that pairs like 'A Christmas Carol' and 'Jingle All the Way', 'A Bend in the Road' and 'The Smoke Jumper' and many other other items in pairs are recommended. The content wise similarity of items in each pair is worth noticing.

4.2 Detecting Cliques in Five Nearest Neighbour Network (5NN)

The Amazon Meta data file contains information not only about the products but also gives a set of 5 similar products for each product. This similarity is based on sub category. We form a 5-NN network from this data. The largest connected component (LCC) is found out and Bron-Kerbosch algorithm [12] is implemented on this LCC to find maximal cliques. It comes up with 3–7 membered cliques. The 5-NN network is made out of similar items and thus the cliques formed share the similarity essence among themselves and thus can be recommended. The number of 3, 4, 5, 6 and 7 membered cliques are 79833, 32211, 7878, 872 and 4 respectively. Thus, we can see with increase in clique size, its frequency decreases.

Products like "Ballroom Dancing", "Much Ado About Ballroom Dancing", "The Complete Idiot's Guide to Ballroom Dancing" are recommended together. Thus, we see that similar products are suggested for co-purchase.

4.3 Collaborative Filtering

In collaborative filtering, predictions about a person's tastes are made from his network of collaboration. The underlying assumption of this approach is that, if a

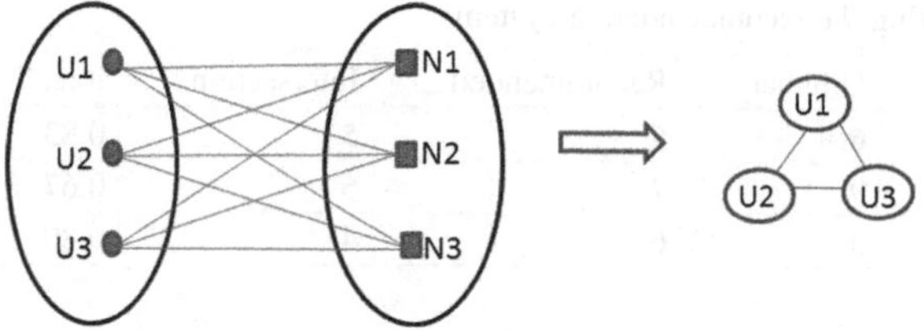

Fig. 7 Edge reduction

person A has the same opinion as that of person B on an issue, then A is more likely to follow B's opinion on a different issue than to follow the opinion of a person chosen randomly. On the contrary, it is very unlikely that two reviewers will have equal taste in every aspect. So, for an item, instead of recommending A's collection for B, it is better to find a neighbour of B with whom it has maximum similarity regarding the sub category of that item. Figure 7 shows a sample edge reduction.

Lemma 1 *For reducing number of edges of a complete bipartite graph* (n_1, n_2, E) *by converting to a one mode network the condition that must be satisfied is* $n_1 - 1 < 2\,n_2$ *or,* $n_2 - 1 < 2\,n_1$

Proof The total number of edges in the bipartite graph (n_1, n_2, E) is $n_1\ n_2$. For reduction the number of edges in the new network should be less than that of the bipartite network. So,

$^{n_1}C_k < n_1\ n_2$ or, $^{n_2}C_k < n_1\ n_2$

On solving the inequalities, the following result is obtained:

$n_1 - 1 < 2n_2$ or, $n_2 - 1 < 2n_1$ □

The bipartite graph consists of two sets one of which is the set of items and the other, the set of reviewers. Here, we have tried to project this two mode network into a one mode network by graphically linking reviewer to reviewer if there exists an item that has been reviewed by both. While transforming the edges are not given direct weights rather a set of data, generic form of which is:

$W(a, b) = \{x, y \mid x$ belongs to common sub categories between a, b and y is the frequency of common occurrence of $x\}$.

Here, Table 2 shows a sample edge. It is clear that these reviewers have a greater match of taste when it comes to the genres *Style*, *Directors* and *Series* in Music, Video and Book respectively.

Table 2 Description of an edge between two reviewers (a, b)

Common subcategories reviewed (x)		Frequency of occurrence (y)
Category	Sub category	
Music	Styles	4153
Video	Directors	1555
Music	Rap	2
Book	Guides and reviews	1

Table 3 Validating the recommendation system

Node number	Original	Recommended	Intersection	Recall	Precision
372,787	6	5	5	0.83	1
255,803	9	7	6	0.67	0.86
392,440	5	6	4	0.80	0.67

4.4 Efficiency of Recommendation System

A recommendation system is useless unless and until a strong evidence of it working fine is provided. Though the transaction table between reviewers and items were not available, different path was chosen to resolve this issue. For each and every time stamp largest connected component (LCC) was found out and a similarity study was made to see how much of our recommendations were present in these LCC.

A Machine Learning Approach Towards Validating: Machine Learning Approach brings a flavor of past experiences while taking current decisions. We have tried to implement the same essence. We have taken two time stamps namely t_1 and t_2 for learning a model of co-purchase predilections and have tested it on the time stamp t_3. The data set provided the edge list of all co-purchase relations between items for time stamp t_3.

Precision and Recall: For a recommender system, precision and recall are the two parameters which give a measure of the accuracy of the system. Precision (i.e. positive predictive value) is the fraction of retrieved instances that are relevant, while recall (also known as sensitivity) is the fraction of relevant instances that are retrieved. Both precision and recall are therefore based on an understanding and measure of relevance shown in Table 3. Here, original refers to the number of items that has been co-purchased with the particular node. Recommended refers to the number of items that our recommendation system has provided for that node. Intersection refers to the number of items common between Original and Recommended.

"The Music of Jerome Kern" (372,787) has high precision and recall value. "Southern Harmony and Musical Companion" (392,440) has high recall value.

5 Conclusion

The study of co-purchase network reveals how human tendencies can shape up co-purchase patterns which bears with it temporal effects. These motifs, if studied carefully can be used to develop strategies to increase sales. We made an attempt to make recommendation systems based on nearest neighbor model and graph topology based collaborative filtering. Since testing recommendations for all the items in the network is computationally expensive, we picked up random samples to reveal how the recommendation system performs for them. We can clearly see that the recommendation can be good for a large part of the dataset but can not guarantee a highly precise recommendation for every item.

References

1. Leskovec, J., Adamic, L.A., Huberman, B.A.: â€œThe dynamics of viral marketing. ACM Trans. Web **1**(1), (2007)
2. Oestreicher-Singer, G., Sundararajan, A.: Linking network structure to e-commerce demand: theory and evidence from amazon.coms copurchase network. In: TPRC 2006. Available in SSRN (2006)
3. Newman, M.: Modularity and community structure in networks. Proc. Nat. Acad. Sci. **103** (23), 85778582 (2006)
4. Clauset, A., Newman, M.E.J., Moore, C.: Finding community structure in very large networks. Phys. Rev. E **70**, 066111 (2004)
5. Luo, F., Wang, J.Z., Promislow, E.: Exploring local community structures in large networks. Web Intell. Agent Syst. **6**(4), 387400 (2008)
6. Srivastava, A.: Motif analysis in the amazon product co-purchasing network. In: CoRR, vol. abs/1012.4050 (2010)
7. Bogdanov, P., Mongiov, M., Singh, A.K.: Mining heavy subgraphs in time-evolving networks. In: Cook, D.J., Pei, J., 0010, W.W., Zaane, O.R., Wu, X. (eds.) ICDM, pp. 8190. IEEE (2011)
8. Lahiri, M., Berger-Wolf, T.Y.: Structure prediction in temporal networks using frequent subgraphs. In: CIDM, p. 3542. IEEE (2007)
9. Wackersreuther, B., Wackersreuther, P., Oswald, A., Bohm, C., Borgwardt, K.M.: Frequent subgraph discovery in dynamic networks. In: Proceedings of the Eighth Workshop on Mining and Learning with Graphs, ser. MLG 10, pp. 155–162. ACM, New York (2010)
10. Lahiri, M., Berger-Wolf, T.Y.: Periodic subgraph mining in dynamic networks. Knowl. Inf. Syst. **24**(3), 467497 (2010)
11. Han, J., Pei, J., Yin, Y., Mao, R.: Mining frequent patterns without candidate generation: a frequent-pattern tree approach. Data Min. Knowl. Disc. **8**(1), 5387 (2004)
12. Bron, C., Kerbosch, J.: Algorithm 457: finding all cliques of an undirected graph. Commun. ACM **16**(9), 575–577 (1973)
13. Basuchowdhuri, P., Shekhawat, M.K., Saha, S.K.: Analysis of Product Purchase Patterns in a Co-purchase Network. EAIT (2014)

Fingerprint Template Protection Using Multiple Spiral Curves

Munaga V.N.K. Prasad, Jaipal Reddy Anugu and C.R. Rao

Abstract In this paper we proposed a method for generating the cancelable fingerprint template using spiral curves by constructing contiguous right angled triangles using the invariant distances between reference minutia and every other minutiae in fingerprint image, then projecting onto a 4D space, features transformed using DFT. The proposed approach experimented by using the FVC database. The approach attains the primary needs diversity, revocability, security of biometric system. Performance is calculated using metrics GAR, FAR, and EER.

Keywords Fingerprint template generation · Contiguous right angled triangles · Projection · Transformation · Security

1 Introduction

Biometrics gained its popularity in secure authentication in comparison with traditional based methods of remembering tokens and passwords [1]. Biometrics traits are inextricably bound to individual's identity [2] and need not to remember. However, biometrics will be same forever, once compromised can not be reissued or canceled [1], thus violating user privacy. If biometrics lost everything lost. Individuals face threat to their physical existence as they can not be kept away from theft. This has raised problems and challenges in security and regarding protection of one's identity. Hence, there should be a biometric technique that ensures security and privacy. Any

M.V.N.K. Prasad (✉) · J.R. Anugu
Institute for Development and Research in Banking Technology, Hyderabad, India
e-mail: mvnkprasad@idrbt.ac.in

J.R. Anugu
e-mail: jaipalreddy51@gmail.com

J.R. Anugu · C.R. Rao
School of Computer and Information Sciences, University of Hyderabad, Hyderabad, India
e-mail: crrcs@uohyd.ernet.in

A. Nagar et al. (eds.), *Proceedings of 3rd International Conference on Advanced Computing, Networking and Informatics*, Smart Innovation, Systems and Technologies 43, DOI 10.1007/978-81-322-2538-6_61

cancelable biometric method must meet the following characteristics [1] (i) Diversity, (ii) Revocability, (iii) Non-invertibility and (iv) Performance.

There are four categories of attacks [3] at sensor module, interface module, software module and database module. Irrevocability of template makes it more dangerous and violates user's privacy. Therefore, need of biometric technologies to provide increased security and privacy protection. The template protection schemes are broadly categorized into cancelable biometrics and biometric cryptosystem [3]. Cancelable biometrics is about to transform the features into irreversible template. The two approaches, namely biometric salting (blending user specific information like password or token with biometric data to generate a new template) and non-invertible transforms (transform original biometric data into irreversible template) are under cancelable biometrics. On other side, Biometric cryptosystem serves by securing cryptographic key with help of biometric data (key binding) or generating cryptographic key (key generation) [3] from biometric data.

The rest of paper is organized as follows: In Sect. 2, discussion regarding the related work done on cancelable biometrics. Section 3 explains about the proposed model. Section 4 explains the experimental setup and analysis of the model in terms of the performance characteristics of cancelable biometrics. Conclusions are discussed in Sect. 5.

2 Literature Review

Reconstruction of original fingerprint image from minutiae set was proved [4]. Later a many methods have come for reconstruction of fingerprint image from minutiae [5–7]. Because of feasibility to inversion, it is not secure to keep the original fingerprint features as biometric template. As an alternative layers of protection could be applied to original fingerprint to convert into new form. But, performance is generally degraded when transformations applied [8]. Preserving the performance while applying transformations being a challenging task.

In literature, there exists direct minutiae transformation and indirect minutiae transformation. In first case the original location and orientation are taken directly for further use; invariant features taken in later case. For feature transformation, [8] proposed first method to generate biometric template using non-invertible transform functions, namely Cartesian, polar and surface-folding transformation. Though the three transformations claimed to be non-invertible, later an approach [9] reveals the invertibility of surface folding transformation if parameters and transformed template are known.

In method [10], a 3D array taken, for each reference minutia other minutiae are translated and rotated. Each cell is marked as '1' if it contains more than one minutiae falling in each cell, otherwise '0'. For same key scenario the performance degraded significantly. A method alignment free fingerprint template [11], the calculated invariant feature set is given as input to user specific transformation function to derive parameters, which can be used to generate cancelable template. Generation of revocable fingerprint template using minutiae triplet [12]. Similarly, revocable fingerprint template generation [13] by taking four invariant features from minutia pair and then using histogram binning [14].

Pair polar coordinate based scheme [15] explores the relative relationship of minutiae in a rotation and shift-free pair-polar framework. Non-inversion is attained through many-to-one mapping relation. The method based on densely infinite-to-one mapping (DITOM) technique [16] elaborates the three features, then quantized, hashed, and binarized using histogram binning. Method based on curtailed circular convolution [17] where generation of random sequence using user specific key, L-point Discrete Fourier Transform (DFT)s of bit-string and random sequence independently, then taking inverse DFTs on the product of DFTs generated. Considered the points as template after removing first p-1 points.

The work of multiline code (MLC) [18] enhanced [19] by taking mean distance of minutiae falling in each semicircle region along the lines. Performed quantization to get bit-string, then permutation of resulted bit-string using user key. Multiline neighboring relation [20] by constructing rectangles with different orientations, followed by determining invariant distances and relative orientations, then projecting on to 2D plane, then transformation using DFT to get template.

3 Proposed Method

The steps involved in proposed method are as follows:

1. Construction of spiral curves.
2. Projection on to 4D space and bit-sting generation.
3. Feature transformation.
4. Matching.

3.1 Construction of Spiral Curve

Fingerprint feature extraction from impression. The minutiae set $N = [x_i, y_i, \theta_i]_1^k$, where k represents minutiae count in fingerprint image. Choose a reference minutia from the set, N.

1. Find out the rotation and translation invariant distance between reference minutia, (x_r, y_r, θ_r) and the every other minutiae (x_i, y_i, θ_i) in image.

$$\begin{bmatrix} x_i^T \\ y_i^T \end{bmatrix} = \begin{bmatrix} \cos\theta_r & -\sin\theta_r \\ \sin\theta_r & \cos\theta_r \end{bmatrix} \begin{bmatrix} (x_i - x_r) \\ -(y_i - y_r) \end{bmatrix} \tag{1}$$

Based on the calculated values x_i^T and y_i^T, distance is evaluated $d_{ri} = \sqrt{x_i^{T2} + y_i^{T2}}$.

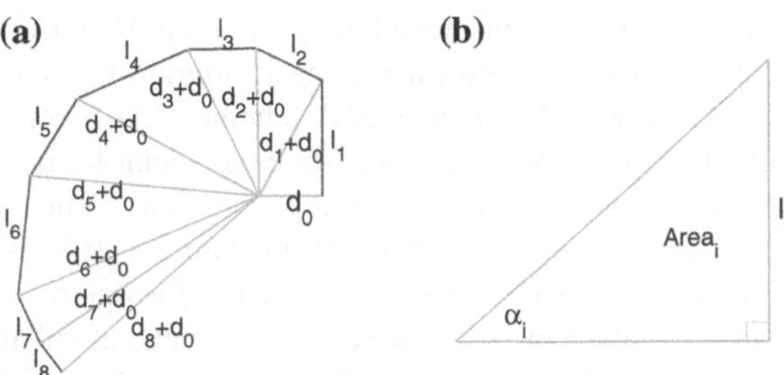

Fig. 1 **a** Construction of spiral curve for reference minutia. **b** Metrics in triangle

2. Sort the distances in ascending order, take n number of least distances to draw contiguous right angled triangles taking distances as hypotenuses of triangles [21]. For first triangle, the base distance is taken as specific value d_0 and the same is added to every distance as shown in Fig. 1a.
3. The other side of triangle, l_i as shown in Fig. 1a is calculated using Pythagoras theorem. We keep the distance l_i, angle at vertex as center α_i, area A_i of each triangle as shown in Fig. 1b and the actual orientation of corresponding minutiae θ_i as feature set for further use.
4. For each reference minutia, (x_r, y_r, θ_r) we represent the feature set as $L_r = [[l_{r1}, \theta_1, \alpha_{r1}, A_{r1}], [l_{r2}, \theta_2, \alpha_{r2}, A_{r2}], \ldots [l_{rn}, \theta_n, \alpha_{rn}, A_{rn}]]$, where $l_{ij}, \alpha_{ij}, A_{ij}$ indicates the metrics taken for each triangle, θ_j indicates the orientation of corresponding minutia from which we have taken invariant distance to reference minutia, n is the number of minutiae taken after sorting the distances.
5. Repeat the steps 1–4 for each other minutiae in minutia set, N of a fingerprint. Thus fingerprint template contains $L = [L_1, L_2, L_3, \ldots, L_k]$ where k refers minutiae count in fingerprint image.

Here, in the feature set we are storing metrics of each triangle and orientations of minutiae. From the feature set we can not find the position and orientation of minutia in the fingerprint image.

3.2 Projection on to 4D Space and Bit-Sting Generation

We generate bit-string by using space based quantization. Each L_r is a vector of order 4, $L_r = (l_{ij}, \theta_j, \alpha_{ij}, A_{ij})$, can be plotted on a 4D-space by taking distance, orientation, angle and area along 4 axes in space with ranges in [0 λ], [0 360], [0 90] and [0 Δ] respectively, where λ is maximum distance and Δ is maximum area. The cells of 4D-space are partitioned into sizes cx, cy, cz and cw along axes [10]. The number of cells in the plane are $A \times B \times C \times D$ where $A = \left\lfloor \frac{maximum\ distance}{cx} \right\rfloor$, $B = \left\lfloor \frac{360}{cy} \right\rfloor$, $C = \left\lfloor \frac{90}{cz} \right\rfloor$ and $D = \left\lfloor \frac{maximum\ area}{cw} \right\rfloor$. Here $\lfloor . \rfloor$ represents the floor function. The template L is mapped to 4D space. Then we will know which cell contains which points on the plane using the Eq. 2.

$$\begin{bmatrix} x_i \\ y_i \\ z_i \\ w_i \end{bmatrix} = \begin{bmatrix} \lfloor l_{ij}/cx \rfloor \\ \lfloor \theta_j/cy \rfloor \\ \lfloor \alpha_{ij}/cz \rfloor \\ \lfloor A_{ij}/cw \rfloor \end{bmatrix} \tag{2}$$

where x_i, y_i, z_i and w_i indicate the x, y, z and w indices on 4D space, cx, cy, cz and cw represents the dimensions of each cell. By visiting each and every cell of 4D space we will get a binary string by taking '1' for one or more point falling onto a cell, otherwise '0'. So the length of bit-string will be I = $A \times B \times C \times D$.

3.3 Feature Transformation

The generated bit-string need to be transformed into non-invertible template for protection. Apply Discrete Fourier Transform on bit-string (H_w) to transform into a complex vector. Performing I-point DFT on H_w, we get frequency domain complex vector, F_i into $I \times 1$ vector. $F = [F_0, F_1, F_2, \ldots, F_{I-1}]$. Multiply the user specific random matrix R with complex vector F to transform into non-invertible template T [16]. $[T]_{p\times 1} = [R]_{p\times q}[F]_{q\times 1}$. Here q = I, the size of bit-string and $p<q$. Thus T is a transformed vector of order $p \times 1$. While at verification we use the same user key to generate the random matrix and then to generate transformed template.

3.4 Matching

Local matching In Local matching, we compare template locally by matching each spiral curve data in enrolled template with every spiral curve data in query template. Fingerprint transformed templates at the time of enrollment and query are represented as $E = [E_1, E_2, E_3, \ldots, E_m]$ and $Q = [Q_1, Q_2, Q_3, \ldots, Q_n]$. Then the distance between both of E_i and Q_j is calculated using Eq. 3

$$d(E_i, Q_j) = \frac{\|E_i - Q_j\|_2}{\|E_i\|_2 + \|Q_j\|_2} \tag{3}$$

where $\|.\|_2$ indicates 2-norm or euclidean norm. Then the matching score can be found using $S(E_i, Q_j) = 1 - d(E_i, Q_j)$. The matching score will come in range of [0 1]. Matching is done according to [19, 20]. To prevent double matching we re-evaluate similarity matrix. $S(E_i, Q_j) = S(E_i, Q_j)$ if $S(E_i, Q_j)$ is maximum for $j \in [1, m]$ and $i \in [1, n]$, otherwise 0.

Global matching In global matching we take the maximum similarity score of the template comparison using Eq. 4. Here Ψ indicates the count of non-zero values in $S(E_i, Q_j)$, m and n are minutiae counts in respective enrolled and query templates. If score is 1, refers exact match and 0 refers mismatch.

$$MS = \frac{\sum_{i=1}^{m} \sum_{j=1}^{n} S(E_i, Q_j)}{\Psi} \quad (4)$$

4 Experimental Results and Analysis

For experimental testing database collected at Fingerprint Verification Competition: FVC2002. Each DB1, DB2, DB3 contains 100 users with 8 samples each. We took 2 impressions per each user. Neurotechnology Verifinger SDK used to extract features. False Rejection Rate (FRR), Genuine Acceptance Rate (1-FRR), False Acceptance Rate (FAR), Equal Error Rate (EER) are measures of performance [1]. FRR is the ratio of total false rejections to total identification attempts or the probability of rejecting a similar image as impostor. FAR is the ratio of total false acceptances to total identification attempts or the probability of accepting dissimilar image as genuine. EER is the value when FRR and FAR are equal. Genuine Acceptance Rate is accepting similar image as genuine.

In the method we have taken minutiae points as center for drawing spiral curves instead of singular points [21] which may not be available in all images. Considered only 4 smallest distances instead of all distances for each reference minutia to reduce complexity. In quantization after fine tuning cx, cy, cz and cw are fixed. In different key scenario we got EER value 0 %. For same key scenario EER values are shown in Table 1. Figure 2 refers the error rate for FVC2002. Receiver Operating Characteristic (ROC) curve for FVC2002 DB1, DB2 and DB3 as in Fig. 3, shows that low recognition rate for DB3 relative to DB1 and DB2 because of low quality images. The proposed model ensures the revocability and diversity by changing uscr key to generate multiple templates which can not be matched. Using different keys 100 transformed templates are generated from a same image on FVC2002 DB2 and matched with enrolled image to find pseudo-imposter

Table 1 Equal error rate for same key scenario

Methods	FVC2002		
	DB1	DB2	DB3
Ahmed et al. [15]	9	6	27
Yang et al. [22]	–	13	–
Jin et al. [13]	5.19	5.65	–
Proposed method	7.85	5.29	17.55

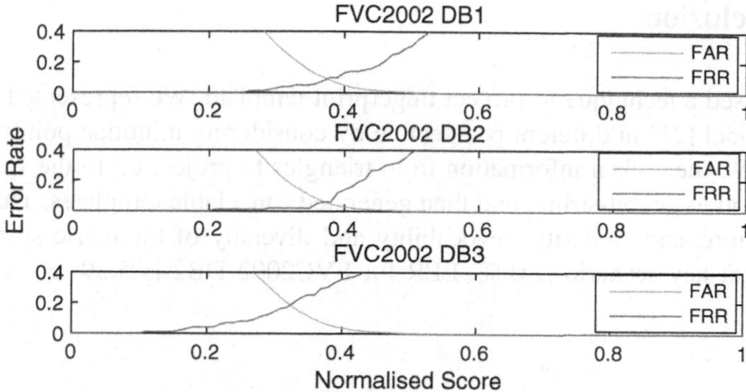

Fig. 2 Equal error rate for FVC2002 same key

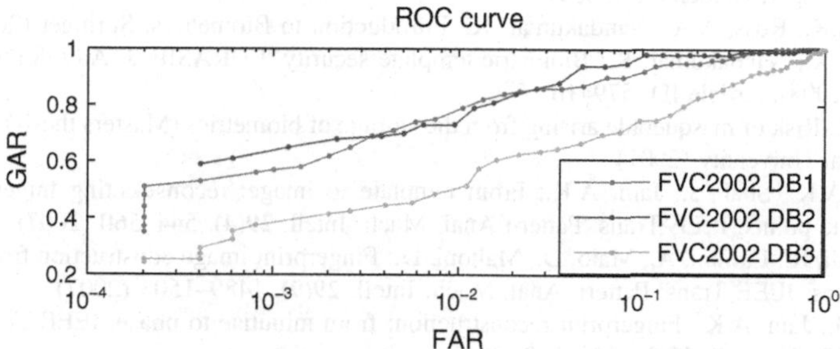

Fig. 3 ROC curve for same key for FVC2002

distribution. As shown in Fig. 4 pseudo-imposter distribution is clearly separated from genuine distribution. In security perspective it is inconceivable to regress original image from a stolen template. Even if adversary knows the spiral curves of each reference minutia, the position of reference minutia and orientation can not be found to try the possibilities to put neighboring minutiae around it.

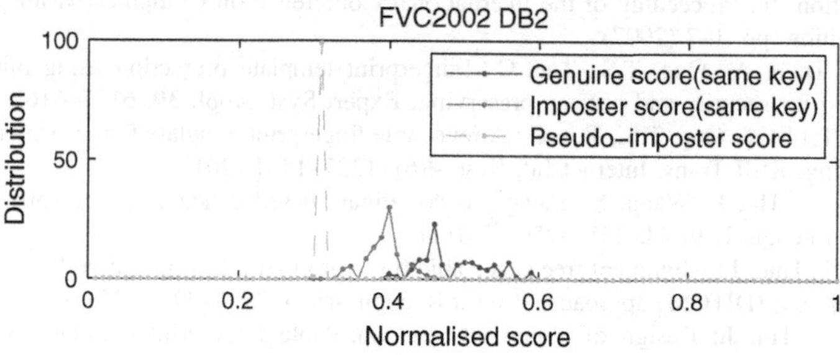

Fig. 4 Genuine, imposter and pseudo-imposter distributions for FVC2002 DB2

5 Conclusion

We proposed a technique to protect fingerprint template. We represented the spiral curves model [21] in different perspective by considering minutiae points as center to shell. We then taken information from triangles to project on to the space based quantization to get bit-string, and then generated cancelable templates. This method meets requirements security, revocability and diversity of biometric system. EER for different key scenario is 0 %. EER for FVC2002 DB2 is 5.29 %.

References

1. Maltoni, D., Maio, D., Jain, A.K., Prabhakar, S.: Handbook of Fingerprint Recognition, 2nd edn. Springer, Heidelberg (2009)
2. Jain, A.K., Ross, A.A., Nandakumar, K.: Introduction to Biometrics. Springer (2011)
3. Jain, A.K., Nandakumar, K.: Biometric template security, EURASIP. J. Adv. Signal Process. 1–17 (2008). Article ID: 579416
4. Hill, C.: Risk of masquerade arising from the storage of biometrics (Masters thesis), Australian National University (2001)
5. Ross, A.K., Shah, J., Jain, A.K.: From template to image: reconstructing fingerprint from minutiae points. IEEE Trans. Pattern Anal. Mach. Intell. **29**(4), 544–560 (2007)
6. Cappelli, R., Lumini, A., Maio, D., Maltoni, D.: Fingerprint image construction from standard templates. IEEE Trans. Pattern Anal. Mach. Intell. **29**(9), 1489–1503 (2007)
7. Feng, J., Jain, A.K.: Fingerprint reconstruction: from minutiae to phase. IEEE Trans. Pattern Anal. Mach. Intell. **33**(2), 209–223 (2011)
8. Ratha, N.K., Chikkerur, S., Connell, J.H., Bolle, R.M.: Generating cancelable fingerprint templates. IEEE Tans. Pattern Anal. Match. Intell. **29**(4), 561–572 (2007)
9. Fenq, Q., Su, F., Cai, A., Zhao, F.F.: Cracking cancelable fingerprint template of Ratha. In: International Symposium on Computer Science and Computational Technology (ISCSCT08), vol. 2, pp. 572–575 (2008)
10. Lee, C., Kim, J.: Cancelable fingerprint templates using minutiae based bit strings. J. Netw. Comput. Appl. **33**, 236–246 (2010)
11. Lee, C., Choi, J.Y., Toh, K.A., Lee, S., Kim, J.: Alignment-free cancelable fingerprint template based on local minutia information. IEEE Trans. Syst. Man. Cybern.-Part B: Cybern. **37**(4), 980–992 (2007)
12. Farooq, F., Bolle, R.M., Jea, T.Y., Ratha, N.K.: Anonymous and revocable fingerprint recognition. In: Proceeding of the International Conference on Computer Vision and Pattern Recognition, pp. 1–7 (2007)
13. Jin, Z., Teoh, A., Ong, T.S., Tee, C.: Fingerprint template protection using minutia-based bit-string for security and privacy preserving. Expert Syst. Appl. **39**, 6157–6167 (2012)
14. Jin, Z., Teoh, A., Ong, T.S., Tee, C.: A revocable fingerprint template for security and privacy preserving. KSII Trans. Internet Inf. Syst. **4**(6), 1327–1341 (2010)
15. Ahmed, T., Hu, J., Wang, S.: Pair-polar coordinate based cancelable fingerprint templates. Pattern Recogn. Lett. **44**, 2555–2564 (2011)
16. Wang, S., Hue, J.: Alignment free cancelable fingerprint template design: a densely infinite to one mapping (DITOM) approach. Pattern Recogn. **45**, 4129–4137 (2012)
17. Wang, S., Hu, J.: Design of alignment-free cancelable fingerprint templates via curtailed circular convolution. Pattern Recogn. **47**, 1321–1329 (2014)

18. Wong, W.J., Wong, M.L.D., Kho, Y.H.: Multiline code: a low complexity revocable fingerprint template for cancelable biometrics. J. Cent. South Univ. **20**, 1292–1297 (2013)
19. Wong, W.J., Teoh, A.B.J., Wong, M.L.D., Kho, Y.H.: Enhanced multiline code for minutiae based fingerprint template protection. Pattern Recogn. Lett. **34**, 1221–1229 (2013)
20. Prasad, M.V.N.K., Kumar, C.S.: Fingerprint template protection using multiline neighboring relation. Expert Syst. Appl. **41**, 6114–6122 (2014)
21. Moujahdi, C., Bebis, G., Ghouzali, S., Rziza, M.: Fingerprint shell secure representation of fingerprint template. Pattern Recogn. Lett. **45**, 189–196 (2014)
22. Yang, H., Jiang, X., Kot, A.C.: Generating secure cancelable fingerprint templates using local and global features. In: 2nd IEEE ICCSIT, pp. 645–649 (2009)

Rhythm and Timbre Analysis for Carnatic Music Processing

Rushiraj Heshi, S.M. Suma, Shashidhar G. Koolagudi, Smriti Bhandari and K.S. Rao

Abstract In this work, an effort has been made to analyze rhythm and timbre related features to identify *raga* and *tala* from a piece of Carnatic music. *Raga* and *Tala* classification is performed using both rhythm and timbre features. Rhythm patterns and rhythm histogram are used as rhythm features. Zero crossing rate (ZCR), centroid, spectral roll-off, flux, entropy are used as timbre features. Music clips contain both instrumental and vocals. To find similarity between the feature vectors T-Test is used as a similarity measure. Further, classification is done using Gaussian Mixture Models (GMM). The results shows that the rhythm patterns are able to distinguish different *ragas* and *talas* with an average accuracy of 89.98 and 86.67 % respectively.

Keywords Carnatic music · *Raga* · *Tala* · Gaussian Mixture Models · Rhythm · Timbre · T-test

R. Heshi (✉) · S. Bhandari
Walchand College of Engineering, Sangli, India
e-mail: rushiraj.heshi@walchandsangli.ac.in

S. Bhandari
e-mail: smriti.bhandari@yahoo.com

S.M. Suma · S.G. Koolagudi
National Institute of Technology, Karnataka Surathkal, India
e-mail: suma.cs13f02@nitk.edu.in

S.G. Koolagudi
e-mail: koolagudi@nitk.edu.in

K.S. Rao
Indian Institute of Technology, Kharagpur, India
e-mail: ksrao@sit.iitkgp.ernet.in

A. Nagar et al. (eds.), *Proceedings of 3rd International Conference on Advanced Computing, Networking and Informatics*, Smart Innovation, Systems and Technologies 43, DOI 10.1007/978-81-322-2538-6_62

1 Introduction

Since large collection of multimedia data is available in digital form, we need to identify ways to make those collections accessible to users. Efficient indexing criteria makes retrieving/accessing an easy task. The techniques for indexing/accessing music are categorized into 3 types. The first two kinds are meta-data and text based access, use meta data and user provided tags for retrieval. However, these tags need not be correct always. Where as in content based Music Indexing and Retrieval (MIR), music signal is analyzed for deciding the genre, hence it is more effective than the previous two techniques. The fundamental concepts of Carnatic music are *raga* (melodic scales) and *tala* (rhythmic cycles). A *raga* in Carnatic music prescribes a set of rules for building a melody very similar to the western concept of mode [1]. *Raga* tells about the set of notes used and the way in which these notes are rendered. Technically a note is a fundamental frequency component of a music signal defined using starting and ending time [2]. A *tala* refers to a fixed time cycle or a metre, set for a particular composition, which is built from groupings of beats [1]. *Tala* has cycles of a defined number of beats and rarely changes within a song. Since *raga* and *tala* are the fundamental concepts, extracting these information from music signal helps in building efficient MIR systems.

In this work, an attempt has been made to analyze different features like rhythm and timbre for classification of *raga* and *tala*. Rhythm is the pattern of regular or irregular pulses caused in music by the occurrence of strong and weak melodic and harmonic beats [3]. Timbre describes those characteristics of sound, which allow the ear to distinguish sounds that have the same pitch and loudness which is related to the melody of the music [3].

Rest of the paper is organized as follows. A brief review of the past works and the issues are discussed in Sect. 2. The features extracted and classifier used are explained in Sect. 3. Section 4 explains the experiments conducted and the analysis of results. Section 5 concludes the work with some future research directions.

2 Related Work

In this section, different feature extraction approaches towards audio retrieval and classification have been discussed. Many of the works have used pitch derivatives as features for *raga* identification since pitch feature is related to melody of the music. In [4], Hidden Markov Model (HMM) is used for the identification of Hindustani ragas. The proposed method uses note sequence as a feature. Many micro-tonal variations present in the ICM make note transcription a challenging task even for a monophonic piece of music. Two heuristics namely Hillpeak heuristics and Note duration one try to overcome these variations. The limitation of this work is limited data set as it contains only two *ragas* and considerably lower accuracy of note transcription. Similar work has been carried out by Arindam et al. [5] using manual note transcription. The HMM evaluated for this sequence claimed to achieve 100 % accuracy, if the given note sequence is correct. However, it is difficult to achieve high accuracy in ICM transcription because of the micro-tonal variations and improvisations. P. Kirthika et al. introduced an audio mining technique based on *raga* and emphasized importance of *raga* in audio classification [6]. Individual notes are used as features. Pitch and timbre indices are considered for classification. Koduri et al. presented *raga* recognition techniques based on pitch extraction methods and KNN is used for classification [7]. Property of the tonic note with pitch of highest mean and least variation shown in pitch histogram, is used for identification of tonic pitch value [8]. Using Semi-Continuous Gaussian Mixture Model (SC-GMM), tonic frequency and *raga* of the musical piece are identified. Only 5 *Sampurna ragas* are used for validating this system. In [9], Rhythm patterns and Rhythm histogram are used as a feature for identifying and tagging songs. GMM is used for classification. In [10], timbre features such as spectral centroid, spectral roll-off, spectral flux, low energy features, MFCC and rhythmic features are used for classification of *raga* and *tala*.

From the literature, it is evident that many of the works have used features such as pitch and its derivatives and set of note information for identifying *raga* and *tala*. In this work features other than pitch and note information are analyzed for classification.

3 Methodology

Figure 1 shown below represents the activities done while implementing the idea. Rhythm and timbre related features are extracted from each frame of the music clip. GMMs are trained using these features to model the training music clips on the basis if their rhythm and timbre. Further trained models are used to classify unknown test clips.

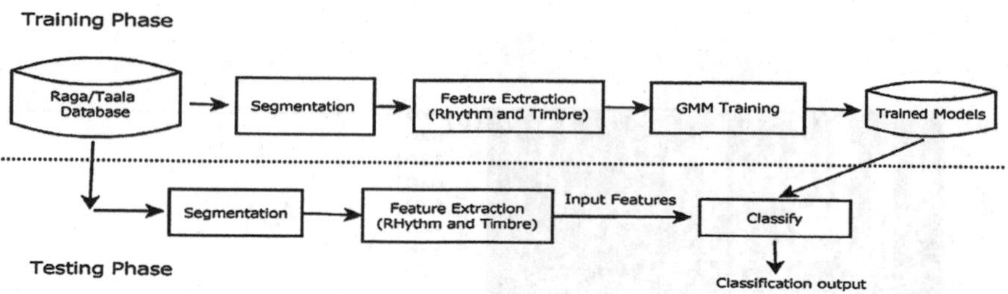

Fig. 1 Schematic diagram to classify *raga* and *tala*

3.1 Features Extraction

Rhythm Features: Rhythm patterns and Rhythm histograms are extracted from the first 60 frames of the given music piece [3]. It results in a 24 × 60 matrix where 24 represents the critical bands of the Bark scale and 60 represents the number of frames (shown in Fig. 2a). The x-axis represents the rhythm frequency up to 10 Hz and the y-axis represents bark scale of 24 critical bands. From the rhythm patterns, rhythm histogram is obtained by adding the values in each frequency bin in the rhythm pattern. This results in a 60 dimensional vector representing the "rhythmic energy" of the corresponding modulation frequencies. In Fig. 2b x-axis represents the rhythm frequency up to 10 Hz and the y-axis represents the magnitude of respective frequency.

Timbre Features: Timbre related features such as ZCR, centroid, roll-off, flux, entropy are extracted from the signal. ZCR is the number of times signal crosses x-axis. Centroid determines frequency area around which most of the signal energy concentrates. Centroid is calculated using Eq. 1.

$$C_t = (M_t[n] * n)/M_t[n] \tag{1}$$

where $M_t[n]$ is the magnitude of the Fourier transform of frame t and frequency bin n. Roll-off is used for finding out frequency such that certain fraction of total energy is contained below that particular frequency. The spectral roll-off is defined as the frequency R_t below which 85 % of the magnitude distribution is concentrated and is calculated using Eq. 2.

$$M_t[n] = 0.85 * M_t[n] \tag{2}$$

The spectral flux is a measure of the amount of local spectral change. Flux is the distance between spectrum of two successive frames. It is calculated using Eq. 3.

$$F_t = (N_t[n] - N_t - 1[n]) \tag{3}$$

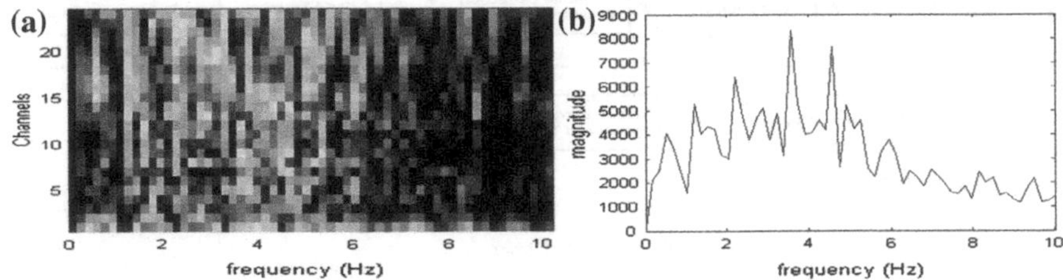

Fig. 2 Rhythm features extracted from the music signal. **a** Rhythm patterns. **b** Rhythm histograms

where $N_t[n]$ and $N_t - 1[n]$ are the normalized magnitude of the Fourier transform of the current frame t, and the previous frame $t - 1$, respectively. Entropy is used to calculate randomness of the signal.

$$H(X) = -p(x_i) \log bp(xi) \tag{4}$$

where $p(x_i)$ is the probability mass function of outcome x_i. These features extracted from the signal are initially checked for similarity for each *raga* and *tala* class. Further are used for classification task.

3.2 Classifier

T-Test: Before developing a classifier model, the T-test is performed to determine whether the means of two groups are statistically different from each other. The result of test is 0 or 1. The output 0 implies T-Test is passed and there is no significant difference between two feature vectors. If output is 1 then T-test does not pass and there is significant difference between two feature vectors.

GMM: GMM is a mixture of Gaussian Distributions. Probability density function for mixture of Gausses is a linear combination of individual PDFs. A GMM is constructed for each class (*raga/tala*). Expectation Maximization algorithm is used for training GMM. In testing phase, the highest probability value (greater than 0.5) is used to decide the output class.

4 Experimentation and Results

4.1 Database

Two different audio datasets are collected for 10 *raga* and 10 *tala* considered for the study are given in Table 1. The music clips include both monophonic and polyphonic music and are rendered by different male and female singers. The dataset consists of 400 clips (20 clips in each type of *raga* or *tala*).

4.2 Performance Evaluation

6 sets of experiments are performed to evaluate the proposed method. Initial four experiments are conducted using T-test to validate rhythm and timbre on *raga* and *tala* datasets. Each music clip is compared with all the other music clips and the

Table 1 Database: list of *ragas* and *talas* used

Raga id	Raga name	Number of clips	Tala id	Tala name	Number of clips
1	Abhogi	20	1	Adi	20
2	Hamsadwani	20	2	Adi (2 kalai)	20
3	Hari Khambhoji	20	3	Desh Adi	20
4	Hindolam	20	4	Khanda Chapu	20
5	Kalyani	20	5	Khanda Ekam	20
6	Malyamarutha	20	6	Misra Chapu	20
7	Mayamalavagowla	20	7	Misra Jhampa	20
8	Mohanam	20	8	Rupakam	20
9	Nattai	20	9	Thisra Triputa	20
10	Shankarabharanam	20	10	Triputa	20

similarity value (0 or 1) is recorded. The percentage of music clips that matches with the same class of clips is calculated. The values in the Table 2 show that rhythm features have better similarity than timbre features. Hence, rhythm features are considered for classification of *raga* and *tala*. Rhythm features of 60 dimensions from 14 music clips are used for GMM training and 6 music clips are used for testing from each *raga* and *tala*. The results in Table 3 show that the rhythm features are useful in classification and hence may be used as secondary features along with pitch related features for *raga* and *tala* identification.

Table 2 Results of similarity test for rhythm and timbre features

		Raga/tala id									
		1	2	3	4	5	6	7	8	9	10
Similarity for ragas (%)	Rhythm features	100	66.6	66.6	87.5	88.8	66.6	66.6	50	50	66.6
	Timbre features	50	55.5	33.3	37.5	33.3	55.5	33.3	37.5	25	33
Similarity for talas (%)	Rhythm features	66.6	100	100	88.8	100	66.6	100	66.6	100	66.6
	Timbre features	33.3	50	33.3	33.3	100	33.3	50	55.5	100	33.3

Table 3 Accuracy of classification of *raga* and *tala* using rhythm features

		Raga/tala id									
		1	2	3	4	5	6	7	8	9	10
Accuracy of classification (%)	Raga	100	66.6	100	66.6	66.6	100	100	100	100	100
	Tala	66.6	100	100	66.6	100	66.6	100	66.6	100	100

5 Summary and Conclusion

In this work, analysis of rhythm and timbre features for classification of *raga* and *tala* of Carnatic music has been done. From the experiments, it is found the rhythm features are able to distinguish *raga* and *tala* better than timbre features. The average accuracy of 89.98 and 86.67 % is achieved for classification of *raga* and *tala* respectively using GMM classifier. Even though the results obtained are promising, it cannot be generalized since it is validated using a small data set. As a future work, combination of rhythm and pitch related features shall be used for *raga* and *tala* classification. MIR systems for music recommendation shall be developed using these features.

References

1. Agarwal, P., Karnick, H., Raj, B.: A comparative study of indian and western music forms. In: International Society for Music Information Retrieval (2013)
2. Klapuri, A., Davy, M.: Signal Processing Methods for Music Transcription. Springer, New York Inc., Secaucus (2006)
3. Orio, N., Piva, R.: Combining timbric and rhythmic features for semantic music tagging. In: International Society for Music Information Retrieval (2013)
4. Pandey, G., Mishra, C., Ipe, P.: Tansen: a system for automatic raga identification pp. 1350–1363 (2003)
5. Bhattacharjee, A., Srinivasan, N.: Hindustani raga representation and identification: a transition probability based approach. IJMBC **2**(1–2), 66–91 (2011)
6. P.Kirthika, Chattamvelli, R.: A review of raga based music classification and music information retrieval (mir). IEEE conference on signal processing (2012)
7. Koduri, G., Gulati, S., Rao, P.: A survey of raaga recognition techniques and improvements to the state-of-the-art. SMC (2011)
8. Ranjani, H.G., Arthi, S., Sreenivas, T.V.: Carnatic music analysis: shadja, swara identification and raga verification in alapana using stochastic models. In: IEEE Workshop on Applications of Signal Processing to Audio and Acoustics, pp. 29–32 (2011)
9. Tzanetakis, G., Cook, P.: Musical genre classification of audio signals. IEEE Transactions on Speech Audio process. **10**(5) (2002)
10. Christopher, R., Kumar, P., Chandy, D.: Audio retrieval using timbral feature. In: IEEE International Conference on Emerging Trends in Computing, Communication and Nanotechnology (2013)

Repetition Detection in Stuttered Speech

Pravin B. Ramteke, Shashidhar G. Koolagudi and Fathima Afroz

Abstract This paper mainly focuses on detection of repetitions in stuttered speech. The stuttered speech signal is divided into isolated units based on energy. Mel-frequency cepstrum coefficients (MFCCs), formants and shimmer are used as features for repetition recognition. These features are extracted from each isolated unit. Using Dynamic Time Warping (DTW) the features of each isolated unit are compared with those subsequent units within one second interval of speech. Based on the analysis of scores obtained from DTW a threshold is set, if the score is below the set threshold then the units are identified as repeated events. Twenty seven seconds of speech data used in this work, consists of 50 repetition events. The result shows that the combination of MFCCs, formants and shimmer can be used for the recognition of repetitions in stuttered speech. Out of 50 repetitions, 47 are correctly identified.

Keywords MFCCs · Formants · Shimmer · Jitter · Dynamic time warping

1 Introduction

Stuttering is a speech disorder which disrupts the normal flow of a speech by involuntary disfluencies. Different types of stutter disfluencies are: 1. Word repetitions 2. Part-word repetitions 3. Prolongation (sound or phoneme pronunciation is stretched) 4. Broken words [1]. The processing and quantification of stuttered speech helps in many medical applications, such as treating psychological disorders, anxiety

P.B. Ramteke (✉) · S.G. Koolagudi · F. Afroz
National Institute of Technology Karnataka, Surathkal 575 025, Karnataka, India
e-mail: ramteke0001@gmail.com

S.G. Koolagudi
e-mail: koolagudi@nitk.ac.in

F. Afroz
e-mail: fathimaNITK@gmail.com

A. Nagar et al. (eds.), *Proceedings of 3rd International Conference on Advanced Computing, Networking and Informatics*, Smart Innovation, Systems and Technologies 43, DOI 10.1007/978-81-322-2538-6_63

related problems those cause stammering, therapy to improve speaking fluencies by reducing stuttering and so on. The most common approach of stuttering assessment is to count the occurrence of types of disfluencies. Main difficulties of manual evaluations of these disfluencies are need of continuous human expert attention, careful recording of number and types of disfluencies, considerable amount of experts time being spent and also the evaluations made by different experts are highly subjective and poorly agree with the other similar evaluations [2]. Hence, the automatic detection of stuttering disfluencies greatly helps Speech Language Pathologists (SLPs) to analyze stuttering patterns, plan appropriate assessment program and monitor the progress during the period of treatment.

The repetition events are more prominent in stuttered speech compared to the other types of disfluencies [3]. Each repetition event have similar acoustic features, therefore it is possible to automatically diagnose and process stuttering patterns. It is one of the key parameters in assessing the stuttering events objectively. This paper mainly focuses on automatic detection of repetition events using energy, heuristics and DTW.

Rest of the paper is arranged as follows. Section 2 critically evaluates some of the past research reported in this area. Methodology of incorporating this idea is discussed in detail in Sect. 3. Results and their analysis is carried out in Sect. 4. Paper concludes with Sect. 5 followed by some important references.

2 Literature Review

Many researches have focused on detection of different types of stammering events in speech signal. MFCCs are claimed to be robust in recognition tasks related to the human voice as they mostly represent phonetic and vocal cavity information. Hence many studies have considered MFCCs as the primary features for the recognition of stuttered events with different classifiers [4]. In [5], MFCCs are used as features for the recognition of syllable repetition. The decision logic is implemented by perceptron based on the score given by score matching. The work claims 83 % accuracy. In [6], SVM classifier with unimodal and multimodal kernel functions are trained using 13 MFCC features to recognizes fluent and disfluent segments in speech. The SVM classifier with unimodal kernel functions achieves 96.13 % accuracy, while with multimodal kernel functions 96.4 % accuracy is claimed. LPCCs are used as features to recognize prolongations and repetitions in stuttered speech. Linear discriminant analysis (LDA) and k-nearest neighbor (k-NN) algorithms used for classification is claimed to achieve accuracy of 89.77 % [7]. In [8], LPC, LPCC and weighted linear prediction cepstral coefficients (WLPCC) are used as a features for recognizing the repetitions and prolongations. k-NN and LDA are used for the classification with WLPCCs claimed to achieve better recognition accuracy of 97.06 % when compared to LPCCs and LPCs (95.69 % and 93.14 % respectively).

From the literature, it is observed that many of the works have focused on the classification of repetition and prolongation events in a stuttered speech using same machine learning and classification algorithms. For training and testing manually separated repetition and prolongation patterns are considered, hence these approaches may not be feasible for real time evaluation. Hence this work focuses on recognition of repetition events by comparing each isolated utterance with the subsequent utterances within the specified interval of time. This approach works on basic heuristics and is suitable for real time applications.

3 Methodology

In this paper, process of repetition recognition is divided into four stages: segmentation, feature extraction, feature comparison and decision making shown in Fig. 1. The following subsections explain framework in detail.

3.1 Segmentation

In the proposed approach the long utterance containing the complete sentence is broken into smaller segments based on average energy, which may be a word or a stammered semiword or syllable [9].

3.2 Feature Extraction

The spectral and prosodic features are used individually and in combination for the recognition of repetition events. In this work 13 MFCC features are extracted as these features approximate the human auditory response more closely and claimed to be robust in recognition tasks [10]. This fact is main motivation to use MFCC features for the detection of repetitive events. Each phoneme is pronounced by the unique articulation of the vocal tract, hence there is a significant difference in their formant frequencies [11]. So, formants may be considered for the speech recognition. In this approach, 3 formant frequencies are considered for the repetition

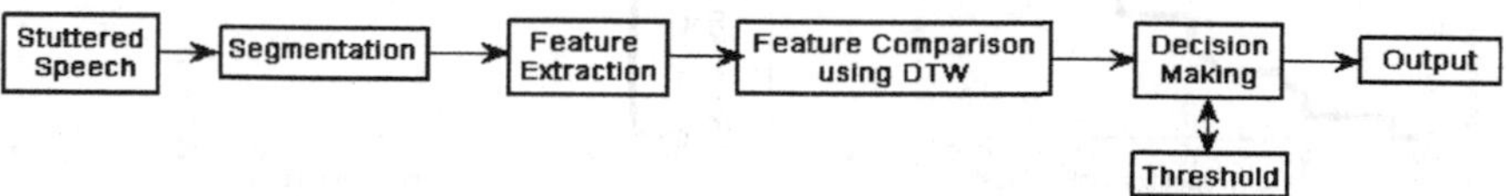

Fig. 1 Proposed framework for recognition of repetitions in stuttered speech

recognition. There is a slight variation in two consecutive pitch periods and their respective amplitudes which are represented as features called as jitter and shimmer respectively [12]. The shimmer can be used for recognition of repetition events since energy is almost similar for each repetition event.

3.3 Feature Comparison and Decision Making

In this paper the DTW based feature comparison is done to evaluate the similarity between different isolated events. In general DTW is an approach used for measuring similarity between two temporal sequences, if one sequence may be warped non-linearly by stretching or shrinking on to the other (e.g. time series) [13]. The procedure to compute DTW is as below:

Consider two time series Q and C, of length n and m respectively, where $Q = q_1, q_2, \ldots, q_i, \ldots, q_n$ and $C = c_1, c_2, \ldots, c_j, \ldots, c_m$. These two sequences are aligned as n-by-m matrix. The (i^{th}, j^{th}) element of the matrix contains the distance $d(q_i, c_j)$ between the two points q_i and c_j. Then, the absolute distance between the values of two sequences is computed using the Euclidean distance:

$$d(q_i, c_j) = \sqrt{(q_i - c_j)^2} \qquad (1)$$

Each matrix element (i,j) corresponds to the alignment between the points q_i and c_j. Then, accumulated distance for two sequences is measured by:

$$D(i,j) = \min[D(i-1, j-1), D(i-1, j), D(i, j-1)] + d(i,j) \qquad (2)$$

The resultant value of $D(i,j)$ gives the optimal match between the two sequences. The value of $D(i,j)$ is small when signals or sequences are more similar else returns the larger distance value. This approach is employed to compare the feature vectors extracted from isolated units.

Figure 2a shows the comparison of feature vectors of repetitive units i.e. reference unit "ऊस" ("oos") and test unit "ऊस" ("oos") aligned on vertical and horizontal axis respectively. Using this approach the 24 feature vectors for reference

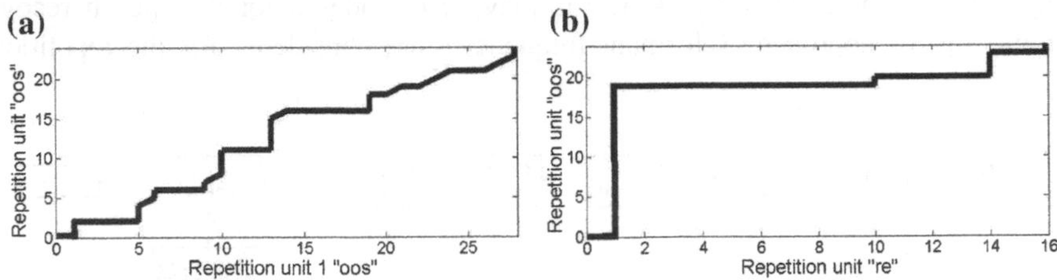

Fig. 2 Comparison of feature vectors of isolated units using DTW. **a** Repetitive speech units "ऊस ऊस" ("oos oos"). **b** Non-repetitive speech units "ऊस रे" ("oos re")

unit and 28 feature vectors for test unit are compared. For this comparison the path almost follows a diagonal of distance matrix and distance values obtained is 14.2435. The path shown in the Fig. 2b shows the minimum distance path computed over the non-repetitive units: test unit "रे" ("re") and reference unit "ऊस" ("oos"). Comparison of 24 reference and 16 test feature vectors gives distance values 19.45 and the path does not follow the diagonal. From this it is observed that there is a significant difference in the path and distance values obtained using DTW feature comparison for repetitive and non-repetitive speech units.

A heuristic approach is employed for the identification of repetition events. The distance values for the repeated units are small compared to that of non-repeated units. Hence, the distance values obtained for the repeated units and non-repeated units are analyzed and the threshold is empirically set to repeated and non-repeated units. If the distance is below the set threshold then the units are classified as repetition events.

4 Results and Discussion

In this section, the results obtained using the combinations of different features are presented. The database consists a recordings of single speaker in Hindi language. There are total 98 repetition events in 30 min of recording. Out of 98 repetitions, the speech data containing 50 repetition events is considered for this study. The features are extracted from each isolated unit using the frame size of 25 ms with frame shift of 10 ms.

Table 1 shows the number of repetition correctly detected using different combination of features. For 17-dimensional (13 MFCCs + 3 formants + 1 shimmer) feature vector, out of 50 repetitions 47 are correctly recognized along with the 19 false positive repetition for threshold 23.0 which is the real concern to be addressed.

Table 1 Results of repetition detection for different combinations of features

Sr. no	Features	Threshold	Total no. of repetitions	Correctly identified repetitions	Misclassified repetitions	False positive repetitions
1	MFCCs (13)	19	50	37	13	19
2	MFCCs (13) + Formant Energy (3)	24	50	47	3	22
3	MFCCs (13) + Formant Energy (3) + Shimmer (1)	23	50	47	3	19

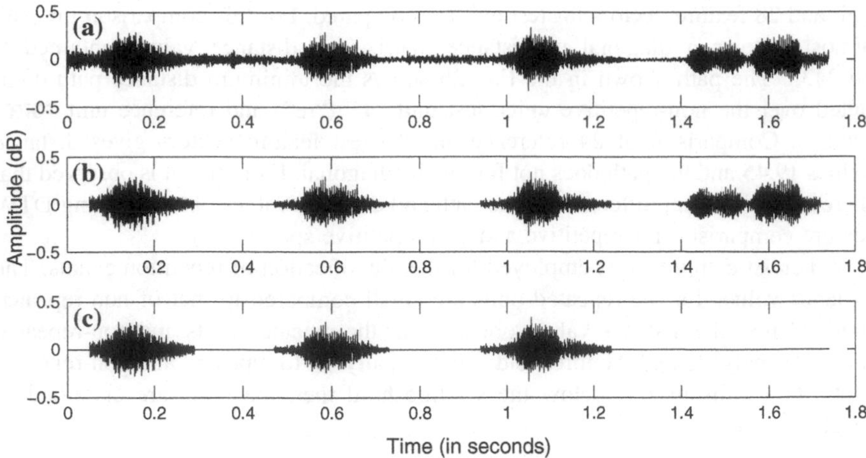

Fig. 3 **a** Input stuttered speech signal. **b** Isolated utterances from stuttered speech signal. **c** Output speech signal consists of 3 utterances of a repetition event.

Figure 3 shows the output of repetition recognition approach applied on a small sample of stuttered speech "ऊस ऊस ऊस पूरे" ("oos oos oos poore"). Figure 3a shows the stuttered speech signal containing 3 repetitions of same units. The units isolated based on energy are shown in Fig. 3b. Figure 3c shows that the 3 repetitions of the same unit "ऊस" ("oos") are correctly identified with the threshold distance of 23.0.

5 Summary and Conclusion

This paper presents a simple heuristic approach for recognition of repetition in a stuttered speech. Spectral and prosodic features are used for the task. A heuristic approach is used for the classification of units into repeated events. The result shows that the combination of MFCCs, formants and shimmer can be used for the characterization of repetitions in stuttered speech. Out of 50 repetitions 47 are correctly identified yielding to an accuracy of 94 % along with 19 false positive detection.

From the result, it may be observed that the false positive units are more in repetition recognition. Hence, this work can further be aimed at reduction of false positive units. Also, this work may be extended to detect the prolongation and interjection, as significant difference in the durations of syllable with prolongation and interjection is observed compared to their normal pronunciation.

References

1. Riper, V.: The Nature of Stuttering. Prentice Hall, New Jersey (1971)
2. Kully, D., Boerg, E.: An investigation of inter-clinic agreement in the identification of fluent and stuttered syllables. J. Fluency Disord. **13**, 309–318 (1988)
3. Conture, E.G.: Stuttering Englewood cliffs, New Jersey: Prentice-Hall, 2nd edn. (1990)
4. Zhang, J., Dong, B., Yan, Y.: A computer-assist algorithm to detect repetitive stuttering automatically. In: International Conference on Asian Language Processing, pp. 249–252 (2013)
5. Ravikumar, K.M., Balakrishna, R., Rajagopal, R., Nagaraj, H.C.: Automatic detection of syllable repetition in read speech for objective assessment of stuttered disfluencies. Proce. World Acad. Sci. **2**, 220–223 (2008)
6. Palfy, J., Pospichal, J.: Recognition of repetitions using support vector machines. In: Signal Processing Algorithms, Architectures, Arrangements, and Applications Conference Proceedings (SPA), 2011, pp. 1–6 (2011)
7. Chee, L.S., Ai, O.C., Hariharan, M., Yaacob, S.: Automatic detection of prolongations and repetitions using LPCC. In: 2009 International Conference for Technical Postgraduates (TECHPOS). pp. 1–4 (2009)
8. Ai, O.C., Hariharan, M., Yaacob, S., Chee, L.S.: Classification of speech dysfluencies with MFCC and LPCC features. J. Med. Syst. **39**, 2157–2165 (2012)
9. Ying, G.S., Mitchell, C.D., Jamieson, L.H.: Endpoint detection of isolated utterances based on a modified teager energy measurement. International Conference on Acoustics, Speech and Signal Processing, vol. 2, pp. 732–735 (1993)
10. James, L.: MFCC tutorial. http://practicalcryptography.com/miscellaneous/machine-learning/guide-mel-frequency-cepstral-coefficients-mfccs/
11. Welling, L., Ney, H.: Formant estimation for speech recognition. IEEE Transactions on Speech Audio Processing, vol. 6, pp. 36–48 (1998)
12. Li, X., Tao, J., Johnson, M.T., Soltis, J., Savage, A., Kirsten, M.L., Newman, J.D.: Stress and emotion classification using Jitter and Shimmer features. In: IEEE International Conference on Acoustics, Speech and Signal Processing, 2007, vol. 4., pp. IV–1081. IEEE (2007)
13. Keogh, E., Ratanamahatana, C.A.: Exact indexing of dynamic time warping. Knowl. Inf. Syst. **7**, 358–386 (2005)

Removal of Artifacts from Electrocardiogram Using Efficient Dead Zone Leaky LMS Adaptive Algorithm

T. Gowri, P. Rajesh Kumar, D.V.R. Koti Reddy and Ur Zia Rahman

Abstract The ability to extract high resolution and valid ECG signals from contaminated recordings is an important subject in the biotelemetry systems. During ECG acquisition several artefacts strongly degrades the signal quality. The dominant artefacts encountered in ECG signal such as Power Line Interference, Muscle Artefacts, Baseline Wander, Electrode Motion Artefacts; and channel noise generated during transmission. The tiny features of ECG signal are masked due to these noises. To track random variations in noisy signals, the adaptive filter is used. In this paper, we proposed Dead Zone Leaky Least Mean Square algorithm, Leaky Least Mean Froth algorithm and Median Leaky LMS algorithms to remove PLI and EM artefacts from ECG signals. Based on these algorithms, we derived some sign based algorithms for less computational complexity. We compare the proposed algorithms with LMS algorithm, which shows better performance in weight drift problem, round off error and low steady state error. The simulation results show that Dead Zone Leaky LMS algorithm gives good correlation factor and SNR ratio.

Keywords Adaptive filter · Artifacts · ECG · LMS algorithm · Noise cancellation

---

T. Gowri (✉)
Department of ECE, GIT, GITAM University, Visakhapatnam 530045, A.P., India
e-mail: gowri3478@yahoo.com

P. Rajesh Kumar
Department of ECE, AUCE, Andhra University, Visakhapatnam 530003, A.P., India
e-mail: rajeshauce@gmail.com

D.V.R. Koti Reddy
Department of Institute of Technology, AUCE, Andhra University, Visakhapatnam 530003, A.P., India
e-mail: rkreddy_67@yahoo.co.in

U.Z. Rahman
Department of ECE, KL University, Guntur 522502, A.P., India
e-mail: mdzr55@gmail.com

A. Nagar et al. (eds.), *Proceedings of 3rd International Conference on Advanced Computing, Networking and Informatics*, Smart Innovation, Systems and Technologies 43, DOI 10.1007/978-81-322-2538-6_64

1 Introduction

The World health organization reports that, ever year, millions of people in this industrialized world suffer cardiovascular disease (CVDs) [1] because they are not treated promptly. When the patient is far from specialist help, biotelemetry is an effective tool for the diagnosis of cardiac abnormalities, there a doctor analyzes the signal and decides on what action to be taken, the decision sent to the patient site for immediate action [2, 3].

When we acquiring ECG signal in clinical laboratory, the signal sometimes corrupted by different types of artefacts. The commonly occurred artefacts are Power Line Interference (PLI) [4], Baseline Wander (BW) [5], Electrode Motion (EM) and Muscle Artefacts (MA). These artefacts strongly affect the signal quality and are very important for clinical monitoring, diagnosis and making their efficient cancellation imperative. Therefore the separation of pure quality ECG signal from background noise corrupted ECG signal is a great importance for examination. So by using filtering techniques we have to distinct valid signal can be separated from the undesired artefacts.

Filtering is used to process a signal either to enhance the Signal to Noise Ratio (SNR) or to eliminate certain types of noises. In biomedical signal analysis when the input signal and artifacts are both stationary and their statistical characteristics are approximately known, and then an optimal filter like a Wiener filter or a matched filter can be used. In practical cases like biomedical signal analysis, a priori information is not available or when the signal or noise is non-stationary; then optimal design is not possible [6–8]. In such a case adaptive filter has an ability to adjust their weight coefficients based on the incoming signal.

Thakor and Zhu [9] presented an adaptive recurrent filter to remove muscle noise using LMS algorithm, which is useful method to acquire the impulse response of the normal QRS complex. The cancellations of baseline wanders, power line interferences by using an efficient finite impulse response and state space recursive least square filter with a reduced number of taps are described in [10–12]. Manuel et al. [13] developed an ECG signal enhancement method based on the empirical mode decomposition for non stationary signals.

In this paper, we propose three algorithms: Dead Zone Leaky LMS (DZLLMS), Leaky Least Mean Fourth (LLMF) and Leaky Median LMS (LMLMS). These algorithms are derived from LMS algorithms. The efficiency of the proposed algorithms are taken by the removal of noises from ECG signals under unconventional conditions such as ECG signal with PLI and EM artefacts. The simulation results shows that DZLLMS algorithm gives the better elimination of noises present in the ECG signal and also high correlation factor.

2 Some Variants of LMS Adaptive Algorithms

The adaptive filter structure is shown in the Fig. 1. Let the reference input to the filter assigned as **x(m)**. The tap inputs x(m), x(m − 1), x(m − 2), …. x(m − N + 1) forms the elements of the N-by-1 tap input vector **x(m)**, Correspondingly, the weight vector represented as **w(m)** = $[w_0, w_1, \ldots, w_{N-1}]^T$. The output y(m) is the FIR filter inner product of reference input and weight vector.

The error e(m) is defined as the difference between the desired response d(m) and the actual response y(m) i.e. e(m) = d(m) − y(m). The updated weight equation for the LMS algorithm is as follows.

$$\mathbf{w}(m+1) = \mathbf{w}(m) + \mu e(m)\mathbf{x}(\mathbf{m}). \quad (1)$$

The convergence of the LMS algorithm and mean square value mainly depends on the step size parameter μ, its value is chosen according to steepest decent phenomena is $2/\lambda_{max}$, where λ_{max}, is the maximum value in the input auto correlation matrix.

In ECG signal processing under critical conditions some of the samples in the ECG signal becomes zero, i.e., the excitation is inadequate. At these samples, the weights are varies drastically. The fluctuations in weights is called weight drift problem. For the tap weight vector in (1), we place small leakage factor γ, then in that algorithm the weight drift problem can be minimized [14, 15]. This algorithm is known as the leaky LMS (LLMS) algorithm. The weight update equation for the LLMS algorithm is as follows.

$$\mathbf{w}(m+1) = (1 - \mu\gamma)\mathbf{w}(m) + \mu e(m)\mathbf{x}(m) \quad (2)$$

In the above recursion (2), the product μγ is much closer to zero.

Sometimes small values of the error signal e(m) may causes disturbances or noises, which results numerical instability. In ECG noise cancellers, these small errors and large errors causes additional filtering operations, which lead to delay for making decision. To avoid these additional computations we need to set a threshold to the error value. In signal processing applications the Dead Zone LMS (DZLMS)

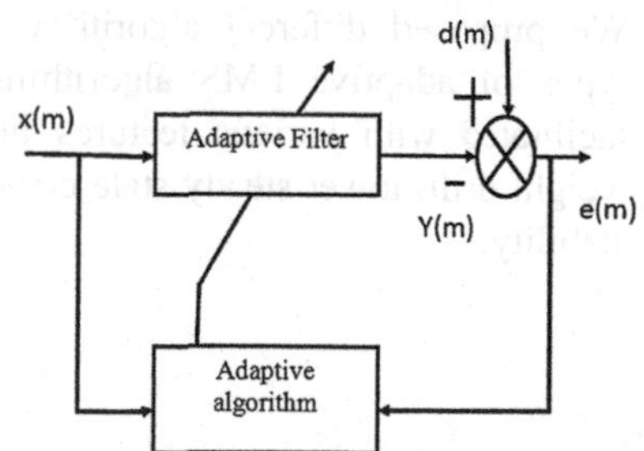

Fig. 1 Adaptive filter structure

is used to diminish the problems of round-off errors. This algorithm uses dead zone nonlinearity [16] and is defined as follows.

$$g\{z\} = \begin{cases} z - t, z > t > 0, \\ 0, -t < z < t, \\ z + t, z < -t. \end{cases} \tag{3}$$

where 't' is a threshold value, and is applied to the error signal in (1). The weight update equation for DZLMS is as follows.

$$\mathbf{w}(m + 1) = \mathbf{w}(m) + \mu g\{e(m)\}\mathbf{x}(m). \tag{4}$$

Adaptive filters based on higher order statistics performs better mean square estimation than LMS algorithm in particular scenarios. Least-Mean Fourth (LMF) algorithm [17] is based on the minimization of the fourth order moment of the output estimation error. The weight recursion for the LMF algorithm is given by

$$\mathbf{w}(m + 1) = \mathbf{w}(m) + \mu e^3(m)\mathbf{x}(m). \tag{5}$$

LMF algorithm exhibits lower steady state error than the conventional LMS algorithm due to the fact that the excess mean-square error of the LMS algorithm is dependent only on the second order moment of the noise.

In ECG signal processing, the amplitude of the P wave increases due to abnormal heart rhythm. Smoothing the noisy gradient components using a non-linear filter is a best remedy for this problem. The median function is used to reject single occurrence of large spikes of noise. This modification leads to Median LMS (MLMS) algorithm and the weight update equation is as follows.

$$\mathbf{w}(m + 1) = \mathbf{w}(m) + \mu \cdot \mathrm{med}_L[e(m)x(m), e(m - 1)x(m - 1)\ldots e(m - N)x(m - M)] \tag{6}$$

3 Proposed Various Adaptive Noise Cancellers for Cardiac Signal Enhancement

We proposed different algorithms by combining different features of different types of adaptive LMS algorithms as discussed above. These algorithms are facilitated with various features like diminish the problem of round off error, weight drift, lower steady state error, smoothes the impulsive noise with increased stability.

3.1 Dead Zone Leaky Least Mean Square Algorithm

By introducing leaky factor in DZLMS algorithm the resultant hybrid algorithm becomes as Dead Zone Leaky LMS (DZLLMS). This algorithms facilitates the noise canceller with the advantages of diminish the problem of round off error, weight drift can be avoided and stability increases. After incorporating these merits with reference to (2) and (4), the weight recursion is as follows.

$$\mathbf{w}(m+1) = (1-\mu\gamma)\mathbf{w}(m) + \mu g\{e(m)\}\mathbf{x}(m). \tag{7}$$

3.2 Leaky Least Mean Fourth Algorithm

In order to achieve lower steady state error and increasing stability, compared to the conventional LMS algorithm, we combine the features of LMF and LLMS described by (2) and (5). The resultant algorithm is Leaky Least Mean Fourth (LLMF) algorithm and mathematically it can be written as follows.

$$\mathbf{w}(m+1) = (1-\mu\gamma)\mathbf{w}(m) + \mu e^3(m)\mathbf{x}(m). \tag{8}$$

3.3 Median Leaky Least Mean Square Algorithm

In order to nullifying impulsive noise and increasing stability, we combine the features of MLMS and LLMS described by (2) and (6). The resultant algorithm is Median Leaky Least Mean Square (MLLMS) algorithm and its mathematical weight update equation can be written as follows.

$$\mathbf{w}(m+1) = (1-\mu\gamma)\mathbf{w}(m) + \mu.\mathrm{med}_L[e(m)\mathbf{x}(m), e(m-1)\mathbf{x}(m-1)\ldots e(m-L)\mathbf{x}(m-L)] \tag{9}$$

4 Extensions to Sign Based Realizations of Adaptive Filters

In the following, we propose new signum based algorithms which makes the used of signum function. We derive different new algorithms from the above DZLLMS, LLMF, MLLMS algorithms. The signum (polarity) function can be applied to the error or data or both. By applying signum function to the algorithm, the multiply and accumulate operations [18, 19] are reduced, and also computation time required to evaluate that algorithm operation reduced. Sign based treatment gives the reduction of computational complexity of the filter and hence suitable for biotelemetry applications. The signum function is defined as follows.

$$\mathrm{Sign}\{\mathrm{p(m)}\} = \begin{cases} 1 : \mathrm{p(m)} > 0, \\ 0 : \mathrm{p(m)} = 0, \\ -1 : \mathrm{p(m)} < 0. \end{cases} \tag{10}$$

In the weight updated equations, the data vector **x**(m) is replaced by Sign **x**(m), the algorithm is called as Signed-Regressor (SR) algorithm; it is also called clipping the input data. Sign Error algorithm or simply Sign (S) algorithm is obtained by replacing error e(m) by with its Sign e(m) in the weight update equations. Because of the replacement of e(m), the implementation of recursion becomes simple than the normal LMS algorithm recursion. In the updated weight equation the data vector **x**(m), error e(m) is replaced by Sign x(m) and Sign e(m), then this algorithm is called Sign-Sign LMS (SSLMS) algorithm. This SSLMS also called as zero forcing LMS because of zero multiplications in the implementation. The sign based techniques can be applied for the above all algorithms to derive the algorithms: SRDZLLMS, SDZLLMS, SSDZLLMS, SRLLMF, SLLMF, SSLLMF, SRMLLMS, SMLLMS, and SSMLLMS.

5 Convergence Characteristics

The convergence curves for various LMS variant adaptive algorithms discussed in the previous sections are shown in Fig. 2. These curves are plotted between MSE and number of iterations. MSE is calculated for each sample for 4000 iterations and the average value is taken for the characterization.

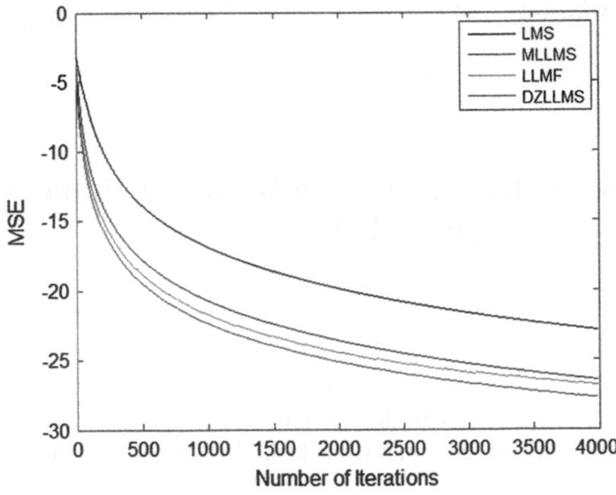

Fig. 2 Convergence curves of LMS based adaptive algorithms

Table 1 Computational complexity of various LMS based algorithms

S. no.	Algorithm	Multiplications	Additions
1	LMS	N + 1	N + 1
2	DZLLMS	N + 3	N + 1
3	LLMF	N + 3	N + 1
4	MLLMS	N + 3	N + 1

These curves are obtained during the adaptive power line interference (60 Hz) cancellation using various adaptive noise cancellers individually with a adaptive filter of length 5, random variance of 0.01 and step size 0.01. From Fig. 2 it is clear that, among all the algorithms DZLLMS outperforms. The performance of MLLMS and LLMF are comparable to each other and superior to LMS. The computational complexity for various LMS based adaptive filters are shown in Table 1.

6 Results and Discussion

To test the ability of the various gradient based adaptive algorithms and their sign based realizations discussed in this chapter we performed various experiments on real ECG signals with various artifacts. For our research work we have taken real ECG records from the physiobank MIT-BIH arrhythmia database [20]. The records are digitized with a sampling frequency of 360 Hz per channel, with a resolution of 11 bit over a range of 10 mV. In this database totally 48 ECG records available, each record, recorded with 30 min long duration. The 22 ECG records are collected from women, 25 records from men with different ages, and record numbers 201, 202 are collected from the same person.

In our simulation purpose we collected six ECG records from MIT-BIH database, and calculated Signal to Noise Ratio (SNR) in decibels, correlation factor, and observed the Excess MSE using different adaptive filters with MATLAB software. The output graphs shown here for data 105 record only, because of space limitation, rest are compared using table form. The number of samples is taken on x-axis and amplitude on y-axis for all figures. In our experiments, we have considered two dominant artefacts, namely PLI and EM.

The ECG signal is corrupted with a power line noise of amplitude 1 mV, with 60 Hz frequency. This corrupted ECG signal is applied at input signal and sampled with a frequency of 200 Hz. The reference signal is synthesized sinusoidal noise generated in the noise generator; and filter output is recovered signal. Four filter structures are designed using LMS, DZLLMS, LLMF and MLLMS algorithms. The simulation results for PLI cancellation using these algorithms are shown in Fig. 3.

The performance of the various algorithms for the removal of PLI is measured in terms of SNR, and is tabulated in Table 2. From Table 2 it is clear that DZLLMS algorithm achieves maximum average SNR over the dataset is 8.9847 dB, where as

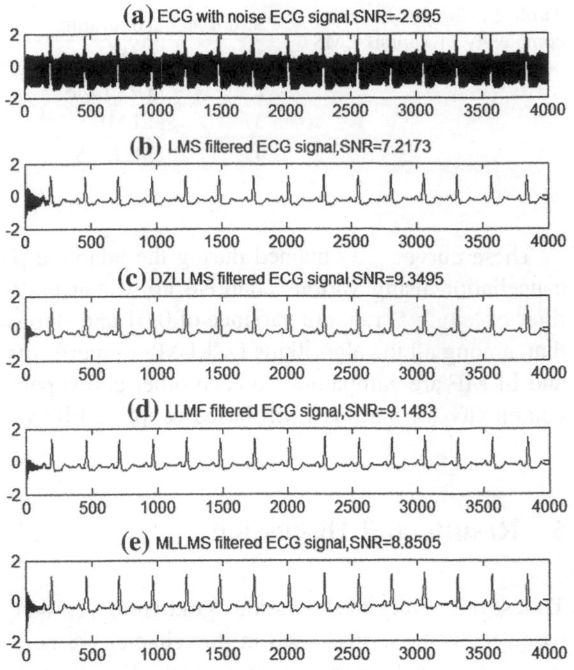

Fig. 3 PLI cancellation using different LMS based adaptive algorithms

Table 2 SNR contrast of LMS based adaptive algorithms for PLI removal

S. no	Rec. no	Before	LMS	DZLLMS	LLMF	MLLMS
1	100	−2.9191	7.0122	9.3848	8.6663	9.1262
2	101	−2.8062	7.0785	9.3327	8.8901	8.9067
3	102	−3.9981	5.7999	7.8636	7.1561	8.0451
4	103	−2.5193	7.3346	9.122	3.4126	9.0657
5	104	−2.9763	6.9617	8.8576	8.3986	9.2303
6	105	−2.6951	7.2159	9.348	9.1486	8.8495
Average SNR			6.9000	8.9847	7.6120	8.8705

MLLMS gets 8.8705 dB, LLMF gets 7.6120 dB and LMS gets 6.9000 dB, these values correlate with the convergence characteristics shown in Fig. 2. To enjoy the less computational complexity the LMS variant adaptive algorithms are combined with signed algorithms and results several algorithms as discussed above. Using these algorithms various ANCs are developed and tested for adaptive PLI removal as shown in Table 3. The sign regressor algorithm gets better elimination of noise. The correlation between desired signal and practical output response compared to derived algorithms is shown in Table 4.

Table 3 SNR contrast of sign based adaptive algorithms for PLI removal

S. no.	Algorithm	Rec. no. 100	Rec. no. 101	Rec. no. 102	Rec. no. 103	Rec. no. 104	Rec. no. 105	Avg SNR
1	SRLMS	8.6338	8.6225	7.5949	8.8461	8.5185	8.6439	8.4766
2	SLMS	5.8667	6.7402	6.6192	7.9602	7.4863	7.0099	6.9740
3	SSLMS	4.8457	6.9779	6.2527	7.7331	6.984	6.7625	6.5926
4	SRDZLLMS	8.9139	8.9578	7.6823	8.7732	8.44	8.9069	8.6123
5	SDZLLMS	8.2388	8.2405	7.4878	8.5106	8.357	8.2853	8.1866
6	SSDZLLMS	7.8969	8.0629	7.2207	8.2789	8.0818	8.1151	7.9427
7	SRLLMF	8.3034	8.542	6.9629	4.2175	8.1486	8.6033	7.4629
8	SLLMF	4.5327	6.3849	6.1621	6.9967	6.902	6.8711	6.3082
9	SSLLMF	4.114	5.9168	6.2493	7.6741	7.1098	6.8111	6.3125
10	SRMLLMS	7.1242	7.5206	6.6655	7.6039	7.3624	7.4196	7.2827
11	SMLLMS	4.5024	6.1029	6.224	7.2474	7.202	7.1454	6.4040
12	SSMLLMS	4.578	5.3369	4.7954	6.019	5.5528	5.729	5.3351

Table 4 Correlation factor analysis for removal of PLI

Rec. no	LMS	DZLLMS	LLMF	MLLMS
100	0.921	0.974	0.962	0.970
101	0.943	0.981	0.976	0.976
102	0.922	0.968	0.956	0.973
103	0.974	0.990	0.931	0.989
104	0.966	0.988	0.986	0.989
105	0.972	0.991	0.990	0.987
Avg Cor	0.949	0.982	0.966	0.980

From MIT-BIH normal sinus rhythm database and noise stress test [21] the real EM noise is collected for testing of the signal using different filters in non-stationary conditions. In this database totally 18 subjects there, with no significant arrhythmias. For this real EM noise we added SNR of 1.25 dB, with a random variance noise of 0.01, because when we transmit a signal form source point to destination point for diagnosis the channel noise added which is here taken as random noise. The real EM noise has a sampling frequency of 360 Hz so that it is anti-aliased and re-sampled to 128 Hz in order to match the ECG sampling rate. Various filter structures are applied for removing of this noise, the simulation results for all LMS based ANCs are shown in Fig. 4. The SNR for various filtering methods is shown in Table 5. From this table it is clear that LLMF effectively filters EM noise, it got an average SNRI of 5.308 dB, DZLLMS gets 5.275 dB, MLLMS gets 4.911 dB and LMS get 4.236 dB. The excess mean square error (EMSE) characteristics for the removal of PLI noise and EM noise as shown in Figs. 5 and 6.

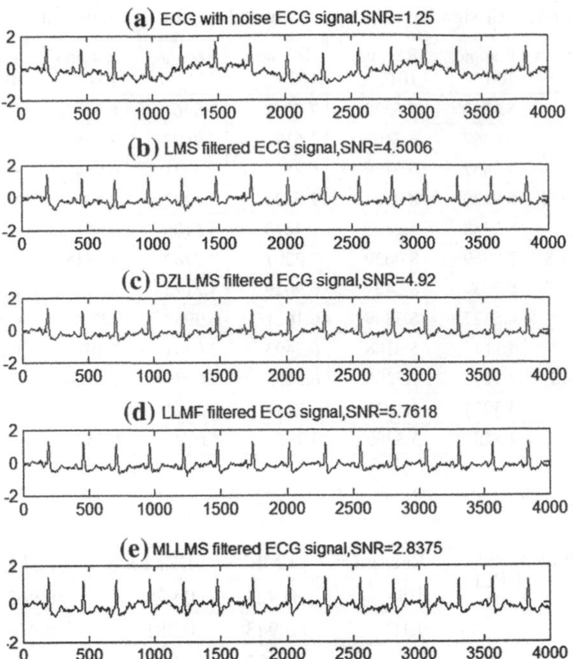

Fig. 4 EM artifact removal using various LMS based adaptive algorithms

Table 5 SNR contrast of proposed adaptive algorithms for EM removal

Rec. no	LMS	DZL LMS	L LMF	ML LMS
100	3.541	6.093	5.957	4.453
101	3.528	5.075	5.711	4.419
102	4.883	6.890	6.489	5.075
103	4.534	5.844	4.595	5.131
104	4.433	2.828	3.337	5.224
105	4.501	4.920	5.763	5.164
Avg SNR	4.236	5.275	5.308	4.911

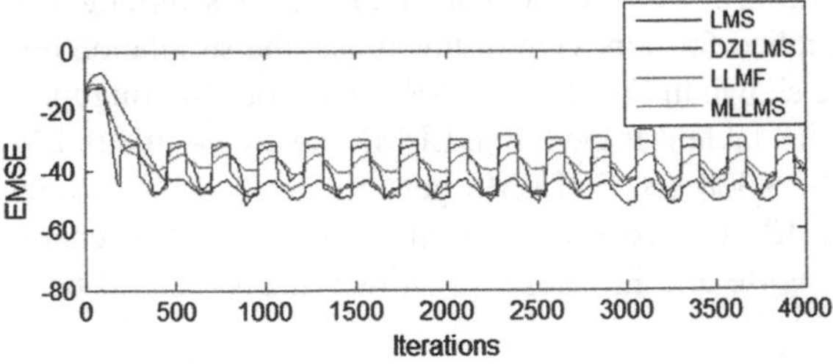

Fig. 5 EMSE characteristics for PLI reduction using different adaptive algorithms

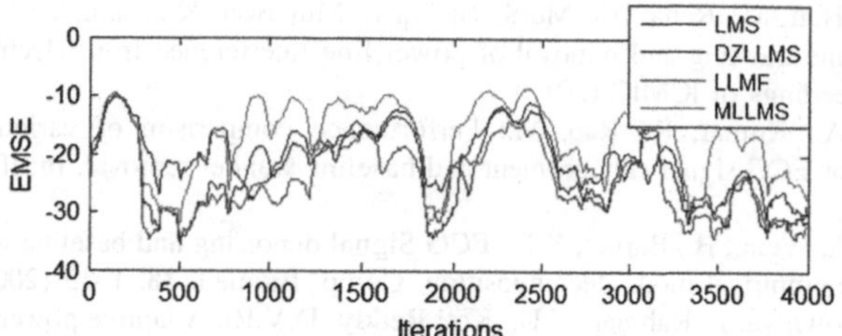

Fig. 6 EMSE characteristics for EM reduction using different adaptive algorithms

7 Conclusion

In this paper we developed weight update variants of LMS based adaptive algorithms are DZLLMS, LLMF and MLLMS. The signed algorithms are derived based on these three algorithms for less number of multiplications and additions. To evaluate the performance of the various ANCs we have plotted the convergence characteristics, computational complexity, SNR and correlation factor. Among the three LMS based algorithms DZLLMS algorithm performs better in the non stationary noise removal and among the sign based versions sign regressor version of algorithms performs better reduction of noise with less computational complexity.

References

1. World Health Organization Report.: Preventing chronic diseases a vital investment. World Health Organization, Geneva (2005). http://www.who.int/chp/chronic_disease_report/en/
2. Derya, E.Ü.: Clinical Technologies: Concepts, Methodologies, Tools and Applications. Telemedicine and Biotelemetry for E-Health Systems. IGI Publications, Berlin (2011)
3. McMurray, J.J., Pfeffer, M.A.: Heart failure. Lancet **365**(9474), 1877–1889 (2005)
4. Suzanna, M., Martens, M., Mischi, M., Oei, S.G., Bergmans, J.W.M.: An improved adaptive power line interference canceller for electrocardiography. IEEE Trans. Biomed. Eng. **53**(11), 2220–2231 (1996)
5. Sayadi, O., Shamsollahi, M.B.: Multiadaptive bionic wavelet transform: application to ECG denoising and baseline wandering reduction. Eurasip J. Adv. Signal Process. **2007** (Article ID 41274), 11 (2007)
6. Widrow, B., Glover, J., McCool, J.M., Kaunitz, J., Williams, C.S., Hearn, R.H., Zeidler, J.R., Dong, E., Goodlin, R.: Adaptive noise cancelling: principles and applications. Proc. IEEE **63**, 1692–1716 (1975)
7. Patrick, S.H.: A comparison of adaptive and non adaptive filters for reduction of power line interference in the ECG. IEEE Trans. Biomed. Eng. **43**(1), 105–109 (1996)
8. Haykin, S.: Adaptive filter theory. Prentice-Hall, Eaglewood Clirs (1986)
9. Thakor, N.V., Zhu, Y.S.: Applications of adaptive filtering to ECG analysis: noise cancellation and arrhythmia detection. IEEE Trans. Biomed. Eng. **38**(8), 785–794 (1991)
10. Van Alste, J.A., Schilder, T.S.: Removal of base-line wander and power-line interference from the ECG by an efficient FIR filter with a reduced number of taps. IEEE Trans. Biomed. Eng. **32**(12), 1052–1060 (1985)

11. Razzaq, N., Butt, M., Rahat Ali, Md.S., Sadiq, I., Munawar, K., Zaidi, T.: Self tuned SSRLS filter for online tracking and removal of power line interference from electrocardiogram. In: U2013 Proceedings of lCMIC (2013)
12. Rehman, S.A., Kumar, R., Rao, M.: Performance comparison of various adaptive filter algorithms for ECG signal enhancement and baseline wander removal. In; IEEE-ICCCNT'12 (2012)
13. Manuel, B.V., Weng, B., Barner, K.E.: ECG Signal denoising and baseline wander correction based on the empirical mode decomposition. Comp. Biomed. **38**, 1–13 (2008)
14. Gowri, T., Sowmya, I., Rahman, Z.U., Koti Reddy, D.V.R.: Adaptive power line interference removal from Cardiac signals using Leaky based normalized higher order filtering techniques. In: IEEE 1st International Conference on AIMS, 259–263 (2013)
15. Tobias, O.J., Seara, R.: Leaky Delayed LMS Algorithm: stochastic analysis for Gaussian data and delay modeling error. IEEE Trans. Signal Process. **52**(6), 1596–1606 (2004)
16. Heiss, M.: Error-minimizing dead zone for basis function networks. IEEE Trans. Neural Netw. **I**(6), 1503–1506 (1996)
17. Walach, E., Widrow, B.: The least mean fourth adaptive algorithm and its family. IEEE Trans. Inf. Theory **30**, 275–283 (1984)
18. Gowri, T., Rajesh Kumar, P., Koti Reddy, D.V.R.: An efficient variable step size Least Mean Square Adaptive Algorithm used to enhance the quality of electrocardiogram signal. Adv. Signal Process. Intell. Recogn. Syst. AISC **264**, 463–475 (2014)
19. Rahman, Z.U.Md., Ahamed, S.R., Koti Reddy, D.V.R.: Efficient and simplified adaptive noise cancellers for ECG sensor based remote health monitoring. IEEE Sens. J. **12**(3), 566–573 (2012)
20. http://www.physionet.org/physiobank/database/mitdb/. MIT-BIH Arrhythmia Database
21. http://www.physionet.org/physiobank/database/nsrdb/. The MIT-BIH Normal Sinus Rhythm Database

Detecting Tampered Cheque Images Using Difference Expansion Based Watermarking with Intelligent Pairing of Pixels

Mahesh Tejawat and Rajarshi Pal

Abstract Image based cheque clearing systems enable faster clearing of cheques in banking system. But the paying bank should receive an unaltered image of a genuine cheque so that its decision does not support a case of fraud. The method, proposed in this paper, successfully detects tampered cheque images. The proposed method is based on a fragile and reversible watermarking technique, namely difference expansion based watermarking. But the major problem with these kinds of reversible watermarking techniques is that the pixel values in the watermarked image often fall outside the dynamic range of the image. This paper demonstrates how intelligent pairing of pixels can solve this problem. Therefore, the revised difference expansion based watermarking (based on intelligent pixel pairing) successfully detects tampered images of cheques.

Keywords Document tamper detection · Cheque fraud detection · Watermarking · Reversible watermarking

1 Introduction

The paying bank must receive an unaltered version of the cheque image. No one should be able to commit frauds by maliciously tampering the contents of the cheque image such as payee name, amount, etc. According to the specifications of Cheque Truncation System (CTS) [1], which is an image based cheque clearing system, the transfer of cheque image from the presenting bank to the paying bank

M. Tejawat (✉)
University of Hyderabad, Hyderabad, India
e-mail: mahesh.griet@gmail.com

R. Pal
Institute for Development and Research in Banking Technology (IDRBT), Hyderabad, India
e-mail: iamrajarshi@yahoo.co.in

A. Nagar et al. (eds.), *Proceedings of 3rd International Conference on Advanced Computing, Networking and Informatics*, Smart Innovation, Systems and Technologies 43, DOI 10.1007/978-81-322-2538-6_65

through an intermediate clearing house is secured by applying asymmetric key encryption in phases. But an end-to-end (i.e., from capturing system of the presenting bank to the paying bank) encryption cannot be adopted. Because, the content of the image has to be accessed at various processing nodes (including the clearing house). Therefore, an unencrypted image of the cheque is available at these processing nodes. This leaves the image vulnerable to malicious tampering. Hence, detection of such a tampering of a cheque image is necessary.

Image forensic techniques mainly focus on identifying tampered images based on reflection [2], shading and shadows [3]. The work in [4] proposes to detect manipulations in the images of signs and billboards based on the deviations in projecting texts on a planar surface. Tampering a cheque image can be as small as modifying '1' to '4' in the amount in figure field and "one" to "four" in corresponding texts (amount in words). So, there is no question of artifacts arising from reflection, shadow or lighting. Moreover, this type of modification is also not related to projecting texts on a planar surface. Therefore, such image forensic techniques [2–4] are not fruitful to detect tampered images of cheques.

Fragile image watermarking based solutions can be used to detect these tampering of cheque images. The watermark is embedded in the image. The watermarked cheque image is, then, communicated to the paying bank. The paying bank extracts the watermark from the received image. Mismatch between the extracted and the known watermark suggests the event of tampering. The work in [5] adopts this strategy based on the watermarking proposed in [6]. But according to this method, the presenting bank must send both the cover and the watermarked images to the paying bank. This is a drawback of this method.

The work in [7] has applied the difference expansion based watermarking scheme [8] to detect tampering of cheque images. But the problem with this difference expansion based scheme [8], as well as any other reversible watermarking, is that the watermark pixel values may fall outside the range [0,255]. When a watermarked pixel value becomes higher than 255, an overflow occurs. Similarly, the case of watermarked pixel value becoming less than 0 is known as underflow. The difference expansion based scheme, as suggested in [8], cannot insert the watermark bits in pixel pairs where there is a chance of occurring these problems (dubbed as non-changeable). A solution to this has been proposed in [7], where suitable methods of inserting watermark at non-changeable pixel pairs have been devised.

This paper introduces the concept of intelligent pairing of pixels in this context such that non-changeable cases do not arise at all. It helps in inserting watermark by expanding the difference between the pixel values without causing overflow and underflow problems. Therefore, the proposed intelligent pairing of pixels enables to apply difference expansion based watermarking in cheque images. The proposed method provides another alternative (apart from the method in [7]) to detect fraudulent tampering of cheque images.

The rest of the paper is organized as follows: In Sect. 2, a brief overview of difference expansion based watermarking is explained as this forms the baseline of the proposed method. Section 3 proposes the intelligent grouping of pixels to overcome underflow and overflow problems. In Sect. 4, the complete scenario of inserting watermark and detection of tampered cheque images is described. Section 5 presents the experimental results. Finally, Sect. 6 draws the conclusion of this paper.

2 Difference Expansion Based Reversible Watermarking

This section discusses the basic philosophy of difference expansion based watermarking (proposed in [8]) as it forms the basis of this paper. One bit watermark is inserted in a pair of 8-bit gray scale pixel values (*x*, *y*) where $0 \leq x,\ y \leq 255$. The basic steps of this insertion algorithm are as follows:

The integer average l and the difference h of the pair of values are calculated using:

$$l = \left\lfloor \frac{x+y}{2} \right\rfloor, \quad h = x - y \tag{1}$$

One watermark bit b is embedded in the difference value as:

$$h' = 2 \times h + b \tag{2}$$

where $b = 0$ or 1.

The watermarked pixel values are calculated using:

$$x' = l + \left\lfloor \frac{h'+1}{2} \right\rfloor, \quad y' = l - \left\lfloor \frac{h'}{2} \right\rfloor \tag{3}$$

As the watermarked image is also an 8-bit gray scale image, the intensity values should be in the range of [0,255].

$$0 \leq x',\ y' \leq 255$$

The difference h is multiplied by 2 while inserting watermark b using Eq. (2) in-order to make the scheme reversible. But due to this multiplication of the difference between the pair of pixel values, in certain cases, the watermarked pixel values may not fall in this range. For example, let two pixel values in the original cheque image be $x = 254$ and $y = 240$. Also let us assume that the watermark bit $b = 0$. Applying Eqs. (1), (2), and (3), the watermarked pixel values become $x' = 261$ and $y' = 233$. In this example, x' falls outside the dynamic range of a 8-bit gray scale image and hence, causes an overflow.

Expandable Difference Values As suggested in [8], the difference value h is expandable for the average value l if and only if

$$h' = |2 \times h + b| \leq \min(2(255 - l),\ 2l + 1) \tag{4}$$

for both $b = 0$ and 1.

Changeable Difference Values As suggested in [8], the difference value h is changeable for the average value l if and only if

$$h' = \left|2 \times \left\lfloor \frac{h}{2} \right\rfloor + b\right| \leq \min(2(255 - l),\ 2l + 1) \tag{5}$$

for both $b = 0$ and 1.

If the difference value cannot be categorized either as expandable or changeable value, then watermark can not be embedded in the difference.

3 Proposed Intelligent Grouping of Pixels

As discussed earlier, difference expansion based watermarking [8] may cause overflow and underflow problems. An approach to avoid this problem has been proposed in [7] and also discussed in Sect. 1. In this section, another method to mitigate this problem has been proposed using intelligent pairing of pixels. Equation (4) suggests following two points: (1) Small differences are expandable, whereas large differences are not expandable. (2) Whether a difference h is small (or large) enough to be (or not to be) expanded is determined by the positioning of the integer average l of the concerned pair of pixel values with respect to two extreme values 0 and 255. These observations lead to an idea that an intelligent pairing of pixels will no more cause overflow and underflow problems. This pairing is carried out based on a group division of intensity values.

Let the universe of intensity values be denoted by

$$U = \{i | 0 \leq i \leq 255\}$$

Let the smallest integer in the set U be assigned in a variable y. Therefore, initially, $y = 0$. The set of values of x which satisfy the Eq. (4) for the assumed values of y is determined. It suggests that the pair $(x = 0,\ y = 0)$ will not cause overflow and underflow problems. So, they can be paired together. Equation (4) indicates that pairing of any value other than zero with a zero value causes underflow. Let this set be denoted by G_1. It can be seen that $G_1 = \{0\}$.

Next, U is modified to $U - G_1$. Again the smallest integer in the set U is considered to be y. The subset of values in this set, which satisfies the Eq. (4) for x, is determined. Let this subset is denoted by G_2. In this case $G_2 = \{1, 2, 3\}$. It suggests

Table 1 Division of intensity values into groups

Groups	Intensity values
G_1	0
G_2	1, 2, 3
G_3	4, 5, … 12
G_4	13, 14, … 39
G_5	40, 41, … 120
G_6	121, 122, … 210
G_7	211, 212, … 240
G_8	241, 242, … 250
G_9	251, 252, 253
G_{10}	254
G_{11}	255

that the values in G_2 can only be paired with the values in G_2. Otherwise, underflow problem arises. Once again the set is reduced to $U = U - G_2$. This procedure is repeated until $U = \phi$, i.e., the null set. At this point, all integers from 0 to 255 have been put under any one of the groups G_i. This experiment divides the integers into 11 groups as shown in Table 1.

This procedure ensures that there is no common element in the sets G_i and G_j (where i and j are not equal). On the other word, $G_i \cap G_j = \phi$. Moreover, $\bigcup_{i=1}^{11} G_i$ includes all integers from 0 to 255. A pair of values (x, y) satisfying Eq. (4) suggests that this pair is expandable, i.e., watermark bit can be inserted without causing an overflow and underflow problems. So, selection of x and y values from the same group ensures that these problems can be avoided.

4 Detection of Tampered Cheque Images

A method to detect tampered cheque images is depicted in this section. It applies the difference expansion based watermarking in every pair of pixels in the important regions of the cheque image. But an intelligent pairing of these pixels (as proposed in previous section) distinguishes this scheme from earlier approaches [7]. Embedding of watermark in the selected pair of pixels is same as that of difference expansion method (Eqs. (1)–(3)). This section describes the steps of inserting watermark in the important regions of a cheque image.

4.1 Insertion of Watermark

This section describes the method of inserting watermark in important regions of the cheque image without causing overflow and underflow (the watermarked pixel values will remain in the range [0,255]). The insertion of watermark algorithm comprises of the following steps:

Identifying Important Regions of the Cheque Image Frauds are carried out in a cheque image by altering the contents at important regions. The seven important regions of a cheque image are identified to contain (a) date, (b) payee name, (c) amount in words, (d) amount in numbers, (e) account number, (f) signature, and (g) MICR code.

A template based identification of these important portions of the image is adopted. A template is a binary image of same size as the cheque image, where pixels in important regions are indicated with a value 1, and other pixels are having a value 0. A sample cheque image is shown in Fig. 1. Important regions as identified by an appropriate template are shown in Fig. 2. The format of a cheque varies from bank to bank. Therefore, distinct templates need to be designed for each bank.

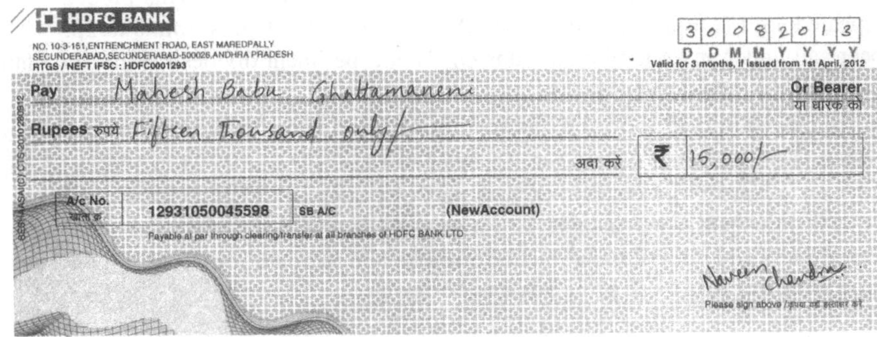

Fig. 1 Original cheque image

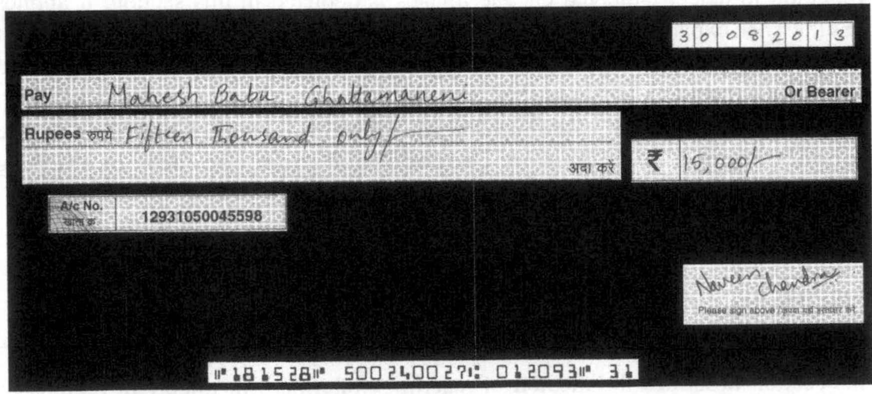

Fig. 2 Seven important regions of original cheque image

Moreover, the same bank may have various formats of cheques. Different templates have to be designed for each of these formats. Based on the format and the issuing bank of the cheque, an appropriate template is invoked to identify the important portions of the cheque.

Intelligent Pairing of Pixels Section 3 discusses how intensity values are placed into eleven groups. Now pixels in the seven important regions of the cheque image are assigned into these eleven groups based on their intensity values. Pixels belonging to the same group are traversed in a row-major sequence and watermark is inserted into each successive pair of pixels in that traversal.

An example shown in Fig. 3 explains this procedure. Let us assume that the red rectangle in the original image (in Fig. 3) shows the pixels belonging to the important region and outside of it are the pixels belonging to un-important region. Now pixels of the important region are assigned into 11 groups based on their intensity values as shown in Table 1 (i.e., group information). For example, the

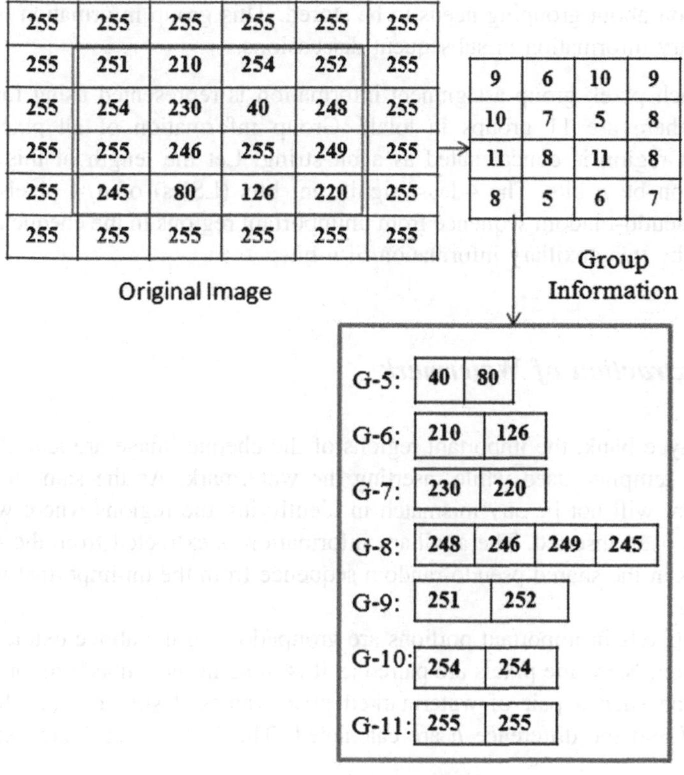

Fig. 3 Example of pairing of pixels

top-left pixel in the red rectangle having intensity value 251 is assigned to group G_9 as suggested by Table 1. Then the pixels belonging to each group are traversed in row-major order and successive pixels are teamed up. For example, according to Fig. 3, there are four pixels belonging to G_8. A row-major traversal forms the pairs as $(248, 246)$ and $(249, 245)$.

Embedding Watermark Successive pixels of a group are paired to insert 1-bit of watermark in the difference of each pair of pixel values. The length of the binary watermark signal is equal to half of the length of total number of pixels present in the seven important regions. Pixels from each of these groups are traversed in row major order and one-bit of watermark is inserted at each pair of pixels using difference expansion based scheme using Eqs. (1)–(3).

Intelligent pairing of pixels depends upon the grouping of intensity values as shown in Table 1. Each pair of pixel values from the important regions in the original cheque image (x, y) is transformed to $x^{'}$ and $y^{'}$ in the watermarked image.

Storing Group Assignment Information In order to retrieve the correct watermark, the knowledge of pairing of pixels during embedding is required. So, information about grouping needs to be stored. This group information is referred as auxiliary information in subsequent discussions.

For each pixel, group assignment information is represented using four binary bits (as there are 11 groups in total). Group information of all pixels in the important region is concatenated as a bit string. Let the length of this auxiliary information be n bits. The 4-least significant bits (LSBs) of $n/4$ pixels selected using a pseudo-random sequence from unimportant regions in the cheque image are replaced by this auxiliary information of n bits.

4.2 *Extraction of Watermark*

At the payee bank, the important regions of the cheque image are identified using the same template used while inserting the watermark. As the same template is used, there will not be any mismatch in identifying the regions where watermark bits have been inserted. The auxiliary information is extracted from the 4LSBs of the pixels in the shared pseudo-random sequence from the un-important regions of the image.

Next, pixels in important portions are grouped using the above extracted group information. Now, the pixels are paired as it is done in the embedding process. Let x and y be such a pair of watermarked pixel values. Using Eq. (1) the integer average l and the difference h are calculated. The LSB bit of h are extracted to obtain the watermark bit b, and new difference value h' is calculated using the following Eq. (6):

$$h' = \left\lfloor \frac{h-b}{2} \right\rfloor \tag{6}$$

The extracted pixel pairs (x', y') are generated from the newly obtained difference h' and the integer average l using Eq. (3).

4.3 Detecting Tampered Cheques

The paying bank receives the watermarked image. Paying bank extracts the sequence of watermark bits from the watermarked image and compares the extracted watermark with the inserted watermark which is known to it. If the watermark is retrieved correctly, then the paying bank is sure that the received cheque image is unaltered, else someone has tampered the cheque image and paying bank refuses to clear the cheque.

5 Experimental Results

Figure 1 shows an original cheque image of HDFC Bank. A sequence of 1's and 0's of length as half of the total number of pixels in the important regions is taken as the watermark. The important regions of the cheque are identified using the template as shown in Fig. 2. The watermark is embedded in the cheque image as described in Sect. 4.1. Figure 4 shows the obtained watermarked image from the original cheque image (Fig. 1). From this watermarked image, the watermark is correctly retrieved.

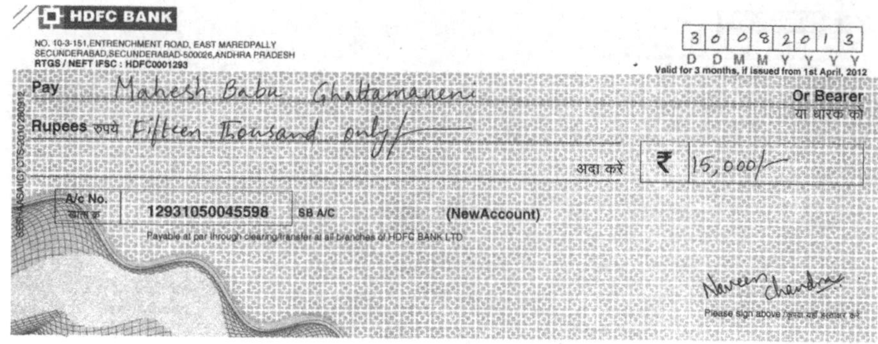

Fig. 4 Watermarked cheque image of the image in Fig. 1

Moreover, the important regions of the cheque are completely restored (Fig. 5). The un-important regions cannot be completely restored because LSB replacement (which is carried out in several pixels in these regions) is irreversible. But, it can be concluded from Figs. 1 and 5 that this change in un-important regions is very nominal. This change is limited in the LSBs of few pixels in unimportant regions. Therefore, this is not perceivable normally.

Figure 6 shows an image which is obtained by tampering the watermarked image in Fig. 4. The payee name has been modified in this example. The inserted sequence of watermark bit could not be retrieved from this modified image. This is because modification has destroyed the watermark signal present in the watermarked image.

Tests are carried out on a cheque image database having 44 cheques from various banks. Moreover, this dataset also contains $44 \times 7 = 308$ tampered images (each genuine cheque image has been tampered in seven different ways). In all above cases, the proposed method differentiates between genuine and tampered images.

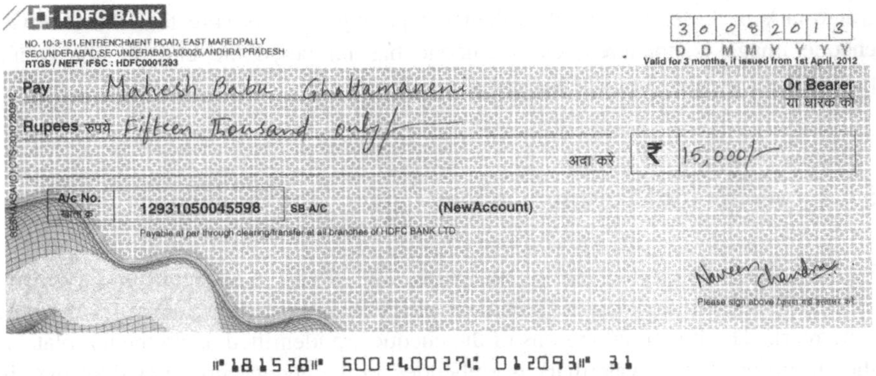

Fig. 5 Restored cheque image from the watermarked image in Fig. 4

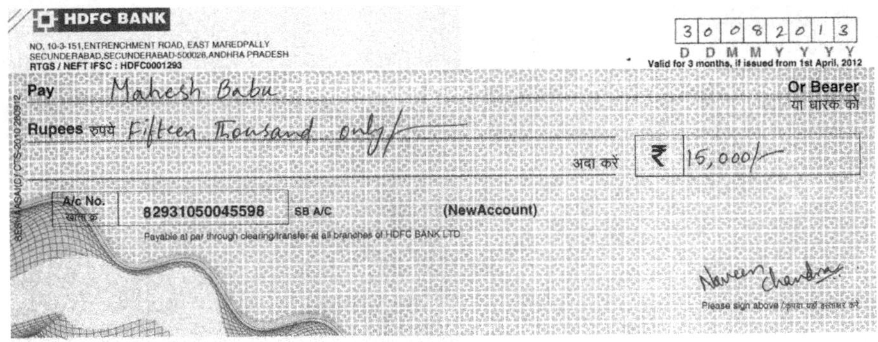

Fig. 6 Tampered cheque image obtained from the image in Fig. 4

6 Conclusion

The proposed method enables the paying bank to detect whether a received image is genuine or not. Thus, this method helps in preventing frauds which are committed by tampering of cheque images. Therefore, this method enhances the security of the cheque clearing process.

A difference expansion [8] based watermarking technique has been adopted for this purpose. But this difference expansion based watermarking suffers due to underflow and overflow problems. An intelligent pairing based solution has been proposed to overcome this problem. This is the key contribution of this paper. The work reported in [7] also detects tampered cheque images without any failure. But the proposed work in this paper is another way of addressing the problem. Instead of categorizing pairs of pixels as in [7], this method attempts to cure the root cause of the underflow and overflow problems. When the difference between the pair of pixel values are more compared to the distance of the values from either 0 or 255, then underflow and overflow take place. Therefore, intelligently pairing up pixels solves the problem.

References

1. NCR Corporation and RBI/NPCI.: CTS clearing house interface—specifications (2010)
2. O'Brien, J., Farid, H.: Exposing photo manipulation with inconsistent reflections. ACM Trans. Graph. **1**, **4**(1–4), 11 (2012)
3. Kee, E., O'Brien, J., Farid, H.: Exposing photo manipulation from shading and shadows. ACM Trans. Graph. **33**, 168–185 (2014)
4. Conotter, V., Boato, G., Farid, H.: Detecting photo manipulations on signs and billboards. In: Proceedings of 17th IEEE International Conference on Image Processing, pp. 1741–1744 (2010)
5. Rajender, M., Pal, R.: Detection of manipulated cheque images in cheque truncation system using mismatch in pixels. In: Proceedings of 2nd International Conference on Business and Information Management, pp. 30–35 (2014)
6. Mukherjee, D.P., Maitra, S., Acton, S.T.: Spatial domain digital watermarking of multimedia objects for buyer authentication. IEEE Trans. Multimedia **6**, 1–15 (2004)
7. Kota, S., Pal, R.: Detecting tampered cheque images in cheque truncation system using difference expansion based watermarking. In; Proceedings of IEEE International Conference on Advance Computing, pp. 1041–1047 (2014)
8. Tian, J.: Reversible data embedding using difference expansion. IEEE Trans. Circuits Syst. Video Technol. **13**, 890–896 (2003)

Printed Odia Digit Recognition Using Finite Automaton

Ramesh Kumar Mohapatra, Banshidhar Majhi and Sanjay Kumar Jena

Abstract Odia digit recognition (ODR) is one of the intriguing areas of research topic in the field of optical character recognition. This communication is an attempt to recognize printed Odia digits by considering their structural information as features and finite automaton with output as recognizer. The sample data set is created for Odia digits, and we named it as Odia digit database (ODDB). Each image is passed through several precompiled standard modules such as binarization, noise removal, segmentation, skeletonization. The image thus obtained is normalized to a size of 32×32 2D image. Chain coding is used on the skeletonised image to retrieve information regarding number of end points, T-joints, X-joints and loops. It is observed that finite automaton is able to classify the digits with a good accuracy rate except the digits ୧, ୨, and ୭.. We have used the correlation function to distinguish between, ୧, ୨, and ୭.. For our experiment we have considered some poor quality degraded printed documents. The simulation result records 96.08 % overall recognition accuracy.

Keywords ODR · Finite automaton · Correlation · Chain coding

R.K. Mohapatra (✉) · B. Majhi · S.K. Jena
Department of Computer Science and Engineering, National Institute of Technology, Rourkela 769008, India
e-mail: mohapatrark@nitrkl.ac.in

B. Majhi
e-mail: bmajhi@nitrkl.ac.in

S.K. Jena
e-mail: skjena@nitrkl.ac.in

A. Nagar et al. (eds.), *Proceedings of 3rd International Conference on Advanced Computing, Networking and Informatics*, Smart Innovation, Systems and Technologies 43, DOI 10.1007/978-81-322-2538-6_66

1 Introduction

A correct interpretation of digits, characters, and symbols from printed documents and other materials, is very much important in the field of document image analysis. The Optical digit recognition (ODR) is a mechanism of automatic recognition of numerals from scanned and digitized papers and other documents. Albeit it is a challenging task, it contributes gigantically to the furtherance of the automation process and enhances the man-machine interface in many practical applications [1, 2] such as:

(a) Helping blind people in reading,
(b) Automatic text entry into the computer,
(c) Preserving historical documents in digitized form, and
(d) Automatic sorting of postal mails through ZIP/PIN code, bank cheques, and other documents.

Odia is the official language of the state Odisha (Former Orissa) accounting for over 40 million people and the second official language of Jharkhand (another state in India). The Odia script arose from the Kalinga script that is the descendant of the Brahmi script of ancient India. It is the sixth classical language among many Indian languages. Now-a-days, automatic analysis and precise recognition of on-line and off-line optical characters [3] is one of the human necessities and intriguing area of research. The Odia digits and their corresponding English numerals are shown in Fig. 1.

Based on the feature vector matching, Akiyama and Hagita [4] have proposed a multi-font character recognition system and have done the experiment for a large set of perplexed written documents. The suggested system is adequate to read various types of preprint documents with an accuracy of 94.8 %. Chaudhuri and Pal [5] have proposed a system to recognize characters from the printed Bangla script with the constraint that the document should have Linotype font text. They claimed an overall accuracy of about 96 % for the text printed on plain paper. In [6] Chaudhuri et al. have described a strategy for the recognition of printed Oriya script and arrogated an overall accuracy of 96.3 % including both Odia text and numerals. They examined their proposed system on an extensive set of published Oriya material. To the best of our knowledge, we observed that a numerable amount of work has been carried out until now for the recognition of printed Odia digits.

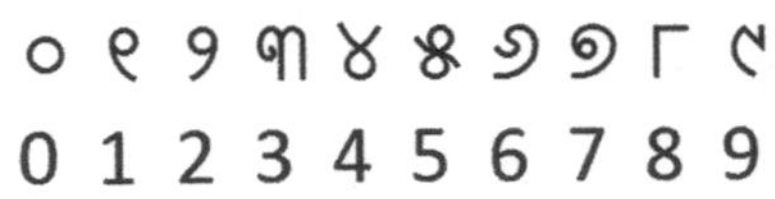

୦ ୧ ୨ ୩ ୪ ୫ ୬ ୭ ୮ ୯
0 1 2 3 4 5 6 7 8 9

Fig. 1 Sample Odia digit data set and its corresponding English digits

The paper is organized as follows. In Sect. 2 the proposed model for the recognition of printed Odia digit and the associated preprocessing methods are discussed. The observational results are presented in Sect. 3 followed by concluding remark in Sect. 4.

2 Proposed Model for Printed Odia Digits

In the Fig. 2, we delineated the steps for the recognition of printed Odia numerals. In our proposed methodology, a database containing the images of printed Odia digits (ODDB) has been maintained, which are collected from various printed documents. It contains printed digits of different shapes, sizes, and fonts. Any test image is first pre-processed using some conventional techniques. This module will perform a series of operations such as binarization followed by the line, word, character segmentations [7], and thinning to make the image manageable for the next phase. Each segmented character is skeletonised using some morphological operations. Use of Freeman's encoding [8, 9] over skeletonized image is explained in Sect. 2.2. From this sequence, we generate a string of length four, $\langle a_1 a_2 a_3 a_4 \rangle$, where a_1, a_2, a_3 and a_4 represent the number of end points, T-joints, X-joints, and loops respectively. All these information are concatenated to generate a string of length four as shown in Table 1. String thus obtained is fed to a finite automaton with output for classification. If the number of endpoints is one, then the image is passed to the correlation function for classification.

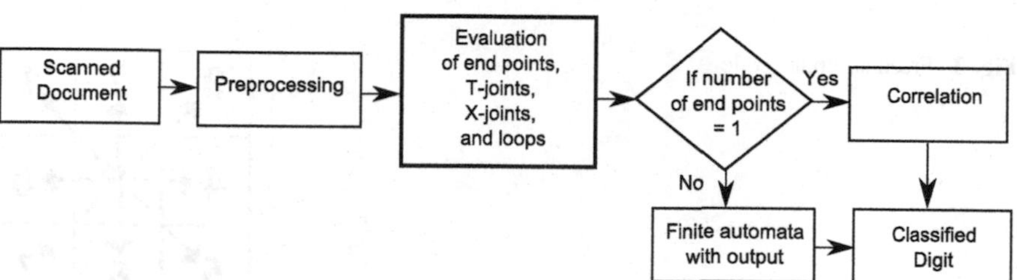

Fig. 2 Proposed model for printed Odia digit recognition

Table 1 Structural information for Odia digits

Odia digits	EPs	TJs	X Js	Loops	String	Odia digits	EPs	TJs	X Js	Loops	String
୦	0	0	0	1	0001	୫	2	0	1	2	2012
୧	1	1	0	1	1101	୬	2	0	0	0	2000
୨	1	1	0	1	1101	୭	1	1	0	1	1101
୩	3	1	0	0	3100	୮	2	1	0	0	2100
୪	2	0	1	1	2011	୯	2	2	0	0	2200

2.1 Pre-processing

Basic morphological operations [10] are applied to perform the tasks, such as binarization, segmentation, thinning, resizing, on the scanned document. We exploit standard methods for this purpose. We utilized the adaptive binarization technique [7] and segmentation to get the isolated digits. The segmented image is slenderized to single pixel width and then it is normalized to a size of 32 × 32.

2.2 Chain Coding

The proposed scheme extracts one string of strokes from the thinned object of the character using Freeman encoding method (see Fig. 3). Except the digit zero all other digits must have at least one endpoint. The extraction is done by starting from the endpoint and tracing the direction from one pixel to another. Finally, the image is represented by a series of numbers from the set {0, 1, 2, 3, 4, 5, 6, 7}. Each number represents the transition between two consecutive boundary pixels. We have considered the digit ୪ and its skeletonized image as shown in Figs. 4 and 5 respectively. To demonstrate how to calculate the parameters such as number of endpoints (EPs), T-joints (TJs), *X*-Joints (*X* Js) and loops, the encoding sequence is obtained experimentally for the digit ୪ which is given by 0 7 7 7 7 7 7 7 7 7 7 1 1 1 0 1 2 1 1 1 2 1 0 0 4 4 5 6 5 5 5 6 5 4 5 5 6 7 7 7 7 7 6 6 6 6 6 5 6 5 4 4 5 4 4 4 3 3 3 3 2 2 2 2 2 1 1 1 1 1 2 3 3 3 3 3 3 3 3 3 4. The sequence generated by the chain code is analyzed to have information on the number of endpoints (EPs), T-joints (TJs),

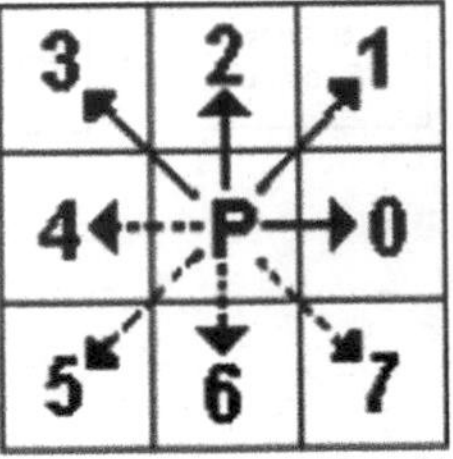

Fig. 3 Freeman chain code

Fig. 4 segmented image of the letter Four

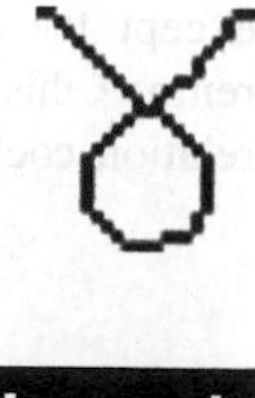

Fig. 5 Skeletonised image of the letter Four

Fig. 6 End points and X-point of the letter 4

X-Joints (X Js) and loops. In Fig. 6 the two end points (top-left and top-right) along with one X-joint (middle) are shown.

2.3 Recognition Phase

For the recognition of printed numerals from any scanned document, first we do the pre-processing to get a normalized thinned image of size 32 × 32. From the Freeman chain coding we get the string as mentioned in the Table 1. In our proposed methodology, we used the finite automaton with output, as recognizer. The state transition diagram is shown in the Fig. 7 where the initial state is Q_0 and the states with two concentric circles are the accepting states with certain output. The equivalent Algorithm 1 gives a step-by-step process of recognizing the digits by the finite automaton with output. It clearly shows that all digits are recognizable

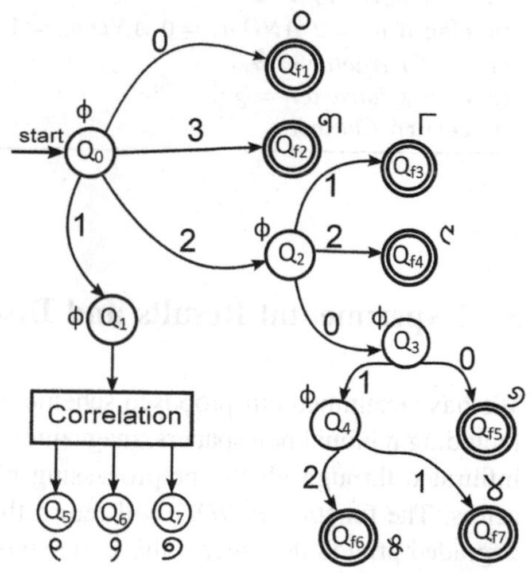

Fig. 7 State diagram for digit recognition

except the digits ୧, ୨, and ୭ because they represent the same string. So, to remove this ambiguity we use the correlation method. The two-dimensional correlation coefficients are calculated as in the Eq. 1,

$$r = \frac{\sum_m \sum_n (A_{mn} - \hat{A})(B_{mn} - \hat{B})}{\sqrt{\left(\sum_m \sum_n (A_{mn} - \hat{A})\right)^2 - \left(\sum_m \sum_n (B_{mn} - \hat{B})\right)^2}}, \tag{1}$$

where $\hat{A}$ = mean of A and $\hat{B}$ mean of B. The function FIND (img) will return the number of end points, T-joints, X-joints, and loops respectively. The function $FUNC_{corr}()$ used in the Algorithm 1, will return the character with maximum correlation value.

Algorithm 1: RECOGNITION OF ODIA DIGITS

Input: A test image (segmented/normalized) img for recognition.
Output: The value $Character$ is returned

$\langle a_1 a_2 a_3 a_4 \rangle$ = Call FIND(img)
if $a_1 = 0$ **then**
  $Character$ = ୦
else if $a_1 = 3$ **then**
  $Character$ = ୩
else if $a_1 = 1$ **then**
  $Character$ = Call $FUNC_{corr}()$
else if $a_1 = 2$ AND $a_2 = 1$ **then**
  $Character$ = ୮
else if $a_1 = 2$ AND $a_2 = 2$ **then**
  $Character$ = ୯
else if $a_1 = 2$ AND $a_2 = 0$ AND $a_3 = 0$ **then**
  $Character$ = ୬
else if $a_1 = 2$ AND $a_2 = 0$ AND $a_3 = 1$ AND $a_4 = 1$ **then**
  $Character$ = ୪
else if $a_1 = 2$ AND $a_2 = 0$ AND $a_3 = 1$ AND $a_4 = 2$ **then**
  $Character$ = ୫
else $Character = \phi$
return $Character$

3 Experimental Results and Discussions

We have examined our proposed scheme on a large set of printed Odia documents, including regional newspapers, magazines, articles, and books. Each document was infiltrated through all the preprocessing phase to have manageable isolated characters. The function *FIND*() will return the string $\langle a_1 a_2 a_3 a_4 \rangle$. Due to some badly degraded printed documents the finite automaton with output is able to classify all

Table 2 Recognition rate of individual Odia digits

Odia digits	Number of occurrence	Number of samples unclassified	Recognition rate (%)
୦	233	2	99.14
୧	246	19	92.27
୨	156	14	91.02
୩	212	3	98.58
୪	286	5	98.25
୫	223	2	99.10
୬	202	10	95.05
୭	313	23	92.65
୮	185	2	98.91
୯	219	7	95.85

the digits with good accuracy except the digits ୧ , ୨ , and ୭ because they represent the same string. To distinguish among these we use the correlation function. We consider 50 such documents where the number of occurrences of each digits along with their misclassification rate is mentioned in the Table 2. For the sake of understanding we have taken a printed document with good quality printing, where it contains ୧ , ୨ , and ୭ along with some other digits and Odia text. We run the procedure $FUNC_{corr}()$ for the test image where digits with identical string are present. The output is depicted in the Fig. 8 with a bounding box. The overall recognition rate is 96.08 %.

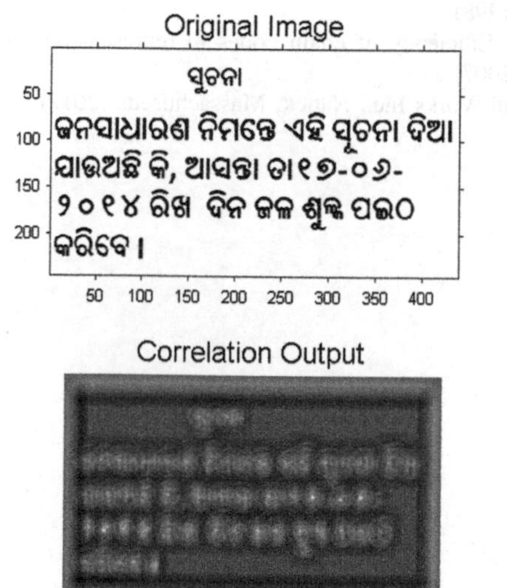

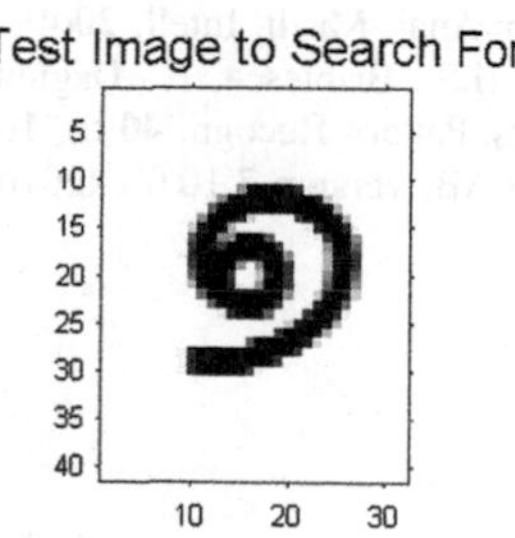

Template Image Found in Original Image

ସୂଚନା

ଜନସାଧାରଣ ନିମନ୍ତେ ଏହି ସୂଚନା ଦିଆ
ଯାଉଅଛି କି, ଆସନ୍ତା ତା୧୭-୦୬-
୨୦୧୪ ରିଖ ଦିନ ଜଳ ଶୁଳ୍କ ପଇଠ
କରିବେ ।

Fig. 8 Recognizing the digit seven

4 Conclusion

Here, we have suggested a scheme for printed Odia digit recognition using finite automaton. The obtained overall recognition accuracy of our system is about 96.08 %, which is better than other techniques suggested in this paper. Certainly, there is a scope for improvement as far as accuracy is concerned. The proposed scheme requires to be tested on perplexed printed documents and degraded historic documents.

References

1. Trier, Ø.D., Jain, A.K., Taxt, T.: Feature extraction methods for character recognition-a survey. Pattern Recogn. **29**(4), 641–662 (1996)
2. Pal, U., Chaudhuri, B.B.: Indian script character recognition: a survey. Pattern Recogn. **37**(9), 1887–1899 (2004)
3. Nikolaou, N., Makridis, M., Gatos, B., Stamatopoulos, N., Papamarkos, N.: Segmentation of historical machine-printed documents using adaptive run length smoothing and skeleton segmentation paths. Image Vis. Comput. **28**(4), 590–604 (2010)
4. Akiyama, T., Hagita, N.: Automated entry system for printed documents. Pattern Recogn. **23**(11), 1141–1154 (1990)
5. Chaudhuri, B.B., Pal, U.: A complete printed Bangla OCR system. Pattern Recogn. **31**(5), 531–549 (1998)
6. Chaudhuri, B.B., Pal, U., Mitra, M.: Automatic recognition of printed Oriya script. Sadhana **27**(1), 23–34 (2002)
7. Louloudis, G., Gatos, B., Pratikakis, I., Halatsis, C.: Text line and word segmentation of handwritten documents. New Frontiers in handwriting recognition. Pattern Recogn. **42**(12), 3169–3183 (2009)
8. Zingaretti, P., Gasparroni, M., Vecci, L.: Fast chain coding of region boundaries. IEEE Trans. Pattern Anal. Mach. Intell. **20**(4), 407–415 (1998)
9. Cruz, H.S., Bribiesca, E., Dagnino, R.M.R.: Efficiency of chain codes to represent binary objects. Pattern Recogn. **40**(6), 1660–1674 (2007)
10. MATLAB. version 7.10.0 (R2010a). The MathWorks Inc., Natick, Massachusetts (2012)

Author Index

A. Nagar et al. (eds.), *Proceedings of 3rd International Conference on Advanced Computing, Networking and Informatics*, Smart Innovation, Systems and Technologies 43, DOI 10.1007/978-81-322-2538-6

Printed in the United States
By Bookmasters